Birkhäuser Advanced Texts
Basler Lehrbücher

Edited by
Herbert Amann, University of Zürich
Steven G. Krantz, Washington University
Shrawan Kumar, University of North Carolina at Chapel Hill

Emmanuele DiBenedetto

Real Analysis

Birkhäuser
Boston • Basel • Berlin

Emmanuele DiBenedetto
Department of Mathematics
Vanderbilt University
Nashville, TN 37240
U.S.A.

Library of Congress Cataloging-in-Publication Data

DiBenedetto, Emmanuele.
   Real Analysis / Emmanuele DiBenedetto.
     p. cm. — (Birkhäuser advanced texts)
   Includes bibliographical references and index.
   ISBN 0-8176-4231-5 (acid-free paper) — ISBN 3-7643-4231-5 (Basel : acid-free paper)
   1. Mathematical analysis. I. Title. II. Series.

   QA300.D46 2001
   515–dc21
                          2001052752
                          CIP

AMS Classification Codes: 03E04, 03E10, 03E20, 03E25, 26A03, 26A09, 26A12, 26A15, 26A16,
26A21, 26A27, 26A30, 26A42, 26A45, 26A46, 26A48, 26A51, 26B05, 26B15, 26B20, 26B25,
26B30, 26B35, 26B40, 26E10, 28A05,28A10, 28A12, 28A15, 28A20, 28A25,28A33, 28A35,
28A50, 28A75,28A78,31B05, 31B10, 35C15, 35E05, 40A05,40A10, 41A10, 42B25, 42B35,
46A03, 46A22, 46A30, 46A32, 46B03, 46B07, 46B10, 46B25, 46B45, 46C05, 46C15,46E05,
46E10, 46E15, 46E35, 46F05, 46F10, 54A05, 54A10, 54A20, 54A25, 54B05, 54B10, 54B15,
54C05, 54C30, 54D05, 54D10, 54D30, 54D45, 54D60, 54D65, 54E35, 54E40, 54E45, 54E50,
54E52

Printed on acid-free paper
© 2002 Birkhäuser Boston

Birkhäuser

ISBN 0-8176-4231-5   SPIN 10798100
ISBN 3-7643-4231-5

Reformatted from the author's files by John Spiegelman, Philadelphia, PA.
Printed and bound by Hamilton Printing, Rensselaer, NY.
Printed in the United States of America.

9 8 7 6 5 4 3 2 1
A member of BertelsmannSpringer Science+Business Media GmbH

Contents

Preface

This book is a self-contained introduction to real analysis assuming only basic notions on limits of sequences in $\mathbb{R}^N$, manipulations of series, their convergence criteria, advanced differential calculus, and basic algebra of sets.

The passage from the setting in $\mathbb{R}^N$ to abstract spaces and their topologies is gradual. Continuous reference is made to the $\mathbb{R}^N$ setting, where most of the basic concepts originated.

The first seven chapters contain material forming the backbone of a basic training in real analysis. The remaining two chapters are more topical, relating to maximal functions, functions of bounded mean oscillation, rearrangements, potential theory, and the theory of Sobolev functions.

Even though the layout of the book is theoretical, the entire book and the last chapters in particular concern applications of mathematical analysis to models of physical phenomena through partial differential equations.

The preliminaries contain a review of the notions of countable sets and related examples. We introduce some special sets, such as the Cantor set and its variants, and examine their structure. These sets will be a reference point for a number of examples and counterexamples in measure theory (Chapter II) and in the Lebesgue differentiability theory of absolute continuous functions (Chapter IV). This initial chapter also contains a brief collection of the various notions of *ordering*, the Hausdorff maximal principle, Zorn's lemma, the well-ordering principle, and their fundamental connections.

These facts keep appearing in measure theory (Vitali's construction of a Lebesgue nonmeasurable set), topological facts (Tychonov's theorem on the compactness of the product of compact spaces; existence of Hamel bases), and functional analysis (Hahn–Banach theorem; existence of maximal orthonormal bases in Hilbert spaces).

Chapter I is an introduction to those basic topological issues that hinge upon analysis or that are, one way or another, intertwined with it. Examples include Urysohn's lemma and the Tietze extension theorem, characterization of compactness and its relation to the Bolzano–Weierstrass property, structure of the compact sets in $\mathbb{R}^N$, and various properties of semicontinuous functions defined on compact sets. This analysis of compactness concerns the structure of the compact subsets of the space of continuous functions (Chapter IV) and the characterizations of the compact subsets of the spaces $L^p(E)$ for all $1 \leq p < \infty$ (Chapter V).

The Tychonov theorem is proved, keeping in mind its application in the proof of the Alaoglu theorem on the weak* compactness of closed balls in a linear, normed space.

We introduce the notions of linear, topological vector spaces and of linear maps and functionals and their relation to boundedness and continuity.

The discussion turns quickly to metric spaces, their topology and their structure. Examples are drawn mostly from spaces of continuous or continuously differentiable functions or integrable functions. The notions and characterizations of compactness are rephrased in the context of metric spaces. This is preparatory to characterizing the structure of compact subsets of $L^p(E)$.

The structure of complete metric spaces is analyzed through Baire's category theorem. This plays a role in subsequent topics, such as an indirect proof of the existence of nowhere-differentiable functions (Chapter IV), in the structure of Banach spaces (Chapter VI), and in questions of completeness and noncompleteness of various topologies on $C_o^\infty(E)$ (Chapter VII).

Chapter II is a modern account of measure theory. The discussion starts from the structure of open sets in $\mathbb{R}^N$ as sequential coverings to construct measures and a brief introduction to the algebra of sets. Measures are constructed from outer measures by the Carathéodory process. The process is implemented in specific examples such as the Lebesgue–Stieltjes measures in $\mathbb{R}$ and the Hausdorff measure. The latter seldom appears in introductory textbooks in real analysis. We have chosen to present it in some detail because it has become, in the past two decades, an essential tool in studying the fine properties of solutions of partial differential equations and systems. The Lebesgue measure in $\mathbb{R}^N$ is introduced directly starting from the Euclidean measure of cubes, rather than regarding it, more or less abstractly, as the N-product of the Lebesgue measure on $\mathbb{R}$. In $\mathbb{R}^N$ we distinguish between Borel sets and Lebesgue-measurable sets, by cardinality arguments and concrete counterexamples.

For general measures, emphasis is put on necessary and sufficient criteria of measurability in terms of $\mathcal{G}_\delta$ and $\mathcal{F}_\sigma$. In this we have in mind the operation of measuring a set as an approximation process. From the applications point of view one would like to approximate the measure of a set by the measure of measurable sets containing it and measurable sets contained into it. The notion is further expanded in the theory of Radon measures and their regularity properties.

It is also further expanded into the covering theorems although these represent an independent topic in their own right. The Vitali covering theorem is presented in its proof due to Banach. The Besicovitch covering is presented by emphasizing its value for general Radon measures in $\mathbb{R}^N$. For both we stress the measure-theoretical nature of the covering as opposed to the notion of covering a set by inclusion.

Coverings have made possible an understanding of the local properties of solutions of partial differential equations, chiefly the Harnack inequality for non-negative solutions of elliptic equations. For this reason, in the "Problems and Complements" of this chapter, we have included various versions of the Vitali and Besicovitch covering theorems.

Chapter III introduces the Lebesgue integral. The theory is preceded by the notions of measurable functions, convergence in measure, Egorov's theorem on selecting almost-everywhere convergent subsequences from sequences convergent in measure, and Lusin's theorem characterizing measurability in terms of quasi-continuity. This theorem is given relevance as it relates to measurability and local behavior of measurable functions. It is also a concrete application of the necessary and sufficient criteria of measurability of the previous chapter.

The integral is constructed starting from nonnegative simple functions by the Lebesgue procedure. Emphasis is placed on convergence theorems and the Vitali theorem on the absolute continuity of the integral. The Peano–Jordan and Riemann integrals are compared to the Lebesgue integral by pointing out differences and analogies.

The theory of product measures and the related integral is developed in the framework of the Carathéodory construction by starting from measurable rectangles. This construction provides a natural setting for the Fubini–Tonelli theorem on multiple integrals.

Applications are provided ranging from the notion of convolution, the convergence of the Marcinkiewicz integral, to the interpretation of an integral in terms of the distribution function of its integrand.

The theory of measures is completed in this chapter by introducing the notion of signed measure and by proving Hahn's decomposition theorem. This leads to other natural notions of decompositions such as the Jordan and Lebesgue decomposition theorems.

It also naturally suggests other notions of comparing two measures such as the absolute continuity of a measure ν with respect to another measure μ. It also suggests representing ν, roughly speaking, as the integral of μ by the Radon–Nikodým theorem.

Relating two measures finds application in the Besicovitch–Lebesgue theorem, presented in the next chapter, and connecting integrability of a function to some of its local properties.

Chapter IV is a collection of applications of measure theory to issues that were at the root of modern analysis. What does it mean for a function of one real variable to be differentiable? When can one compute an integral by the fundamental theorem of calculus? What does it mean to take the derivative on an integral?

These issues motivated a new way of measuring sets and the need for a new notion of integral.

The discussion starts from functions of bounded variation in an interval and their Jordan's characterization as the difference of two monotone functions. The notion of differentiability follows naturally from the definition of the four Dini numbers. For a function of bounded variation, its Dini numbers, regarded as functions, are measurable. This is a remarkable fact due to Sierpinski and Banach.

Functions of bounded variations are almost-everywhere differentiable. This is a celebrated theorem of Lebesgue. It uses in an essential way Vitali's covering theorem of Chapter I.

We introduce the notion of absolutely continuous functions and discuss similarities and differences with respect to functions of bounded variation. The Lebesgue theory of differentiating an integral is developed in this context. A natural related issue is that of the density of a Lebesgue-measurable subset of an interval. Almost every point of a measurable set is a density point for that set. The proof uses a remarkable theorem of Fubini on differentiating, term by term, a series of monotone functions.

Similar issues for functions of N real variables are far more delicate. We present the theory of differentiating a measure ν with respect to another μ by precisely identifying such a derivative in terms of the singular part and the absolutely continuous part of μ with respect to ν. The various decompositions of measures of Chapter III find their natural application here along with the Radon–Nikodým theorem.

The pivotal point of the theory is the Besicovitch–Lebesgue theorem asserting that the limit of the integral of a measurable function f when the domain of integration shrinks to a point x actually exists for almost all x and equals the value of f at x. The shrinking procedure is achieved by using balls centered at x, and the measure can be any Radon measure. This is the strength of the Besicovitch covering theorem. We discuss the possibility of replacing balls with domains that are, roughly speaking, comparable to a ball.

As a consequence, almost every point of an N-dimensional Lebesgue-measurable set is a density point for that set.

The final part of the chapter contains an array of facts of common use in real analysis. These include basic facts on convex functions of one variable and their almost-everywhere double differentiability. In the "Problems and Complements," we introduce the Legendre transform and indicate the main properties and features.

We present the Ascoli–Arzelá theorem, keeping in mind a description of compact subsets of spaces of continuous functions.

We also include a theorem of Kirzbraun and Pucci extending bounded, continuous functions in a domain into bounded, continuous functions in the whole $\mathbb{R}^N$ with the *same* upper bound and the *same* concave modulus of continuity. This theorem does not seem to be widely known.

The final part of the chapter contains a detailed discussion of the Stone–Weierstrass theorem. We present first the Weierstrass theorem (in N dimensions) as a pure fact of approximation theory. The polynomials approximating a continuous function f in the sup-norm over a compact set are constructed explicitly by means of the Bernstein polynomials. The Stone theorem is then presented as a way of identifying the structure of a class of functions that can be approximated by polynomials.

Chapter V introduces the theory of L^p spaces for $1 \leq p \leq \infty$. The basic inequalities of Hölder and Minkowski are introduced and used to characterize the norm and related topology of these spaces. A discussion is provided to identify elements of $L^p(E)$ as equivalence classes.

We also introduce the $L^p(E)$ spaces for $0 < p < 1$ and the related topology. We establish that there are no convex open sets except $L^p(E)$ itself and the empty set.

We then turn to questions of convergence in the sense of $L^p(E)$ and their completeness (Riesz–Fisher theorem) as well as issues of separating such spaces by simple functions. The latter serves as a tool in the notion of weak convergence of sequences of functions in $L^p(E)$. Strong and weak convergence are compared and basic facts relating weak convergence and convergence of norms are stated and proved.

The "Problems and Complements" section contains an extensive discussion comparing the various notions of convergence.

We introduce the notion of functional in $L^p(E)$ and its boundedness and continuity and prove the Riesz representation theorem, characterizing the form of all the bounded linear functionals in $L^p(E)$ for $1 \leq p < \infty$. This proof is based on the Radon–Nikodým theorem and as such is measure-theoretical in nature.

We present a second proof of the same theorem based on the topology of L^p. The open balls that generate the topology of $L^p(E)$ are *strictly* convex for $1 < p < \infty$. This fact is proved by means of the Hanner and Clarkson inequalities, which, while technical, is of interest in their own right.

The Riesz representation theorem permits one to prove that if E is a Lebesgue-measurable set in $\mathbb{R}^N$, then $L^p(E)$ for $1 \leq p < \infty$, are separable. It also permits one to select weakly convergent subsequences from bounded ones. This fact holds in general, reflexive, separable Banach spaces (Chapter VI). We have chosen to present it independently as part of L^p theory. It is our point of view that a good part of functional analysis draws some of its key facts from concrete spaces such as spaces of continuous functions, the L^p space, and the spaces ℓ_p.

The remainder of the chapter presents some technical tools regarding $L^p(E)$ for E a Lebesgue-measurable set in $\mathbb{R}^N$, to be used in various parts of the later chapters. These include the continuity of the translation in the topology of $L^p(E)$, the Friedrichs mollifiers, and the approximation of functions in $L^p(E)$ with $C^\infty(E)$ functions. It also includes a characterization of the compact subsets in $L^p(E)$.

Chapter VI is an introduction to those aspects of functional analysis closely related to the Euclidean spaces $\mathbb{R}^N$, the spaces of continuous functions defined on some open set $E \subset \mathbb{R}^N$, and the spaces $L^p(E)$. These naturally suggest the notion of finite-dimensional and infinite-dimensional normed spaces. The difference between the two is best characterized in terms of the compactness of their closed unit ball. This is a consequence of a beautiful counterexample of Riesz.

The notions of maps and functionals is rephrased in terms of the norm topology. In $\mathbb{R}^N$ one thinks of a linear functional as an affine function whose level sets are hyperplanes through the origin. Much of this analogy holds in general normed spaces with the proper rephrasing.

Families of pointwise equibounded maps are proven to be uniformly equibounded as an application of Baire's category theorem.

We also briefly consider special maps such as those generated by Riesz potentials (estimates of these potentials are provided in Chapter III) and related Fredholm integral equations.

Proofs of the classical open mapping theorem and closed graph theorem are presented as a way of inverting continuous maps to identify isomorphisms out of continuous linear maps.

The Hahn–Banach theorem is viewed in its geometrical aspects of separating closed convex sets in a normed space and of "drawing" tangent planes to a convex set.

These facts all play a role in the notion of weak topology and its properties. Mazur's theorem on weak and strong closure of convex sets in a normed space is related to the weak topology of the $L^p(E)$ spaces. These provide the main examples, as convexity is explicit through Clarkson's inequalities.

The last part of the chapter gives an introduction to Hilbert spaces and its geometrical aspects through the parallelogram identity. We present the Riesz representation theorem of functionals through the inner product. The notion of basis is introduced and its cardinality is related to the separability of a Hilbert space. We introduce orthonormal systems and indicate the main properties (Bessel's inequality) and some construction procedures (Gram–Schmidt). The existence of a complete system is a consequence of the Hausdorff maximum principle. We also discuss various equivalent notions of completeness.

Chapter VII is about spaces of real-valued, continuous functions, differentiable functions, infinitely differentiable functions with compact support in some open set $E \subset \mathbb{R}^N$, and weakly differentiable functions. Together with the $L^p(E)$ spaces, these are among the backbone spaces of real analysis.

We prove the Riesz representation theorem for continuous functions of compact support in $\mathbb{R}^N$. The discussion starts from positive functionals and their representation. Radon measures are related to positive functionals and bounded, signed Radon measures are related to bounded linear functionals. Analogous facts hold for the space of continuous functions with compact support in some open set $E \subset \mathbb{R}^N$.

We then turn to making precise the notion of a topology for $C_o^\infty(E)$. Completeness and noncompleteness are related to metric topologies in a constructive way. We introduce the Schwartz topology and the notion of continuous maps and functionals with respect to such a topology. This leads to the theory of distributions and its related calculus (derivatives, convolutions, etc., of distributions).

Their relation to partial differential equations is indicated through the notion of fundamental solution. We compute the fundamental solution for the Laplace operator also in view of its applications to potential theory (Chapter VIII) and to Sobolev inequalities (Chapter IX).

The notion of weak derivative in some open set $E \subset \mathbb{R}^N$ is introduced as an aspect of the theory distributions. We outline their main properties and state and prove the (by now classical) Meyers–Serrin theorem. Extension theorems and approximation by smooth functions defined in domains larger than E are provided. This leads naturally to a discussion of the smoothness properties of ∂E for these approximations and/or extensions to take place (cone property, segment property, etc.).

We present some calculus aspects of weak derivatives (chair rule, approximations by difference quotients, etc.) and turn to a discussion of $W^{1,\infty}(E)$ and its relation to Lipschitz functions. For the latter we conclude the chapter by stating and proving the Rademacher theorem.

Chapter VII is a collection of topics of common use in real analysis and its applications. First is the Wiener version of the Vitali covering theorem (commonly referred to as the "simple version" of Vitali's theorem). This is applied to the notion of maximal function, its properties, and its related strong-type L^p estimates for $1 < p < \infty$. Weak estimates are also proved and used in the Marcinkiewicz interpolation theorem. We prove the by-now-classical Calderón–Zygmund decomposition theorem and its applications to the space functions of bounded mean oscillation (BMO) and the Stein–Fefferman L^p estimate for the sharp maximal function.

The space of BMO is given some emphasis. We give the proof of the John–Nirenberg estimate and provide its counterexample. We have in mind here the limiting case of some potential estimates (later in the chapter) and the limiting Sobolev embedding estimates (Chapter IX).

We introduce the notion of rearranging the values of functions and provide their properties and the related notion of equimeasurable function. The discussion is for functions of one real variable. Extensions to functions of N real variables are indicated in the "Problems and Complements."

The goal is to prove the Riesz convolution inequality by rearrangements. The several existing proofs (Riesz, Zygmund, Hardy–Littlewood–Polya) all use, one way or another, the symmetric rearrangement of an integrable function.

We have reproduced here the proof of Hardy–Littlewood–Polya as it appears in their monograph [24]. In the process we need to establish Hardy's inequality, of interest in its own right.

The Riesz convolution inequality is presented in several of its variants, leading to an N-dimensional version of it, through an application of the continuous version of the Minkowski inequality.

Besides the intrinsic interest of these inequalities, what we have in mind here is to recover some limiting cases of potential estimates and their related Sobolev embedding inequalities.

The final part of the chapter introduces the Riesz potentials and their related L^p estimates, including some limiting cases. These are on the one hand based on the previous Riesz convolution inequality, and, on the other hand, on Trudinger's version of the BMO estimates for particular functions arising as potentials.

Chapter IX provides an array of embedding theorems for functions in Sobolev spaces. Their importance to analysis and partial differential equations cannot be underestimated. Although good monographs exist [1, 39] I have found it laborious to extract the main facts, listed in a clean manner and ready for applications.

We start from the classical Gagliardo–Nirenberg inequalities and proceed to Sobolev inequalities. We have made an effort to trace, in the various embedding inequalities, how the smoothness of the boundary enters in the estimates. For example, whenever the cone condition is required, we trace back in the various constants the dependence on the height and the angle of the cone. We present the Poincaré inequalities for bounded, convex domains E and trace the dependence of the various constants on the "modulus of convexity" of the domain through the ratio of the radius of the smallest ball containing E and the largest ball contained in E. The limiting case $p = N$ of the Sobolev inequality builds on the limiting inequalities for the Riesz potentials, and it is preceded by an introduction to Morrey spaces and their connection to BMO.

The characterization of the compact subsets of $L^p(E)$ (Chapter V) is used to prove Reillich's theorem on compact Sobolev inequalities.

We introduce the notion of trace of function in $W^{1,p}(\mathbb{R}^N \times \mathbb{R}^+)$ on the hyperplane $x_{N+1} = 0$. Through a partition of unity and a local covering this provides the notion of trace of functions in $W^{1,p}(E)$ on the boundary ∂E, provided such a boundary is sufficiently smooth. Sharp inequalities relating functions in $W^{1,p}(E)$ with the integrability and regularity of their traces on ∂E are established in terms of fractional Sobolev spaces. Such inequalities are first established for E being a half-space and ∂E a hyperplane, and then extended to general domains E with sufficiently smooth boundary ∂E. In the "Problems and Complements," we characterize functions f defined and integrable on ∂E as traces on ∂E of functions in some Sobolev spaces $W^{1,p}(E)$. The relation between p and the order of integrability of f on ∂E is shown to be sharp. For special geometries, such as a ball, the inequality relating the integral of the traces and the Sobolev norm can be made explicit. This is indicated in the "Problems and Complements."

The last part of the chapter contains a newly established *multiplicative* Sobolev embedding for functions in $W^{1,p}(E)$ that do not necessarily vanish on ∂E. The open set E is required to be convex. Its value is in its applicability to the asymptotic behavior of solutions to Neumann problems related to parabolic partial differential equations.

Acknowledgments

This book has grown out of courses and topics in real analysis that I have taught over the years at Indiana University Bloomington, Northwestern University, the University of Rome Tor Vergata, Italy, and Vanderbilt University.

My thanks go to the numerous students who have pointed out misprints and imprecise statements. Among these, let me thank in particular John Renze, Ethan Pribble, Lan Yueheng, Kamlesh Parwani, Ronnie Sadka, Marco Battaglini, Donato Gerardi, Zsolt Macskasi, Tianhong Li, Todd Fisher, Liming Feng, Mikhail Perepelitsa, Lucas Bergman, Derek Bruff, David Peterson, Hannah Callender, Elizabeth Carver, and Yulya Babenko.

Special thanks go to Michael O'Leary, David Diller, Giuseppe Tomassetti, and Gianluca Bonuglia. The material of the last three chapters results from topical seminars organized over the years with these former students.

I am indebted to Allen Devinatz, Edward Nussbaum, Ethan Devinatz, Haskell Rhosental, and Juan Manfredi for providing me with some counterexamples and for helping me to make precise some facts related to unbounded linear functionals in normed spaces.

I would like also to thank Henghui Zou, James Serrin, Avner Friedman, Craig Evans, Robert Glassey, Herbert Amann, Enrico Magenes, Giorgio Talenti, Gieri Simonett, Vincenzo Vespri, Paolo Bisegna, and Mike Mihalik for reading, at various stages, portions of the manuscript and for providing valuable critical comments.

The input of Daniele Andreucci has been crucial, and it needs to be singled out. He read the entire manuscript and made critical remarks and suggestions. He has also worked out in detail a large number of the problems and suggested some of his own. I am very much indebted to him.

My notes were conceived as a book in 1994 while I was teaching topics in real analysis at the School of Engineering of the University of Rome, Tor Vergata, Italy.

Special thanks go to Franco Maceri, former Dean of the School of Engineering of the University of Rome, for his vision of mathematics as the natural language of applied sciences and for fostering that vision.

I learned real analysis from Carlo Pucci at the University of Florence, Italy in 1974–1975. My view of analysis and these notes are influenced by Pucci's teaching. According to his way of thinking, every theorem had to be motivated and had to go along with examples and counterexamples, i.e., had to withstand a scientific "critique."

Preliminaries

1   Countable sets

A set E is countable if it can be put in one-to-one correspondence with a subset of the natural numbers $\mathbb{N}$. Every subset of a countable set is countable.

Proposition 1.1. *The set $\mathcal{S}_E$ of the finite sequences of elements of a countable set E is countable.*

Proof. Let $\{2, 3, 5, 7, 11, \ldots, m_j, \ldots\}$ denote the sequence of prime numbers. Every positive integer n has a unique factorization of the type

$$n = 2^{\alpha_1} 3^{\alpha_2} \cdots m_j^{\alpha_j}, \tag{1.1}$$

where the sequence $\{\alpha_1, \ldots, \alpha_j\}$ is finite and the α_i are nonnegative integers. Now let $\sigma \equiv \{e_1, e_2, \ldots, e_j\}$ be an element of $\mathcal{S}_E$. Since E is countable, to each element e_i there corresponds a unique positive integer α_i. Thus to σ there corresponds the unique positive integer given by (1.1). $\qquad\square$

Corollary 1.2. *The set of pairs $\{m, n\}$ of integer numbers is countable.*

Corollary 1.3. *The set $\mathbf{Q}$ of the rational numbers is countable.*

Proof. The rational numbers can be put in one-to-one correspondence with a subset of the ratios $\frac{m}{n}$ for two integers m, n with $n \neq 0$. $\qquad\square$

Proposition 1.4. *The union of a countable collection of countable sets is countable.*

Proof. Let $E_1, E_2, \ldots, E_m, \ldots$ be a countable collection of countable sets. Since each of the E_j is countable, we list their elements as

$$
\begin{aligned}
E_1 &\equiv \{a_{11} \quad a_{12} \quad a_{13} \quad \cdots \quad a_{1n} \quad \cdots \}, \\
E_2 &\equiv \{a_{21} \quad a_{22} \quad a_{23} \quad \cdots \quad a_{2n} \quad \cdots \}, \\
E_3 &\equiv \{a_{31} \quad a_{32} \quad a_{33} \quad \cdots \quad a_{3n} \quad \cdots \}, \\
&\cdots \\
E_m &\equiv \{a_{m1} \quad a_{m2} \quad a_{m3} \quad \cdots \quad a_{mn} \quad \cdots \}, \\
&\cdots
\end{aligned}
$$

Thus the elements of $\bigcup E_n$ are in one-to-one correspondence with a subset of the ordered pairs $\{m, n\}$ of natural numbers. $\square$

Proposition 1.5 (Cantor[1]). *The interval* $[0, 1]$ *is not countable.*

Proof. The proof uses Cantor's diagonal process. If $[0, 1]$ were countable, its elements could be listed as the countable collection

$$
x_j = 0.a_{1j}a_{2j}\cdots a_{mj}\cdots, \qquad j = 1, 2, \ldots,
$$

where a_{mj} are integers from 0 to 9. Now set $a_j = 1$ if a_{jj} is even and $a_j = 2$ if a_{jj} is odd. Then, the element $x = 0.a_1a_2\cdots$ is in $[0, 1]$ and is different from any one of the $\{x_j\}$. $\square$

2   The Cantor set

Divide the closed interval $[0, 1]$ into three equal subintervals and remove the central open interval $I_1 = (\frac{1}{3}, \frac{2}{3})$ so that

$$
[0, 1] - I_1 = \left[0, \frac{1}{3}\right] \bigcup \left[\frac{2}{3}, 1\right].
$$

Subdivide each of these intervals into three equal parts and remove their central open interval. If I_2 is the set that has been removed,

$$
I_2 = \left(\frac{1}{3^2}, \frac{2}{3^2}\right) \bigcup \left(\frac{7}{3^2}, \frac{8}{3^2}\right);
$$

$$
[0, 1] - \left(I_1 \bigcup I_2\right) = \left[0, \frac{1}{3^2}\right] \bigcup \left[\frac{2}{3^2}, \frac{3}{3^2}\right] \bigcup \left[\frac{6}{3^2}, \frac{7}{3^2}\right] \bigcup \left[\frac{8}{3^2}, 1\right].
$$

We subdivide each of the closed intervals making up $[0, 1] - (I_1 \bigcup I_2)$ into three equal subintervals and remove their central open interval. If I_3 is the set that has

---

[1]G. Cantor, De la puissance des ensembles parfaits de points, *Acta Math.*, **4** (1884), 381–392.

been removed,

$$I_3 = \left(\frac{1}{3^3}, \frac{2}{3^3}\right) \bigcup \left(\frac{7}{3^3}, \frac{8}{3^3}\right) \bigcup \left(\frac{19}{3^3}, \frac{20}{3^3}\right) \bigcup \left(\frac{25}{3^3}, \frac{26}{3^3}\right);$$

$$[0, 1] - \left(I_1 \bigcup I_2 \bigcup I_3\right)$$

$$= \left[0, \frac{1}{3^3}\right] \bigcup \left[\frac{2}{3^3}, \frac{3}{3^3}\right] \bigcup \left[\frac{6}{3^3}, \frac{7}{3^3}\right] \bigcup \left[\frac{8}{3^3}, \frac{9}{3^3}\right]$$

$$\bigcup \left[\frac{18}{3^3}, \frac{19}{3^3}\right] \bigcup \left[\frac{20}{3^3}, \frac{21}{3^2}\right] \bigcup \left[\frac{24}{3^3}, \frac{25}{3^3}\right] \bigcup \left[\frac{26}{3^3}, 1\right].$$

Proceeding in this fashion we define a sequence of disjoint open sets I_n, each being the finite, disjoint union of open intervals and satisfying[2]

$$\text{meas}(I_n) = \frac{2^{n-1}}{3^n} \quad \text{and} \quad \sum \text{meas}(I_n) = 1. \tag{2.1}$$

The Cantor set $\mathcal{C}$ is the set that remains after removing, the union of the I_n out of $[0, 1]$, i.e.,

$$\mathcal{C} = [0, 1] - \bigcup I_n. \tag{2.2}$$

The set $\mathcal{C}$ is closed and each of its points is an accumulation point of the extremes of the intervals I_n. Thus $\mathcal{C}$ coincides with the set of all its accumulation points. Set

$$\mathcal{E} \equiv \left\{ \begin{array}{c} \text{the collection of all sequences } \{\varepsilon_n\}, \\ \text{the numbers } \varepsilon_n \text{ are either 0 or 1} \end{array} \right\}. \tag{2.3}$$

Then every element $x \in \mathcal{C}$ can be represented as[3]

$$x = \sum \frac{2}{3^j} \varepsilon_j \quad \text{for some sequence } \{\varepsilon_n\} \in \mathcal{E}. \tag{2.4}$$

Every element of $\mathcal{C}$ is associated with one and only one sequence $\{\varepsilon_n\} \in \mathcal{E}$ by the representation formula (2.4). For example,

$$\frac{1}{3} \longleftrightarrow \{0, 1, 1, 1, \dots, 1, \dots\}; \qquad \frac{2}{3} \longleftrightarrow \{1, 0, 0, 0, \dots, 0, \dots\};$$

$$\frac{1}{9} \longleftrightarrow \{0, 0, 1, 1, \dots, 1, \dots\}; \qquad \frac{2}{9} \longleftrightarrow \{0, 1, 0, 0, \dots, 0, \dots\};$$

$$\frac{7}{9} \longleftrightarrow \{1, 0, 1, 1, \dots, 1, \dots\}; \qquad \frac{8}{9} \longleftrightarrow \{1, 1, 0, 0, \dots, 0, \dots\}.$$

Vice versa, any such sequence identifies by (2.4) one and only one element of $\mathcal{C}$. The set of all sequences in $\mathcal{E}$ has the cardinality of the real numbers in $[0, 1]$

---

[2] The measure here is meant to be the Euclidean measure of intervals $(a, b) \subset \mathbb{R}$, i.e., $\text{meas}(a, b) = b - a$.

[3] See Sections 2.1 and 2.2 of the Problems and Complements.

being their binary representation. Thus $\mathcal{C}$ has the cardinality of $\mathbb{R}$ and therefore is uncountable. It also follows from (2.4) that the Cantor set could be defined alternatively as the set of those numbers in $[0, 1]$ whose ternary expansion has only the digits 0 and 2. The two definitions are equivalent.

3   Cardinality

Two sets X and Y have the same *cardinality* if there exists a one-to-one map f from X onto Y. In such a case one writes $\text{card}(X) = \text{card}(Y)$.

If X is finite, then $\text{card}(X)$ is the number of elements of X.

The formal inequality $\text{card}(X) \leq \text{card}(Y)$ means that there exists a one-to-one map from X into Y. In particular, in $X \subset Y$, then $\text{card}(X) \leq \text{card}(Y)$. The formal inequality $\text{card}(X) \geq \text{card}(Y)$ means that there exists a one-to-one map from X onto Y. In particular, in $X \supset Y$, then $\text{card}(X) \geq \text{card}(Y)$.

Proposition 3.1 (Schröder–Bernstein). *Assume that*

$$\text{card}(X) \leq \text{card}(Y) \quad and \quad \text{card}(X) \geq \text{card}(Y).$$

Then $\text{card}(X) = \text{card}(Y)$.

Proof. Let f be a one-to-one function from X into Y and let g be a one-to-one function from Y into X. Partition X into the disjoint union of three sets X_o, X_1, X_2 by the following iterative procedure. If x is not in the range of g we say that $x \in X_1$. If x is in the range of g, form $g^{-1}(x) \in Y$. If $g^{-1}(x)$ is not in the range of f, the process terminates and we say that $x \in X_2$. Otherwise, form $f^{-1}(g^{-1}(x))$. If such an element is not in the range of g, the process terminates and we say that $x \in X_1$. Proceeding in this fashion, either the process can be continued indefinitely or it terminates. If it terminates with an element not in the range of g we say that the starting element is in X_1. If it terminates with an element not in the range of f we say that the starting x is in X_2. If it can be continued indefinitely we say that $x \in X_o$.

The three sets X_o, X_1, and X_2 are disjoint and $X = X_o \bigcup X_1 \bigcup X_2$. Similarly, $Y = Y_o \bigcup Y_1 \bigcup Y_2$, where the sets Y_j, $j = 0, 1, 2$ are constructed similarly.

By construction, f is a bijection from X_o onto Y_o and from X_1 onto Y_2. Similarly, g is a bijection from Y_1 onto X_2. Thus the map $h : X \to Y$ defined by

$$h(x) = \begin{cases} f(x) & \text{if } x \in X_o \bigcup X_1, \\ g^{-1}(x) & \text{if } x \in X_2 \end{cases}$$

is a one-to-one map from X onto Y. □

Corollary 3.2. $\text{card}(X) = \text{card}(Y)$ *if and only if* $\text{card}(X) \leq \text{card}(Y)$ *and* $\text{card}(X) \geq \text{card}(Y)$.

The formal strict inequality $\text{card}(X) < \text{card}(Y)$ means that any one-to-one function $f : X \to Y$ is not a surjection; i.e., roughly speaking, X contains strictly fewer elements than Y. For example, $\text{card}(\mathbb{N}) < \text{card}(\mathbb{R})$.

3.1   Some examples. A set X has the cardinality of $\mathbb{N}$ if it can be put in a one-to-one correspondence with $\mathbb{N}$. In such a case, one writes $\mathrm{card}(X) = \mathrm{card}(\mathbb{N})$. For example, $\mathrm{card}(\mathbf{Z}) = \mathrm{card}(\mathbf{Q}) = \mathrm{card}(\mathbb{N})$.

A set X has the cardinality of $\mathbb{R}$ if it can be put in a one-to-one correspondence with $\mathbb{R}$. In such a case one writes $\mathrm{card}(X) = \mathrm{card}(\mathbb{R})$. For example, if $\mathcal{C}$ is the Cantor set, $\mathrm{card}(\mathcal{C}) = \mathrm{card}(\mathbb{R})$.

For a positive integer m, denote by

$$X^m = \underbrace{X \times X \times \cdots \times X}_{m \text{ times}}$$

the collection of all m-tuples $(x_1, x_2, \ldots, x_m)$ of elements of X. Also, denote by 2^X the set of all subsets of a set X. Thus $2^{\mathbb{N}}$ is the collection of all subsets of $\mathbb{N}$ and $2^{\mathbb{R}}$ is the collection of all subsets of $\mathbb{R}$.

Proposition 3.3. *For all positive integers m,*[4]

$$\mathrm{card}(\mathbb{N}^m) = \mathrm{card}(\mathbb{N}), \qquad \mathrm{card}(2^{\mathbb{N}}) = \mathrm{card}(\mathbb{R}). \tag{3.1}$$

Moreover, for any X,

$$\mathrm{card}(2^X) > \mathrm{card}(X). \tag{3.2}$$

Proof. The first part of (3.1) follows from Proposition 1.1. To prove the second, let A be a nonempty subset of $\mathbb{N}$. Such a set consists of an increasing sequence, finite or infinite, of positive integers, say, for example,

$$A = \{m_1, m_2, \ldots, m_n, \ldots\}.$$

Label by zero the elements of $\mathbb{N} - A$, and by 1 the elements of A, and keep their ordering within $\mathbb{N}$. This process uniquely identifies the sequence

$$\varepsilon_A = \{0, \ldots, 1_{m_1}, \ldots, 0, \ldots, 1_{m_2}, \ldots\}.$$

This is a sequence of the type of $\mathcal{E}$ introduced in (2.3). If $A = \emptyset$, we associate with $\emptyset$ the sequence $\varepsilon_{\emptyset} \in \mathcal{E}$ whose elements are all zero. Vice versa, any sequence $\{\varepsilon_n\} \in \mathcal{E}$ identifies uniquely, by the inverse process, one and only one element of $2^{\mathbb{N}}$. Thus $2^{\mathbb{N}}$ is in one-to-one correspondence with $\mathcal{E}$, which in turn is in one-to-one correspondence with $\mathbb{R}$.

Statement (3.2) is proved by establishing that no function $f : X \to 2^X$ can be a surjection. Let f be any such function and set[5]

$$A_f = \{x \in X \mid x \notin f(x)\}.$$

If $f : X \to 2^X$ were a surjection, there would exist $y \in X$ such that $f(y) = A_f$. By the definition of A_f, such a y can be neither in A_f nor in $X - A_f$. $\square$

---

[4]Here we are interested only in the cardinality of powers of $\mathbb{N}$ and $\mathbb{R}$, to estimate the cardinality of the Borel sets, versus the cardinality of the Lebesgue-measurable sets (Section 14 of the Problems and Complements of Chapter II).

[5]This is a generalized Cantor diagonalization process. The set A_f could be empty.

Corollary 3.4. *Given any nonempty set X, there exists a set Y containing X and of strictly larger cardinality.*

Corollary 3.5. $\mathrm{card}(\mathbb{R}) < \mathrm{card}(2^{\mathbb{R}})$.

4   Cardinality of some infinite Cartesian products

For a positive integers m, the set X^m is the collection of all m-tuples of elements of X. Any such m-tuple $(x_1, \ldots, x_m)$ can be regarded as a function from the first m integers $(1, \ldots, m)$ into X.

By analogy, $X^{\mathbb{N}}$ is defined as the collection of all sequences $\{x_n\}$ of elements of X. Any such sequence $\{x_n\}$ can be regarded as a function from $\mathbb{N}$ into X.

For example, $\mathbb{R}^{\mathbb{N}}$ is the collection of all sequences of real numbers or, equivalently, is the collection of all functions $f : \mathbb{N} \to \mathbb{R}$.

The product space $[0, 1]^{\mathbb{N}}$ is called the Hilbert cube and is the collection of all sequences $\{x_n\}$ of elements in $[0, 1]$.

Let $\{0, 1\}$ denote any set consisting of only two elements, say, for example, 0 and 1. Then $\{0, 1\}^{\mathbb{N}}$ is in one-to-one correspondence with the Cantor set. Therefore,

$$\mathrm{card}(\{0, 1\}^{\mathbb{N}}) = \mathrm{card}(\mathbb{R}). \tag{4.1}$$

If A is any set, countable or not, the Cartesian product X^A is defined to be the collection of all functions $f : A \to X$. For example, $(0, 1)^{(0, 1)}$ is the collection of all functions $f : (0, 1) \to (0, 1)$. Likewise, $\mathbb{R}^{\mathbb{R}}$ is the set of all real-valued functions defined in $\mathbb{R}$.

Proposition 4.1. $\mathrm{card}(2^{\mathbb{N}} \times 2^{\mathbb{N}}) = \mathrm{card}(2^{\mathbb{N}}) = \mathrm{card}(\mathbb{R})$.

Proof. We exhibit a one-to-one correspondence between $(2^{\mathbb{N}} \times 2^{\mathbb{N}})$ and $2^{\mathbb{N}}$. For any two subsets A and B of $\mathbb{N}$, set

$$g(A) = \bigcup_{n \in A} \{2n\}, \qquad h(B) = \bigcup_{m \in B} \{2m - 1\}. \tag{4.2}$$

Then for any pair of sets $A, B \in 2^{\mathbb{N}}$, define

$$f(A, B) = g(A) \bigcup h(B) = C \in 2^{\mathbb{N}}. \tag{4.3}$$

Every element $(A, B) \in (2^{\mathbb{N}} \times 2^{\mathbb{N}})$ is mapped into one and only one element of $2^{\mathbb{N}}$. Vice versa, given any $C \in 2^{\mathbb{N}}$, separating its even and odd numbers identifies in a unique manner two sets A and B in $2^{\mathbb{N}}$, and hence a unique element $(A, B) \in (2^{\mathbb{N}} \times 2^{\mathbb{N}})$ such that (4.3) holds. $\square$

Corollary 4.2. *For every positive integer m,*

$$\mathrm{card}(\mathbb{R}^m) = \mathrm{card}(\mathbb{R}). \tag{4.4}$$

Proof. Assume first that $m = 2$. Since there is a one-to-one correspondence between $\mathbb{R}$ and $2^{\mathbb{N}}$,

$$\text{card}(\mathbb{R}^2) = \text{card}(\mathbb{R} \times \mathbb{R}) = \text{card}(2^{\mathbb{N}} \times 2^{\mathbb{N}}) = \text{card}(\mathbb{R}).$$

For general $m \in \mathbb{N}$, the statement follows by induction.   □

Proposition 4.3. *Let X, Y, and Z be any triple of sets. Then*

$$\text{card}(X^{Y \times Z}) = \text{card}\{(X^Y)^Z\}.$$

Proof. Let $f \in X^{Y \times Z}$ so that $f(y, z) \in X$ for all pairs $(y, z) \in Y \times Z$. For a fixed $z \in Z$, set

$$h_z(y) = f(y, z). \tag{4.5}$$

This gives an element of X^Y. Thus as z ranges over Z, the map $h_z(\cdot)$ uniquely identifies a function from Z into X^Y. Conversely, any element $h_z \in (X^Y)^Z$ uniquely identifies an element $f \in X^{Y \times Z}$ by formula (4.5).   □

Corollary 4.4. $\text{card}(\mathbb{R}^{\mathbb{N}}) = \text{card}(\mathbb{R})$.

Proof. Since $\mathbb{R}$ is in one-to-one correspondence with $\{0, 1\}^{\mathbb{N}}$,

$$\text{card}(\mathbb{R}^{\mathbb{N}}) = \text{card}[(\{0, 1\}^{\mathbb{N}})^{\mathbb{N}}] = \text{card}(\{0, 1\}^{\mathbb{N} \times \mathbb{N}})$$
$$= \text{card}(\{0, 1\}^{\mathbb{N}}) = \text{card}(\mathbb{R})$$

since $\mathbb{N} \times \mathbb{N}$ and $\mathbb{N}$ are in one-to-one correspondence.   □

Corollary 4.5. $\text{card}(\mathbb{N}^{\mathbb{N}}) = \text{card}(\mathbb{R})$.

Proof. An element of $\mathbb{N}^{\mathbb{N}}$ is a sequence of elements of $\mathbb{N}$. In particular, $\mathbb{N}^{\mathbb{N}}$ contains those sequences containing only two fixed elements of $\mathbb{N}$, say, for example, $\{1, 2\}$. Therefore, $\{1, 2\}^{\mathbb{N}} \subset \mathbb{N}^{\mathbb{N}}$ and

$$\text{card}(\mathbb{N}^{\mathbb{N}}) \geq \text{card}(\{1, 2\}^{\mathbb{N}}) = \text{card}(\mathbb{R}).$$

On the other hand, $\mathbb{N}^{\mathbb{N}}$ is contained in $\mathbb{R}^{\mathbb{N}}$. Therefore,

$$\text{card}(\mathbb{N}^{\mathbb{N}}) \leq \text{card}(\mathbb{R}^{\mathbb{N}}) = \text{card}(\mathbb{R}).$$

The conclusion now follows from Proposition 3.1.   □

5   Orderings, the maximal principle, and the axiom of choice

A relation $\prec$ on a set X is a *partial ordering* of X if it is transitive (i.e., $x \prec y$ and $y \prec z$ implies $x \prec z$) and antisymmetric (i.e., $x \prec y$ and $y \prec x$ implies $x = y$).

The relation $\leq$ of *less than or equal to* is a partial ordering of $\mathbb{R}$. The *set inclusion* $\subseteq$ partially orders the set 2^X of all subsets of X.

A partial ordering $\prec$ on a set X is a *linear ordering* on X if for any two elements $x, y \in X$, either $x \prec y$ or $y \prec x$. The relation $\leq$ is a linear ordering on $\mathbb{R}$, whereas $\subseteq$ is not a linear ordering on 2^X.

Let X be a set partially ordered by a relation $\prec$ and let $E \subset X$. An *upper bound* of a subset $E \subset X$ is an element $x \in X$ such that $y \prec x$ for all $y \in E$. If $x \in E$, then x is a *maximal element* of E.

A linearly ordered subset $E \subset X$ is *maximal* if any linearly ordered subset of X is contained in E.

Hausdorff Maximal Principle. *Every partially ordered set contains a maximal linearly ordered subset. Partial and linear order are meant with respect to the same ordering $\prec$.*

The Hausdorff maximal principle is taken as an *axiom*. The following maximality and ordering statements, i.e., *Zorn's lemma*, the *axiom of choice*, and the *well-ordering principle*, will be proven as a consequence of the Hausdorff maximal principle. It can be proven that they are equivalent, i.e., that any one of them, taken as an *axiom*, implies the remaining three.[6]

Zorn's Lemma. *Let X be a partially ordered set such that every linearly ordered subset has an upper bound. Then X has a maximal element.*

Proof. Let M be the maximal linearly ordered set claimed by the maximal principle. An upper bound for M is a maximal element of X. $\qquad\square$

The Axiom of Choice. *Let $\{X_\alpha\}$ be a collection of nonempty sets, as the index α ranges over some set A. There exists a function f defined in A, such that $f(\alpha) \in X_\alpha$.*

Equivalently, one may choose an element x_α out of each set X_α. If the sets of the collection $\{X_\alpha\}$ are disjoint, one might select one and only one representative out of each X_α.

Proof. A function φ that maps the elements β of a subset $B \subset A$ into elements of X_β may be regarded as a set of ordered pairs

$$\varphi = \bigcup \{(\beta, x_\beta) | \beta \in B \subset A \text{ and } x_\beta \in X_\beta\}$$

---

[6]See [27. pp. 33–36].

such that no two of such pairs have the same first coordinate. The set Φ of all such φ is partially ordered by inclusion. Let Ψ be a linearly ordered subset of Φ and set

$$\psi = \bigcup \{\varphi | \varphi \in \Psi\}.$$

Such a ψ is a union of pairs (α, x_α) for $\alpha \in A$ and $x_\alpha \in X_\alpha$. Since Ψ is linearly ordered, any two such pairs (α, x_α) and (β, x_β) must belong to some $\varphi \in \Psi$. Therefore, $\alpha \neq \beta$ since φ is a function. This implies that ψ is a function in Φ and is an upper bound for Ψ. Thus every linearly ordered subset of Φ has an upper bound. By Zorn's lemma, Φ has a maximal element f.

To prove the axiom, it remains to show that the domain of f is A. If not, we may select an element $\alpha \in A - \text{dom}\{f\}$ and associate with it an element $x_\alpha \in X_\alpha$, since X_α is not empty. Then the function

$$f' = f \bigcup (\alpha, x_\alpha)$$

is in Φ and $f \subset f'$, against the maximality of f. $\qquad\square$

Corollary 5.1. *Let X be a set. There exists a function $f : 2^X \to X$ such that $f(E) \in E$ for every $E \subset X$, i.e., one may choose an element out of every subset of X.*

6   Well-ordering

A linear ordering $\prec$ is a *well-ordering* of X if every subset of X has a first element. The set $\mathbb{N}$ of positive integers is well-ordered by $\leq$. The set $\mathbb{R}$ is not well-ordered by the same ordering.

Well-Ordering Principle. *Every set X can be well-ordered; i.e., there exists a relation $\prec$ that well-orders X.*[7]

Proof. Let $f : 2^X \to X$ be a function, as in Corollary 5.1, whose existence is guaranteed by the axiom of choice. Set $x_1 = f(X)$ and

$$x_n = f\left(X - \bigcup_{j=1}^{n-1} x_j\right) \quad \text{for } n \geq 2. \tag{6.1}$$

The sequence $\{x_n\}$ can be given the ordering of $\mathbb{N}$ and, as such, is well-ordered. A well-ordering for X is constructed by rendering transfinite such a process.

Let D be a subset of X and let $\prec$ be a linear ordering defined on D. A subset $E \subset D$ is a *segment* relative to $\prec$ if for any $x \in E$, all the elements $y \in D$ such that $y \prec x$ belong to E.

---

[7]E. Zermelo, Neuer Beweis für die Wohlordnung, *Math. Ann.*, **65** (1908), 107–128.

The segments of $\{x_n\}$ relative to the ordering induced by $\mathbb{N}$ are the sets of the form $\{x_1, x_2, \ldots, x_m\}$ for some $m \in \mathbb{N}$. The union and intersection of two segments is a segment. The empty set is a segment relative to any linear ordering $\prec$.

Denote by $\mathcal{F}$ the family of *linear* orderings $\prec$ defined on subsets D of X and satisfying the following:

If $E \subset D$ is a segment, then the first element of $(D - E)$ is $f(X - E)$.   (*)

Such a family is not empty since the ordering of $\mathbb{N}$ on the domain $D = \{x_n\}$ is in $\mathcal{F}$.

Lemma 6.1. *Every element of $\mathcal{F}$ is a well-ordering on its domain.*

Proof. Let $D \subset X$ be the domain of a linear ordering $\prec \in \mathcal{F}$. For a nonempty subset $A \subset D$, let

$$E = \{y \in D | y \prec x \text{ for all } x \in A \text{ and } y \neq x\}.$$

Then the first element of A is the first element of $(D - E)$, and the latter is $f(X - E)$ since E is a segment.   □

Lemma 6.2. *Let $\prec_1$ and $\prec_2$ be two elements in $\mathcal{F}$ with domains D_1 and D_2. Then one of the two domains, say, for example, D_1, is a segment for the other, say, for example, D_2, with respect to the corresponding ordering $\prec_2$. Moreover, $\prec_1$ and $\prec_2$ coincide on such a segment.*

Proof. Let E denote the set of all x such that

$$\{y \in D_1 | y \prec_1 x\} = \{y \in D_2 | y \prec_2 x\}$$

and such that $\prec_1$ and $\prec_2$ agree on these two sets. By construction, E is a segment for both D_1 and D_2 with respect to their orderings.

The set E must coincide with at least one of D_1 and D_2. If not, the element $f(X - E)$ is the first element of both $(D_1 - E)$ and $(D_2 - E)$. By the definition of E, such an element is in E. This is, however, a contradiction since $f(X - E)$ is not in E.   □

Let D_o be the union of the domains of the elements of $\mathcal{F}$. Also let $\prec_o$ be that ordering on D_o that coincides with the ordering $\prec$ in $\mathcal{F}$ on its domain D. By Lemma 6.2, this is a linear ordering on D_o and satisfies requirement (*) of the class $\mathcal{F}$. Therefore, by Lemma 6.1, it is a well-ordering on D_o. It remains to show that $D_o = X$. Consider the set

$$D'_o = D_o \bigcup \{f(X - D_o)\}$$

and the ordering $\prec'_o$ that coincides with $\prec_o$ on D_o and by which $f(X - D_o)$ follows any element of D_o. One checks that $\prec'_o$ satisfies requirements (a) and (b) of the class $\mathcal{F}$ and its domain is D'_o. Therefore, $D_o = D'_o$. However, this is a contradiction unless $(X - D_o) = \emptyset$.   □

6.1   The first uncountable. Let X be an uncountable set, well-ordered by the ordering $\prec$. Without loss of generality, we may assume that X has an upper bound, i.e., that there exists some $x^* \in X$ such that $x \prec x^*$ for all $x \in X - \{x^*\}$.

Indeed, if not, we may add to X an element x^*, and on $X \bigcup \{x^*\}$ define an ordering $\prec^*$ to coincide with $\prec$ on X and by which $x \prec x^*$ for all $x \in X$.

For $x \in X$, set

$$E_x = \{y \in X | y \prec x\}.$$

If E_x is a finite set, then x is called a *finite ordinal*. Now consider the set

$$E^o = \{x \in X | E_x \text{ is infinite}\}.$$

Since X is infinite and has an upper bound, $E^o \neq \emptyset$. Since X is well-ordered by $\prec$, the set E^o has a least element denoted by ω. The set E_ω is infinite, and for each $x \prec \omega$, the set E_x is finite. For this reason, ω is referred to as the *first infinite ordinal*.

The set E_ω is the set of *finite ordinals* and is in one-to-one correspondence with the set $\mathbb{N}$ of the natural numbers ordered with the natural ordering of $\mathbb{N}$.

If E_x is a countable set, then x is called a *countable ordinal*. Set

$$E^1 = \{x \in X | E_x \text{ is uncountable}\}.$$

Since X is uncountable and has an upper bound, $E^1 \neq \emptyset$. Since X is well-ordered by $\prec$, the set E^1 has a least element denoted by Ω. The set E_Ω is uncountable, and for each $x \prec \Omega$ the set E_x is countable. For this reason, Ω is referred to as the *first uncountable ordinal*.

Let $\mathbb{R}$ be well-ordered by $\prec$. The cardinality of E_ω is denoted by $\aleph_o$. The cardinality of E_Ω is denoted by $\aleph_1$. The inclusion $E_\Omega \subset \mathbb{R}$ implies $\aleph_1 \leq \text{card}(\mathbb{R})$. The *continuum hypothesis* states that the cardinality of $\mathbb{R}$ is $\aleph_1$; i.e., roughly speaking, there are no sets, in the sense of cardinality, between E_Ω and $\mathbb{R}$.

PROBLEMS AND COMPLEMENTS

1   COUNTABLE SETS

1.1   A real number x is called *algebraic* if it is the root of a polynomial with rational coefficients, i.e., if there exists an $(n + 1)$-tuple $\{a_o, a_1, a_2, \ldots, a_n\}$ of rational numbers such that

$$a_o + a_1 x + a_2 x^2 + \cdots + a_n x^n = 0.$$

The collection of all the algebraic numbers is countable.

1.2   The collection of the prime numbers is countable.

2 THE CANTOR SET

2.1   Let $s \geq 2$ be a positive integer and let $\{\sigma_n\}$ be a sequence of positive integers that can take only the values $1, \ldots, (s-1)$. Any $x \in [0, 1]$ has the representation

$$x = \sum \frac{\sigma_n}{s^n}, \quad \text{where } \{\sigma_n\} \text{ is one such sequence.} \tag{2.1c}$$

The sequence $\{\sigma_n\}$ that identifies x is unique except if x is of the form $x = r/s^n$ for some positive integer r. In the latter case there are exactly 2 sequences that identify x. Conversely, if $\{\sigma_n\}$ is any such sequence, then (2.1c) converges to a number $x \in [0, 1]$. For $s = 2, 3, 10$, this gives the binary, ternary, or decimal representation of x.

2.2   Let $\mathcal{L}_n$ and $\mathcal{R}_n$ denote, respectively, the set of all the left- and right-hand points of the 2^{n-1} intervals removed out of $[0, 1]$, at the nth step of the process. For example,[8]

$$\mathcal{L}_1 = \left\{ \frac{1}{3} \right\}; \quad \mathcal{L}_2 = \left\{ \frac{1}{3^2}, \frac{7}{3^2} \right\}; \quad \mathcal{L}_3 = \left\{ \frac{1}{3^3}, \frac{7}{3^3}, \frac{19}{3^3}, \frac{25}{3^3} \right\}; \ldots;$$

$$\mathcal{R}_1 = \left\{ \frac{2}{3} \right\}; \quad \mathcal{R}_2 = \left\{ \frac{2}{3^2}, \frac{8}{3^2} \right\}; \quad \mathcal{R}_3 = \left\{ \frac{2}{3^3}, \frac{8}{3^3}, \frac{20}{3^3}, \frac{26}{3^3} \right\}; \ldots.$$

Also let $\mathcal{L}_o = \{1\}$ and $\mathcal{R}_o = \{0\}$. Then

$$\mathcal{C} \supset \bigcup \left(\mathcal{L}_n \bigcup \mathcal{R}_n \right) \quad \text{(strict inclusion).}$$

The elements of $\mathcal{L}_n$ and $\mathcal{R}_n$ can be constructed recursively as

$$\mathcal{L}_n = \left\{ \text{numbers of the form } r + \frac{1}{3^n} \text{ for } r \in \bigcup_{j=0}^{n-1} \mathcal{R}_j \right\};$$

$$\mathcal{R}_n = \left\{ \text{numbers of the form } r + \frac{2}{3^n} \text{ for } r \in \bigcup_{j=0}^{n-1} \mathcal{R}_j \right\}.$$

It follows that if $\beta_n \in \mathcal{R}_n$, there exists a finite sequence of positive integers

$$j_1 < j_2 < \cdots < j_{n-1} \leq (n-1)$$

such that

$$\beta_n = \frac{2}{3^{j_1}} + \frac{2}{3^{j_2}} + \cdots + \frac{2}{3^{j_{n-1}}} + \frac{2}{3^n}.$$

---

[8]This description of $\mathcal{L}_n$ and $\mathcal{R}_n$ is needed in the construction of the Cantor ternary function in Section 5.3 of the Problems and Complements of Chapter IV.

Therefore, β_n has a ternary expansion with only digits 0 and 2 of the form

$$\beta_n = \{0, \ldots, 2_{j_1}, 0, \ldots, 2_{j_2}, 0, \ldots, 2_{j_{n-1}}, 0, \ldots, 2_n, 0_{n+1}, 0, 0, \ldots\}.$$

Likewise, an element $\alpha_n \in \mathcal{L}_n$ is of the form

$$\alpha_n = \frac{2}{3^{j_1}} + \frac{2}{3^{j_2}} + \cdots + \frac{2}{3^{j_{n-1}}} + \frac{1}{3^n}.$$

The ternary expansion of α_n has only digits 0 and 2 and is of the form

$$\alpha_n = \{0, \ldots 2_{j_1}, 0, \ldots, 2_{j_2}, 0, \ldots, 2_{j_{n-1}}, 0, \ldots, 1_n, 0_{n+1}, 0, 0, \ldots\}$$
$$= \{0, \ldots 2_{j_1}, 0, \ldots, 2_{j_2}, 0, \ldots, 2_{j_{n-1}}, 0, \ldots, 0_n, 2_{n+1}, 2, 2, \ldots\}.$$

If (α_n, β_n) is an interval removed out of $[0, 1]$ at the nth step of the process, then α_n and β_n have the same ternary expansion up to the terms of order $(n-1)$. The term of order n is 2 for the right-hand point and 0 for the left-hand point. The remaining terms in the expansion of α_n are all 2 and those in the expansion of β_n are all 0.

2.3   A GENERALIZED CANTOR SET OF POSITIVE MEASURE. Let $\alpha \in (0, 1)$, and out of the interval $[0, 1]$ remove the central open interval

$$I_1^\alpha = \left(\frac{1}{2} - \frac{\alpha}{6}, \frac{1}{2} + \frac{\alpha}{6}\right) \quad \text{of length } \frac{\alpha}{3}.$$

Out of each of the remaining two intervals $(0, \frac{1}{2} - \frac{\alpha}{6})$ and $(\frac{1}{2} + \frac{\alpha}{6}, 1)$, remove the two central intervals each of length $\frac{1}{9}\alpha$ and denote by I_2^α their union. Proceeding in this fashion, define sequences of open sets I_n^α, each being the finite union of disjoint open intervals and satisfying

$$\text{meas}(I_n^\alpha) = \alpha \frac{2^{n-1}}{3^n} \quad \text{and} \quad \sum \text{meas}(I_n^\alpha) = \alpha. \tag{2.2c}$$

The generalized Cantor set is the set that remains after removing the union of the I_n^α out of $[0, 1]$; i.e.,

$$\mathcal{C}_\alpha = [0, 1] - \bigcup_n I_n^\alpha. \tag{2.3c}$$

2.4   A GENERALIZED CANTOR SET OF MEASURE ZERO. Fix $\delta \in (0, \frac{1}{2})$. From $[0, 1]$ remove the open middle interval I_0 of length $(1 - 2\delta)$ so that

$$[0, 1] - I_0 = J_{1,1} \bigcup J_{1,2},$$

where

$$J_{1,1} = [0, \delta], \qquad J_{1,2} = [1 - \delta, 1], \qquad I_o = (\delta, 1 - \delta).$$

Next, out of $J_{1,1}$ and $J_{1,2}$, remove the open middle intervals $I_{1,1}$ and $I_{1,2}$, each of length $\delta - 2\delta^2$, and denote by I_1 their union. Thus

$$I_1 = I_{1,1} \bigcup I_{1,2}, \qquad \text{meas}(I_1) = 2\delta - 4\delta^2,$$

where

$$I_{1,1} = (\delta^2, \delta - \delta^2), \qquad I_{1,2} = (1 - (\delta - \delta^2), 1 - \delta^2).$$

The set that remains is

$$[0, 1] - \left(I_0 \bigcup I_1 \right) = J_{2,1} \bigcup J_{2,2} \bigcup J_{2,3} \bigcup J_{2,4},$$

where

$$J_{2,1} = [0, \delta^2], \qquad\qquad J_{2,2} = [\delta - \delta^2, \delta];$$
$$J_{2,3} = [1 - \delta, 1 - (\delta - \delta^2)], \qquad J_{2,4} = [1 - \delta^2, 1].$$

From this,

$$\text{meas}\left(I_0 \bigcup I_1 \right) = 1 - 4\delta^2 \quad \text{and} \quad \sum_{j=1}^{4} \text{meas}(J_{2,j}) = 4\delta^2.$$

Next, out of each $J_{2,j}$, remove the open middle interval $I_{2,j}$ of length $\delta^2 - 2\delta^3$ and denote by I_3 their union. Thus

$$I_3 = \bigcup_{j=1}^{4} I_{2,j}, \quad \text{and} \quad \text{meas}(I_3) = 4\delta^2 - 8\delta^3.$$

The set that remains is

$$[0, 1] - \bigcup_{i=0}^{3} I_i = \bigcup_{j=1}^{2^3} J_{3,j},$$

where $J_{3,j}$ are closed disjoint subintervals of $[0, 1]$, each of length δ^3.

Proceeding in this fashion, we generate sequences of open sets I_n, each being the finite union of disjoint open intervals and of disjoint closed intervals $J_{m,n} \subset [0, 1]$ such that

$$[0, 1] - \bigcup_{i=0}^{n} I_i = \bigcup_{j=1}^{2^n} J_{n,j}.$$

Moreover,

$$\sum_{i=0}^{n} \text{meas}(I_i) = 1 - 2^n \delta^n \qquad\qquad (2.4c)$$

and

$$\sum_{j=1}^{2^n} \text{meas}(J_{n,j}) = 2^n \delta^n. \tag{2.5c}$$

The generalized Cantor set $\mathcal{C}_\delta$ is defined as

$$\mathcal{C}_\delta = [0, 1] - \bigcup_{i=0}^{\infty} I_i. \tag{2.6c}$$

For $\delta = \frac{1}{3}$ this coincides with the Cantor set $\mathcal{C}$ introduced in (2.2). From (2.6c), it follows that $\text{meas}(\mathcal{C}_\delta) = 0$ for all $\delta \in (0, \frac{1}{2})$.

Remark 2.1c. For each $n \in \mathbb{N}$ the Cantor set $\mathcal{C}_\delta$ is covered by the union of the closed intervals $J_{n,j}$ for $j = 1, 2, \ldots, 2^n$.

2.5   PERFECT SETS. A nonempty set $E \subset \mathbb{R}^N$ is *perfect* if it coincides with the set of its accumulation points. The Cantor set is perfect. The set of the rational numbers is not perfect.

Proposition 2.1c. *A perfect set is uncountable.*

Proof. Let E be perfect and assume, by contradiction, that is countable, i.e., that $E = \bigcup x_n$, for a sequence $\{x_n\}$ of elements in $\mathbb{R}^N$. Pick $y_1 \in (E - x_1)$. Since $\text{dist}\{x_1, y_1\} > 0$, there exists a closed cube Q_1, centered at y_1, that avoids x_1. Since E is perfect and Q_1 is closed and bounded,

$$Q_1 \bigcap E \text{ is compact and avoids } x_1.$$

Next, remove x_2 out of E. The point y_1, as an element of E, is an accumulation point of elements in $(E - x_2)$. Therefore, the intersection

$$\mathring{Q}_1 \bigcap (E - x_2)$$

is not empty and we may pick an element y_2 out of it. Since y_2 is different than x_1 and x_2, there exists a closed cube Q_2 centered at y_2, contained in Q_1, and not containing x_1 and x_2. For such a cube,

$$Q_1 \bigcap E \supset Q_2 \bigcap E \text{ is compact and avoids } x_1 \text{ and } x_2.$$

Proceeding in this fashion, we construct a nested sequence of closed, bounded sets $Q_n \bigcap E$ satisfying the following:

$$Q_{n-1} \bigcap E \supset Q_n \bigcap E \text{ is compact and avoids } x_1, \ldots, x_n.$$

Thus $\bigcap (Q_n \bigcap E) = \emptyset$. However, this is impossible (see Section 2.6 below). $\square$

2.6   Let $\{E_n\}$ be a sequence of nonempty, closed sets in $\mathbb{R}^N$ such that $E_{n+1} \subset E_n$. If E_{n_o} is bounded for some index n_o, then $\bigcap E_n \neq \emptyset$. If all the E_n are unbounded, their intersection might be empty.

I

Topologies and Metric Spaces

1 Topological spaces

Let X be a set. A collection $\mathcal{U}$ of subsets of X defines a *topology* on X if

(i) the empty set $\emptyset$ and X belong to $\mathcal{U}$;

(ii) the union of any collection of sets in $\mathcal{U}$ is in $\mathcal{U}$;

(iii) the intersection of finitely many elements of $\mathcal{U}$ is in $\mathcal{U}$.

The pair $\{X; \mathcal{U}\}$, i.e., X endowed with the topology generated by $\mathcal{U}$, is a topological space. The elements $\mathcal{O}$ of $\mathcal{U}$ are the *open* sets of X. A set C in X is *closed* if $(X - C)$ is open. The empty set $\emptyset$ and X are both open and closed. It follows from the definitions that the finite union of closed sets is closed and the intersection of any collection of closed sets is closed.

An open neighborhood of a set $A \subset X$ is any open set $\mathcal{O}$ that contains A. In particular, an open neighborhood of a singleton $x \in X$ is any open set $\mathcal{O}$ such that $x \in \mathcal{O}$. A subset $\mathcal{O} \subset X$ is open if and only if it is an open neighborhood of any of its points.

A point $x \in A$ is an *interior* point of A if there exists an open set $\mathcal{O}$ such that $x \in \mathcal{O} \subset A$. The interior of A is the set of all its interior points. A set $A \subset X$ is open if and only if it coincides with its interior.

A point x is a point of *closure* of A if every open neighborhood of x intersects A. The closure $\overline{A}$ of A is the set of all the points of closure of A. A set A is closed if and only if $A = \overline{A}$. It follows that A is closed if and only if it is the intersection of all closed sets containing A.

Let $\{x_n\}$ be a sequence of elements of X. A point $x \in X$ is a *cluster* point for the sequence $\{x_n\}$ if every open set $\mathcal{O}$ containing x contains x_n for infinitely many n.

The sequence $\{x_n\}$ converges to x if for every open set $\mathcal{O}$ containing x, there exists a positive integer $m(\mathcal{O})$ depending on $\mathcal{O}$, such that $x_n \in \mathcal{O}$ for all $n \geq m(\mathcal{O})$.

Thus limit points for $\{x_n\}$ are cluster points. The converse is false. Indeed, there exist sequences $\{x_n\}$ with a cluster point x_o such that no subsequence of $\{x_n\}$ converges to x_o (see Section 1.11 of the Problems and Complements).

Proposition 1.1 (Cauchy). *A sequence $\{x_n\}$ of elements of a topological space $\{X; \mathcal{U}\}$ converges to x if and only if every subsequence $\{x_{n'}\} \subset \{x_n\}$ contains in turn a subsequence $\{x_{n''}\} \subset \{x_{n'}\}$ converging to x.*

Let A and B be subsets of X. The set B is *dense* in A if $A \subset \overline{B}$. If also $B \subset A$, then $\overline{A} = \overline{B}$.

The space $\{X; \mathcal{U}\}$ is *separable* if it contains a countable dense set.

Let X_o be a subset of X. The collection $\mathcal{U}$ induces a topology on X_o by the family

$$\mathcal{U}_o = \left\{ \mathcal{O} \bigcap X_o \,|\, \mathcal{O} \in \mathcal{U} \right\}.$$

The pair $\{X_o; \mathcal{U}_o\}$ is a topological subspace of $\{X; \mathcal{U}\}$.

A subspace of a separable topological space need not be separable (see Section 4.9 of the Problems and Complements).

Let $\{X; \mathcal{U}\}$ and $\{Y; \mathcal{V}\}$ be any two topological spaces. A function $f : X \to Y$ is continuous at a point $x \in X$ if for every open set $O \in \mathcal{V}$ containing $f(x)$, there is an open set $\mathcal{O} \in \mathcal{U}$ containing x and such that $f(\mathcal{O}) \subset O$. A function $f : X \to Y$ is continuous if it is continuous at every $x \in X$. This implies that f is continuous if and only if the preimage of every open set is open, i.e., if for every open set $O \in \mathcal{V}$, the set $f^{-1}(O)$ is an open set $\mathcal{O} \in \mathcal{U}$. Equivalently, f is continuous if and only if the preimage of a closed set is closed.

The restriction of a continuous function $f : X \to Y$ to a subset $X_o \subset X$ is continuous with respect to the induced topology of $\{X_o; \mathcal{U}_o\}$.

A *homeomorphism* between $\{X; \mathcal{U}\}$ and $\{Y; \mathcal{V}\}$ is a continuous one-to-one function f from X onto Y, with continuous inverse f^{-1}.

If $f : X \to Y$ is a homeomorphism, then $f(\mathcal{O}) \in \mathcal{V}$ for all $\mathcal{O} \in \mathcal{U}$.

Two homeomorphic topological spaces are equivalent in the sense that the elements of X are in one-to-one correspondence with the elements of Y, and the open sets making up the topology of $\{X; \mathcal{U}\}$ are in one-to-one correspondence with the open sets making up the topology of $\{Y; \mathcal{V}\}$.

The collection 2^X of all subsets of X generates a topology on X called the *discrete topology*. Every function f from $\{X; 2^X\}$ into a topological space $\{Y; \mathcal{V}\}$ is continuous.

The *trivial topology* on X is that for which the only open sets are X and $\emptyset$. The closure of any point $x \in X$ is X. All the continuous real-valued functions defined on X are constant.

As a shorthand notation, we denote by X a topological space whenever a topology $\mathcal{U}$ is clear from the context or whenever the specification of a topology $\mathcal{U}$ is immaterial.

By a *real-valued* function f defined on some $\{X; \mathcal{U}\}$, we mean a function from $\{X; \mathcal{U}\}$ into $\mathbb{R}$ endowed with the Euclidean topology.

1.1 Hausdorff and normal spaces. A topological space $\{X; \mathcal{U}\}$ is a *Hausdorff* space if it separates points, i.e., if for any two distinct points x, $y \in X$, there exist disjoint open sets $\mathcal{O}_x$ and $\mathcal{O}_y$ such that $x \in \mathcal{O}_x$ and $y \in \mathcal{O}_y$.

Proposition 1.2. *Let* $\{X; \mathcal{U}\}$ *be a Hausdorff topological space. Then the points* $x \in X$ *are closed.*[1]

Proof. Every point $y \in (X - x)$ is contained in some open set contained in $(X - x)$. Since $(X - x)$ is the union of all such open sets, it is open. Thus $\{x\}$ is closed. □

A topological space $\{X; \mathcal{U}\}$ is *normal* if it separates closed sets, i.e., if for any two disjoint closed sets C_1 and C_2, there exist disjoint open sets $\mathcal{O}_1$ and $\mathcal{O}_2$ such that $C_1 \subset \mathcal{O}_1$ and $C_2 \subset \mathcal{O}_2$.

A normal space need not be Hausdorff. For example, the trivial topology is not Hausdorff but it is normal. However, if in addition the singletons $\{x\}$ are closed, then a normal space is Hausdorff. The converse is false, as there exist Hausdorff spaces that do not separate closed sets (see Section 1.19 of the Problems and Complements).

2 Urysohn's lemma

Lemma 2.1 (Urysohn[2]). *Let* $\{X; \mathcal{U}\}$ *be normal. Given any two closed, disjoint sets A and B in X, there exists a continuous function $f : X \to [0, 1]$ such that*

$$f(x) = \begin{cases} 0 & \text{for } x \in A, \\ 1 & \text{for } x \in B. \end{cases}$$

Proof. We may assume that neither A nor B is empty. Indeed, if, for example, $A = \emptyset$, the function $f \equiv 1$ satisfies the conclusion of the lemma.

Let t denote rational diadic numbers in $[0, 1]$, i.e., of the form

$$t = \frac{m}{2^n}, \quad m = 0, 1, \ldots, 2^n, \quad n = 0, 1, \ldots.$$

For each such t we construct an open set $\mathcal{O}_t$ in such a way that the family $\{\mathcal{O}_t\}$ satisfies

$$\mathcal{O}_o \supset A, \quad \mathcal{O}_1 = (X - B), \quad \text{and} \quad \overline{\mathcal{O}}_\tau \subset \mathcal{O}_t \quad \text{whenever } \tau < t. \tag{2.1}$$

---

[1]The converse is false. See Section 4.2 of the Problems and Complements.
[2]P. Urysohn, Über die Mächtigkeit der zusammenhängenden Mengen, *Math. Ann.*, **94** (1925), 290.

Since $\{X; \mathcal{U}\}$ is normal, there exists an open set $\mathcal{O}_o$ containing A and whose closure is contained in $(X - B)$.

For $n = 0$ and $m = 0$, we select such an open set $\mathcal{O}_o$. For $n = 0$ and $m = 1$, we select $\mathcal{O}_1$ as in (2.1).

To $n = 1$ and $m = 0, 1, 2$, there correspond sets $\mathcal{O}_o, \mathcal{O}_{\frac{1}{2}}, \mathcal{O}_1$. The first and last have been selected and we select set $\mathcal{O}_{\frac{1}{2}}$ so that

$$\overline{\mathcal{O}}_o \subset \mathcal{O}_{\frac{1}{2}} \subset \overline{\mathcal{O}}_{\frac{1}{2}} \subset \mathcal{O}_1.$$

To $n = 2$ and $m = 0, 1, 2, 3, 4$ there correspond open sets $\mathcal{O}_{\frac{m}{2^2}}$ of which only $\mathcal{O}_{\frac{1}{4}}$ and $\mathcal{O}_{\frac{3}{4}}$ have to be selected. Since $\{X; \mathcal{U}\}$ is normal, there exist an open set $\mathcal{O}_{\frac{1}{4}}$ such that

$$\overline{\mathcal{O}}_o \subset \mathcal{O}_{\frac{1}{4}} \subset \overline{\mathcal{O}}_{\frac{1}{4}} \subset \mathcal{O}_{\frac{1}{2}}.$$

Analogously, there exists an open set $\mathcal{O}_{\frac{3}{4}}$ such that

$$\overline{\mathcal{O}}_{\frac{1}{2}} \subset \mathcal{O}_{\frac{3}{4}} \subset \overline{\mathcal{O}}_{\frac{3}{4}} \subset \mathcal{O}_1.$$

Proceeding by induction, if the open sets $\mathcal{O}_{m/2^{n-1}}$ have been selected, we choose the sets $\mathcal{O}_{m/2^n}$ by first observing that the ones corresponding to m even have already been selected. Therefore, we have only to choose those corresponding to m odd. For any such m fixed, the open sets $\mathcal{O}_{(m-1)/2^n}$ and $\mathcal{O}_{(m+1)/2^n}$ have been selected. Since $\{X; \mathcal{U}\}$ is normal, there exists an open set $\mathcal{O}_{m/2^n}$ such that

$$\overline{\mathcal{O}}_{\frac{m-1}{2^n}} \subset \mathcal{O}_{\frac{m}{2^n}} \subset \overline{\mathcal{O}}_{\frac{m}{2^n}} \subset \mathcal{O}_{\frac{m+1}{2^n}}.$$

Define $f : X \rightarrow [0, 1]$ by setting $f(x) = 1$ for $x \in B$ and

$$f(x) = \inf\{t | x \in \mathcal{O}_t\} \quad \text{for } x \in (X - B).$$

By construction, $f(x) = 0$ on A and $f(x) = 1$ on B. It remains to prove that f is continuous. From the definition of f, it follows that for all $s \in (0, 1]$,

$$[f < s] = \bigcup_{t < s} \mathcal{O}_t \quad \text{and} \quad [f \leq s] = \bigcap_{t > s} \mathcal{O}_t.$$

Therefore, $[f < s]$ is open. On the other hand by the last part of (2.1), $\overline{\mathcal{O}}_\tau \subset \mathcal{O}_t$ whenever $\tau < t$. Therefore,

$$[f \leq s] = \bigcap_{t > s} \mathcal{O}_t = \bigcap_{t > s} \overline{\mathcal{O}}_t,$$

and $[f \leq s]$ is closed. $\qquad\qquad\qquad\qquad\qquad\qquad\qquad\qquad\qquad\qquad\square$

Corollary 2.2. *A Hausdorff space $\{X; \mathcal{U}\}$ is normal if and only if, for every pair of closed disjoint subsets C_1 and C_2 of X, there exists a continuous function $f : X \rightarrow \mathbb{R}$ such that $f = 1$ on C_1 and $f = 0$ on C_2.*

3 The Tietze extension theorem

Theorem 3.1 (Tietze[3]). *Let $\{X; \mathcal{U}\}$ be normal. A continuous function f from a closed subset C of X into $\mathbb{R}$ has a continuous extension on X, i.e., there exists a continuous real-valued function f_* defined on the whole X, such that $f = f_*$ on C. Moreover, if f is bounded, i.e., if*

$$|f(x)| \leq M \quad \text{for all } x \in C \text{ for some } M > 0,$$

then f_* satisfies the same bound.

Proof. Assume first that f is bounded and that $M \leq 1$. We will construct a sequence of real-valued, continuous functions $\{g_n\}$, defined on the whole X, such that for all $n \in \mathbb{N}$,

$$|g_n(x)| \leq \frac{1}{3}\left(\frac{2}{3}\right)^{n-1} \quad \text{for all } x \in X \tag{3.1}$$

and

$$\left| f(x) - \sum_{j=1}^{n} g_j(x) \right| \leq \left(\frac{2}{3}\right)^n \quad \text{for all } x \in C. \tag{3.2}$$

Assuming the sequence $\{g_n\}$ has been constructed, by virtue of (3.1), the series $\sum g_n$ is uniformly convergent in X and $|\sum g_n| \leq 1$. Therefore, the functions

$$f_n = \sum_{j=1}^{n} g_j$$

are continuous and form a sequence $\{f_n\}$, uniformly convergent on X to a continuous function f_*.[4]

From (3.2), it follows that $f = f_*$ on C. It remains to construct the sequence $\{g_n\}$.

Since f is continuous, the two sets $[f \leq -\frac{1}{3}]$ and $[f \geq \frac{1}{3}]$ are closed and disjoint. By the Urysohn lemma, there exists a continuous function

$$g_1 : X \to \left[-\frac{1}{3}, \frac{1}{3}\right]$$

such that

$$g_1 = \frac{1}{3} \quad \text{on} \quad \left[f \geq \frac{1}{3}\right] \quad \text{and} \quad g_1 = -\frac{1}{3} \quad \text{on} \quad \left[f \leq -\frac{1}{3}\right].$$

---

[3]H. Tietze, Über Funktionen, die auf einer abgeschlossenen Menge stetig sind, *J. Math.*, **145** (1915), 9–14.

[4]See Section 1.8 of the Problems and Complements.

By construction,

$$|f(x) - g_1(x)| \leq \frac{2}{3} \quad \text{for all } x \in C.$$

The function $h_1 = (f - g_1)$ is continuous and bounded on C. The two sets $[h_1 \leq -\frac{2}{9}]$ and $[h_1 \geq \frac{2}{9}]$ are closed and disjoint. Therefore, by the Urysohn lemma, there exists a continuous function

$$g_2 : X \to \left[-\frac{2}{9}, \frac{2}{9} \right]$$

such that

$$g_2 = \frac{2}{9} \quad \text{on} \left[h_1 \geq \frac{2}{9} \right] \quad \text{and} \quad g_2 = -\frac{2}{9} \quad \text{on} \left[h_1 \leq -\frac{2}{9} \right].$$

By construction,

$$|h_1 - g_2| = |f - (g_1 + g_2)| \leq \frac{4}{9} \quad \text{on } C.$$

The sequence $\{g_n\}$ is constructed inductively by this procedure. This proves Tietze's theorem if f is bounded and $|f| \leq 1$. If f is bounded and $|f| \leq M$, the conclusion follows by replacing f with f/M. If f is unbounded, set

$$f_o = \frac{f}{1 + |f|}.$$

Since $f_o : C \to \mathbb{R}$ is continuous and bounded, it has a continuous extension $g_o : X \to [-1, 1]$. In particular,

$$g_o(x) = \frac{f(x)}{1 + |f(x)|} \in (-1, 1) \quad \text{for all } x \in C.$$

This implies that the set $[|g_o| = 1]$ is closed and disjoint from C. By the Urysohn lemma, there exists a continuous function $\eta : X \to \mathbb{R}$ such that $\eta = 1$ on C and $\eta = 0$ on $[|g_o| = 1]$. The function

$$g = \frac{\eta g_o}{1 - \eta |g_o|} : X \to \mathbb{R},$$

is continuous and coincides with f on C. $\qquad\square$

4   Bases, axioms of countability, and product topologies

A family of open sets $\mathcal{B}$ is a *base* for the topology of $\{X; \mathcal{U}\}$ if for every open set $\mathcal{O}$ and every $x \in \mathcal{O}$, there exists a set $B \in \mathcal{B}$ such that $x \in B \subset \mathcal{O}$.

A collection $\mathcal{B}_x$ of open sets is a base at x if for each open set $\mathcal{O}$ containing x, there exists $B \in \mathcal{B}_x$ such that $x \in B \subset \mathcal{O}$.

Thus if $\mathcal{B}$ is a base for the topology of $\{X; \mathcal{U}\}$, it is also a base for each of the points of X. More generally, $\mathcal{B}$ is a base if and only if it contains a base for each of the points of X.

A set $\mathcal{O}$ is open if and only if for each $x \in \mathcal{O}$ there exists $B \in \mathcal{B}$ such that $x \in B \subset \mathcal{O}$.

Let $\mathcal{B}$ be a base for $\{X; \mathcal{U}\}$. Then

(i) every $x \in X$ belongs to some $B \in \mathcal{B}$;

(ii) for any two given sets B_1 and B_2 in $\mathcal{B}$ and every $x \in B_1 \cap B_2$, there exists some $B_3 \in \mathcal{B}$ such that $x \in B_3 \subset B_1 \cap B_2$.

The notion of base is induced by the presence of a topology generated by $\mathcal{U}$ on X. Conversely, if a collection of sets $\mathcal{B}$ satisfies (i) and (ii), then it permits one to construct a topology on X for which $\mathcal{B}$ is a base.

Proposition 4.1. *Let $\mathcal{B}$ be a collection of sets in X satisfying* (i) *and* (ii). *There exists a collection $\mathcal{U}$ of subsets of X, which generates a topology on X, for which $\mathcal{B}$ is a base.*

Proof. Let $\mathcal{U}$ consist of the empty set $\emptyset$ and the collection of all subsets $\mathcal{O}$ of X, such that for every $x \in \mathcal{O}$ there exists an element $B \in \mathcal{B}$ such that $x \in B \subset \mathcal{O}$. Such a collection is not empty since $X \in \mathcal{U}$.

It follows from the definition that the union of any collection of elements in $\mathcal{U}$ remains in $\mathcal{U}$. Moreover, $\mathcal{U}$ contains the empty set $\emptyset$ and X.

Let $\mathcal{O}_1$ and $\mathcal{O}_\in$ be any two elements of $\mathcal{U}$ with nonempty intersection. For every $x \in \mathcal{O}_1 \cap \mathcal{O}_2$ there exist sets B_1, B_2, B_3 in $\mathcal{B}$ such that $B_i \subset \mathcal{O}_i, i = 1, 2$ and

$$x \in B_3 \subset B_1 \cap B_2 \subset \mathcal{O}_1 \cap \mathcal{O}_2.$$

Therefore, $\mathcal{O}_1 \cap \mathcal{O}_2 \in \mathcal{U}$. This implies that the collection $\mathcal{U}$ generates a topology on X for which $\mathcal{B}$ is a base. □

A topological space $\{X; \mathcal{U}\}$ satisfies the *first axiom of countability* if each point $x \in X$ has a countable base.

The space $\{X; \mathcal{U}\}$ satisfies the *second axiom of countability* if there exists a countable base for its topology.

Proposition 4.2. *Every topological space satisfying the second axiom of countability is separable.*

Proof. Let $\{\mathcal{O}\}$ be a countable base for the topology of $\{X; \mathcal{U}\}$. For each $i \in \mathbb{N}$ select an element $x_i \in \mathcal{O}_i$. This generates a countable, dense subset of $\{X; \mathcal{U}\}$. □

4.1   Product topologies. Let $\{X_1; \mathcal{U}_1\}$ and $\{X_2; \mathcal{U}_2\}$ be two topological spaces. The *product topology* $\mathcal{U}_1 \times \mathcal{U}_2$ on the Cartesian product $X_1 \times X_2$ is constructed by considering the collection $\mathcal{B}$ of all products

$$\mathcal{O}_1 \times \mathcal{O}_2, \quad \text{where } \mathcal{O}_i \in \mathcal{U}_i, \quad i = 1, 2.$$

These are called the *open rectangles* of the product topology. First, one verifies that they form a base in the sense of (i) and (ii). Then the product topological space

$$\{X_1 \times X_2; \mathcal{U}_1 \times \mathcal{U}_2\}.$$

is constructed by the procedure of Proposition 4.1. The symbol $\mathcal{U}_1 \times \mathcal{U}_2$ means the collection $\mathcal{U}$ of sets in $X_1 \times X_2$, constructed by the procedure of Proposition 4.1.

If $\{X_1; \mathcal{U}_1\}$ and $\{X_2; \mathcal{U}_2\}$ are Hausdorff spaces, then the topological product space is a Hausdorff space.

Let $X_1 \times X_2$ be endowed with the product topology $\mathcal{U}_1 \times \mathcal{U}_2$. Then the projections

$$\pi_j : X_1 \times X_2 \longrightarrow X_j, \quad j = 1, 2,$$

are continuous. Moreover, $\mathcal{U}_1 \times \mathcal{U}_2$ is the weakest topology on $X_1 \times X_2$ for which such projections are continuous. The procedure can be iterated to construct the product topology on the product of n topological spaces $\{X_i; \mathcal{U}_i\}_{i=1}^n$. Such a topology is the weakest topology for which the projections

$$\pi_j : \prod_{i=1}^n X_i \longrightarrow X_j, \quad j = 1, \ldots, n,$$

are continuous. More generally, given an infinite family of topological spaces $\{X_\alpha; \mathcal{U}_\alpha\}_{\alpha \in A}$, the product topology $\prod \mathcal{U}_\alpha$ on the Cartesian product $\prod X_\alpha$ is constructed as the weakest topology for which the projections

$$\pi_\beta : \prod X_\alpha \longrightarrow X_\beta$$

are continuous for all $\beta \in A$. Such a topology must contain the collection

$$\mathcal{B} = \left\{ \begin{array}{c} \text{finite intersections of the inverse images} \\ \pi_\alpha^{-1}(\mathcal{O}_\alpha) \text{ for } \alpha \in A \text{ and } \mathcal{O}_\alpha \in \mathcal{U}_\alpha \end{array} \right\}.$$

One verifies that $\mathcal{B}$ satisfies the conditions (i) and (ii) of a base. Then the product topology is generated, starting from such a base, by the procedure of Proposition 4.1.[5]

Let f be a function defined on A such that $f(\alpha) \in X_\alpha$ for all $\alpha \in A$.

The collection $\{f(\alpha)\}$ can be identified with a point in $\prod X_\alpha$. Conversely, any point $x \in \prod X_\alpha$ can be identified with one such function. If $X_\alpha = X$ for some set X and all $\alpha \in A$, then $\prod X_\alpha$ is denoted by X^A and it is identified with the set of all functions defined on A and with values in X. If $A = \mathbb{N}$, then $X^{\mathbb{N}}$ is the set of all sequences $\{x_n\}$ of elements of X.

---

[5]A justification for such a construction is in Remark 8.1.

5   Compact topological spaces

A collection $\mathcal{F}$ of open sets $\mathcal{O}$ is an *open covering* of X if every $x \in X$ is contained in some open set $\mathcal{O} \in \mathcal{F}$. The covering is countable if it contains countably many elements, and it is finite if it contains a finite number of open sets.

It might occur that X is covered by a subfamily $\mathcal{F}'$ of elements of $\mathcal{F}$. Such a subfamily, if it exists, is an open subcovering of X relative to $\mathcal{F}$.

The topological space $\{X; \mathcal{U}\}$ is *compact* if every open covering $\mathcal{F}$ contains a finite subcovering $\mathcal{F}'$. A set $X_o \subset X$ is compact if $\{X_o; \mathcal{U}_o\}$ is a compact topological space.

A collection $\mathcal{G}$ of closed subsets of X has the *finite intersection property* if the elements of any finite subcollection have nonempty intersection.

Let $\mathcal{F}$ be an open covering for X and let $\mathcal{G}$ be the collection of the complements of the elements in $\mathcal{F}$. The elements of $\mathcal{G}$ are closed, and if X is compact, $\mathcal{G}$ does not have the finite intersection property.

More generally, X is compact if and only if every collection of closed subsets of X with empty intersection does not have the finite intersection property.

Proposition 5.1.

(i) $\{X; \mathcal{U}\}$ *is compact if and only if every collection $\mathcal{G}$ of closed sets with the finite intersection property has nonempty intersection.*

(ii) *Let E be a closed subset of a compact space $\{X; \mathcal{U}\}$. Then E is compact.*

(iii) *Let E be a compact subset of a Hausdorff topological space. Then E is closed.*[6]

(iv) *The continuous image of a compact set is compact.*

Proof. Part (i) follows from the previous remarks. To prove (ii), let $\mathcal{F}$ be any open covering for E. Then the collection $\{\mathcal{F}, (X - E)\}$ is an open covering for X. From this we may extract a finite subcovering $\mathcal{F}'$ for X which, by possibly removing $(X - E)$, gives a finite subcovering for E.

Turning to (iii), let $y \in (X - E)$ be fixed. Since X is a Hausdorff space, for every $x \in E$ there exist disjoint open sets $\mathcal{O}_x$ and $\mathcal{O}_{x,y}$ separating x and y. The collection $\{\mathcal{O}_x\}$ forms an open cover for E, from which we extract a finite one, say, for example, $\{\mathcal{O}_{x_1}, \ldots, \mathcal{O}_{x_n}\}$ for some positive integer n. The intersection $\bigcap_{j=1}^{n} \mathcal{O}_{x_j, y}$ is open and does not intersect E. Therefore, every element y of the complement of E contains an open neighborhood not intersecting E. Thus E is closed.

To establish (iv), let $\{X; \mathcal{U}\}$ be compact and let $\{Y; \mathcal{V}\}$ be the image of a continuous function f from X onto Y. Given an open covering $\{\Phi\}$ of Y, the collection

$$\mathcal{F} \equiv \{f^{-1}(\Phi) | \Phi \in \{\Phi\}\}$$

---

[6]The assumption that $\{X; \mathcal{U}\}$ be Hausdorff cannot be removed. Indeed, if $\mathcal{U}$ is the trivial topology on X, every proper subset of X is compact and not closed.

is an open covering for X from which we may extract a finite one

$$\{f^{-1}(\Phi_1), f^{-1}(\Phi_2), \ldots, f^{-1}(\Phi_n)\}.$$

Then the finite collection $\{\Phi_1, \ldots, \Phi_n\}$ covers Y.   □

A topological space $\{X; \mathcal{U}\}$ is *locally compact* if for each $x \in X$ there exists an open set $\mathcal{O}$ containing x and such that $\overline{\mathcal{O}}$ is compact. For example, $\mathbb{R}^N$ endowed with the Euclidean topology is locally compact but not compact.

5.1   Sequentially compact topological spaces. A topological space $\{X; \mathcal{U}\}$ is *countably compact* if every countable open covering of X contains a finite sub-covering. If $\{X; \mathcal{U}\}$ is compact it is also countably compact. The converse is false (see Section 5.7 of the Problems and Complements).

A topological space $\{X; \mathcal{U}\}$ has the Bolzano–Weierstrass property if every infinite sequence $\{x_n\}$ of elements of X has at least one cluster point. A topological space $\{X; \mathcal{U}\}$ is *sequentially compact* if every infinite sequence $\{x_n\}$ of elements of X has a convergent subsequence. Thus if $\{X; \mathcal{U}\}$ is sequentially compact, it has the Bolzano–Weierstrass property. The converse is false.[7]

Proposition 5.2.

(i) *The continuous image of a countably compact space is countably compact.*

(ii) $\{X; \mathcal{U}\}$ *is countably compact if and only if every countable family $\mathcal{G}$ of closed sets with the finite intersection property has nonempty intersection.*

(iii) $\{X; \mathcal{U}\}$ *has the Bolzano–Weierstrass property if and only if it is countably compact.*

(iv) *If $\{X; \mathcal{U}\}$ is sequentially compact, then it is countably compact.*

(v) *If $\{X; \mathcal{U}\}$ is countably compact and if it satisfies the first axiom of countability, then it is sequentially compact.*

Proof. Parts (i) and (ii) follow from the definitions and the proof of Proposition 5.1. To establish (iii), we use the characterization of countable compactness stated in (ii). Let $\{X; \mathcal{U}\}$ be countably compact and let $\{x_n\}$ be a sequence of elements of X. The closed sets

$$B_n \equiv \text{closure of } \{x_n, x_{n+1}, \ldots, \}$$

satisfy the finite intersection property and therefore have nonempty intersection. Any element $x \in \bigcap B_n$ is a cluster point for $\{x_n\}$.

Conversely, let $\{X; \mathcal{U}\}$ satisfy the Bolzano–Weierstrass property and let $\{B_n\}$ be a countable collection of closed subsets of X, with the finite intersection property. Since for all $n \in \mathbb{N}$ the intersection $\bigcap_{j=1}^{n} B_j$ is nonempty, we may select an

---

[7] By Proposition 5.2(v), a counterexample can be constructed starting from a space $\{X; \mathcal{U}\}$ that does not satisfy the first axiom of countability. See Section 5.7 of the Problems and Complements.

element x_n out of it. The sequence $\{x_n\}$ has at least one cluster point x, which by construction belongs to the intersection of all B_n.

Part (iv) follows from (iii) since sequential compactness implies the Bolzano–Weierstrass property.

To prove (iv), let $\{x_n\}$ be an infinite sequence of elements of X and let x be in the closure of $\{x_n\}$.

Since $\{X; \mathcal{U}\}$ satisfies the first axiom of countability, there exists a nested countable collection of open sets $\mathcal{O}_m \supset \mathcal{O}_{m+1}$, each containing x and an element x_m out of the sequence $\{x_n\}$.

The subsequence $\{x_m\}$ converges to x. $\qquad\qquad\qquad\qquad\qquad\qquad\square$

Proposition 5.3. *Let $\{X; \mathcal{U}\}$ satisfy the second axiom of countability. Then every open covering of X contains a countable subcovering.*

Proof. Let $\{B_n\}$ a countable collection of open sets that forms a base for the topology of $\{X; \mathcal{U}\}$, and let $\mathcal{F}$ be an open covering of X. To each B_n we associate one and only one open set $\mathcal{O}_n \in \mathcal{F}$ that contains it. The countable collection $\mathcal{O}_n$ is a countable subcovering of $\mathcal{F}$. $\qquad\qquad\qquad\qquad\qquad\qquad\square$

Corollary 5.4. *For spaces satisfying the second axiom of countability, compactness, countable compactness, and sequential compactness are equivalent.*

6 Compact subsets of $\mathbb{R}^N$

Let E be a subset of $\mathbb{R}^N$. We regard E as a topological space with the topology inherited from the Euclidean topology of $\mathbb{R}^N$.

Proposition 6.1. *The closed interval $[0, 1]$ is compact.*

Proof. Let $\{I_\alpha\}$ be a collection of open intervals covering $[0, 1]$ and set

$$\mathcal{E} = \bigcup \left\{ \begin{array}{l} x \in [0, 1] \text{ such that the closed interval } [0, x] \text{ is} \\ \text{covered by finitely many elements out of } \{I_\alpha\} \end{array} \right\}.$$

Let $c = \sup\{x | x \in \mathcal{E}\}$. Since 0 belongs to some I_α, we have $0 < c \leq 1$. Such an element c is covered by some open set $I_{\alpha_c} \in \{I_\alpha\}$, and therefore there exist $\varepsilon > 0$ such that

$$(c - \varepsilon, c + \varepsilon) \subset I_{\alpha_c}.$$

By the definition of c, the interval $[0, c - \varepsilon]$ is covered by finitely many open sets $\{I_1, I_2, \ldots, I_n\}$ out of $\{I_\alpha\}$. Augmenting such a finite collection with I_{α_c} gives a finite covering of $[0, c + \varepsilon]$. Thus if $c < 1$, it is not the supremum of the set $\mathcal{E}$. $\quad\square$

Proposition 6.2. *The closed interval $[0, 1]$ has the Bolzano–Weierstrass property.*

Proof. Let $\{x_n\}$ be an infinite sequence of elements of $[0, 1]$ without a cluster point in $[0, 1]$. Then each of the open intervals $(x - \varepsilon, x + \varepsilon)$ for $x \in [0, 1]$ and $\varepsilon > 0$ contains at most finitely many elements of $\{x_n\}$. The collection of all such intervals forms an open covering of $[0, 1]$, from which we may select a finite one. This would imply that $\{x_n\}$ is finite. $\square$

Corollary 6.3. *Every sequence in* $[0, 1]$ *has a convergent subsequence.*

Corollary 6.4. *A bounded, closed subset* $E \subset \mathbb{R}^N$ *has the Bolzano–Weierstrass property.*

Proof. Let $\{x_n\}$ be a sequence of elements in E and represent each of the x_n in terms of its coordinates, i.e.,

$$x_n = (x_{1,n}, x_{2,n}, \ldots, x_{N,n}).$$

Since $\{x_n\}$ is bounded, each of the sequences $\{x_{j,n}\}$ is contained in some closed interval $[a_j, b_j]$. Out of $\{x_{1,n}\}$ we extract a convergent subsequence $\{x_{1,n_1}\}$. Then out of $\{x_{2,n_1}\}$ we extract a convergent subsequence $\{x_{2,n_2}\}$. Proceeding in this fashion, the sequence $\{x_{n_N}\}$ has a limit. Since E is closed, such a limit is in E. $\square$

Proposition 6.5 (Borel[8]). *Let* E *be a bounded, closed subset of* $\mathbb{R}^N$. *Then every open covering* $\mathcal{U}$ *of* E *contains a finite subcovering* $\mathcal{U}'$.

Proof. By Proposition 5.3, we may assume the covering is countable, say, for example, $\mathcal{U} = \{\mathcal{O}_n\}$. We claim that

$$E \subset \bigcup_{i=1}^{m} \mathcal{O}_i \quad \text{for some } m \in \mathbb{N}.$$

Indeed, if not, we may select for each positive integer n, an element

$$x_n \in E - \bigcup_{i=1}^{n} \mathcal{O}_n,$$

and select, out of the sequence $\{x_n\}$, a subsequence $\{x_{n'}\}$ convergent to some $x \in E$. Since the collection $\{\mathcal{O}_n\}$ covers E, there exists an index m such that $x \in \mathcal{O}_m$. Thus $x_{n'} \in \mathcal{O}_m$ for infinitely many n'. $\square$

Proposition 6.6 (Heine–Borel). *Every compact subset of* $\mathbb{R}^N$, *endowed with the Euclidean topology, is closed and bounded.*

Proof. Let $E \subset \mathbb{R}^N$ be compact. Since $\mathbb{R}^N$, endowed with the Euclidean topology, is a Hausdorff space, E is closed by Proposition 5.1(iii).

The collection of balls $\{B_n\}$ centered at the origin and radius $n \in \mathbb{N}$ is an open covering for E. Since E is contained in the union of a finite subcovering, it is bounded. $\square$

Theorem 6.7. *A subset* E *of* $\mathbb{R}^N$ *is compact if and only if is bounded and closed.*

---

[8] E. Borel, Théorie des fonctions, *Ann. École Norm. Ser.* 3, **12** (1895), 9–55. Borel's proof assumes that the covering is countable. The extension to general coverings was observed by F. Riesz (F. Riesz, Sur un théoréme de E. Borel, *C. R. Acad. Sci. Paris*, **144** (1905), 224–226).

7 Continuous functions on countably compact spaces

Let f be a map from a topological space $\{X; \mathcal{U}\}$ into $\mathbb{R}$, and for $t \in \mathbb{R}$, set

$$[f < t] = \{x \in X | f(x) < t\}.$$

The sets $[f \leq t]$, $[f \geq t]$, and $[f > t]$ are defined analogously.

A map f from a topological space $\{X; \mathcal{U}\}$ into $\mathbb{R}$ is *upper semicontinuous* if $[f < t]$ is open for all $t \in \mathbb{R}$. A map $f : X \to \mathbb{R}$ is *lower semicontinuous* if $[f > s]$ is open for all $s \in \mathbb{R}$. A map $f : X \to \mathbb{R}$ is continuous if and only if it is both upper and lower semicontinuous.

Theorem 7.1 (Weierstrass–Baire). *Let $\{X; \mathcal{U}\}$ be countably compact and let $f : X \to \mathbb{R}$ be upper semicontinuous. Then f is bounded above in X, and it achieves its maximum in X.*

Proof. The collection of sets $\{[f < n]\}$ is a countable open covering of X from which we extract a finite one, say, for example,

$$[f < n_1], [f < n_2], \dots, [f < n_N].$$

Then $f \leq \max\{n_1, \dots, n_N\}$. Thus f is bounded above. Next, let f_o denote the supremum of f on X. The sets $[f \geq f_o - \frac{1}{n}]$ are closed and form a family with the finite intersection property. Since $\{X; \mathcal{U}\}$ is countably compact, their intersection is nonempty. Therefore, there is an element $x_o \in [f \geq f_o - \frac{1}{n}]$ for all $n \in \mathbb{N}$. By construction, $f(x_o) = f_o$. $\square$

Corollary 7.2.

  (i) *A continuous real-valued function from a countably compact topological space $\{X; \mathcal{U}\}$ takes its maximum and minimum in X.*

  (ii) *A continuous real-valued function from a countably compact topological space $\{X; \mathcal{U}\}$ is uniformly continuous.*

Theorem 7.3 (Dini). *Let $\{X; \mathcal{U}\}$ be countably compact and let $\{f_n\}$ be a sequence of upper semicontinuous, real-valued functions such that $f_{n+1} \leq f_n$ for all $n \in \mathbb{N}$ and converging pointwise in X to a lower semicontinuous function f. Then $\{f_n\} \to f$ uniformly in X.*[9]

Proof. By possibly replacing f_n with $(f_n - f)$, we may assume that $\{f_n\}$ is a decreasing sequence of upper semicontinuous functions converging to zero pointwise in X.

For every $\varepsilon > 0$, the collection of open sets $[f_n < \varepsilon]$ covers X and we extract a finite cover, say, for example, up to a possible reordering

$$[f_{n_1} < \varepsilon], [f_{n_2} < \varepsilon], \dots, [f_{n_\varepsilon} < \varepsilon], \quad n_1 < n_2 \cdots < n_\varepsilon.$$

---

[9]See also Sections 7.5–7.7 in the Problems and Complements.

Since $\{f_n\}$ is decreasing $[f_{n_\varepsilon} < \varepsilon] = X$. Thus $f_n(x) < \varepsilon$ for all $x \in X$ and all $n \geq n_\varepsilon$.   □

8   Products of compact spaces

Theorem 8.1 (Tychonov[10]). *Let $\{X_\alpha; \mathcal{U}_\alpha\}$ be a family of compact spaces. Then $\prod X_\alpha$ endowed with the product topology is compact.*

The proof is based on showing that every collection of closed sets with the finite intersection property has nonempty intersection.

Lemma 8.2. *Let $\{X; \mathcal{U}\}$ be a topological space and let $\mathcal{G}_o$ be a collection of subsets of X with the finite intersection property. There exists a maximal collection $\mathcal{G}$ of subsets of X with the finite intersection property and containing $\mathcal{G}_o$; i.e., if $\mathcal{G}'$ is another collection of subsets of X with the finite intersection property and containing $\mathcal{G}_o$, then $\mathcal{G}' \equiv \mathcal{G}$. Moreover, the finite intersection of elements in $\mathcal{G}$ is in $\mathcal{G}$ and each subset of X that intersects every set of $\mathcal{G}$ is in $\mathcal{G}$.[11]*

Proof. The family of all collections of sets with the finite intersection property and containing $\mathcal{G}_o$ is partially ordered by inclusion so that by the Hausdorff principle, there is a maximal linearly ordered subfamily $\mathcal{F}$. We claim that $\mathcal{G}$ is the union of all the collections in $\mathcal{F}$.

Any n-tuple $\{E_1, \ldots, E_n\}$ of elements of $\mathcal{G}$ belongs to at most n collections $\mathcal{G}_j$. Since $\{\mathcal{G}_j\}$ is linearly ordered, there is a collection $\mathcal{G}_n$ that contains the others. Therefore, $E_i \in \mathcal{G}_n$ for all $i = 1, 2, \ldots, n$, and since $\mathcal{G}_n$ has the finite intersection property, $\bigcap E_i \neq \emptyset$.

Thus $\mathcal{G}$ has the finite intersection property. The maximality of $\mathcal{G}$ follows by its construction. The collection $\mathcal{G}'$ of all finite intersections of sets in $\mathcal{G}$ contains $\mathcal{G}_o$ and has the finite intersection property. Therefore, $\mathcal{G}' = \mathcal{G}$ by maximality.

Let E be a subset of X that intersects all the sets in $\mathcal{G}$. Then the collection $\mathcal{G} \bigcup \{E\}$ has the finite intersection property and contains $\mathcal{G}_o$. Therefore, $E \in \mathcal{G}$ by maximality.   □

Proof of Tychonov's theorem. Let $\mathcal{G}_o$ be a collection of closed sets in $\prod_\alpha X_\alpha$ with the finite intersection property and let $\mathcal{G}$ be the maximal collection constructed in Lemma 8.2.

While the sets in $\mathcal{G}_o$ are closed, the elements of $\mathcal{G}$ need not be closed.

We will establish that the intersection of the closure of all elements in $\mathcal{G}$ is nonempty. For each $\alpha \in A$, let $\mathcal{G}_\alpha$ be the collection of the projection of $\mathcal{G}$ into X_α, i.e.,

$$\mathcal{G}_\alpha = \{\text{collection of } \pi_\alpha(E) | E \in \mathcal{G}\}.$$

---

[10]A. Tychonov, Über einen Funktionenräum, *Math. Ann.*, **111** (1935), 762–766.

[11]It is not claimed here that the elements of $\mathcal{G}$ are closed.

The sets in $\mathcal{G}_\alpha$ need not be closed nor open. However, since $\mathcal{G}$ has the finite intersection property in $\prod X_\alpha$, the collection $\mathcal{G}_\alpha$ has the finite intersection property in X_α. Therefore, the collection of their closures in $\{X_\alpha, \mathcal{U}_\alpha\}$,

$$\overline{\mathcal{G}}_\alpha = \{\text{collection of } \overline{\pi_\alpha(E)} | E \in \mathcal{G}\},$$

has nonempty intersection, since each of $\{X_\alpha; \mathcal{U}_\alpha\}$ is compact.

Select an element

$$x_\alpha \in \bigcap \{\overline{\pi_\alpha(E)} | E \in \mathcal{G}\} \subset X_\alpha.$$

We claim that the element $x \in \prod X_\alpha$, whose α-coordinate is x_α, belongs to the closure of all sets in $\mathcal{G}$.

Let $\mathcal{O}$ be a set, open in the product topology, that contains x. By the construction of the product topology, there exists finitely many indices $\alpha_1, \alpha_2, \ldots, \alpha_n$ and finitely many sets $\mathcal{O}_{\alpha_j}$, open in X_{α_j}, such that

$$x \in \bigcap_{j=1}^{n} \pi_{\alpha_j}^{-1}(\mathcal{O}_{\alpha_j}) \subset \mathcal{O}.$$

For each j, the projection x_{α_j} belongs to $\mathcal{O}_{\alpha_j}$. Since x_{α_j} belongs to the closure of all sets in $\mathcal{G}_{\alpha_j}$, the open set $\mathcal{O}_{\alpha_j}$ intersects all the sets in $\mathcal{G}_{\alpha_j}$. Therefore, $\pi_{\alpha_j}^{-1}(\mathcal{O}_{\alpha_j})$ intersects all the sets in $\mathcal{G}$ and, by Lemma 8.2, belongs to $\mathcal{G}$.

Likewise, the finite intersection $\bigcap_{j=1}^{n} \pi_{\alpha_j}^{-1}(\mathcal{O}_{\alpha_j})$ intersects every element in $\mathcal{G}$, and therefore it belongs to $\mathcal{G}$. Thus an arbitrary open set $\mathcal{O}$ containing x intersects all the sets in $\mathcal{G}$, and therefore x belongs to the closure of all such sets. $\qquad\square$

Remark 8.1. Tychonov's theorem provides a motivation for defining the topology on a product space $\prod X_\alpha$ as the weakest topology for which all the projection maps π_α are continuous. Indeed, if all the topological spaces $\{X_\alpha; \mathcal{U}_\alpha\}$ are Hausdorff, the product topology is also a Hausdorff topology. But then any topology stronger than the product topology would violate Tychonov's theorem.[12]

9 Vector spaces

A *linear space* consists of a set X, whose elements are called *vectors*, and a field $\mathcal{F}$, whose elements are called *scalars*, endowed with operations of sum and multiplication by scalars

$$+ : X \times X \to X, \qquad \bullet : \mathcal{F} \times X \to X$$

satisfying the addition laws

---

[12] See Proposition 5.1c and Section 5.3 of the Problems and Complements.

$$x + y = y + x,$$
$$(x + y) + z = x + (y + z) \quad \text{for all } x, y, z \in X;$$

there exists $\Theta \in X$ such that $x + \Theta = x$ for all $x \in X$;

for all $x \in X$, there exists $-x \in X$ such that $x + (-x) = \Theta$

and the scalar multiplication laws

$$\lambda(x + y) = \lambda x + \lambda y \quad \text{for all } x, y \in X;$$
$$\lambda(\mu x) = (\lambda \mu) x \quad \text{for all } \lambda, \mu \in \mathcal{F};$$
$$(\lambda + \mu) x = \lambda x + \mu x \quad \text{for all } \lambda, \mu \in \mathcal{F};$$
$$1x = x, \quad \text{where 1 is the unit element of } \mathcal{F}.$$

It follows that $\lambda \Theta = \Theta$ for all $\lambda \in \mathcal{F}$, and if 0 is the zero-element of $\mathcal{F}$, then $0x = \Theta$ for all $x \in X$. Also, for all $x, y \in X$ and $\lambda \in \mathcal{F}$,

$$(-1)x = -x, \qquad x - y = x + (-y), \qquad \lambda(x - y) = \lambda x - \lambda y.$$

A nonempty subset $X_o \subset X$ is a linear *subspace* of X if it is closed under the inherited operations of sum and multiplication by scalars. The largest linear subspace of X is X itself, and the smallest is the null space $\{\Theta\}$.

A *linear combination* of an n-tuple of vectors $\{x_1, \ldots, x_n\}$ is an expression of the form

$$y = \sum_{j=1}^{n} \lambda_j x_j, \quad \text{where } \{\lambda_1, \ldots, \lambda_n\} \text{ is an } n\text{-tuple of scalars.}$$

If $X_o \subset X$, the *linear span* of X_o is the set of all linear combinations of elements of X_o. It is a linear space, and it is the smallest linear subspace of X containing X_o or *spanned* by X_o.

An n-tuple $\{\mathbf{e}_1, \ldots, \mathbf{e}_n\}$ of vectors is *linearly independent* if

$$\sum_{j=1}^{n} \lambda_j \mathbf{e}_j = 0 \quad \text{implies that} \quad \lambda_j = 0 \quad \text{for all } j = 1, \ldots, n.$$

A linear space X is of dimension n if it contains an n-tuple of linearly independent vectors whose span is the whole X. Any such n-tuple, say, for example, $\{\mathbf{e}_1, \ldots, \mathbf{e}_n\}$ is a *basis* in the sense that given $x \in X$ there exists an n-tuple of scalars

$$\{\lambda_1, \ldots, \lambda_n\} \quad \text{such that} \quad x = \sum_{j=i}^{n} \lambda_j \mathbf{e}_j.$$

For each $x \in X$, the n-tuple $\{\lambda_1, \ldots, \lambda_n\}$ is uniquely determined by the basis $\{\mathbf{e}_1, \ldots, \mathbf{e}_n\}$.

While $\mathcal{F}$ could be any field, we will consider $\mathcal{F} = \mathbb{R}$ and call X the vector space over the reals.

Let A and B be subsets of a linear space X and let $\alpha, \beta \in \mathbb{R}$. Define the set operation

$$\alpha A + \beta B \equiv \bigcup \{\alpha a + \beta b \mid a \in A, b \in B\}.$$

One verifies that the sum is commutative and associative, i.e., that

$$A + B = B + A \quad \text{and} \quad A + (B + C) = (A + B) + C.$$

Moreover,

$$A + \left(B \bigcup C\right) = (A + B) \bigcup (A + C).$$

However, $A + A \neq 2A$ and $A - A \neq \{\Theta\}$.

9.1   Convex sets. A convex combination of two elements $x, y \in X$ is an element of the form

$$tx + (1 - t)y, \quad \text{where } t \in [0, 1].$$

As t ranges over $[0, 1]$ this describes the *line segment* of extremities x and y. The convex combination of n elements $\{x_1, \ldots, x_n\}$ of X is an element of the form

$$\sum_{j=1}^{n} \alpha_j x_j, \quad \text{where } \alpha_j \geq 0 \text{ and } \sum \alpha_j = 1.$$

A set $A \subset X$ is convex if for any pair $x, y \in A$, the elements $tx + (1 - t)y$ belong to A for all $t \in [0, 1]$; equivalently if the line segment of extremities x and y belongs to A.

The *convex hull* $c(A)$ of a set $A \subset X$ is the smallest convex set containing A. It can be characterized as either the intersection of all the convex sets containing A or the set of all convex combinations of n-tuples of elements in A for any n.

The intersection of convex sets is convex; the union of convex sets need not be convex. Linear subspaces of X are convex.

9.2   Linear maps and isomorphisms. Let X and Y be linear spaces over $\mathbb{R}$. A map $T : X \to Y$ is linear if

$$T(\lambda x + \mu y) = \lambda T(x) + \mu T(y) \quad \text{for all } x, y \in X \text{ and } \lambda, \mu \in \mathbb{R}.$$

The image of T is $T(X) \subset Y$ and the kernel of T is $\ker\{T\} = T^{-1}\{0\}$.

Since T is linear, $T(X)$ is a linear subspace of Y and $\ker\{T\}$ is a linear subspace of X.

A linear map $T : X \to Y$ is an *isomorphism* between X and Y if it is one-to-one and onto. The inverse of an isomorphism is an isomorphism, and the composition of two isomorphisms is an isomorphism.

If X and Y are finite dimensional and isomorphic, then they have the same dimension.

10 Topological vector spaces

A vector space X over $\mathbb{R}$ endowed with a topology $\mathcal{U}$ is a *topological vector space* if the operations of sum and the multiplication by scalars,

$$+ : X \times X \to X, \qquad \bullet : \mathbb{R} \times X \to X,$$

are continuous with respect to the product topologies of $X \times X$ and $\mathbb{R} \times X$.

Fix $x_o \in X$. The translation by x_o is defined by

$$T_{x_o}(x) = x_o + x \quad \text{for all } x \in X.$$

Likewise, for a fixed $\lambda \in \mathbb{R} - \{0\}$, the dilation by λ is defined by

$$D_\lambda(x) = \lambda x \quad \text{for all } x \in X.$$

If $\{X; \mathcal{U}\}$ is a topological vector space over $\mathbb{R}$, the maps T_{x_o} and D_λ are homeomorphisms from $\{X; \mathcal{U}\}$ onto itself. In particular, if $\mathcal{O}$ is open, then $x + \mathcal{O}$ is open for all fixed $x \in X$.

Any topology with such a property is *translation invariant*.[13]

Let $\{X; \mathcal{U}\}$ be a topological vector space. If $\mathcal{B}_\Theta$ is a base at the zero element Θ of X, then for any fixed $x \in X$, the collection $\mathcal{B}_x \equiv x + \mathcal{B}_\Theta$ forms a base for the topology $\mathcal{U}$ at x. Thus a base $\mathcal{B}_\Theta$ at Θ determines the topology $\mathcal{U}$ on X.

If the elements of the base $\mathcal{B}_\Theta$ are convex, the topology of $\{X; \mathcal{U}\}$ is called *locally convex*.[14]

An open neighborhood of the origin $\mathcal{O}$ is symmetric if $\mathcal{O} = -\mathcal{O}$.

The next remarks imply that the topology of a topological vector space, while not necessarily locally convex, is, roughly speaking, ball-like and, while not necessarily Hausdorff, is, roughly speaking, close to being Hausdorff.

Proposition 10.1. *Let $\{X; \mathcal{U}\}$ be a topological vector space. Then we have the following:*

(i) *The topology $\mathcal{U}$ is generated by a symmetric base $\mathcal{B}_\Theta$.*

(ii) *If $\mathcal{O}$ is an open neighborhood of the origin, then $X = \bigcup_{\lambda \in \mathbb{R}} \lambda \mathcal{O}$.*

(iii) *$\{X; \mathcal{U}\}$ is Hausdorff if and only if the points are closed.*

(iv) *$\{X; \mathcal{U}\}$ is Hausdorff if and only if $\bigcap \{\mathcal{O} \in \mathcal{B}_\Theta\} = \{\Theta\}$.*

---

[13]This notion can be used to construct a vector topological space $\{X; \mathcal{U}\}$ for which the sum is not continuous. It suffices to construct a vector space endowed with a topology that is not translation invariant. For an example of a linear, topological vector space for which the product by scalars is not continuous, see Section 10.4 of the Problems and Complements.

[14]An example of a topological vector space with a nonlocally convex topology is in Section 6 of Chapter V.

Proof. The continuity of the multiplication by scalars implies that if $\mathcal{O}$ is open, $\lambda\mathcal{O}$ also is open for all $\lambda \in \mathbb{R} - \{0\}$. If $\Theta \in \mathcal{O}$, then $\Theta \in \lambda\mathcal{O}$ for all $|\lambda| \leq 1$. In particular, if $\mathcal{O}$ is an open neighborhood of the origin, $-\mathcal{O}$ also is an open neighborhood of the origin. The set $A \equiv -\mathcal{O} \cap \mathcal{O}$ is an open neighborhood of the origin and is symmetric since $A = -A$. One verifies that the collection of such symmetric sets is a base $\mathcal{B}_\Theta$ at the origin for the topology of $\{X; \mathcal{U}\}$.

To prove (ii), fix $x \in X$ and let $\mathcal{O}$ be an open neighborhood of Θ. Since $0 \cdot x = \Theta$, by the continuity of the product by scalars, there exist $\varepsilon > 0$ and an open neighborhood $\mathcal{O}_x$ of x such that $\lambda \cdot y \in \mathcal{O}$ for all $|\lambda| < \varepsilon$ and all $y \in \mathcal{O}_x$. Thus $\delta \cdot x \in \mathcal{O}$ for some $0 < |\delta| < \varepsilon$ and $x \in \delta^{-1}\mathcal{O}$.

The direct part of (iii) follows from Proposition 1.1. For the converse, assume that Θ and $x \in (X - \Theta)$ are closed. Then there exists an open set $\mathcal{O}$ containing the origin Θ and not containing x. Since $\Theta + \Theta = \Theta$ and the sum $+ : (X \times X) \to X$ is continuous, there exist two open sets $\mathcal{O}_1$ and $\mathcal{O}_2$ such that $\mathcal{O}_1 + \mathcal{O}_2 \subset \mathcal{O}$. Set

$$\mathcal{O}_o = \mathcal{O}_1 \cap \mathcal{O}_2 \cap (-\mathcal{O}_1) \cap (-\mathcal{O}_2).$$

Then

$$\mathcal{O}_o + \mathcal{O}_o \subset \mathcal{O} \quad \text{and} \quad \mathcal{O}_o \cap (x + \mathcal{O}_o) = \emptyset.$$

The last statement is a consequence of (iii). $\square$

Proposition 10.2. *Let $\{X; \mathcal{U}\}$ and $\{Y; \mathcal{V}\}$ be topological vector spaces. A linear map $T : X \to Y$ is continuous if and only if it is continuous at the origin Θ of X.*

Proof. Since T is linear, $T(\Theta) = \theta \in Y$, where θ is the origin of Y. Let $O \in \mathcal{V}$ be an open set containing θ. By assumption $T^{-1}(O)$ is an open set containing Θ. Let $x \in X$ be fixed. An open set in Y that contains $T(x)$ is of the form $T(x) + O$, where O is an open set containing θ. The preimage $T^{-1}(T(x) + O)$ contains the open set $x + T^{-1}(O)$. $\square$

10.1 Boundedness and continuity. Let $\{X; \mathcal{U}\}$ be a topological vector space. A subset $E \subset X$ is bounded if for every open neighborhood $\mathcal{O}$ of the origin Θ, there exists $\mu > 0$ such that $E \subset \lambda\mathcal{O}$ for all $\lambda > \mu$.

A map T from a topological vector space $\{X; \mathcal{U}\}$ into a topological vector space $\{Y; \mathcal{V}\}$ is *bounded* if it maps bounded subsets of X into bounded subsets of Y. An example of an unbounded linear map between two topological vector spaces is in Section 15. Further examples are in Sections 3.4 and 3.5 of the Problems and Complements of Chapter VI.

Proposition 10.3. *A linear, continuous map T from a topological vector space $\{X; \mathcal{U}\}$ into a topological vector space $\{Y; \mathcal{V}\}$ is bounded.*[15]

---

[15]For general topological vector spaces $\{X; \mathcal{U}\}$ and $\{Y; \mathcal{V}\}$, the converse is false; i.e., linearity and boundedness do not imply continuity of T. See Section 10.3 of the Problems and Complements for a counterexample. However, the converse is true for linear, bounded maps T between *metric* vector spaces, as stated in Proposition 14.2.

Proof. Let $E \subset X$ be bounded. For every open neighborhood O of the origin θ of Y, open in the topology of $\{Y; \mathcal{V}\}$, the inverse image $T^{-1}(O)$ is an open neighborhood of the origin Θ, open in the topology of $\{X; \mathcal{U}\}$. Since E is bounded, there exists some $\delta > 0$ such that $E \subset \delta T^{-1}(O)$. Therefore, $T(E) \subset \delta O$.   □

11   Linear functionals

If the target space Y is the field $\mathbb{R}$ endowed with the Euclidean topology, the linear map $T : X \to \mathbb{R}$ is called a *functional* on $\{X; \mathcal{U}\}$.

A linear functional $T : \{X; \mathcal{U}\} \to \mathbb{R}$ is bounded in a neighborhood of Θ if there exists an open set $\mathcal{O}$ containing Θ and a positive number k such that $|T(x)| < k$ for all $x \in \mathcal{O}$.

Proposition 11.1. *Let $T : \{X; \mathcal{U}\} \to \mathbb{R}$ be a not-identically-zero linear functional. Then*

(i) *if T is bounded in a neighborhood of the origin, then T is continuous;*

(ii) *if $\ker\{T\}$ is closed, then T is bounded in a neighborhood of the origin;*

(iii) *T is continuous if and only if $\ker\{T\}$ is closed;*

(iv) *T is continuous if and only if it is bounded in a neighborhood of the origin.*

Proof. Let $\mathcal{O}$ be an open neighborhood of the origin such that $|T(x)| \leq k$ for all $x \in \mathcal{O}$. For every $\varepsilon \in (0, k)$, the preimage of the open interval $(-\varepsilon, \varepsilon)$ contains the open sets $\lambda \mathcal{O}$ for all $0 < \lambda < \varepsilon/k$. Thus T is continuous at the origin and therefore continuous by Proposition 10.2.

Turning to (ii), if $\ker\{T\}$ is closed, there exist $x \in X$ and some open neighborhood $\mathcal{O}$ of Θ such that $(x + \mathcal{O}) \cap \ker\{T\} = \emptyset$. By Proposition 10.1(i), we may assume that $\mathcal{O}$ is symmetric and that $\lambda \mathcal{O} \subset \mathcal{O}$ for all $|\lambda| \leq 1$. This implies that $T(\mathcal{O})$ is a symmetric interval about the origin of $\mathbb{R}$. If such an interval is bounded, there is nothing to prove. If such an interval coincides with $\mathbb{R}$, then there exists $y \in \mathcal{O}$ such that $T(y) = T(x)$. Thus $(x - y) \in \ker\{T\}$ and $(x + \mathcal{O}) \cap \ker\{T\}$ is not empty since $y \in \mathcal{O}$. The contradiction proves (ii).

To prove (iii), observe that the origin $\{0\}$ of $\mathbb{R}$ is closed. Therefore, if T is continuous, $T^{-1}(0) = \ker\{T\}$ is closed.

The remaining statements follow from (i) and (ii).   □

12   Finite-dimensional topological vector spaces

The next proposition asserts that an n-dimensional Hausdorff topological vector space can only be given up to a homeomorphism, the Euclidean topology of $\mathbb{R}^n$.

Proposition 12.1. *Let $\{X; \mathcal{U}\}$ be an n-dimensional Hausdorff topological vector space over $\mathbb{R}$. Then $\{X; \mathcal{U}\}$ is homeomorphic to $\mathbb{R}^n$ equipped with the Euclidean topology.*

Proof. Given a basis $\{\mathbf{e}_1, \ldots, \mathbf{e}_n\}$ for $\{X; \mathcal{U}\}$,

$$X = \bigcup \left\{ \sum_{i=1}^{n} \lambda_i \mathbf{e}_i \,|\, (\lambda_1, \ldots, \lambda_n) \right\}.$$

The representation map

$$\mathbb{R}^n \ni (\lambda_1, \ldots, \lambda_n) \longrightarrow T(\lambda_1, \ldots, \lambda_n) = \sum_{i=1}^{n} \lambda_i \mathbf{e}_i \in X$$

is linear, one-to-one, and onto. Let $\mathcal{O}$ be an open neighborhood of the origin in X, which we may assume to be symmetric and such that $\alpha \mathcal{O} \subset \mathcal{O}$ for all $|\alpha| \leq 1$. By the continuity of the sum and multiplication by scalars, the preimage $T^{-1}(\mathcal{O})$ contains an open ball about the origin of $\mathbb{R}^n$. Thus T is continuous at the origin and therefore continuous.

To show that T^{-1} is continuous assume first that $n = 1$. In such a case $T(\lambda) = \lambda \mathbf{e}$ for some $\mathbf{e} \in (X - \Theta)$. The kernel of the inverse map $T^{-1} : X \to \mathbb{R}$ consists of only the zero element $\{\Theta\}$, which is closed since X is Hausdorff. Therefore, T^{-1} is continuous by Proposition 11.1(iii).

Proceeding by induction, assume that the representation map T is a homeomorphism between $\mathbb{R}^m$ and any m-dimensional Hausdorff space for all $m = 1, 2, \ldots, (n-1)$. Thus, in particular, any $(n-1)$-dimensional Hausdorff space is closed.

The inverse of the representation map has the form

$$X \ni x \longrightarrow T^{-1}(x) = (\lambda_1(x), \ldots, \lambda_{n-1}(x), \lambda_n(x)).$$

Each of the n maps $\lambda_j(\cdot) : X \to \mathbb{R}$ is a linear functional on X whose null space is an $(n-1)$-dimensional subspace of X. Such a subspace is closed by the induction hypothesis. Thus each of the $\lambda_j(\cdot)$ is continuous. $\square$

Corollary 12.2. *Every finite-dimensional subspace of a Hausdorff topological vector space is closed.*

If $\{X; \mathcal{U}\}$ is n-dimensional and not Hausdorff, it is not homeomorphic to $\mathbb{R}^n$. An example is $\mathbb{R}^N$ with the trivial topology.

12.1   Locally compact spaces. A topological vector space $\{X; \mathcal{U}\}$ is locally compact if there exists an open neighborhood of the origin whose closure is compact.

Proposition 12.3. *Let $\{X; \mathcal{U}\}$ be a Hausdorff, locally compact topological vector space. Then X is of finite dimension.*

Proof. Let $\mathcal{O}$ be an open neighborhood of the origin, whose closure is compact. We may assume that $\mathcal{O}$ is symmetric and $\lambda \mathcal{O} \subset \mathcal{O}$ for all $|\lambda| \leq 1$. There exist at most finitely many points $x_1, x_2, \ldots, x_n \in \mathcal{O}$ such that

$$\overline{\mathcal{O}} \subset \left(x_1 + \tfrac{1}{2}\mathcal{O} \right) \bigcup \left(x_2 + \tfrac{1}{2}\mathcal{O} \right) \bigcup \cdots \bigcup \left(x_n + \tfrac{1}{2}\mathcal{O} \right). \tag{12.1}$$

The space $Y = \text{span}\{x_1, \ldots, x_n\}$ is a closed, finite-dimensional subspace of X. From (12.1),

$$\frac{1}{2}\mathcal{O} \subset Y + \frac{1}{4}\mathcal{O}.$$

Therefore,

$$\mathcal{O} \subset Y + \frac{1}{2}\mathcal{O} \subset 2Y + \frac{1}{4}\mathcal{O} = Y + \frac{1}{4}\mathcal{O}.$$

Thus by iteration,

$$\mathcal{O} \subset \bigcap\left(Y + \frac{1}{2^n}\mathcal{O}\right) = \overline{Y} = Y.$$

This implies that $\lambda\mathcal{O} \subset Y$ for all $\lambda \in \mathbb{R}$. Thus by Proposition 10.1(ii),

$$X = \bigcup \lambda\mathcal{O} \subset Y \subset X. \qquad \square$$

The assumption that $\{X; \mathcal{U}\}$ be Hausdorff cannot be removed. Indeed, any $\{X; \mathcal{U}\}$ with the trivial topology is compact, and hence locally compact. However, it is not Hausdorff and, in general, is not of finite dimension.

13   Metric spaces

A *metric* on a nonvoid set X is a function $d : X \times X \to \mathbb{R}^+$ satisfying the following properties:

(i) $d(x, y) \geq 0$ for all pairs $(x, y) \in X \times X$.

(ii) $d(x, y) = 0$ if and only if $x = y$.

(iii) $d(x, y) = d(y, x)$ for all pairs $(x, y) \in X \times X$.

(iv) $d(x, y) \leq d(x, z) + d(y, z)$ for all $x, y, z \in X$.

This last requirement is called the *triangle inequality*.

The pair $\{X; d\}$ is a metric space. Denote by

$$B_\rho(x) = \{y \in X | d(y, x) < \rho\}$$

the open ball centered at x and of radius $\rho > 0$. The collection $\mathcal{B}$ of all such balls satisfies conditions (i) and (ii) of Section 4 and therefore, by Proposition 4.1, generates a topology $\mathcal{U}$ on $\{X; d\}$, called metric topology, for which $\mathcal{B}$ is a base.

The notions of open or closed sets can be given in terms of the elements of $\mathcal{B}$. In particular, a set $\mathcal{O} \subset X$ is open if for every $x \in \mathcal{O}$, there exists some $\rho > 0$ such that $B_\rho(x) \subset \mathcal{O}$.

A point x is a point of closure for a set E if $B_\varepsilon(x) \cap E \neq \emptyset$ for all $\varepsilon > 0$. A set E is closed if and only if it coincides with the set of all its points of closure. In particular, points are closed.

Let $\{x_n\}$ be a sequence of elements of X. A point $x \in X$ is a cluster point for $\{x_n\}$ if for all $\varepsilon > 0$, the open ball $B_\varepsilon(x)$ contains x_n for infinitely many n.

The sequence $\{x_n\}$ converges to x if for every $\varepsilon > 0$ there exists n_ε such that $d(x, x_n) < \varepsilon$ for all $n \geq n_\varepsilon$.

The sequence $\{x_n\}$ is a Cauchy sequence if for every $\varepsilon > 0$ there exists an index n_ε such that $d(x_n, x_m) \leq \varepsilon$ for all $m, n \geq n_\varepsilon$.

A metric space $\{X; d\}$ is *complete* if every Cauchy sequence $\{x_n\}$ of elements of X converges to some element $x \in X$.

13.1 Separation and axioms of countability. The distance between two subsets A, B of X is defined by

$$d(A, B) = \inf_{x \in A; y \in B} d(x, y).$$

Proposition 13.1. *Let A be any subset of X. The function $x \to d(A, x)$ is continuous in $\{X; d\}$.*

Proof. Let $x, y \in X$ and $z \in A$. By requirement (iv) of a metric,

$$d(z, x) \leq d(x, y) + d(z, y).$$

Taking the infimum of both sides for $z \in A$ gives

$$d(A, x) \leq d(x, y) + d(A, y).$$

Interchanging the roles of x and y yields

$$|d(A, x) - d(A, y)| \leq d(x, y). \qquad \square$$

If E_1 and E_2 are two disjoint closed subsets of $\{X; d\}$, then the two sets

$$\mathcal{O}_1 = \{x \in X \,|\, d(x, E_1) < d(x, E_2)\},$$
$$\mathcal{O}_2 = \{x \in X \,|\, d(x, E_2) < d(x, E_1)\}$$

are open and disjoint. Moreover, $E_1 \subset \mathcal{O}_1$ and $E_2 \subset \mathcal{O}_2$. Thus every metric space is normal. In particular, every metric space is Hausdorff.

Every metric space satisfies the first axiom of countability. Indeed, the collection of balls $B_\rho(x)$ as ρ ranges over the rational numbers of $(0, 1)$ is a countable base for the topology at x.

Proposition 13.2. *A metric space $\{X; d\}$ is separable if and only if it satisfies the second axiom of countability.*[16]

---

[16]An example of nonseparable metric space is in Section 18.1 of Chapter V. See also Section 18.2 of the Problems and Complements of Chapter V.

Proof. Let $\{X; d\}$ be separable and let A be a countable, dense subset of $\{X; d\}$. The collection of balls centered at points of A and with rational radius forms a countable base for the topology of $\{X; d\}$. The converse follows from Proposition 4.2.   □

Corollary 13.3. *Every subset of a separable metric space is separable.*

Proof. Let $\{x_n\}$ be a countable dense subset. For a pair of positive integers (m, n), consider the balls $B_{1/m}(x_n)$ centered at x_n and radius $1/m$. If Y is a subset of X, the ball $B_{1/m}(x_n)$ must intersect Y for some pair (m, n). For any such pair, select an element $y_{n,m} \in B_{1/m}(x_n) \bigcap Y$. The collection of such $y_{m,n}$ is a countable, dense subset of Y.   □

13.2   Equivalent metrics. From a given metric d on X, one can generate other metrics. For example, one might set

$$d_o(x, y) = \frac{d(x, y)}{1 + d(x, y)}.   \tag{13.1}$$

One verifies that d_o satisfies the requirements (i)–(iii). To verify that d_o satisfies (iv), it suffices to observe that the function

$$t \longrightarrow \frac{t}{1 + t} \quad \text{for } t \geq 0$$

is nondecreasing. Thus d_o is a new metric on X and generates the metric space $\{X; d_o\}$.

Starting from the Euclidean metric in $\mathbb{R}^N$, one may introduce a new metric by

$$d_*(x, y) = \left| \frac{x}{1 + |x|} - \frac{y}{1 + |y|} \right|, \quad x, y \in \mathbb{R}^N.   \tag{13.2}$$

More generally, the same set X can be given different metrics, say, for example, d_1 and d_2, to generate metric spaces $\{X; d_1\}$ and $\{X; d_2\}$.

Two metrics d_1 and d_2 on the same set X are equivalent if they generate the same topology. Equivalently, d_1 and d_2 are equivalent if they define the same open sets. In such a case, the identity map between $\{X; d_1\}$ and $\{X; d_2\}$ is a homeomorphism.

13.3   Pseudometrics. A function $d : (X \times X) \to \mathbb{R}$ is a pseudometric if it satisfies all but condition (ii) of the requirements of being a metric. For example,

$$d(x, y) = ||x| - |y||$$

is a pseudometric on $\mathbb{R}$. The open balls $B_\rho(x)$ are defined as for metrics and generate a topology on X, called the pseudometric topology. The space $\{X; d\}$ is a pseudometric space. The statements of Propositions 13.1 and 13.2 and Corollary 13.3 continue to hold for pseudometric spaces.

14   Metric vector spaces

Let $\{X; d\}$ and $\{Y; \eta\}$ be metric spaces. The notion of continuity of a function from X into Y can be rephrased in terms of the metrics η and d. Precisely, a function $f : \{X; d\} \to \{Y; \eta\}$ is continuous at some $x \in X$ if and only if for every $\varepsilon > 0$, there exists $\delta = \delta(\varepsilon, x)$ such that

$$\eta\{f(x), f(y)\} < \varepsilon \quad \text{whenever } d(x, y) < \delta. \tag{14.1}$$

The function f is continuous if it is continuous at each $x \in X$.

A function $f : \{X; d\} \to \{Y; y\}$ is uniformly continuous if the choice of δ in (14.1) depends on ε and is independent of $x \in X$.

A homeomorphism f between two metric spaces $\{X; d\}$ and $\{Y; \eta\}$ is *uniform* if the map $f : X \to Y$ is one-to-one and onto, and if it is *uniformly* continuous and has uniformly continuous inverse. There exist homeomorphisms between metric spaces that are not uniform.[17]

An isometry between $\{X; d\}$ and $\{Y; \eta\}$ is a homeomorphism f between $\{X; d\}$ and $\{Y; \eta\}$ that preserves distances, i.e., such that

$$\eta\{f(x), f(y)\} = d(x, y) \quad \text{for all } x, y \in X.$$

It follows from the definition that an isometry is a uniform homeomorphism between $\{X; d\}$ and $\{Y; \eta\}$.

Let $\{X_1; d_1\}$ and $\{X_2; d_2\}$ be metric spaces. The product metric $(d_1 \times d_2)$ on the Cartesian product $(X_1 \times X_2)$ is defined by

$$(d_1 \times d_2)\{(x_1, x_2), (y_1, y_2)\} = d_1(x_1, y_1) + d_2(x_2, y_2)$$

for all $x_1, y_1 \in X_1$ and $x_2, y_2 \in X_2$. One verifies that the topology generated by $(d_1 \times d_2)$ on $(X_1 \times X_2)$ coincides with the product topology of $\{X_1; d_1\}$ and $\{X_2; d_2\}$.

If X is a *vector* space, then $\{X; d\}$ is a topological vector space if the operations of sum and product by scalars,

$$+ : X \times X \longrightarrow X, \qquad \bullet : \mathbb{R} \times X \longrightarrow X,$$

are continuous with respect to the topology generated by d on X and the topology generated by $(d \times d)$ on $X \times X$.

A metric d on a vector space X is translation invariant if

$$d(x + z, y + z) = d(x, y) \quad \text{for all } x, y, z \in X.$$

If d is translation invariant, then the metric d_o of (13.1) is translation invariant. The metric d_* in (13.2) is not translation invariant.

Proposition 14.1. *If d on a vector space X is translation invariant, then the sum $+ : (X \times X) \to X$ is continuous.*

---

[17]For example, $f : (0, \infty) \to (0, 1)$ by $x \to \frac{1}{x}$.

Proof. It suffices to show that

$$X \times X \ni (x, y) \longrightarrow x + y$$

is continuous at an arbitrary point $(x_o, y_o) \in X \times X$. From the definition of product topology,

$$
\begin{aligned}
d(x + y, x_o + y_o) &= d(x - x_o, y_o - y) \\
&\leq d(x - x_o, \Theta) + d(y_o - y, \Theta) \\
&= d(x, x_o) + d(y, y_o) \\
&= (d \times d)\{(x, y), (x_o, y_o)\}. \qquad \square
\end{aligned}
$$

Translation-invariant metrics generate translation-invariant topologies. There exist non-translation-invariant metrics that generate translation-invariant topologies.

Remark 14.1. The topology generated by a metric on a vector space X need not be locally convex. A counterexample is in Corollary 6.2 of Chapter V.

Remark 14.2. In general, the notion of a metric on a vector space X does not imply, alone, any continuity statement of the operations of sum or product by scalars. Indeed, there exist metric spaces for which both operations are discontinuous.

To construct examples, let $\{X; d\}$ be a metric vector space and let h be a discontinuous bijection from X onto itself. Setting

$$d_h(x, y) \overset{\text{def}}{=} d(h(x), h(y)) \tag{14.2}$$

defines a metric in X. The bijection h can be chosen in such a way that for the metric vector space $\{X; d_h\}$ the sum and the multiplication by scalars are both discontinuous.[18]

14.1   Maps between metric spaces. The notion of maps between metric spaces and their properties, is inherited from the corresponding notions between topological vector spaces. In particular, Propositions 10.1–10.3 and 11.1 continue to hold in the context of metric spaces. However, for metric spaces, Proposition 10.3 admits a converse.

Proposition 14.2. *Let $\{X; d\}$ and $\{Y; \eta\}$ be metric vector spaces. A bounded linear map $T : X \to Y$ is continuous.*

Proof. For any ball $\mathcal{B}_r$ in $\{Y; \eta\}$, of radius r and centered at the origin of Y, there exists a ball $\mathcal{B}_\rho$ in $\{X; d\}$, centered at the origin of X such that $\mathcal{B}_\rho \subset T^{-1}(\mathcal{B}_r)$. If not, for all $\delta > 0$ the ball $\delta^{-1} B_1$ is not contained in $T^{-1}(\mathcal{B}_r)$. Thus $T(B_1)$ is not contained in $\delta \mathcal{B}_r$ for any $\delta > 0$ against the boundedness of T. The contradiction implies T is continuous at the origin and, by linearity, T is continuous everywhere. $\qquad \square$

---

[18]This construction was suggested by Ethan Devinatz. See Section 14 of the Problems and Complements for a choice of h.

15 Spaces of continuous functions

Let E be a subset of $\mathbb{R}^N$, denote by $C(E)$ the collection of all continuous functions $f : E \to \mathbb{R}$, and set

$$d(f, g) = \sup_{x \in E} |f(x) - g(x)|, \quad f, g \in C(E). \tag{15.1}$$

If E is compact, this defines a metric in $C(E)$ by which $C(E)$ turns into a metric vector space. The metric in (15.1) generates a topology in $C(E)$ called the topology of *uniform convergence*.

Cauchy sequences in $C(E)$ converge uniformly to a continuous function in E. In this sense, $C(E)$ is *complete*.

If E is compact, $C(E)$ is separable.[19]

If E is open, a function $f \in C(E)$, while bounded on every compact subset of E, in general, is not bounded in E. Let $\{E_n\}$ be a collection of bounded open sets invading E, i.e., $\overline{E}_n \subset E_{n+1}$ for all n, and $E = \bigcup E_n$. For every $f, g \in C(E)$, set

$$d_n(f, g) = \sup_{x \in \overline{E}_n} |f(x) - g(x)|.$$

Each d_n, while a metric in $C(\overline{E}_n)$, is a pseudometric in $C(E)$. Setting

$$d(f, g) = \sum \frac{1}{2^n} \frac{d_n(f, g)}{1 + d_n(f, g)} \tag{15.2}$$

define a metric in $C(E)$ by which $\{C(E); d\}$ is a metric vector space.

A sequence $\{f_n\}$ of functions in $C(E)$ converges to $f \in C(E)$ in the metric (15.2) if and only if $\{f_n\} \to f$ uniformly on every compact subset of E.

Cauchy sequences in $C(E)$ converge uniformly over compact subsets of E, to a function in $C(E)$. In this sense, the space $C(E)$ with the topology generated by the metric (15.2) is complete.

Denote by $\mathcal{L}^1(E)$ the collection of functions in $C(E)$ whose Riemann integral over E is finite. Since $\mathcal{L}^1(E)$ is a linear subspace of $C(E)$, it can be given the metric (15.2) and the corresponding topology. This turns $\mathcal{L}^1(E)$ into a metric vector space. The linear functional

$$T(f) = \int_E f \, dx : \mathcal{L}^1(E) \longrightarrow \mathbb{R}$$

is unbounded and hence discontinuous. As an example let $E = (0, 1)$. The functions $f_n(t) = t^{\frac{1}{n} - 1}$ are all in $\mathcal{L}^1(0, 1)$ and the sequence $\{f_n\}$ is bounded in the topology of (15.2) since $d(f_n, 0) \leq 1$. However, $T(f_n) = n$.

The linear functional

$$T(f) = \int_E f \, dx : C(\overline{E}) \longrightarrow \mathbb{R}$$

is bounded and hence continuous.

---

[19]See Corollary 16.2 of Chapter IV.

15.1   Spaces of continuously differentiable functions. Let E be an open subset of $\mathbb{R}^N$ and denote by $C^1(E)$ the collection of all continuously differentiable functions $f : E \to \mathbb{R}$.

Denote by $C^1(\overline{E})$ the collection of functions in $C^1(E)$ whose derivatives f_{x_j}, $j = 1, \ldots, N$, admit a continuous extension to $\overline{E}$. For $f, g \in C^1(\overline{E})$, formally set[20]

$$d(f, g) = \sup_{x \in \overline{E}} |f(x) - g(x)| + \sum_{j=1}^{N} \sup_{x \in \overline{E}} |f_{x_j}(x) - g_{x_j}(x)|. \qquad (15.3)$$

If E is bounded, so that $\overline{E}$ is compact, this defines a metric in $C^1(\overline{E})$ by which $C^1(\overline{E})$ turns into a metric vector space.

Cauchy sequences in $C^1(\overline{E})$ converge to functions in $C^1(\overline{E})$. Therefore, $C^1(\overline{E})$ is complete.

The space $C^1(\overline{E})$ can also be given the metric (15.1). This turns $C^1(\overline{E})$ into a metric space. The topology generated by such a metric in $C^1(\overline{E})$ is the same as the topology that $C^1(E)$ inherits as a subspace of $C(\overline{E})$. With respect to such a topology, $C^1(\overline{E})$ is not complete.

The linear map

$$T(f) = f_{x_j} : C^1(\overline{E}) \longrightarrow C(\overline{E}) \quad \text{for a fixed } j \in \{1, \ldots, N\}$$

is bounded, and hence continuous, if $C^1(\overline{E})$ is given the metric (15.3). It is unbounded and hence discontinuous if $C^1(\overline{E})$ is given the metric (15.1).

As an example, let $\overline{E} = [0, 1]$. The functions $f_n(t) = t^n$ are in $C^1[0, 1]$ for all $n \in \mathbb{N}$ and the sequence $\{f_n\}$ is bounded in $C[0, 1]$ since $d(f_n, 0) = 1$. However, $T(f_n) = nt^{n-1}$ is unbounded in $C[0, 1]$.

If E is open and $f \in C^1(E)$, the functions f and f_{x_j}, $j = 1, \ldots, N$, while bounded on every compact subset of E, in general, are not bounded in E. A metric in $C^1(E)$ can be introduced along the lines of (15.2).

16   On the structure of a complete metric space

Let $\{X; \mathcal{U}\}$ be a topological space. A set $E \subset X$ is *nowhere dense* in X, if $\overline{E}^c$ is dense in X. If E is nowhere dense, then $\overline{E}$ also is nowhere dense.

A closed set E is nowhere dense if and only if it does not contain any open set. If E is nowhere dense, for any open set $\mathcal{O}$ the complement $\mathcal{O} - \overline{E}$ must contain an open set. Indeed, if not, $\overline{E}$ would contain the open set $\mathcal{O}$.

If E is nowhere dense and open, $\overline{E} - E$ is nowhere dense. If E is nowhere dense and closed, $E - \overset{\circ}{E}$ is nowhere dense.

A finite subset of $[0, 1]$ is nowhere dense in $[0, 1]$.

---

[20]The extensions of f_{x_j} to $\overline{E}$ are again denoted by f_{x_j}.

The Cantor set is nowhere dense in $[0, 1]$. Such a set is the *uncountable* union of nowhere dense sets.

The rationals are not nowhere dense in $[0, 1]$. However, they are the *countable* union of nowhere dense sets in $[0, 1]$. Thus the uncountable union of nowhere dense sets might be nowhere dense and the countable union of nowhere dense sets might be dense.

A set $E \subset X$ is said to be *meager*, or of the *first category*, if it is the countable union of nowhere dense sets. A set that is not of first category is said to be of the *second category*.

The complement of a set of first category is called a *residual* or *nonmeager* set.

The rationals in $[0, 1]$ are a set of first category. The Cantor set is of first category in $[0, 1]$.

A metric space $\{X; d\}$ is *complete* if every Cauchy sequence $\{x_n\}$ of elements of X converges to some element $x \in X$. An example of noncomplete metric space is the set of the rationals in $[0, 1]$ with the Euclidean metric.

Every metric space can be completed as indicated in Section 16.3 of the Problems and Complements. The completion of the rationals are the real numbers.

The *Baire category theorem* asserts that a complete metric space cannot be the countable union of nowhere dense sets, much the same way as $[0, 1]$ is not the union of the rationals.

Theorem 16.1 (Baire[21]). *A complete metric space is of second category.*

Proof. If not, there exists a countable collection $\{E_n\}$ of nowhere dense subsets of X, such that $X = \bigcup E_n$. Pick $x_o \in X$ and consider the open ball $B_1(x_o)$ centered at x_o and radius 1. Since E_1 is nowhere dense in X, the complement $B_1(x_o) - \overline{E}_1$ contains an open set. Select an open ball $B_{r_1}(x_1)$, such that

$$\overline{B}_{r_1}(x_1) \subset B_1(x_o) - \overline{E}_1 \subset \overline{B}_1(x_o).$$

The selection can be done so that $r_1 < \frac{1}{2}$. Since E_2 is nowhere dense the complement $B_{r_1}(x_1) - \overline{E}_2$ contains an open set so that we may select an open ball $B_{r_2}(x_2)$ such that

$$\overline{B}_{r_2}(x_2) \subset B_{r_1}(x_1) - \overline{E}_2 \subset \overline{B}_{r_1}(x_1).$$

The selection can be done so that $r_2 < \frac{1}{3}$. Proceeding in this fashion generates a sequence of points $\{x_n\}$ and a family of balls $\{B_{r_n}(x_n)\}$ such that

$$\overline{B}_{r_{n+1}}(x_{n+1}) \subset \overline{B}_{r_n}(x_n), \quad r_n \leq \frac{1}{n+1}$$

and

$$\overline{B}_{r_n}(x_n) \bigcap \bigcup_{j=1}^{n} \overline{E}_n = \emptyset \quad \text{for all } n.$$

[21]R. Baire, Sur les fonctions de variables réelles, *Ann. Mat.*, 3-3 (1899), 1–124.

The sequence $\{x_n\}$ is Cauchy, and we let x denote its limit. Such a limit x is an element of X since $\{X; d\}$ is complete. Now the element x must belong to all the closed ball $\overline{B}_{r_n}(x_n)$ and does not belong to any of the $\overline{E}_n$. Thus $x \notin \bigcup \overline{E}_n$ and $X \neq \bigcup E_n$. $\qquad\qquad\qquad\square$

Corollary 16.2. *A complete metric space $\{X; d\}$ does not contain open subsets of first category.*

Theorem 16.3 (Banach–Steinhaus[22]). *Let $\{X; d\}$ be a complete metric space and let $\mathcal{F}$ be a family of continuous, real-valued functions defined in X. Assume that the functions $f \in \mathcal{F}$ are pointwise equibounded, i.e., that for all $x \in X$, there exists a positive number $F(x)$ such that*

$$|f(x)| \leq F(x) \quad \text{for all } f \in \mathcal{F}. \tag{16.1}$$

Then there exists a nonempty open set $\mathcal{O} \in X$ and a positive number F such that

$$|f(x)| \leq F \quad \text{for all } f \in \mathcal{F} \quad \text{and all } x \in \mathcal{O}. \tag{16.2}$$

Thus if the functions of the family $\mathcal{F}$ are pointwise equibounded in X, they are uniformly equibounded within some open subset of X. For this reason, the theorem is also referred to as the *uniform boundedness principle*.

Proof. For $n \in \mathbb{N}$, let $E_{n,f}$ and E_n be subsets of X defined by

$$E_{n,f} = \{x \in X \,|\, |f(x)| \leq n\}, \qquad E_n = \bigcap_{f \in \mathcal{F}} E_{n,f}.$$

The sets $E_{n,f}$ are closed, since the functions f are continuous. Therefore, in addition, the sets E_n are closed. Since the functions f are pointwise equibounded, for each $x \in X$ there exists some integer n such that $|f(x)| \leq n$ for all $f \in \mathcal{F}$. Therefore, each $x \in X$ belongs to some E_n, i.e., $X = \bigcup E_n$.

Since $\{X; d\}$ is complete, by the Baire category theorem, at least one of the E_n must not be nowhere dense. Since E_n is closed, it must contain a nonempty open set $\mathcal{O}$. Such a set satisfies (16.2) with $F = n$. $\qquad\qquad\qquad\square$

The Baire category theorem and related category arguments are remarkable as they afford function-theoretical conclusions from purely topological information.

17 Compact and totally bounded metric spaces

Since a metric space satisfies the first axiom of countability, sequential compactness, countable compactness, and the Bolzano–Weierstrass property all coincide.[23]

---

[22]S. Banach and H. Steinhaus, Sur le principe de la condensation de singularités, *Fund. Math.*, 9 (1927), 50–61.
[23]See Proposition 5.2(iii)–(v).

A metric space $\{X; d\}$ is *totally bounded* if for each $\varepsilon > 0$ there exists a finite collection of elements of X, say $\{x_1, x_2, \ldots, x_m\}$ for some positive integer m depending upon ε, such that X is covered by the union of the balls $B_\varepsilon(x_i)$ of radius ε and centered at x_i.

A finite sequence $\{x_1, \ldots, x_m\}$ with such a property is called a finite ε-net for X.

Proposition 17.1. *A countably compact metric space* $\{X; d\}$ *is totally bounded.*

Proof. Proceeding by contradiction, assume that there exists some $\varepsilon > 0$ for which there is no finite ε-net. Then for a fixed $x_1 \in X$, the ball $B_\varepsilon(x_1)$ does not cover X and we choose $x_2 \in X - B_\varepsilon(x_1)$. The union of the two balls $B_\varepsilon(x_1)$ and $B_\varepsilon(x_2)$ does not cover X, and we select

$$x_3 \in X - B_\varepsilon(x_1) \bigcup B_\varepsilon(x_2).$$

Proceeding in this fashion generates a sequence of points $\{x_n\}$ at mutual distance of at least ε. Such a sequence cannot have a cluster point, thus contradicting the Bolzano–Weierstrass property. $\qquad\square$

Corollary 17.2. *A countably compact metric space is separable.*

Proof. For positive integers m and n, let $E_{n,m}$ be the finite ε-net of X corresponding to $\varepsilon = \frac{1}{n}$. The union $\bigcup E_{n,m}$ is a countable subset of X which is dense in X. $\quad\square$

If $\{X; d\}$ is separable, every open covering of X has a countable subcovering. Therefore, countable compactness implies compactness.[24] Thus for metric spaces, all the various notions of compactness are equivalent.

We next examine the relation between compactness and total boundedness.

If $\{X; d\}$ is compact, it is also totally bounded. Indeed, having fixed $\varepsilon > 0$, the balls $B_\varepsilon(x)$ centered at all points of X form an open covering of X, from which one may extract a finite one.

It turns out that total boundedness implies compactness, provided the metric space $\{X; d\}$ is complete.

Proposition 17.3. *A totally bounded and complete metric space* $\{X; d\}$ *is sequentially compact.*

Proof. Let $\{x_n\}$ be a sequence of elements of X. The proof consists of selecting a Cauchy sequence $\{x_{n'}\} \subset \{x_n\}$. Since $\{X; d\}$ is complete, such a Cauchy subsequence would then converge to some $x \in X$, thereby establishing that $\{X; d\}$ is sequentially compact.

Fix $\varepsilon = \frac{1}{2}$ and determine a corresponding $\frac{1}{2}$-net, say, for example,

$$\{y_{1,1}, y_{1,2}, \ldots, y_{1,m_1}\} \quad \text{for some positive integer } m_1.$$

---

[24] See Proposition 5.3 and Corollary 5.4.

The union of the balls $B_{\frac{1}{3}}(y_{1,j})$ for $j = 1, 2, \ldots, m_1$ covers X. Therefore, at least one of these balls, say, for example, $B_{\frac{1}{3}}(y_{1,j})$, contains infinitely many elements of $\{x_n\}$. We select these elements and relabel them, to form a sequence $\{x_{n_1}\}$. These elements satisfy

$$d(x_{n_1}, x_{m_1}) < 1.$$

Next, let $\varepsilon = \frac{1}{2^2}$ and determine a corresponding $\frac{1}{4}$-net, say, for example,

$$y_{2,1}, y_{2,2}, \ldots, y_{2,m_2} \quad \text{for some positive integer } m_2.$$

There exists a ball $B_{\frac{1}{4}}(y_{2,j})$ for some $j \in \{1, 2, \ldots, m_2\}$ that contains infinitely many elements of $\{x_{n_1}\}$. We select these elements and relabel them to form a sequence $\{x_{n_2}\}$. These elements satisfy

$$d(x_{n_2}, x_{m_2}) < \frac{1}{2}.$$

Let $h \geq 2$ be a positive integer. If the subsequence $\{x_{n_{h-1}}\}$ has been selected, we let $\varepsilon = 2^{-h}$ and determine a corresponding 2^{-h}-net, say, for example,

$$y_{h,1}, y_{h,2}, \ldots, y_{h,m_h} \quad \text{for some positive integer } m_h.$$

There exists a ball $B_{\frac{1}{2^h}}(y_{h,j})$, for some $j \in \{1, 2, \ldots, m_h\}$ that contains infinitely many elements of $\{x_{n_{h-1}}\}$. We select these elements and relabel them to form a sequence $\{x_{n_h}\}$, whose elements satisfy

$$d(x_{n_h}, x_{m_h}) < \frac{1}{2^{h+1}}.$$

The Cauchy subsequence $\{x_{n'}\}$ is selected by diagonalization out of the sequences $\{x_{n_h}\}$. $\square$

Theorem 17.4. *A metric space $\{X; d\}$ is compact if and only if is totally bounded and complete.*

17.1 Precompact subsets of X. The various notions of compactness and their characterization in terms of total boundedness do not require that $\{X; d\}$ be a vector space. Thus, in particular, they apply to any subset $K \subset X$, endowed with the metric d inherited from $\{X; d\}$, by regarding $\{K; d\}$ as a metric space in its own right.

Proposition 17.5. *A subset $K \subset X$ of a metric space $\{X; d\}$ is compact if and only if it is sequentially compact.*

A subset $K \subset X$ is *precompact* if its closure $\overline{K}$ is compact.

Proposition 17.6. *A subset K of a complete metric space $\{X; d\}$ is precompact if and only if is totally bounded.*

PROBLEMS AND COMPLEMENTS

1 TOPOLOGICAL SPACES

1.1 A countable union of open sets is open. A countable union of closed sets need not be closed.

1.2 A countable intersection of closed sets is closed. A countable intersection of open sets need not be open.

1.3 Let $\mathcal{U}_1$ and $\mathcal{U}_2$ be topologies on X. Then $\mathcal{U}_1 \bigcap \mathcal{U}_2$ is a topology and $\mathcal{U}_1 \bigcup \mathcal{U}_2$ need not be a topology on X.

1.4 The Euclidean topology on $\mathbb{R}$ induces a relative topology on $[0, 1)$. The sets $[0, \varepsilon)$ for $\varepsilon \in (0, 1)$ are open in the relative topology of $[0, 1)$ and not in the original topology of $\mathbb{R}$.

1.5 Let $X = \mathbb{N} \bigcup \{\omega\}$, where ω is the first infinite ordinal. A set $\mathcal{O} \subset X$ is open if it either is any subset of $\mathbb{N}$ or if contains $\{\omega\}$ and all but finitely many elements of $\mathbb{N}$. The collection of all such sets, complemented with $\emptyset$ and X, defines a topology on X. A function $f : X \to \mathbb{R}$ is continuous with respect to such a topology if and only if $\lim f(n) = f(\omega)$.

1.6 Linear combinations of continuous functions are continuous. Let $g : \{X; \mathcal{U}\} \to \{Y; \mathcal{V}\}$ and $f : \{Y; \mathcal{V}\} \to \{Z; \mathcal{Z}\}$ be continuous. Then $f(g) : \{X; \mathcal{U}\} \to \{Z; \mathcal{Z}\}$ is continuous. The maximum or minimum of two real-valued, continuous functions is continuous.

1.7 Let $\mathcal{U}_1$ and $\mathcal{U}_2$ be topologies on X. The topology $\mathcal{U}_1$ is stronger or finer than $\mathcal{U}_2$ if $\mathcal{U}_2 \subset \mathcal{U}_1$, i.e., roughly speaking, if $\mathcal{U}_1$ contains more open sets than $\mathcal{U}_2$; equivalently if the identity map from $\{X; \mathcal{U}_1\}$ onto $\{X; \mathcal{U}_2\}$ is continuous.

1.8 Let $\{f_n\}$ be a sequence of real-valued, continuous functions from $\{X; \mathcal{U}\}$ into $\mathbb{R}$. If $\{f_n\} \to f$ uniformly, then f is continuous.

1.9 Let $C \subset X$ be closed and let $\{x_n\}$ be a sequence of points in C. Every cluster point of $\{x_n\}$ belongs to C.

1.10 Let $f : X \to Y$ be continuous and let $\{x_n\}$ be a sequence in X. If x is a cluster point of $\{x_n\}$, then $f(x)$ is a cluster point of $\{f(x_n)\}$.

1.11 Let X be the collection of pairs (m, n) of nonnegative integers. Any subset of X that does not contain $(0, 0)$ is declared to be open. A set $\mathcal{O}$ that contains $(0, 0)$ is open if and only if for all but a finite number of integers m, the set

$$\left\{ n \in \mathbb{N} \bigcup \{0\} | (m, n) \notin \mathcal{O} \right\}$$

is finite. For a fixed m the collection of (m, n) as n ranges over $\mathbb{N} \bigcup \{0\}$ can be regarded as a *column*.

With this terminology, a set $\mathcal{O}$ containing $(0, 0)$ is open if and only if it contains all but a finite number of elements for all but a finite number of columns. This defines a Hausdorff topology on X. No sequence in X can converge to $(0, 0)$.

The sequence (n, n) has $(0, 0)$ as a cluster point, but no subsequence of (n, n) converges to $(0, 0)$.[25]

1.12 Connected spaces A topological space $\{X; \mathcal{U}\}$ is connected if it is not the union of two disjoint open sets. A subset $X_o \subset X$ is connected if the space $\{X_o; \mathcal{U}_o\}$ is connected.

1.13 The continuous image of a connected space is connected.

1.14 Let $\{A_\alpha\}$ be a family of connected subsets of $\{X; \mathcal{U}\}$ with nonempty intersection, i.e., $\bigcap A_\alpha \neq \emptyset$. Then $\bigcup A_\alpha$ is connected.

1.15 *INTERMEDIATE VALUE THEOREM.* Let f be a real-valued continuous function on a connected space $\{X; \mathcal{U}\}$. Let $a, b \in X$ such that $f(a) < z < f(b)$ for some real number z. There exists $c \in X$ such that $f(c) = z$.

1.16 The discrete topology is a Hausdorff topology. If X is finite, then the discrete topology is the only one for which $\{X; \mathcal{U}\}$ is Hausdorff.

1.17 Let $\{X; \mathcal{U}\}$ be Hausdorff. Then $\{X; \mathcal{U}_1\}$ is Hausdorff for any stronger topology $\mathcal{U}_1$.

1.18 A Hausdorff space $\{X; \mathcal{U}\}$ is normal if and only if, for any closed set C and any open set $\mathcal{O}$ such that $C \subset \mathcal{O}$, there exists an open set O such that $C \subset O \subset \overline{O} \subset \mathcal{O}$.

1.19 SEPARATION PROPERTIES OF TOPOLOGICAL SPACES. A topological space $\{X; \mathcal{U}\}$ is said to be *regular* if points are separated from closed sets, i.e., if for a given closed set $C \subset X$ and x not in C, there exist disjoint open sets $\mathcal{O}_C$ and $\mathcal{O}_x$ such that $C \subset \mathcal{O}_C$ and $x \in \mathcal{O}_x$.

A Hausdorff space is said to be of type T_2. A regular space for which the singletons $\{x\}$ are closed is said to be of type T_3. A normal space for which the singletons $\{x\}$ are closed, is said to be of type T_4.

The separation properties of a topological space $\{X; \mathcal{U}\}$ are classified as follows:

T_0: Points are separated by open sets; i.e., for any two given points $x, y \in X$, there exists an open set containing one of the two points, say, for example, y, but not the other.

T_1: The singletons $\{x\}$ are closed.

T_2: Hausdorff spaces.

---

[25]R. Arens, Note on convergence in topology. *Math. Mag.*, **23** (1950), 229–234.

T_3: Regular $+ T_1$.

T_4: Normal $+ T_1$.

From the definitions, it follows that $T_4 \implies T_3 \implies T_2 \implies T_1 \implies T_0$.

The converse implications are false in general. In particular, T_0 does not imply T_1. For example, the space $X = \{x, y\}$ with the open sets $\emptyset$, X, y is T_0 and not T_1. We have already observed that T_1 does not imply Hausdorff.

Hausdorff, in turn, does not imply normal. Counterexamples are rather specialized.[26]

We will be concerned only with spaces that are Hausdorff and normal.

4   BASES, AXIOMS OF COUNTABILITY, AND PRODUCT TOPOLOGIES

4.1   Let $\{X; \mathcal{U}\}$ satisfy the first axiom of countability and let $A \subset X$. For every $x \in \overline{A}$, there exists a sequence $\{x_n\}$ of elements of A converging to x. For every cluster point y of a sequence $\{x_n\}$ of points in X, there exists a subsequence $\{x_{n'}\} \to y$.

4.2   Let X be infinite and let $\mathcal{U}$ consist of the empty set and the collection of all subsets of X whose complement is finite. Then $\mathcal{U}$ is a topology on X. If X is uncountable, $\{X; \mathcal{U}\}$ does not satisfy the first axiom of countability.

The points are closed but $\{X; \mathcal{U}\}$ is not Hausdorff.

4.3   Let $\mathcal{B}$ be the collection of all intervals of the form $[\alpha, \beta)$. Then $\mathcal{B}$ is a base for a topology $\mathcal{U}$ on $\mathbb{R}$, constructed as in Proposition 4.1. The set $\mathbb{R}$ endowed with such a topology satisfies the first but not the second axiom of countability. The intervals $[\alpha, \beta)$ are both open and closed. This is called the *half-open interval* topology.

The sequence $\{1 - \frac{1}{n}\}$ converges to 1 in the Euclidean topology and not in the half-open interval topology.

4.4   Let X be an uncountable set, well ordered by $\prec$, and let Ω be the first uncountable. Set

$$X_o = \{x \in X | x \prec \Omega\}, \qquad X_1 = X_o \bigcup \{\Omega\}.$$

Consider the collection $\mathcal{B}_o$ of sets

$$
\begin{aligned}
\{x \in X_o | x \prec \alpha\} &\quad \text{for some } \alpha \in X_o, \\
\{x \in X_o | \beta \prec x\} &\quad \text{for some } \beta \in X_o, \\
\{x \in X_o | \alpha \prec x \prec \beta\} &\quad \text{for } \alpha, \beta \in X_o.
\end{aligned}
$$

---

[26]R. H. Sorgenfrey, On the topological product of paracompact spaces, *Bull. Amer. Math. Soc.*, **53** (1947), 631–632; W. T. van Est and H. Freudenthal, Trennung durch stetige Funktionen in topologischen Räumen, *Indag. Math.*, **13** (1951), 305.

Similarly, define a collection of sets $\mathcal{B}_1$, where the various elements are taken out of X_1.

(i) The collection $\mathcal{B}_o$ forms a base for a topology $\mathcal{U}_o$ on X_o. The resulting space $\{X_o; \mathcal{U}_o\}$ satisfies the first but not the second axiom of countability. Moreover, $\{X_o; \mathcal{U}_o\}$ is separable.

(ii) The collection $\mathcal{B}_1$ forms a base for a topology $\mathcal{U}_1$ on X_1. The resulting space $\{X_1; \mathcal{U}_1\}$ does not satisfy the first axiom of countability and is not separable.

4.5   The product of two connected topological spaces is connected.

4.6   The product of a family $\{X_\alpha; \mathcal{U}_\alpha\}$ of Hausdorff spaces is Hausdorff.

4.7   A sequence $\{x_n\}$ of elements of $\prod X_\alpha$ converges to some $x \in \prod X_\alpha$ if and only if the sequences of the projections $\{x_{\alpha,n}\}$ converge to the projections x_α of x.

4.8   The *countable* product of separable topological spaces is separable.

4.9   Let $\{X; \mathcal{U}\}$ satisfy the second axiom of countability. Every topological subspace of X is separable. If $\{X; \mathcal{U}\}$ is separable but it does not satisfy the second axiom of separability, a topological subspace of X might not be separable.

The interval $[0, 1]$ with the half-open interval topology is separable. The Cantor set $\mathcal{C} \subset [0, 1]$ with the inherited topology is not separable.

4.10   THE BOX TOPOLOGY. Let $\{X_\alpha; \mathcal{U}_\alpha\}$ be a family of topological spaces and set

$$\mathcal{B} = \bigcup \left\{ \prod \mathcal{O}_\alpha | \mathcal{O}_\alpha \in \mathcal{U}_\alpha \right\}.$$

Each set in $\mathcal{B}$ is an open rectangle since it is the Cartesian product of open sets in $\mathcal{U}_\alpha$. The collection $\mathcal{B}$ forms a base for a topology in $\prod_\alpha X_\alpha$, called the box topology.

While the projections π_α are continuous with respect to such a topology, the box topology contains, roughly speaking, too many open sets.

As an example let $[0, 1]$ be endowed with the topology inherited from the Euclidean topology on $\mathbb{R}$. Then the Hilbert box $[0, 1]^{\mathbb{N}}$ can be endowed with either the product topology or the box topology. The sequence

$$\{x_n\} = \left\{ \frac{1}{n}, \frac{1}{n}, \dots, \frac{1}{n}, \dots \right\}$$

converges to zero in the product topology and not in the box topology. Indeed, the neighborhood of the origin

$$\mathcal{O} = \prod \left[0, \frac{1}{j} \right)$$

does not contain any of the elements of $\{x_n\}$.

5   COMPACT TOPOLOGICAL SPACES

5.1   A Hausdorff and compact topological space is *regular* (i.e., it separates closed sets from points) and normal.

5.2   If $\{X; \mathcal{U}\}$ is compact, then $\{X; \mathcal{U}_o\}$ is compact for any weaker topology $\mathcal{U}_o \subset \mathcal{U}$. The converse is false.

Proposition 5.1c. *Let $f : \{X; \mathcal{U}\} \to \{Y; \mathcal{V}\}$ be continuous, one-to-one, and onto. If $\{X; \mathcal{U}\}$ is compact and $\{Y; \mathcal{V}\}$ is Hausdorff, then f is a homeomorphism.*

Proof. The inverse f^{-1} is one-to-one and onto, and it is continuous if for every closed set $C \subset X$, the image $f(C)$ is closed.

   If $C \subset X$ is closed, it is compact and its continuous image $f(C)$ is compact and hence closed since $\{Y; \mathcal{V}\}$ is Hausdorff. $\square$

5.3   Let $\{X; \mathcal{U}\}$ be Hausdorff and compact. Then Proposition 5.1c implies that

   (i) $\{X; \mathcal{U}_1\}$ is not compact for any stronger topology $\mathcal{U}_1$;

   (ii) $\{X; \mathcal{U}_o\}$ is not Hausdorff for any weaker topology $\mathcal{U}_o$;

   (iii) if $\{X; \mathcal{U}_1\}$ is compact for a stronger topology $\mathcal{U}_1 \supset \mathcal{U}$, then $\mathcal{U} = \mathcal{U}_1$.

Thus the topological structure of a compact Hausdorff space $\{X; \mathcal{U}\}$ is rigid in the sense that one cannot strengthen its topology without losing compactness and cannot weaken it without losing the separation property.

5.4   Let $\|x\|$ be the Euclidean norm in $\mathbb{R}^N$ and consider the function

$$f(x) = \begin{cases} \dfrac{\max\{|x_1|, \ldots, |x_N|\}}{\|x\|} x & \text{for } x \neq 0, \\ 0 & \text{for } x = 0. \end{cases}$$

The function f maps cubes of edge 2ρ in $\mathbb{R}^N$ onto balls of radius ρ in $\mathbb{R}^N$; it is continuous, one-to-one, and onto. Thus f is a homeomorphism between $\mathbb{R}^N$ equipped with the topology generated by the cubes with faces parallel to the coordinate planes, and $\mathbb{R}^N$ equipped with the topology generated by the balls.

5.5   A space X consisting of more than one point and equipped with the trivial topology is compact and not Hausdorff.

5.6   Let $\{X; \mathcal{U}\}$ be locally compact. A subset $C \subset X$ is closed if and only if $C \cap K$ is closed for every closed compact subset $K \subset X$.

5.7   Let $\{X_o; \mathcal{U}_o\}$ and $\{X_1; \mathcal{U}_1\}$ be the spaces introduced in Section 4.4 of the Problems and Complements. $\{X_o; \mathcal{U}_o\}$ is sequentially compact but not compact. $\{X_1; \mathcal{U}_1\}$ is compact.

5.8   THE ALEXANDROV ONE-POINT COMPACTIFICATION OF $\{X; \mathcal{U}\}$. Let $\{X; \mathcal{U}\}$ be a noncompact Hausdorff topological space. Having fixed $x_* \notin X$, consider the set $X_* = X \bigcup \{x_*\}$ and define a collection of sets $\mathcal{U}_*$ consisting of $\mathcal{U}$, X_*, and all subsets $\mathcal{O}_* \subset X_*$ containing x_* and such that $(X_* - \mathcal{O}_*)$ is compact in $\{X; \mathcal{U}\}$. Then $\mathcal{U}_*$ is a Hausdorff topology on X_*. Moreover, $\{X_*; \mathcal{U}_*\}$ is compact, X is dense in X_*, and the restriction of $\mathcal{U}_*$ coincides with the original topology $\mathcal{U}$ on X.[27]

5.9   The topological space of Section 1.5 of the Problems and Complements is compact. It can be regarded as the Alexandrov one-point compactification of $\mathbb{N}$ equipped with the discrete topology.

5.10   The one-point compactification of $\mathbb{R}^N$ with its Euclidean topology is homeomorphic to the unit sphere in $\mathbb{R}^{N+1}$ by stereographic projection.

5.11   The one-point compactification of $\{X_o; \mathcal{U}_o\}$ in Section 4.4 of the Problems and Complements is $\{X_1; \mathcal{U}_1\}$.

7   CONTINUOUS FUNCTIONS ON COUNTABLY COMPACT SPACES

ON THE WEIERSTRASS–BAIRE THEOREM

7.1   The set of discontinuities of a real-valued function could be as diverse as possible. As an example, consider the functions

$$f(x) = \begin{cases} 1 & \text{if } x \in \mathbf{Q}, \\ -1 & \text{if } x \in [0, 1] - \mathbf{Q}, \end{cases}$$

$$g(x) = \begin{cases} x & \text{if } x \in \mathbf{Q}, \\ -x & \text{if } x \in [0, 1] - \mathbf{Q}. \end{cases}$$

The first is everywhere discontinuous but its absolute value is continuous. The second is continuous only at $x = 0$.

7.2   There exists a function $f : (0, 1) \to \mathbb{R}$ continuous at the irrationals and discontinuous at the rationals of $(0, 1)$.

A rational number $r \in (0, 1]$ can be written as the ratio m/n of two positive integers in *lowest terms*.[28]

Then set

$$f(x) = \begin{cases} \dfrac{1}{n} & \text{if } x \in \mathbf{Q} \bigcap (0, 1], \\ 0 & \text{if } x \in (0, 1] - \mathbf{Q}. \end{cases} \tag{7.1c}$$

---

[27] A. D. Alexandrov, On the extension of a Hausdorff space to an H-closed space, *C. R. Dokl. Acad. Sci. USSR. (N. S.)*, **37** (1942), 118–121.

[28] That is, m and n are the smallest integers for which $r = \frac{m}{n}$. A rational number r is an equivalence class of ratios of the form $\frac{m}{n}$. Out of such an equivalence class, we select the representative in *lowest terms*.

However, there exists no function $f : (0, 1] \to \mathbb{R}$ continuous at the rationals and discontinuous at the irrationals of $(0, 1]$.[29]

The function in (7.1c) is everywhere upper semicontinuous in $(0, 1]$ since, for every $x_o \in (0, 1]$,

$$\limsup_{x \to x_o} f(x) = 0 \le f(x_o).$$

7.3   There exist functions that are everywhere finite in their domain of definition and not bounded in every subset of their domain of definition.

Continue to represent a rational number $r \in (0, 1)$ as the ratio m/n of two positive integers in *lowest terms*. Then set

$$f(x) = \begin{cases} n & \text{if } x \in \mathbf{Q} \bigcap (0, 1), \\ 0 & \text{if } x \in (0, 1) - \mathbf{Q}. \end{cases} \tag{7.2c}$$

Such a function is everywhere finite in $[0, 1]$ and unbounded in every subinterval of $[0, 1]$. Indeed, let $I \subset [0, 1]$ be an interval. If f were bounded in I, then the denominator n of all rational numbers $\frac{m}{n} \in I$ would be bounded. This would imply that there are only finitely many rationals in I.

7.4   There exist real-valued, bounded functions, defined on a compact set, that take neither maxima nor minima.

Continue to represent a rational number $r \in (0, 1)$ as the ratio m/n of two positive integers in *lowest terms*. Then set

$$f(x) = \begin{cases} (-1)^n \dfrac{n}{n+1} & \text{if } x \in \mathbf{Q} \bigcap (0, 1), \\ 0 & \text{if } x \in (0, 1) - \mathbf{Q}. \end{cases} \tag{7.3c}$$

About any point of $(0, 1)$ the values of f are arbitrarily close to ± 1.

The function in (7.3c) is nowhere upper semicontinuous in $[0, 1]$. Indeed, for every $x_o \in [0, 1]$,

$$\limsup_{x \to x_o} f(x) = 1 > f(x_o).$$

Thus the assumption that f be upper semicontinuous cannot be relaxed in the Weierstrass–Baire theorem.

The function in (7.3c) is also nowhere monotone in $[0, 1]$.

ON THE ASSUMPTIONS OF DINI'S THEOREM

7.5   The assumption that the limit function f be lower semicontinuous cannot be removed from Dini's theorem. Indeed, the sequence $\{x^n\}$ for $x \in [0, 1]$ is decreasing and each x^n is continuous in $[0, 1]$, but the limit f is not lower semicontinuous. Accordingly, the convergence $\{x^n\} \to f$ is not uniform in $[0, 1]$.

---

[29] See also Corollary 16.4c of the Problems and Complements.

7.6   The assumption that each of the f_n is upper semicontinuous cannot be removed from Dini's theorem. Set

$$f_n(x) = \begin{cases} 0 & \text{for } x = 0, \\ 1 & \text{for } 0 < x < \dfrac{1}{n}, \\ 0 & \text{for } \dfrac{1}{n} \le x \le 1. \end{cases}$$

The sequence $\{f_n\}$ is decreasing, it converges to zero pointwise in $[0, 1]$, but the convergence is not uniform.

7.7   The requirement that the sequence $\{f_n\}$ be decreasing cannot be removed from Dini's theorem. Set

$$f_n(x) = \begin{cases} 2n^2 x & \text{for } 0 \le x \le \dfrac{1}{2n}, \\ n - 2n^2\left(x - \dfrac{1}{2n}\right) & \text{for } \dfrac{1}{2n} \le x \le \dfrac{1}{n}, \\ 0 & \text{for } \dfrac{1}{n} \le x \le 1. \end{cases}$$

The functions f_n are continuous in $[0, 1]$ and converge to zero pointwise in $[0, 1]$. However, the convergence is not uniform.

9   VECTOR SPACES

9.1   The element $\Theta \in X$ is unique.

9.2   Let A, B, and C be subsets of a vector space X. Then we have the following:

(i) $A \cap B \ne \emptyset$ if and only if $\Theta \in A - B$.

(ii) $A \cap (B + C) \ne \emptyset$ if and only if $B \cap (A - C) \ne \emptyset$; equivalently if and only if $C \cap (A - B) \ne \emptyset$.

(iii) X_o is a subspace of X if and only if $\alpha X_o = X_o$ for all $\alpha \in \mathbb{R}$ and $x + X_o = X_o$ for all $x \in X_o$.

(iv) If X_o and X_1 are linear subspaces of X, then $\alpha X_o + \beta X_1$ is a subspace of X.

(v) If A and B are convex, then $A + B$ is convex and λA is convex for all $\lambda \in \mathbb{R}$.

9.3   **HAMEL BASES.** A collection $\{x_\alpha\}$ of elements of a vector space X is linearly independent if any *finite* subcollection of elements $\{x_\alpha\}$ is linearly independent.

A linearly independent collection $\{x_\alpha\}$ is a *Hamel basis* for a vector space X if $\text{span}\{x_\alpha\} = X$; equivalently if every $x \in X$ has a unique representation as a *finite* linear combination of elements of $\{x_\alpha\}$, i.e.,

$$x = \sum_{j=1}^{m} c_j x_{\alpha_j} \quad \text{for some finite } m, \quad c_j \in \mathbb{R}.$$

Proposition 9.1c. *Every vector space X has a Hamel basis.*

Proof. Let $\mathcal{L}$ be the collection of all subsets of X whose elements are linearly independent. This collection is partially ordered by inclusion. Every linearly ordered subcollection $\{B_\sigma\}$ of $\mathcal{L}$ has an upper bound given by $B = \bigcup B_\sigma$. Indeed, the elements of B are linearly independent since any finitely many of them must belong to some B_σ, and the elements of B_σ are linearly independent.

Therefore, by Zorn's lemma, $\mathcal{F}$ has a maximal element $\{x_\alpha\}$. The elements of $\{x_\alpha\}$ are linearly independent, and every $x \in X$ can be written as a finite linear combination of them. Indeed, if not, the collection $\{x_\alpha, x\}$ belongs to $\mathcal{F}$, contradicting the maximality of $\{x_\alpha\}$. $\qquad\square$

9.4   A Hamel basis of $\mathbb{R}^N$ is the usual Euclidean basis.

9.5   Let ℓ denote the collection of all sequences $\{c_n\}$ of real numbers and consider the countable subcollection of ℓ,

$$\begin{aligned}
\mathbf{e}_1 &= \{1, 0, 0, \ldots, 0_m, 0, \ldots\}, \\
\mathbf{e}_2 &= \{0, 1, 0, \ldots, 0_m, 0, \ldots\}, \\
\cdots &= \cdots \quad \cdots \quad \cdots \quad \cdots \quad \cdots \quad \cdots, \\
\mathbf{e}_m &= \{0, 0, 0, \ldots, 1_m, 0, \ldots\}, \\
\cdots &= \cdots \quad \cdots \quad \cdots \quad \cdots \quad \cdots \quad \cdots.
\end{aligned} \qquad (9.1c)$$

Every $x \in \ell$ can be written as $x = \sum c_n \mathbf{e}_n$. However, $\{\mathbf{e}_n\}$ is not a Hamel basis for ℓ.

9.6   ON THE DIMENSION OF A VECTOR SPACE. If the Hamel basis of a vector space X is of the form $\{x_n\}$ for $n \in \mathbb{N}$, the dimension of X is $\aleph_0$, i.e., the cardinality of $\mathbb{N}$.

More generally, if $\{x_\alpha\}$ for $\alpha \in A$ is a Hamel basis for a vector space X, then the dimension of X is the cardinality of A. This definition of dimension of X is independent of the choice of the Hamel basis.

9.7   Let ℓ_o denote the collection of all sequences of real numbers $\{c_n\}$ with only finitely many nonzero elements. Then (9.1c) is a Hamel basis for ℓ_o and the dimension of ℓ_o is $\aleph_0$.

9.8   Let $\ell[0, 1]$ denote the collection of all sequences $\{c_n\}$ of real numbers in $[0, 1]$. The dimension of $\ell[0, 1]$ is no less than the cardinality of $\mathbb{R}$ since the collection

$$x_\alpha = \{\alpha, \alpha^2, \ldots, \alpha^n, \ldots\}, \quad \alpha \in (0, 1),$$

is linearly independent.

9.9   A vector space with a countable Hamel basis is separable.

9.10   The pair $\{\mathbb{R}; \mathbf{Q}\}$, i.e., the reals $\mathbb{R}$ over the field of the rationals $\mathbf{Q}$, is a vector space. If $x \in \mathbb{R}$ is not an algebraic number, then the elements $\{1, x, x^2, \ldots\}$ are linearly independent.[30] The dimension of $\{\mathbb{R}; \mathbf{Q}\}$ is at least the cardinality of $\mathbb{R}$.

10   TOPOLOGICAL VECTOR SPACES

10.1   Let A and B be subsets of a topological vector space $\{X; \mathcal{U}\}$. Then

(i) if A and B are open, then $\alpha A + \beta B$ is open;

(ii) $\overline{A} + \overline{B} \subset \overline{A + B}$; the inclusion is, in general, strict unless either one of $\overline{A}$ or $\overline{B}$ is compact;

(iii) if $A \subset X$ is convex, then $\overline{A}$ and $\overset{\circ}{A}$ are convex;

(iv) the convex hull of an open set is open.

10.2   If $x \in \mathcal{O} \in \mathcal{B}_{\Theta}$, there exists an open set A such that $x + A \subset \mathcal{O}$.

10.3   The identity map from $\mathbb{R}$ equipped with the Euclidean topology, onto $\mathbb{R}$ equipped with the half-open interval topology of Section 4.3 of the Problems and Complements, is bounded and linear but not continuous.

10.4   Let E be an open set in $\mathbb{R}^N$ and denote by $C(E)$ the linear vector space of all real-valued continuous functions defined in E. In $C(E)$, introduce a topology as follows. For $g \in C(E)$ and $\rho > 0$, stipulate that the set

$$\mathcal{O}_{g,\rho} = \left\{ f \in C(E) \mid \sup_E |f - g| < \rho \right\},$$

is an open neighborhood of g. The collection of such $\mathcal{O}_{g,\rho}$ is a base for a topology in $C(E)$. The sum $+ : C(E) \times C(E) \to C(E)$ is continuous with respect to such a topology. However, the multiplication by scalars $\bullet : \mathbb{R} \times C(E) \to C(E)$ is not continuous.

13   METRIC SPACES

13.1   Properties (i) and (iii) in the definition of a metric follow from (ii) and (iv). Setting $x = y$ in the triangle inequality (iv) and using (ii) gives $2d(x, z) \geq 0$ for all $x, z \in X$. Setting $z = y$ in (iv) gives

$$d(x, y) \leq d(y, x) \quad \text{and, by symmetry,} \quad d(y, x) \leq d(x, y).$$

Thus a metric could be defined as a function $d : (X \times X) \to \mathbb{R}$ satisfying (ii) and (iv).

---

[30] See Section 1.1 of the Problems and Complements of the Preliminaries.

13.2   The identically zero pseudometric generates the trivial topology on X. The function $d(x, y) = 1$ if $x \neq y$ and $d(x, y) = 0$ if $x = y$ is the *discrete* metric on X and generates the discrete topology. With respect to such a metric, the open balls $B_1(x)$ contain only the element x and their closure still coincides with x. Thus

$$\overline{B}_1(x) \neq \{y \in X | d(x, y) \leq 1\}.$$

13.3   The function $(x, y) \to \min\{1; |x - y|\}$ is a metric on $\mathbb{R}$.

13.4   Let $A \subset X$. Then $\overline{A} = \bigcup\{x | d(A, x) = 0\}$.

13.5   A function $f : \{X; d\} \to \{Y; \eta\}$ is continuous at $x \in X$ if and only if $\{f(x_n)\} \to f(x)$ for every sequence $\{x_n\} \to x$.

13.6   Two metrics d_1 and d_2 on X are equivalent if and only if

(i) for every $x \in X$ and every ball $B_\rho^1(x)$ in the metric d_1, there exists a radius $r = r(\rho, x)$ such that the ball $B_r^2(x)$ in the metric d_2 is contained in $B_\rho^1(x)$;

(ii) for every $x \in X$ and every ball $B_r^2(x)$ in the metric d_2, there exists a radius $\rho = \rho(r, x)$ such that the ball $B_\rho^1(x)$ in the metric d_1 is contained in $B_r^2(x)$.

The two metrics are uniformly equivalent if the choices of r in (i) and the choice of ρ in (ii) are independent of $x \in X$. Equivalently, d_1 and d_2 are uniformly equivalent if and only if the identity map between $\{X; d_1\}$ and $\{X; d_2\}$ is a uniform homeomorphism.

13.7   In $\mathbb{R}^N$, the following metrics are uniformly equivalent:

$$d_p(x, y) = \begin{cases} \left(\sum_{i=1}^N |x_i - y_i|^p \right)^{\frac{1}{p}} & \text{for } p \in [1, \infty), \\ \max_{1 \leq i \leq N} |x_i - y_i| & \text{for } p = \infty. \end{cases}$$

The discrete metric in $\mathbb{R}^N$ is not equivalent to any of the metrics d_p.

13.8   The metric d_o in (13.1) is equivalent, but not uniformly equivalent, to the original metric d.

13.9   A metric space $\{X; d\}$ is bounded if there exists an element $\Theta \in X$ and a number $M > 0$ such that $d(x, \Theta) < M$ for all $x \in X$. Boundedness depends only on the metric, and it is neither an intrinsic property of X nor a topological property. In particular, the same set X can be endowed with two equivalent metrics d and d_o in such a way that $\{X; d\}$ is not bounded and $\{X; d_o\}$ is bounded.

13.10   **THE HAUSDORFF DISTANCE OF SETS.**   Let $\{X; d\}$ be a metric space. For $A \subset X$ and $\sigma > 0$, set

$$A_\sigma = \{x \in X | d(x, A) < \sigma\}.$$

The Hausdorff distance of two sets A and B in X is[31]

$$d_{\mathcal{H}}(A, B) = \inf\{\sigma > 0 \text{ such that } A \subset B_{\sigma} \text{ and } B \subset A_{\sigma}\}.$$

If A and B have nonempty intersection, their distance is zero but their Hausdorff distance might be positive.

There exist distinct subsets A and B of X whose Hausdorff distance is zero. Thus $d_{\mathcal{H}}$ is a pseudometric on 2^X and generates the pseudometric space $\{2^X; d_{\mathcal{H}}\}$.

The identity map from $\{X; d\}$ to $\{2^X; d_{\mathcal{H}}\}$ is an isometry.

The topology on $\{2^X; d_{\mathcal{H}}\}$ is generated only by the original metric d via the definition of $d_{\mathcal{H}}$ and not by the topology of $\{X; d\}$. Indeed, there might exist metrics d_1 and d_2 that generate the same topology on X and such that the corresponding Hausdorff distances $d_{1,\mathcal{H}}$ and $d_{2,\mathcal{H}}$ generate different topologies on 2^X. As an example, let $X = \mathbb{R}^+$ endowed with the two equivalent metrics

$$d_1(x, y) = \left| \frac{x}{1+x} - \frac{y}{1+y} \right|, \qquad d_2(x, y) = \min\{1; |x - y|\}.$$

The topologies of $\{2^{\mathbb{R}^+}; d_{1,\mathcal{H}}\}$ and $\{2^{\mathbb{R}^+}; d_{2,\mathcal{H}}\}$ are different. The set of natural numbers $\mathbb{N}$ is a point in $2^{\mathbb{R}^+}$. The ball $B_{\varepsilon}^1(\mathbb{N})$ centered at $\mathbb{N}$ and of radius $\varepsilon \in (0, 1)$, in the topology of $\{2^{\mathbb{R}^+}; d_{1,\mathcal{H}}\}$ contains infinitely many finite subsets of $\mathbb{R}^+$. The ball $B_{\varepsilon}^2(\mathbb{N})$ in the topology of $\{2^{\mathbb{R}^+}; d_{2,\mathcal{H}}\}$ does not contain any finite subset of $\mathbb{R}^+$.

13.11   COUNTABLE PRODUCTS OF METRIC SPACES. Let $\{X_n; d_n\}$ be a countable collection of metric spaces. Then the product topology on $\prod X_n$ coincides with the topology generated by the metric

$$d(x, y) = \sum \frac{1}{2^n} \frac{d_n(x_n, y_n)}{1 + d_n(x_n, y_n)}. \tag{13.1c}$$

This will follow from the two inclusions:

(i) Every open neighborhood $\mathcal{O}_y$ of a point $y \in \prod X_n$, open in the product topology, contains a ball $B_{\varepsilon}(y)$ with respect to the metric in (13.1c).

(ii) Every ball $B_{\varepsilon}(y)$ with respect to the metric in (13.1c) contains an open neighborhood of y, open in the product topology.

Elements $y \in \prod X_n$ are sequences $\{y_n\}$ such that $y_n \in X_n$. For a fixed $y \in \prod X_n$ an open neighborhood $\mathcal{O}_y$ of y, open in the product topology, contains an open set of the form

$$\mathcal{O}_{y,k} = \prod_{n=1}^{k} B_{\varepsilon}^{(n)}(y_n) \quad \text{for some finite } k, \tag{13.2c}$$

where $B_{\varepsilon}^{(n)}(y_n)$ is the ball in $\{X_n; d_n\}$, centered at y_n and of radius ε.

---

[31] See [25, Chapter VIII].

There exists $\delta > 0$ sufficiently small depending on ε and k such that the ball $B_\delta(y)$ in $\prod X_n$ is contained in $\mathcal{O}_y$. Indeed, from

$$\sum \frac{1}{2^n} \frac{d_n(x_n, y_n)}{1 + d_n(x_n, y_n)} < \delta,$$

it follows that the number δ can be chosen so small that

$$d_n(x_n, y_n) < 2^{n+1}\delta \leq \varepsilon \quad \text{for all } n = 1, 2, \ldots, k.$$

Thus $B_\delta(y) \subset \mathcal{O}_y$. Conversely, every ball $B_\delta(y)$ in $\prod X_n$ contains an open set of the form (13.2c). Indeed, let k be a positive integer so large that

$$\sum_{n=k}^{\infty} \frac{1}{2^n} < \frac{\delta}{2}.$$

For such a k fixed the open set in (13.2c) with $\varepsilon = \frac{1}{2}\delta$ is contained in $B_\delta(y)$.

13.12   The countable product of complete metric spaces is complete.

14   METRIC VECTOR SPACES

Referring back to (14.2), the discontinuity of the bijection h is meant with respect to the topology generated by the original metric, whereas the discontinuity of the sum $+ : X \times X \to X$ or the product by scalars $\bullet : \mathbb{R} \times X \to X$ should be proved with respect to the new metric d_h.

14.1   Let $X = \mathbb{R}$, and let d be the usual Euclidean metric. Define

$$h(x) = \begin{cases} 1 & \text{if } x = 0, \\ 0 & \text{if } x = 1, \\ x & \text{otherwise} \end{cases}$$

and let $d_h = d(h)$ be defined as in (14.2).

For $\varepsilon \in (0, 1)$ the ball $B_\varepsilon(0)$, centered at 0 and radius ε, in the new metric d_h consists of the singleton $\{0\}$ and the open ball $B_\varepsilon(1)$, in the original Euclidean metric, of radius ε and centered at 1 from which the singleton $\{1\}$ has been removed.

Likewise the ball $B_\varepsilon(1)$ of radius ε and centered at 1, in the new metric d_h, consists of the singleton $\{1\}$ and the open ball $B_\varepsilon(0)$, in the original Euclidean metric, of radius ε and centered at 0 from which the singleton $\{0\}$ has been removed.

For such a metric d_h, both the sum and the multiplication by scalars are discontinuous.

14.2   The half-open interval topology of Section 4.3 of the Problems and Complements is not metrizable, i.e., there exists no metric on $\mathbb{R}$ that generates the half-open interval topology.

Combine Section 10.3 of the Problems and Complements with Proposition 14.2.

16   ON THE STRUCTURE OF A COMPLETE METRIC SPACE

16.1   Let d_1 and d_2 be two equivalent metrics on the same vector space X. The two metric spaces $\{X: d_1\}$ and $\{X: d_2\}$ have the same topology and the identity map is a homeomorphism. However, the identity map does not preserve completeness. As an example, consider $\mathbb{R}$ with the Euclidean metric and the metric d_o given in (13.1) corresponding to the Euclidean metric.

16.2   Intersection properties of a complete metric space.

Proposition 16.1c (Cantor). *Let $\{X; d\}$ be a complete metric space, and let $\{E_n\}$ be a countable collection of closed subsets of X such that $E_{n+1} \subset E_n$ and* diam$\{E_n\} \to 0$. *Then $\bigcap E_n \neq \emptyset$.*

16.3   COMPLETION OF A METRIC SPACE. Every metric space $\{X; d\}$ can be completed by the following procedure:

(a) First, one defines X' as the set of all the Cauchy sequences $\{x_n\}$ of elements in X and verifies that such a set has the structure of a linear space. Then on X', one defines a distance function

$$(\{x_n\}; \{y_n\}) \longrightarrow d'(\{x_n\}; \{y_n\}) = \lim d(x_n, y_n).$$

Since $\{x_n\}$ and $\{y_n\}$ are Cauchy sequences in X, the sequence $\{d(x_n, y_n)\}$ is a Cauchy sequence in $\mathbb{R}^+$. Thus the indicated limit exists.

Since several pairs of Cauchy sequences might generate the same limit, this is not a metric on X'. One verifies, however, that it is a pseudometric.

(b) In X', introduce an equivalence relation by which two sequences $\{x_n\}$ and $\{y_n\}$ are equivalent if $d'(\{x_n\}; \{y_n\}) = 0$. One verifies that such a relation is symmetric, reflexive, and transitive and therefore generates equivalence classes.

Define X^* as the set of equivalence classes of all Cauchy sequences of $\{X; d\}$. Any such class contains only sequences at zero, mutual pseudodistance.

For any two such equivalence classes x^* and y^*, choose representatives $\{x_n\} \in x^*$ and $\{y_n\} \in y^*$, and set

$$d^*(x^*, y^*) = d'(\{x_n\}; \{y_n\}).$$

One verifies that the definition is independent of the choices of the representative and that d^* defines a metric in X^*.

The original metric space $\{X; d\}$ is embedded into $\{X^*; d^*\}$ by identifying elements of X with elements of X^* as constant Cauchy sequences. Such an embedding is an isometry.

(c) The metric space $\{X^*; d^*\}$ is complete. Let $\{x_j^*\}$ be a Cauchy sequence in $\{X^*; d^*\}$ and select a representative $\{x_{j,n}\}$ out of each equivalence class x_j^*. By construction, any such representative is a Cauchy sequence in $\{X; d\}$. Therefore, for each $j \in \mathbb{N}$, there exists an index n_j such that

$$d(x_{j,n}, x_{j,n_j}) \leq \frac{1}{j} \quad \text{for all } n \geq n_j.$$

By diagonalization, select now the sequence $\{x_{j,n_j}\}$ and verify that it is a Cauchy sequence in $\{X; d\}$. Thus $\{x_{j,n_j}\}$ identifies an equivalence class $x^* \in X^*$. The Cauchy sequence $\{x_n^*\}$ converges to x^* in $\{X^*; d^*\}$.

Finally, the original metric space $\{X; d\}$, with the indicated embedding, is dense in $\{X^*; d^*\}$.

Remark 16.1c. While every metric space can be completed, a deeper problem is that of characterizing the elements of the new space and its metric. A typical example is the completion of the rational numbers into the real numbers.

16.4   SOME CONSEQUENCES OF BAIRE'S THEOREM. The category theorem is equivalent to the following.

Proposition 16.2c. *Let $\{X; d\}$ be a complete metric space. Then a countable collection $\{\mathcal{O}_n\}$ of open dense subsets of X has nonempty intersection.*

16.5   Let $\{X; d\}$ be a complete metric space. Then every closed, proper subset of X of the first category is nowhere dense.

16.6   The countable union of sets of first category is of the first category.

16.7   Let $\{E_n\}$ be a countable collection of closed subsets of a complete metric space $\{X; d\}$ such that $\bigcup E_n = X$. Then $\bigcup \overset{\circ}{E}_n$ is dense in X.

16.8   The rational numbers $\mathbf{Q}$ cannot be expressed as the countable intersection of open intervals.

Proposition 16.3c. *Let $f : \mathbb{R} \to \mathbb{R}$ be continuous on a dense subset E_o of $\mathbb{R}$. Then f is continuous on a set E of the second category.*

Proof. For $x \in (0, 1)$ and $\varepsilon > 0$, set

$$f'(x) = \sup_{\varepsilon > 0} \inf_{|x-y| < \varepsilon} f(y), \qquad f''(x) = \inf_{\varepsilon > 0} \sup_{|x-y| < \varepsilon} f(y).$$

For $n \in \mathbb{N}$, also set

$$\mathcal{O}_n = \left\{ x \in \mathbb{R} | f''(x) - f'(x) < \frac{1}{n} \right\}.$$

The set of continuity of f is the intersection of the $\mathcal{O}_n$. The sets $\mathcal{O}_n$ are open and dense in $\mathbb{R}$, and their complements $\mathcal{O}_n^c$ are nowhere dense in $\mathbb{R}$. Therefore, $\bigcup \mathcal{O}_n^c$ is of the first category in $\mathbb{R}$ and $\bigcap \mathcal{O}_n$ is of the second category in $\mathbb{R}$.   □

Corollary 16.4c. *There exists no function $f : [0, 1] \to \mathbb{R}$ continuous only at the rationals of $[0, 1]$.*[32]

---

[32]See also Section 7.2 of the Problems and Complements.

II

Measuring Sets

1 Partitioning open subsets of $\mathbb{R}^N$

Proposition 1.1 (Cantor[1]). *Every open subset E of $\mathbb{R}$ is the union of a countable collection of pairwise-disjoint open intervals.*

Proof. For $x \in E$, let E_x be the union of all open intervals containing x and contained in E. By construction, E_x is an interval.

If x and y are two distinct elements of E, then either $E_x = E_y$ or $E_x \bigcap E_y = \emptyset$. Indeed, if their intersection is nonempty, their union is an interval containing both x and y and contained in E.

Since E_x is an interval, it contains a rational number. Therefore, the collection of the intervals E_x that are distinct is countable and E is the union of such intervals. $\qquad\square$

An open subset of $\mathbb{R}^N$ cannot be partitioned, in general, into countably many, mutually disjoint open cubes. However, it can be partitioned into a countable collection of *closed* diadic cubes with pairwise-disjoint interior.

Let $\mathbf{q} \equiv (q_1, q_2, \ldots, q_N) \in \mathbf{Z}^N$ denote an N-tuple of integers. Having fixed a positive integer p and and some N-tuples $\mathbf{q}$, we let $Q_{p,\mathbf{q}}$ denote the closed diadic cube

$$Q_{p,\mathbf{q}} \equiv \left\{ x \in \mathbb{R}^N \mid \frac{q_i - 1}{2^p} \le x_i \le \frac{q_i}{2^p}, \quad i = 1, \ldots, N \right\}. \qquad (1.1)$$

---

[1] G. Cantor, Über unendliche, lineare Punktmannigfaltigkeiten, *Math. Ann.*, **20** (1882), 113–121.

The whole $\mathbb{R}^N$ can be partitioned into closed diadic cubes with pairwise-disjoint interior by slicing it with the hyperplanes $\{x_j = q_{\ell_j} 2^{-p}\}$, where for each $j = 1, 2, \ldots, N$, the numbers q_{ℓ_j} range over all integers $\mathbf{Z}$.

By this procedure,

$$\mathbb{R}^N \equiv \bigcup_{\mathbf{q} \in \mathbf{Z}^N} Q_{p,\mathbf{q}} \quad \text{for each fixed } p \in \mathbb{N}.$$

Moreover, any two of these cubes have disjoint interior, i.e.,

$$\overset{\circ}{Q}_{p,\mathbf{q}} \cap \overset{\circ}{Q}_{p,\mathbf{q}'} = \emptyset \quad \text{whenever } \mathbf{q} \neq \mathbf{q}'.$$

Proposition 1.2. *An open set $E \subset \mathbb{R}^N$ is the union of a countable collection of closed diadic cubes with pairwise-disjoint interior.*

Proof. Consider the closed diadic cubes $Q_{1,\mathbf{q}}$. At most countably many of them are contained in E and we denote by $\mathcal{Q}_1$ their union, i.e.,

$$\mathcal{Q}_1 \equiv \left\{ \bigcup Q_{1,\mathbf{q}} | Q_{1,\mathbf{q}} \subset E \right\}.$$

The complement $(E - \mathcal{Q}_1)$ is nonempty and because The union of the $\mathcal{Q}_n$ consist of countably many closed diadic cubes of the type (1.1) with pairwise-disjoint interior. Moreover,

$$E = \bigcup \mathcal{Q}_n.$$

Indeed, since E is open, for every $x \in E$ there exists a closed diadic cube $Q_{p,\mathbf{q}}$ contained in E and of the geometric structure of $\mathcal{Q}_1$ is also open. Therefore, at most countably many of the closed diadic cubes $Q_{2,\mathbf{q}}$ are contained in $(E - \overset{\circ}{\mathcal{Q}}_1)$, and we denote their union by $\mathcal{Q}_2$, i.e.,

$$\mathcal{Q}_2 \equiv \left\{ \bigcup Q_{2,\mathbf{q}} | Q_{2,\mathbf{q}} \subset (E - \overset{\circ}{\mathcal{Q}}_1) \right\}.$$

Proceeding in this fashion, we define inductively

$$\mathcal{Q}_n \equiv \left\{ \bigcup_{\mathbf{q}} Q_{n,\mathbf{q}} | Q_{n,\mathbf{q}} \subset \left(E - \bigcup_{j=1}^{n-1} \overset{\circ}{\mathcal{Q}}_j \right) \right\}, \quad n = 2, 3, \ldots.$$

containing x. Such a cube must be contained in some of the $\mathcal{Q}_n$ for some $n \in \mathbb{N}$. □

Remark 1.1. The decomposition of an open set E into closed diadic cubes is not unique. Having determined one, we reorder and relabel the cubes as $\{Q_n\}$ and write $E = \bigcup Q_n$.

The collection $\{Q_n\}$ can be constructed as in the proof of Proposition 1.2. Another decomposition is in Section 1.1 of the Problems and Complements.

2   Limits of sets, characteristic functions, and σ-algebras

Let $\{E_n\}$ be a countable collection of sets. The upper limit E'' and the lower limit E' of the sequence $\{E_n\}$ are defined as

$$E'' \equiv \bigcap_{n=1}^{\infty} \bigcup_{j=n}^{\infty} E_j = \limsup E_n,$$

$$E' \equiv \bigcup_{n=1}^{\infty} \bigcap_{j=n}^{\infty} E_j = \liminf E_n. \tag{2.1}$$

The sequence $\{E_n\}$ is said to be *convergent* if $E' = E''$, and in such a case we write

$$E'' = E' = \lim E_n. \tag{2.2}$$

For example, such a limit exists if the collection $\{E_n\}$ is *monotone increasing*, i.e., if $E_n \subset E_{n+1}$ for all $n \in \mathbb{N}$, or if $\{E_n\}$ is *monotone decreasing*, i.e., if $E_{n+1} \subset E_n$ for all $n \in \mathbb{N}$. However, for the limit to exist, the sequence $\{E_n\}$ need not be monotone.

It might also occur that a sequence $\{E_n\}$ of nonvoid sets and whose cardinality tends to infinity has a limit, and the limit is the empty set. For example, this occurs if E_n is the set of positive integers between n and $2n$.

The characteristic function χ_E of a set E is defined as

$$\chi_E(x) = \begin{cases} 1 & \text{if } x \in E, \\ 0 & \text{if } x \notin E. \end{cases}$$

It follows from the definition that for a sequence of sets $\{E_n\}$,

$$\chi_{\bigcup_{j=n}^{\infty} E_j} = \sup_{j \geq n} \chi_{E_j}, \qquad \chi_{\bigcap_{j=n}^{\infty} E_j} = \inf_{j \geq n} \chi_{E_j}.$$

Thus the notion of upper and lower limit of a sequence of sets can be equivalently rephrased in terms of upper and lower limit of the corresponding characteristic functions.

A set $E \subset \mathbb{R}^N$ is of the type $\mathcal{F}_\sigma$ if it is the union of a countable collection of closed subsets of $\mathbb{R}^N$. It is of the type $\mathcal{G}_\delta$ if it is the intersection of a countable collection of open subsets of $\mathbb{R}^N$. The set of the rational numbers is of the type $\mathcal{F}_\sigma$ and the set of the irrational numbers is of the type $\mathcal{G}_\delta$.

Similarly, we may define sets of type $\mathcal{F}_{\sigma\delta}$ and $\mathcal{G}_{\delta\sigma}$ as

$$\mathcal{F}_{\sigma\delta} \equiv \{\text{countable intersection of sets of the type } \mathcal{F}_\sigma\},$$

$$\mathcal{G}_{\delta\sigma} \equiv \{\text{countable union of sets of the type } \mathcal{G}_\delta\}.$$

Let X be a set. A collection $\mathcal{A}$ of a subset of X is an *algebra* if it contains the empty set and if the union of any two elements in $\mathcal{A}$ is in $\mathcal{A}$ and the complement of any element in $\mathcal{A}$ is in $\mathcal{A}$. It follows that $X \in \mathcal{A}$ and the *finite* union and intersection of elements in $\mathcal{A}$ remains in $\mathcal{A}$.

The collection $\mathcal{A}$ is a σ-*algebra* of subsets of X if it is an algebra and if the union of a *countable* collection of elements in $\mathcal{A}$ remains in $\mathcal{A}$. This would imply also that the intersection of a countable collection of elements in $\mathcal{A}$ remains in $\mathcal{A}$.

There exist algebras that are not σ-algebras (see Section 2.3 of the Problems and Complements).

The collection 2^X of all subsets of X is a σ-algebra called the *discrete* σ-algebra of subsets of X. The collection $\{X; \emptyset\}$ is a σ-algebra called the *trivial* σ-algebra of subsets of X.

Proposition 2.1. *Given any collection $\mathcal{O}$ of subsets of X, there exists a smallest σ-algebra $\mathcal{A}_o$ that contains $\mathcal{O}$.*

Proof. Let $\mathcal{F}$ be the collection of all the σ-algebras containing $\mathcal{O}$. Such a collection is nonvoid since it contains the discrete σ-algebra. Set

$$\mathcal{A}_o = \bigcap \{\mathcal{A} | \mathcal{A} \in \mathcal{F}\}.$$

If two sets A_1 and A_2 are in $\mathcal{A}_o$, they belong to all the σ-algebras in $\mathcal{F}$. Therefore, their union is in $\mathcal{A}_o$ since it must be in all the $\mathcal{A} \in \mathcal{F}$. Analogously, one proves that the complement of a set in $\mathcal{A}_o$ remains in $\mathcal{A}_o$ and the countable union of sets in $\mathcal{A}_o$ remains in $\mathcal{A}_o$. Therefore, $\mathcal{A}_o$ is a σ-algebra.

If $\mathcal{A}'$ is any σ-algebra containing $\mathcal{O}$, then it must be in the family $\mathcal{F}$. Thus $\mathcal{A}_o \subset \mathcal{A}'$. $\square$

Let $\mathcal{B}$ denote the smallest σ-algebra containing the open subsets of $\mathbb{R}^N$. The elements of $\mathcal{B}$ are the Borel sets and $\mathcal{B}$ is the Borel σ-algebra.

Open and closed subsets of $\mathbb{R}^N$ are Borel sets. Sets of the type $\mathcal{F}_\sigma$, $\mathcal{G}_\delta$, $\mathcal{F}_{\sigma\delta}$, $\mathcal{G}_{\delta\sigma}$, ... are also Borel sets.[2]

3 Measures

Denote by $\mathbb{R}^*$ the set of the *extended* real numbers,

$$\mathbb{R}^* \equiv \{-\infty\} \bigcup \mathbb{R} \bigcup \{+\infty\}$$

with the usual ordering of the formal operations; i.e., for example, for any $c \in \mathbb{R}$,

$$-\infty < c < +\infty, \qquad \pm\infty \pm c = \pm\infty, \qquad (\pm\infty)c = (\pm\infty)\operatorname{sign} c.$$

Let X be a set and let $\mathcal{A}$ be a σ-algebra of subsets of X. A set function μ defined on $\mathcal{A}$ with values on the extended reals $\mathbb{R}^*$ is *countably subadditive* if for a countable collection $\{E_n\}$ of elements of $\mathcal{A}$,

$$\mu \left(\bigcup E_n \right) \leq \sum \mu(E_n).$$

---

[2]See [3] and [25].

The set function μ is *countably additive* if for a countable collection $\{E_n\}$ of *disjoint* elements of $\mathcal{A}$,

$$\mu\left(\bigcup E_n\right) = \sum \mu(E_n) \quad \left(E_i \bigcap E_j = \emptyset \text{ for } i \neq j\right).$$

The triple $\{X, \mathcal{A}, \mu\}$ is a measure space and μ is a measure on X if

(i) the domain of μ is a σ-algebra $\mathcal{A}$;

(ii) μ is nonnegative on $\mathcal{A}$;

(iii) μ is countably additive;

(iv) $\mu(E) < \infty$ for some $E \in \mathcal{A}$.

Proposition 3.1. *Let* $\mu : \mathcal{A} \to \mathbb{R}^*$ *be a measure. Then we have the following:*
- *For every pair* A, B *of elements of* $\mathcal{A}$ *such that* $A \subset B$,

$$\mu(A) \leq \mu(B) \quad and \quad \mu(B - A) = \mu(B) - \mu(A). \tag{3.1}$$

- *For every pair* A, B *of elements of* $\mathcal{A}$,

$$\mu\left(A\bigcup B\right) = \mu(A) + \mu(B) - \mu\left(A\bigcap B\right). \tag{3.2}$$

The set function μ *is countably subadditive. Moreover, for a countable collection* $\{E_n\}$ *of elements of* $\mathcal{A}$,[3]

$$\liminf \mu(E_n) \geq \mu\left(\liminf E_n\right). \tag{3.3}$$

Moreover,

$$\limsup \mu(E_n) \leq \mu(\limsup E_n), \tag{3.4}$$

provided $\mu(\bigcup E_n) < \infty$.[4]

Proof. Write $B = A\bigcup(B - A)$. Since A and $(B - A)$ are disjoint, by (iii),

$$\mu(B) = \mu(A) + \mu(B - A).$$

This proves (3.1) since $\mu(B - A) \geq 0$. To prove (3.2), we write $A \bigcup B$ as a disjoint union, i.e.,

$$A\bigcup B = A\bigcup\left(B - \left(A\bigcap B\right)\right),$$

---

[3]This is a version of Fatou's lemma (see Section 8 of Chapter III).

[4]This assumption cannot be removed as there exist monotone decreasing sequences of sets $\{E_n\}$ such that $\mu(E_n) = \infty$ for all n and $\mu(\lim E_n) = 0$. For such sequences, (3.3) continues to hold and (3.4) fails.

and apply (iii) and (3.1). Let $\{E_n\}$ be a countable collection of sets in $\mathcal{A}$. Set $B_1 = E_1$ and

$$B_n = E_n - \bigcup_{j=1}^{n-1} E_j \quad \text{for } n = 2, 3, \ldots.$$

The sets B_n are mutually disjoint and their union coincides with the union of the E_n. Therefore, by property (iii) of μ and the monotonicity (3.1),

$$\mu\left(\bigcup E_n\right) = \mu\left(\bigcup B_n\right) = \sum \mu(B_n)$$

$$= \sum \mu\left(E_n - \bigcup_{j=1}^{n-1} E_j\right) \le \sum \mu(E_n).$$

Thus μ is countably subadditive.

From the definition of the lower limit of a sequence of sets,

$$\liminf E_n = \bigcup D_n, \quad \text{where } D_n = \bigcap_{j \ge n} E_j.$$

The sets D_n form a monotone increasing family of sets, i.e., $D_n \subset D_{n+1}$. Therefore,

$$\bigcup D_n = D_1 \bigcup \bigcup_n (D_{n+1} - D_n).$$

Since the sets $(D_{n+1} - D_n)$ are disjoint, by (iii) and the monotonicity (3.1),

$$\mu(\liminf E_n) = \mu\left(\bigcup D_n\right) = \mu\left\{D_1 \bigcup \bigcup_n (D_{n+1} - D_n)\right\}$$

$$= \mu(D_1) + \sum \mu(D_{n+1} - D_n)$$

$$= \mu(D_1) + \sum \{\mu(D_{n+1}) - \mu(D_n)\}$$

$$= \lim \mu(D_n) \le \liminf \mu(E_n).$$

This proves (3.3). To prove (3.4), consider the increasing family of sets

$$\bigcup E_n - \bigcup_{j \ge n} E_j$$

and compute

$$\mu\left(\lim\left(\bigcup E_n - \bigcup_{j \ge n} E_j\right)\right) \ge \mu\left(\bigcup E_n - \limsup_{j \ge n} \bigcup E_j\right)$$

$$= \mu\left(\bigcup E_n\right) - \mu\left(\limsup_{j \ge n} \bigcup E_j\right).$$

On the other hand, by (3.3),

$$\mu\left(\lim\left(\bigcup E_n - \bigcup_{j\geq n} E_j\right)\right) \leq \liminf \mu\left(\bigcup E_n - \bigcup_{j\geq n} E_j\right)$$

$$= \mu\left(\bigcup E_n\right) - \limsup \mu\left(\bigcup_{j\geq n} E_j\right).$$

Combining these two inequalities gives

$$\limsup \mu\left(\bigcup_{j\geq n} E_j\right) \leq \mu\left(\limsup \bigcup_{j\geq n} E_j\right),$$

from which (3.4) follows. $\qquad\square$

Remark 3.1. Let $E \in \mathcal{A}$ be a set for which (iv) holds. Then (3.1) implies that $\mu(\emptyset) = \mu(E - E) = 0$.

3.1   Finite, σ-finite, and complete measures. The measure μ is finite if $\mu(X) < \infty$ and it is σ-finite if there exists a countable collection $\{E_n\}$ of subsets of X such that[5]

$$X = \bigcup E_n \quad \text{and} \quad \mu(E_n) < \infty \quad \text{for all } n \in \mathbb{N}.$$

A measure space $\{X, \mathcal{A}, \mu\}$ is complete if every subset of a set of measure zero is in the σ-algebra $\mathcal{A}$. From the monotonicity of μ, it follows that if $\{X, \mathcal{A}, \mu\}$ is complete, the measure of every subset of a set of measure zero is zero. An example of an incomplete measure is in Section 14.3.

3.2   Some examples. Let X be a set. The set function $\mu(\emptyset) = 0$ and $\mu(X) = \infty$ is a measure on the trivial σ-algebra $\{X; \emptyset\}$.

Let X be a set and let E be a subset of X. Define $\mu(E)$ to be the number of elements in E if E is finite and infinity otherwise. The set function μ is a measure on 2^X called the *counting* measure on X.

Let $X = \{x_n\}$ be a sequence and let $\{\alpha_n\}$ be a sequence of nonnegative numbers. The set function

$$\mu(E) = \sum \{\alpha_n | x_n \in E\}$$

is a σ-finite measure on X.

Let X be infinite, and for every subset $E \subset X$, define $\mu(E) = 0$ if E is countable and $\mu(E) = \infty$ otherwise. This defines a measure on 2^X.

The sum of two measures defined on the same σ-algebra is a measure.

---

[5] Section 3.3 of the Problems and Complements might provide an example of a non-σ-finite measure.

Let $\{\mu_n\}$ be a sequence of measures on X defined on the same σ-algebra $\mathcal{A}$. Then $\mu = \sum \mu_n$ is a measure on X defined on $\mathcal{A}$.

Let $\{X, \mathcal{A}, \mu\}$ be a measure space, and for a fixed set $B \in \mathcal{A}$, define

$$\mathcal{A}_B = \left\{ \text{the collection of sets } A \bigcap B \text{ for } A \in \mathcal{A} \right\}.$$

Then $\mathcal{A}_B$ is a σ-algebra and the restriction of μ to $\mathcal{A}_B$ is a measure.

Let $\mathcal{A}$ be the discrete σ-algebra of all subsets of $\mathbb{R}^N$. Having fixed $x \in \mathbb{R}^N$, define

$$\mu(E) = \begin{cases} 1 & \text{if } x \in E, \\ 0 & \text{if } x \notin E. \end{cases} \tag{3.5}$$

One verifies that μ is a measure defined on $2^{\mathbb{R}^N}$. Alternatively, one might define $\mu(E)$ by the formula

$$\mu(E) = \chi_E(x) \quad \text{for all } E \in 2^{\mathbb{R}^N}. \tag{3.5$'$}$$

The measure μ defined by either (3.5) or (3.5)$'$ is called the *Dirac delta-measure* in $\mathbb{R}^N$ with mass concentrated at some x, and it is denoted by δ_x.

4   Outer measures and sequential coverings

A set function μ_e from the subsets of X into $\mathbb{R}^*$ is an *outer measure* if

(i) μ_e is defined on *all* the subsets of X;

(ii) μ_e is nonnegative and $\mu_e(\emptyset) = 0$;

(iii) μ_e is monotone;

(iv) μ_e is countably subadditive.

A collection $\mathcal{Q}$ of subsets of X is a *sequential covering* for X if it contains the empty set and if for every $E \subset X$ there exists a countable collection $\{Q_n\}$ of elements of $\mathcal{Q}$ such that $E \subset \bigcup Q_n$. For example, the collection of the closed cubes in $\mathbb{R}^N$ is a sequential covering for $\mathbb{R}^N$.

Outer measures can be constructed from a sequential covering $\mathcal{Q}$ by the following procedure.

Let λ be a given nonnegative set function defined on $\mathcal{Q}$, taking values on $\mathbb{R}^*$ and such that $\lambda(\emptyset) = 0$. For every $E \subset X$, set

$$\mu_e(E) = \inf \left\{ \sum \lambda(Q_n) | Q_n \in \mathcal{Q} \text{ and } E \subset \bigcup Q_n \right\}. \tag{4.1}$$

It follows from the definition that if $\mu_e(E) < \infty$, for every $\varepsilon > 0$, there exists a countable collection $\{Q_{\varepsilon,n}\}$ of elements of $\mathcal{Q}$ such that

$$E \subset \bigcup Q_{\varepsilon,n} \quad \text{and} \quad \sum \lambda(Q_{\varepsilon,n}) \le \mu_e(E) + \varepsilon. \tag{4.2}$$

Proposition 4.1. *The set function μ_e defined by* (4.1) *is an outer measure.*

Proof. Requirements (i)–(iii) are a direct consequence of the definitions. Let $\{E_n\}$ be a countable collection of subsets of X. In proving (iv), we may assume that $\mu_e(E_n) < \infty$ for all $n \in \mathbb{N}$.

Having fixed an arbitrary $\varepsilon > 0$, for each $n \in \mathbb{N}$, there exists a countable collection $\{Q_{j_n}\}$ of elements of Q such that

$$E_n \subset \bigcup Q_{j_n} \quad \text{and} \quad \sum \lambda(Q_{j_n}) \le \mu_e(E_n) + \varepsilon 2^{-n}.$$

The union of the sets Q_{j_n}, as both n and j_n range over $\mathbb{N}$, covers the union of the E_n. Therefore,

$$\mu_e \left(\bigcup E_n \right) \le \sum_n \sum_{j_n} \lambda(Q_{j_n}) \le \sum \mu_e(E_n) + \varepsilon.$$

By taking $E_n = \emptyset$ for $n > m$ for some $m \in \mathbb{N}$, this also implies that the outer measure is finitely subadditive. $\qquad\square$

The outer measure μ_e generated by the sequential covering Q and the nonnegative set function λ need not coincide with λ on elements of Q. By construction,

$$\mu_e(Q) \le \lambda(Q) \quad \text{for all } Q \in Q, \tag{4.3}$$

and strict inequality might occur.[6]

4.1   The Lebesgue outer measure in $\mathbb{R}^N$. Let Q denote the collection of the diadic cubes in $\mathbb{R}^N$ and let λ be the Euclidean measure of cubes.[7] Then for $E \subset \mathbb{R}^N$,

$$\mu_e(E) = \left\{ \sum \left(\frac{\text{diam } Q_n}{\sqrt{N}} \right)^N \mid E \subset \bigcup Q_n, \ Q_n \in Q \right\} \tag{4.4}$$

defines the Lebesgue outer measure of subsets of $\mathbb{R}^N$.

4.2   The Lebesgue–Stieltjes outer measure. Let $f : \mathbb{R} \to \mathbb{R}$ be monotone increasing and right continuous. For an open interval $(a, b) \subset \mathbb{R}$, define

$$\lambda(a, b) = f(b) - f(a).$$

The collection of open intervals (a, b) forms a sequential covering of $\mathbb{R}$. The corresponding outer measure $\mu_{f,e}$ is the Lebesgue–Stieltjes outer measure on $\mathbb{R}$ generated by f.[8]

---

[6]This happens, for example, for the Lebesgue–Stieltjes outer measure (Section 4.2) and the Hausdorff outer measure (Section 5).

[7]These cubes might be open or closed or neither. It is required that they have the structure (1.1) with some or all or none of the inequalities defining them being strict.

[8]T. J. Stieltjes, Note sur l'intégrale $\int_a^b f(x)G(x)dx$, *Nouv. Ann. Math. Paris Sér.* 3, **7** (1898), 161–171; H. Lebesgue, Sur l'intégrale de Stieltjes et sur les opérations fonctionnelles linéaires, *C. R. Acad. Sci. Paris*, **150** (1910), 86–88.

From the construction, it follows that

$$\mu_{f,e}(a, b] = f(b) - f(a).$$

However, it might occur that

$$\mu_{f,e}(a, b) < f(b) - f(a).$$

Thus for such an outer measure, (4.3) might hold with strict inequality.

5 The Hausdorff outer measure in $\mathbb{R}^N$

For $\varepsilon > 0$, let $\mathcal{E}_\varepsilon$ be the sequential covering of $\mathbb{R}^N$, consisting of all subsets E of $\mathbb{R}^N$ whose diameter is less than ε.[9]

Fix $\alpha > 0$ and set $\lambda(\emptyset) = 0$ and[10]

$$\mathcal{E}_\varepsilon \ni E \longrightarrow \lambda(E) = \operatorname{diam}(E)^\alpha.$$

This defines a nonnegative set function on $\mathcal{E}_\varepsilon$, which, in turn, generates the outer measure

$$\mathcal{H}_{\alpha,\varepsilon}(E) = \inf \left\{ \sum \operatorname{diam}(E_n)^\alpha \,|\, E \subset \bigcup E_n, \ E_n \in \mathcal{E}_\varepsilon \right\}. \tag{5.1}$$

For such an outer measure, the inequality in (4.3) might be strict. Indeed, if Q is a cube of unit edge in $\mathbb{R}^N$,

$$\lambda(Q) = (\sqrt{N})^\alpha \quad \text{and} \quad \mathcal{H}_{\alpha,1}(Q) = 0 \quad \text{for all } \alpha > N.$$

If $\varepsilon' < \varepsilon$, then $\mathcal{H}_{\alpha,\varepsilon} \leq \mathcal{H}_{\alpha,\varepsilon'}$. Therefore, the limit

$$\mathcal{H}_\alpha(E) = \lim_{\varepsilon \to 0} \mathcal{H}_{\alpha,\varepsilon}(E) \tag{5.2}$$

exists and defines a nonnegative set function $\mathcal{H}_\alpha$ on the subsets of $\mathbb{R}^N$.

Proposition 5.1. $\mathcal{H}_\alpha$ *is an outer measure on* $\mathbb{R}^N$. *Moreover,*

$$\begin{aligned}
\mathcal{H}_\alpha(E) < \infty &\quad \text{implies} \quad \mathcal{H}_\beta(E) = 0 \quad \text{for all } \beta > \alpha. \\
\mathcal{H}_\alpha(E) > 0 &\quad \text{implies} \quad \mathcal{H}_\beta(E) = \infty \quad \text{for all } \beta < \alpha.
\end{aligned} \tag{5.3}$$

Finally, for every set $E \subset \mathbb{R}^N$,

$$\frac{1}{N^{N/2}} \mathcal{H}_N(E) \leq \mu_e(E) \leq \frac{\kappa_N}{2^N} \mathcal{H}_N(E), \tag{5.4}$$

where κ_N *is the volume of the unit ball in* $\mathbb{R}^N$ *and* μ_e *is the Lebesgue outer measure in* $\mathbb{R}^N$. *In particular, for* $N = 1$, *the Lebesgue outer measure on* $\mathbb{R}$ *coincides with the Hausdorff outer measure* $\mathcal{H}_1$.

---

[9] For a set $A \subset \mathbb{R}^N$, $\operatorname{diam}(A) = \sup\{|x - y|; x, y \in A\}$.

[10] F. Hausdorff, Dimension und äusseres Mass, *Math. Ann.*, **79** (1919), 157–179.

Proof. The first statement is proved as in Proposition 4.1. Let $\{E_n\}$ be a countable collection of elements of $\mathcal{E}_\varepsilon$ such that $E \subset \bigcup E_n$. For $\beta > \alpha$,

$$\mathcal{H}_{\beta,\varepsilon}(E) \leq \sum \operatorname{diam}(E_n)^\beta \leq \varepsilon^{\beta-\alpha} \sum \operatorname{diam}(E_n)^\alpha.$$

This, in turn, implies that

$$\mathcal{H}_{\beta,\varepsilon}(E) \leq \varepsilon^{\beta-\alpha} \mathcal{H}_{\alpha,\varepsilon}(E).$$

The statements in (5.3) are equivalent.

In proving (5.4), we may assume that both $\mu_e(E)$ and $\mathcal{H}_N(E)$ are finite.

Having fixed $\varepsilon > 0$, there exists a countable collection of diadic cubes $\{Q_n\}$ whose union covers E and

$$\mu_e(E) \geq \sum \left(\frac{\operatorname{diam} Q_n}{\sqrt{N}} \right)^N - \varepsilon.$$

By possibly subdividing the cubes Q_n and using the finite additivity of the Euclidean measure of cubes, we may assume that $\operatorname{diam}(Q_n) < \varepsilon$. Then

$$\mu_e(E) \geq \frac{1}{N^{N/2}} \sum \operatorname{diam}(Q_n)^N - \varepsilon \geq \frac{1}{N^{N/2}} \mathcal{H}_{N,\varepsilon}(E) - \varepsilon$$

for all $\varepsilon > 0$. This proves the lower bound. For the upper bound, having fixed $\varepsilon > 0$, there exists a countable collection $\{E_n\}$ of elements of $\mathcal{E}_\varepsilon$ such that

$$\mathcal{H}_{N,\varepsilon}(E) \geq \sum \operatorname{diam}(E_n)^N - \frac{1}{2}\varepsilon.$$

Each of the E_n is included in a ball B_n of diameter $\operatorname{diam}(E_n)$. Such a ball can be covered by finitely many diadic cubes $\{Q_{n,j}\}_{j=1}^{j_n}$ with pairwise-disjoint interior. Such a covering of E_n can be chosen so that

$$\operatorname{diam}(E_n)^N = \frac{2^N}{\kappa_N}\{\text{volume of } B_n\}$$

$$\geq \frac{2^N}{\kappa_N} \sum_{j=1}^{j_n} \left(\frac{\operatorname{diam} Q_{n,j}}{\sqrt{N}} \right)^N - \frac{1}{2^{n+1}}\varepsilon.$$

Therefore, by (4.4),

$$\mathcal{H}_{N,\varepsilon}(E) \geq \frac{2^N}{\kappa_N} \sum_n \sum_{j=1}^{j_n} \left(\frac{\operatorname{diam} Q_{n,j}}{\sqrt{N}} \right)^N - \varepsilon$$

$$\geq \frac{2^N}{\kappa_N} \mu_e(E) - \varepsilon$$

for all $\varepsilon > 0$. $\qquad\square$

The Hausdorff outer measure is additive on sets that are at mutual positive distance.

Proposition 5.2. *Let E and F be subsets of $\mathbb{R}^N$ such that*

$$\text{dist}\{E; F\} = \delta > 0.$$

Then for all $\alpha > 0$,

$$\mathcal{H}_\alpha\left(E \bigcup F\right) = \mathcal{H}_\alpha(E) + \mathcal{H}_\alpha(F). \tag{5.5}$$

Proof. Since $\mathcal{H}_\alpha$ is subadditive, it suffices to prove (5.5) with equality replaced by $\geq$. Also, we may assume that $\mathcal{H}_\alpha(E \bigcup F)$ is finite; otherwise, the statement is obvious.

Having fixed $\varepsilon < \frac{1}{2}\delta$, there exists a collection of sets $\{G_n\}$ whose union contains $E \bigcup F$ and each of which has diameter less than ε such that

$$\mathcal{H}_{\alpha,\varepsilon}\left(E \bigcup F\right) \geq \sum \text{diam}(G_n)^\alpha - \varepsilon.$$

If a point $x \in E$ is covered by some G_n, such a set does not intersect F. Likewise, if $y \in F$ is covered by G_m, then $G_m \bigcap E = \emptyset$.

Therefore, the collection $\{G_n\}$ can be separated into two subcollections $\{E_n\}$ and $\{F_n\}$. The union of the E_n contains E and the union of the F_n contains F. From this,

$$\mathcal{H}_{\alpha,\varepsilon}\left(E \bigcup F\right) \geq \sum \text{diam}(E_n)^\alpha + \sum \text{diam}(F_n)^\alpha - \varepsilon$$

$$\geq \mathcal{H}_{\alpha,\varepsilon}(E) + \mathcal{H}_{\alpha,\varepsilon}(F) - \varepsilon.$$

The statement follows by letting $\varepsilon \to 0$. $\qquad\qquad\square$

6   Constructing measures from outer measures

Let μ_e be an outer measure on X and let A and E be any two subsets of X. Then the set identity $A = (A \bigcap E) \bigcup (A - E)$ implies

$$\mu_e(A) \leq \mu_e\left(A \bigcap E\right) + \mu_e(A - E). \tag{6.1}$$

Consider the collection $\mathcal{A}$ of those sets $E \subset X$ satisfying (6.1) with equality for all sets $A \subset X$, i.e., the collection of the sets $E \subset X$ such that[11]

$$\mu_e(A) \geq \mu_e\left(A \bigcap E\right) + \mu_e(A - E) \tag{6.2}$$

for all sets $A \subset X$. If $E \in \mathcal{A}$, we say that E is μ_e-measurable.

---

[11] See [6].

Proposition 6.1.

(i) *The empty set is in $\mathcal{A}$.*

(ii) *If $E \in \mathcal{A}$, the complement E^c of E also belongs to $\mathcal{A}$.*

(iii) *If E is a set of outer measure zero, then $E \in \mathcal{A}$.*

(iv) *If E_1 and E_2 belong to $\mathcal{A}$, their union also belongs to $\mathcal{A}$.*

(v) *If E_1 and E_2 belong to $\mathcal{A}$, $E_1 - E_2$ also belongs to $\mathcal{A}$.*

(vi) *If E_1 and E_2 belong to $\mathcal{A}$, $E_1 \cap E_2$ also belongs to $\mathcal{A}$.*

(vii) *Let $\{E_n\}$ be a collection of disjoint sets in $\mathcal{A}$. Then*

$$\mu_e \left(A \cap \bigcup E_n \right) = \sum \mu_e \left(A \cap E_n \right) \quad \text{for every } A \subset X.$$

(viii) *The countable union of sets in $\mathcal{A}$ remains in $\mathcal{A}$.*

Proof. Statement (i) follows from the definition, and (ii) follows from the set identities

$$A \cap E^c = A - E, \qquad A - E^c = A \cap E.$$

Statement (iii) follows from the monotonicity of μ_e and

$$\mu_e \left(A \cap E \right) + \mu_e(A - E) \leq \mu_e(E) + \mu_e(A) \leq \mu_e(A).$$

To prove (iv), let E_1 and E_2 be elements of $\mathcal{A}$ and write (6.2) for the pair A, E_1 and for the pair $(A - E_1)$, E_2; i.e.,

$$\mu_e(A) \geq \mu_e \left(A \cap E_1 \right) + \mu_e(A - E_1),$$
$$\mu_e(A - E_1) \geq \mu_e \left((A - E_1) \cap E_2 \right) + \mu_e((A - E_1) - E_2).$$

Add these inequalities and use the subadditivity of μ_e and the set identities

$$((A - E_1) - E_2) = A - \left(E_1 \bigcup E_2 \right),$$
$$\left((A - E_1) \cap E_2 \right) \bigcup \left(A \cap E_1 \right) = A \cap \left(E_1 \bigcup E_2 \right).$$

This gives

$$\mu_e(A) \geq \mu_e \left(A \cap \left(E_1 \bigcup E_2 \right) \right) + \mu_e \left(A - \left(E_1 \bigcup E_2 \right) \right).$$

Statements (v) and (vi) follow from the set identities

$$E_1 - E_2 = \left(E_1^c \bigcup E_2 \right)^c, \qquad E_1 \cap E_2 = \left(E_1^c \bigcup E_2^c \right)^c.$$

We first establish (vii) for a finite collection of disjoint sets. That is, if

$$B_n = \bigcup_{j=1}^{n} E_j \quad \text{and} \quad E_i \bigcap E_j = \emptyset \quad \text{for } i \neq j,$$

then for every set $A \subset X$,

$$\mu_e \left(A \bigcap B_n \right) = \sum_{j=1}^{n} \mu_e \left(A \bigcap E_j \right).$$

The statement is obvious for $n = 1$. Assuming it holds for n, we show that it continues to hold for $(n + 1)$. By (iv), B_n is in $\mathcal{A}$, and we may write (6.2) for the pair $(A \bigcap B_{n+1})$ and B_n. This gives

$$\mu_e \left(A \bigcap B_{n+1} \right) = \mu_e \left(A \bigcap B_{n+1} \bigcap B_n \right) + \mu_e \left(A \bigcap B_{n+1} - B_n \right)$$
$$= \mu_e \left(A \bigcap B_n \right) + \mu_e \left(A \bigcap E_{n+1} \right).$$

Now let $\{E_n\}$ be a countable collection of disjoint sets in $\mathcal{A}$. By the subadditivity and monotonicity of the outer measure μ_e,

$$\sum \mu_e(A \bigcap E_n) \geq \mu_e \left(A \bigcap \bigcup E_n \right) \geq \mu_e \left(A \bigcap \bigcup_{n=1}^{m} E_n \right)$$
$$= \sum_{n=1}^{m} \mu_e(A \bigcap E_n) \quad \text{for all } m \in \mathbb{N}.$$

To prove (viii), we assume first that the sets of the collection $\{E_n\}$ are mutually disjoint. For every $A \subset X$ and every $m \in \mathbb{N}$,

$$\mu_e(A) = \mu_e \left(A \bigcap \bigcup_{n=1}^{m} E_n \right) + \mu_e \left(A - \bigcup_{n=1}^{m} E_n \right)$$
$$\geq \sum_{n=1}^{m} \mu_e \left(A \bigcap E_n \right) + \mu_e \left(A - \bigcup E_n \right).$$

Letting $m \to \infty$ and using the countable subadditivity of μ_e shows that the union of the E_n satisfies (6.2) and therefore is in $\mathcal{A}$.

For a general countable collection $\{E_n\}$ of elements of $\mathcal{A}$, write

$$D_1 = E_1 \quad \text{and} \quad D_n = E_{n+1} - \bigcup_{j=1}^{n} E_j.$$

The sets D_n are in $\mathcal{A}$ and are mutually disjoint and their union coincides with the union of the E_n.   □

Proposition 6.2. *The restriction of μ_e to $\mathcal{A}$ is a complete measure.*

Proof. By Proposition 6.1, $\mathcal{A}$ is a σ-algebra and the restriction $\mu_e|_{\mathcal{A}}$ satisfies the requirements of a complete measure. In particular, the countable additivity follows from (vii) with A replaced with the union of the E_n, and the completeness follows from (iii). $\qquad\square$

7 The Lebesgue–Stieltjes measure on $\mathbb{R}$

The Lebesgue–Stieltjes outer measure $\mu_{f,e}$ induced by an increasing, right-continuous function $f : \mathbb{R} \to \mathbb{R}$ generates a σ-algebra $\mathcal{A}_f$ of subset of $\mathbb{R}$ and a measure μ_f defined on $\mathcal{A}_f$.

Proposition 7.1. $\mathcal{A}_f$ *contains the Borel sets on $\mathbb{R}$.*

Proof. It suffices to verify that intervals of the type $(\alpha, \beta]$ belong to $\mathcal{A}_f$, i.e., that for every subset $E \subset \mathbb{R}$,

$$\mu_{f,e}(E) \geq \mu_{f,e}\left(E \bigcap (\alpha, \beta]\right) + \mu_{f,e}(E - (\alpha, \beta]). \tag{7.1}$$

Lemma 7.2. *Let λ_f be the Lebesgue–Stieltjes set function defined on open intervals of $\mathbb{R}$, from which $\mu_{f,e}$ is constructed. Then for any open interval I and any interval of the type $(\alpha, \beta]$,*

$$\lambda_f(I) \geq \mu_{f,e}\left(I \bigcap (\alpha, \beta]\right) + \mu_{f,e}(I - (\alpha, \beta]).$$

Proof. Let $I = (a, b)$ and assume that $(\alpha, \beta] \subset (a, b)$. Denote by ε and η positive numbers. By the right-continuity of f and the definition of $\mu_{f,e}$,

$$\begin{aligned}
\lambda_f(I) &= f(b) - f(a) \\
&= \lim_{\varepsilon \to 0}(f(b) - f(\beta + \varepsilon)) \\
&\quad + \lim_{\varepsilon \to 0}\lim_{\eta \to 0}(f(\beta + \varepsilon) - f(\alpha + \eta)) \\
&\quad + \lim_{\eta \to 0}(f(\alpha + \eta) - f(a)) \\
&\geq \mu_{f,e}(\beta, b) + \mu_{f,e}(\alpha, \beta] + \mu_{f,e}(a, \alpha] \\
&\geq \mu_{f,e}(\alpha, \beta] + \mu_{f,e}(I - (\alpha, \beta]).
\end{aligned}$$

The cases when $\alpha \leq a < \beta < b$ and $a \leq \alpha < b < \beta$ are handled in a similar way. $\qquad\square$

Returning to the proof of (7.1), we may assume that $\mu_{f,e}(E)$ is finite. Having fixed $\varepsilon > 0$, there exists a collection of intervals $\{I_n\}$ whose union covers E and

$$
\begin{aligned}
\mu_{f,e}(E) + \varepsilon &\geq \sum \lambda_f(I_n) \\
&\geq \sum \mu_{f,e}\left(I_n \bigcap (\alpha, \beta]\right) + \sum \mu_{f,e}(I_n - (\alpha, \beta]) \\
&\geq \mu_{f,e}\left(\bigcup I_n \bigcap (\alpha, \beta]\right) + \mu_{f,e}\left(\bigcup I_n - (\alpha, \beta]\right) \\
&\geq \mu_{f,e}\left(E \bigcap (\alpha, \beta]\right) + \mu_{f,e}(E - (\alpha, \beta])
\end{aligned}
$$

in view of the countable subadditivity and monotonicity of $\mu_{f,e}$. $\square$

7.1 Borel measures. A measure μ in $\mathbb{R}^N$ is a Borel measure if the σ-algebra of its domain of definition contains the Borel sets. By Proposition 7.1, the Lebesgue–Stieltjes measure on $\mathbb{R}$ is a Borel measure.

8 The Hausdorff measure on $\mathbb{R}^N$

For a fixed $\alpha > 0$, the Hausdorff outer measure $\mathcal{H}_\alpha$ generates a σ-algebra $\mathcal{A}_\alpha$ and a measure μ_α called the Hausdorff measure on $\mathbb{R}^N$.

Proposition 8.1. $\mathcal{A}_\alpha$ *contains the Borel sets in* $\mathbb{R}^N$.

Proof. It suffices to verify that closed sets $E \subset \mathbb{R}^N$ belong to $\mathcal{A}_\alpha$, i.e., that for every subset $A \subset \mathbb{R}^N$,

$$
\mathcal{H}_\alpha(A) \geq \mathcal{H}_\alpha\left(A \bigcap E\right) + \mathcal{H}_\alpha(A - E) \quad \text{for } E \subset \mathbb{R}^N \text{ closed.} \tag{8.1}
$$

We may assume that $\mathcal{H}_\alpha(A)$ is finite. For $n \in \mathbb{N}$, set

$$
E_n = \left\{ x \in \mathbb{R}^N \,|\, \text{dist}\{x; E\} \leq \frac{1}{n} \right\}.
$$

and estimate that

$$
\begin{aligned}
\mathcal{H}_\alpha(A) &= \mathcal{H}_\alpha\left(\left(A \bigcap E_n\right)\bigcup(A - E_n)\right) \\
&\geq \mathcal{H}_\alpha\left(\left(A \bigcap E\right)\bigcup(A - E_n)\right) \\
&= \mathcal{H}_\alpha\left(A \bigcap E\right) + \mathcal{H}_\alpha(A - E_n).
\end{aligned}
$$

The last inequality follows from Proposition 5.2 since

$$
\text{dist}\left\{\left(A \bigcap E\right); (A - E_n)\right\} \geq \frac{1}{n}.
$$

From this,

$$
\mathcal{H}_\alpha(A) \geq \mathcal{H}_\alpha\left(A \bigcap E\right) + \mathcal{H}_\alpha(A - E) - \mathcal{H}_\alpha\left(A \bigcap(E_n - E)\right). \quad \square
$$

Lemma 8.2. $\lim \mathcal{H}_\alpha (A \bigcap (E_n - E)) = 0.$

Proof. For $j = n, (n+1), \ldots,$ set

$$F_j = \left\{ x \in A \Big| \frac{1}{j+1} < \text{dist}\{x; E\} \leq \frac{1}{j} \right\}.$$

Then

$$A \bigcap (E_n - E) = \bigcup_{j=n}^{\infty} F_j$$

and

$$\mathcal{H}_\alpha \left(A \bigcap (E_n - E) \right) \leq \sum_{j=n}^{\infty} \mathcal{H}_\alpha (F_j).$$

The conclusion would follows from this if the series

$$\sum \mathcal{H}_\alpha (F_j)$$

were convergent. To this end, we regroup the F_j into those whose index j is even and those whose index is odd. By construction,

$$\text{dist}\{F_j; F_i\} > 0 \quad \text{if } i \neq j$$

and if either both i and j are even or if both are odd. For any two such sets,

$$\mathcal{H}_\alpha (F_j) + \mathcal{H}_\alpha (F_i) = \mathcal{H}_\alpha \left(F_j \bigcup F_i \right)$$

by virtue of Proposition 5.2. Therefore, for all finite m,

$$\sum_{j=1}^{m} \mathcal{H}_\alpha (F_j) \leq \sum_{h=1}^{m} \mathcal{H}_\alpha (F_{2h}) + \sum_{h=0}^{m} \mathcal{H}_\alpha (F_{2h+1})$$

$$= \mathcal{H}_\alpha \left(\bigcup_{h=1}^{m} F_{2h} \right) + \mathcal{H}_\alpha \left(\bigcup_{h=0}^{m} F_{2h+1} \right)$$

$$\leq 2\mathcal{H}_\alpha (F_1) \leq 2\mathcal{H}_\alpha (A). \qquad \square$$

Corollary 8.3. *For all $\alpha > 0$, the Hausdorff measure $\mathcal{H}_\alpha$ is a Borel measure.*

The definition 5.1, valid for $\alpha > 0$, and the previous remarks suggest that we define $\mathcal{H}_o$ to be the counting measure.

9   Extending measures from semialgebras to σ-algebras

A measure space $\{X, \mathcal{A}, \mu\}$ can be constructed starting from a sequential covering $\mathcal{Q}$ of X and a nonnegative set function $\lambda : \mathcal{Q} \to \mathbb{R}^*$. First, one constructs an outer measure μ_e by the procedure of Section 4, and then by the procedure of Section 6, such an outer measure μ_e generates a σ-algebra $\mathcal{A}$ of subsets of X and a measure μ defined on $\mathcal{A}$.

The elements of the originating sequential covering $\mathcal{Q}$ need not be measurable with respect to the resulting measure μ.[12] Even if the elements of $\mathcal{Q}$ are μ_e-measurable, the set function λ and the outer measure μ_e might disagree on $\mathcal{Q}$.[13]

The next proposition provides sufficient conditions on both $\mathcal{Q}$ and λ for the elements of $\mathcal{Q}$ to be measurable and for λ to coincide with μ_e on $\mathcal{Q}$.

The collection $\mathcal{Q}$ is said to be a *semialgebra* if the following hold:

(i) The intersection of any two elements in $\mathcal{Q}$ is in $\mathcal{Q}$.

(ii) For any two elements Q_1 and Q_2 in $\mathcal{Q}$, the difference $(Q_1 - Q_2)$ is the *finite* disjoint union of elements in $\mathcal{Q}$.

The collection $\mathcal{Q}$ of all parallelepipeds in $\mathbb{R}^N$ whose faces are parallel to the diagonal planes is a semialgebra.[14] The collection of the open intervals on $\mathbb{R}$ is not a semialgebra.

The set function $\lambda : \mathcal{Q} \to \mathbb{R}^*$ is finitely additive on $\mathcal{Q}$ if for any finite collection $\{Q_1, Q_2, \ldots, Q_m\}$ of disjoint elements of $\mathcal{Q}$ whose union is in $\mathcal{Q}$,[15]

$$\lambda \left(\bigcup_{j=1}^m Q_j \right) = \sum_{j=1}^m \lambda(Q_j) \quad \left(\bigcup_{j=1}^m Q_j \in \mathcal{Q} \right). \tag{9.1}$$

The set function λ is countably subadditive on $\mathcal{Q}$ if for any countable collection $\{Q_j\}$ of elements of $\mathcal{Q}$ whose union is in $\mathcal{Q}$,

$$\lambda \left(\bigcup Q_j \right) \leq \sum \lambda(Q_j) \quad \left(\bigcup Q_j \in \mathcal{Q} \right). \tag{9.2}$$

The Euclidean measure of parallelepipeds is finitely additive. The Lebesgue–Stieltjes set function λ_f is finitely additive. The Hausdorff set function is not finitely additive.

---

[12]An example can be constructed a posteriori. Having constructed $\{X, \mathcal{A}, \mu\}$, choose a non-μ-measurable set E and consider the sequential covering $\mathcal{Q}' = \mathcal{Q} \bigcup E$ and the set function λ' defined as λ on $\mathcal{Q}$ and $\lambda(E) = \mu_e(E)$. The pair $\{\mathcal{Q}'; \lambda'\}$ generates the same measure μ.

[13]Examples can be constructed using the Lebesgue–Stieltjes outer measure in $\mathbb{R}$ or the Hausdorff outer measure in $\mathbb{R}^N$.

[14]The parallelepipeds are not required to be closed or open.

[15]The notion of sequential covering $\mathcal{Q}$ does not require that $\mathcal{Q}$ be closed under finite union, even if it is a semialgebra.

Proposition 9.1. *Assume that Q is a semialgebra and that the set function λ is finitely additive on Q. Then $Q \subset A$. If, in addition, λ is countably subadditive, then λ agrees with μ_e on Q.*

Proof. Let $Q \in \mathcal{Q}$ be fixed and select $A \subset X$. If $\mu_e(A) = \infty$, then (6.2) holds with $E = Q$. If $\mu_e(A) < \infty$, having fixed $\varepsilon > 0$, there exists a countable collection $\{Q_{\varepsilon,n}\}$ of elements of Q such that

$$\varepsilon + \mu_e(A) \geq \sum \lambda(Q_{\varepsilon,n}) \quad \text{and} \quad A \subset \bigcup Q_{\varepsilon,n}. \tag{9.3}$$

For each n fixed, write

$$Q_{\varepsilon,n} = \left(Q_{\varepsilon,n} \bigcap Q \right) \bigcup (Q_{\varepsilon,n} - Q).$$

The intersection $Q_{\varepsilon,n} \bigcap Q$ is in Q by virtue of (i), whereas by (ii),

$$(Q_{\varepsilon,n} - Q) = \bigcup_{j_n=1}^{m_n} Q_{j_n},$$

where the Q_{j_n} are disjoint elements of Q. Therefore, each $Q_{\varepsilon,n}$ can be written as the finite union of disjoint elements of Q. By (9.1),

$$\lambda(Q_{\varepsilon,n}) = \lambda \left(Q_{\varepsilon,n} \bigcap Q \right) + \sum_{j_n=1}^{m_n} \lambda(Q_{j_n}). \tag{9.4}$$

The elements of the collection $\{Q_{\varepsilon,n} \bigcap Q\}$ are in Q and their union covers $A \bigcap Q$. Likewise, the elements of the collection $\{Q_{j_n}\}$ as j_n ranges over $\{1, \ldots, m_n\}$ and n ranges over $\mathbb{N}$ are in Q and their union covers $(A - Q)$. Therefore, by the definition of outer measure,

$$\sum_n \lambda(Q_{\varepsilon,n}) = \sum_n \lambda \left(Q_{\varepsilon,n} \bigcap Q \right) + \sum_n \sum_{j_n=1}^{m_n} \lambda(Q_{j_n})$$

$$\geq \mu_e \left(A \bigcap Q \right) + \mu_e(A - Q).$$

Putting this in the first part of (9.3) yields

$$\varepsilon + \mu_e(A) \geq \mu_e \left(A \bigcap Q \right) + \mu_e(A - Q) \quad \text{for all } \varepsilon > 0.$$

This proves that Q is μ_e-measurable.

To prove that $\mu_e(Q) = \lambda(Q)$, we may assume that $\mu_e(Q) < \infty$. Fix an arbitrary $\varepsilon > 0$ and let $\{Q_{\varepsilon,n}\}$ be a countable collection of elements in Q such that

$$\varepsilon + \mu_e(Q) \geq \sum \lambda(Q_{\varepsilon,n}) \quad \text{and} \quad Q \subset \bigcup Q_{\varepsilon,n}. \tag{9.5}$$

From (9.4), for each fixed n, we estimate that

$$\lambda(Q_{\varepsilon,n}) \geq \lambda\left(Q_{\varepsilon,n} \bigcap Q\right).$$

The elements of the collection $\{Q_{\varepsilon,n} \bigcap Q\}$ are in Q by (i). Moreover, their union is in Q since, by the second part of (9.5),

$$Q = \bigcup\left(Q \bigcap Q_{\varepsilon,n}\right).$$

Therefore, by (9.2),

$$\varepsilon + \mu_e(Q) \geq \sum \lambda\left(Q_{\varepsilon,n} \bigcap Q\right) \geq \lambda(Q). \qquad \square$$

Remark 9.1. For the inclusion $Q \subset \mathcal{A}$, it suffices to require that λ be finitely additive on Q. If λ is both finitely additive and countably subadditive on Q, then it is also countably additive on Q.

A set function λ on a semialgebra Q is said to be a *measure on* Q if it satisfies the requirements (i)–(iv) of Section 3. The countable additivity (iii) is required to hold for any countable collection $\{Q_n\}$ of sets in Q whose union is in Q.

Remark 9.2. The assumptions of Proposition 9.1 are verified if λ is a measure on a semialgebra Q. In such a case, $Q \subset \mathcal{A}$ and μ_e agrees with λ on Q. This way the measure μ, the restriction of μ_e to $\mathcal{A}$, can be regarded as an extension of λ from Q to $\mathcal{A}$.

9.1   On the Lebesgue–Stieltjes and Hausdorff measures.

The assumptions of Proposition 9.1 are not satisfied, on different accounts, for either the Lebesgue–Stieltjes measure in $\mathbb{R}$ or the Hausdorff measure in $\mathbb{R}^N$.

For the Lebesgue–Stieltjes measure, Q is the collection of all open intervals $(a, b) \subset \mathbb{R}$. Such a collection is not a semialgebra. The set function λ_f is finitely additive and countably additive. The Lebesgue–Stieltjes measurability of the open intervals must be established by a different argument (Proposition 7.1). One also verifies that $\mu_f(a, b) = \lambda_f(a, b)$

For the Hausdorff measure in $\mathbb{R}^N$, the sequential covering Q is the collection of all subsets of $\mathbb{R}^N$ whose diameter is less than some $\varepsilon > 0$. Such a collection is a semialgebra. However, the set function $\lambda(E) = \text{diam}(E)^\alpha$ is not finitely additive for all α, except for $\alpha = N = 1$. The $\mathcal{H}_\alpha$-measurability of the open sets is established by an independent argument (Proposition 8.1).

10   Necessary and sufficient conditions for measurability

Let $\{X, \mathcal{A}, \mu\}$ be the measure space generated by the pair $\{Q; \lambda\}$, where Q is a semialgebra of subsets of X and λ is a measure on Q.

Denote by $\mathcal{Q}_\sigma$ the collection of all sets that are the countable union of elements of $\mathcal{Q}$. An element $E_\sigma \in \mathcal{Q}_\sigma$ is of the form

$$E_\sigma = \bigcup Q_n, \quad \text{where } Q_n \text{ are elements of } \mathcal{Q}, \tag{10.1}$$

and it can be written as a countable union of disjoint elements of $\mathcal{Q}$ by setting

$$E_\sigma = \bigcup D_n, \tag{10.2}$$

where $D_1 = Q_1$ and

$$D_n = Q_n - \bigcup_{j=1}^{n-1} Q_j = \bigcap_{j=1}^{n-1}(Q_n - Q_j). \tag{10.3}$$

Since $\mathcal{Q}$ is a semialgebra, each of the complements $(Q_n - Q_j)$ is the finite, disjoint union of elements of $\mathcal{Q}$. Therefore, each D_n is the finite, disjoint union of elements of $\mathcal{Q}$.

Also, denote by $\mathcal{Q}_{\sigma\delta}$ the collection of sets that are the countable intersection of elements of $\mathcal{Q}_\sigma$.

Proposition 10.1. *Let $E \subset X$ be of finite outer measure. For every $\varepsilon > 0$, there exists a set $E_{\sigma,\varepsilon} \in \mathcal{Q}_\sigma$ such that*

$$E \subset E_{\sigma,\varepsilon} \quad \text{and} \quad \mu_e(E) \geq \mu(E_{\sigma,\varepsilon}) - \varepsilon. \tag{10.4}$$

Moreover, there exists a set $E_{\sigma\delta} \in \mathcal{Q}_{\sigma\delta}$ such that

$$E \subset E_{\sigma\delta} \quad \text{and} \quad \mu_e(E) = \mu(E_{\sigma\delta}). \tag{10.4}'$$

Proof. By (4.2), for every $\varepsilon > 0$, there exists $E_{\sigma,\varepsilon} \in \mathcal{Q}_\sigma$ of the form (10.1)–(10.3) such that

$$\mu_e(E) + \varepsilon \geq \sum \lambda(Q_n) \geq \sum \mu(D_n)$$
$$= \mu\left(\bigcup D_n\right) = \mu(E_{\sigma,\varepsilon}).$$

This proves (10.4). As a consequence, for each $n \in \mathbb{N}$, there exists a set $E_{\sigma,\frac{1}{n}} \in \mathcal{Q}_\sigma$ such that

$$\mu\left(E_{\sigma,\frac{1}{n}}\right) - \frac{1}{n} \leq \mu_e(E) \leq \mu\left(E_{\sigma,\frac{1}{n}}\right).$$

Thus $E_{\sigma\delta} = \bigcap E_{\sigma,\frac{1}{n}}$ is one possible choice for such a claimed set. $\qquad\square$

Remark 10.1. Proposition 10.1 continues to hold if μ is any measure that agrees with λ on $\mathcal{Q}$.

Proposition 10.2. *Let $\{X, \mathcal{A}, \mu\}$ be the measure space generated by a measure λ on a semialgebra $\mathcal{Q}$. A set $E \subset X$ of finite outer measure is μ-measurable if and only if for every $\varepsilon > 0$, there exists a set $E_{\sigma,\varepsilon}$ of the type $\mathcal{Q}_\sigma$ such that*

$$E \subset E_{\sigma,\varepsilon} \quad and \quad \mu_e(E_{\sigma,\varepsilon} - E) \leq \varepsilon. \tag{10.5}$$

Proof. The necessary condition follows from Proposition 10.1. Indeed, if E is μ-measurable,

$$\mu(E_{\sigma,\varepsilon}) \leq \mu_e(E) + \varepsilon \Longleftrightarrow \mu(E_{\sigma,\varepsilon} - E) \leq \varepsilon.$$

For the sufficient condition, assuming (10.5) holds, we verify that E satisfies (6.2) for all $A \subset X$. Since $E_{\sigma,\varepsilon}$ is μ-measurable,

$$\begin{aligned}
\mu_e(A) &= \mu_e\left(A \bigcap E_{\sigma,\varepsilon}\right) + \mu_e(A - E_{\sigma,\varepsilon}) \\
&\geq \mu_e\left(A \bigcap E\right) + \mu_e\left((A - E) - A\bigcap(E_{\sigma,\varepsilon} - E)\right) \\
&\geq \mu_e\left(A \bigcap E\right) + \mu_e(A - E) - \mu_e\left(A\bigcap(E_{\sigma,\varepsilon} - E)\right) \\
&\geq \mu_e\left(A \bigcap E\right) + \mu_e(A - E) - \varepsilon. \qquad \square
\end{aligned}$$

Proposition 10.3. *Let $\{X, \mathcal{A}, \mu\}$ be a measure space generated by a measure λ on a semialgebra $\mathcal{Q}$.*

A set $E \subset X$ of finite outer measure is μ-measurable if and only if there exists a set $E_{\sigma\delta} \in \mathcal{Q}_{\sigma\delta}$ such that

$$E \subset E_{\sigma\delta} \quad and \quad \mu_e(E_{\sigma\delta} - E) = 0. \tag{10.6}$$

Proof. The necessary condition is contained in Proposition 10.1. For the sufficient condition, assuming 10.6 holds, we verify that E satisfies 6.2 for all $A \subset X$. Since $E_{\sigma\delta}$ is μ-measurable,

$$\begin{aligned}
\mu_e(A) &= \mu_e\left(A \bigcap E_{\sigma\delta}\right) + \mu_e(A - E_{\sigma\delta}) \\
&\geq \mu_e\left(A \bigcap E\right) + \mu_e\left((A - E) - A\bigcap(E_{\sigma\delta} - E)\right) \\
&\geq \mu_e\left(A \bigcap E\right) + \mu_e(A - E). \qquad \square
\end{aligned}$$

11   More on extensions from semialgebras to σ-algebras

Theorem 11.1. *Every measure λ on a semialgebra $\mathcal{Q}$ generates a measure space $\{X, \mathcal{A}, \mu\}$, where $\mathcal{A}$ is a σ-algebra containing $\mathcal{Q}$ and μ is a measure on $\mathcal{A}$, which agrees with λ on $\mathcal{Q}$. Moreover, if $\mathcal{Q}_o$ is the smallest σ-algebra containing $\mathcal{Q}$, the restriction of μ to $\mathcal{Q}_o$ is an extension of λ to $\mathcal{Q}_o$.*

If λ is σ-finite, such an extension is unique.[16]

Proof. There is only to prove the uniqueness whenever λ is σ-finite. Assume that μ_1 and μ_2 are both extensions of λ and let μ_e be the outer measure generated by $\{Q; \lambda\}$.

By the procedure in (10.1)–(10.3), every element of Q_σ can be written as the disjoint union of elements of Q. Since μ_1 and μ_2 agree on Q, they also agree on Q_σ.

Next, we show that both μ_1 and μ_2 agree with the outer measure μ_e on sets of Q_o of finite outer measure.

Let $E \in Q_o$ be of finite outer measure. By (10.4) and Remark 10.1, for every $\varepsilon > 0$, there exists $E_{\sigma,\varepsilon} \in Q_\sigma$ such that

$$\mu_1(E_{\sigma,\varepsilon}) \le \mu_e(E) + \varepsilon.$$

Since $E \subset E_{\sigma,\varepsilon}$ and $\varepsilon > 0$ is arbitrary, this implies

$$\mu_1(E) \le \mu_e(E)$$

for all $E \in Q_o$ of finite outer measure. Also, since μ_e is a measure on Q_o,

$$\mu_e(E_{\sigma,\varepsilon} - E) = \mu_e(E_{\sigma,\varepsilon}) - \mu_e(E)$$
$$= \mu_1(E_{\sigma,\varepsilon}) - \mu_e(E) \le \varepsilon.$$

From this, since both μ_e and μ_1 are measures on Q_o and $E \subset E_{\sigma,\varepsilon}$,

$$\mu_e(E) \le \mu_e(E_{\sigma,\varepsilon}) = \mu_1(E_{\sigma,\varepsilon})$$
$$= \mu_1(E) + \mu_1(E_{\sigma,\varepsilon} - E)$$
$$\le \mu_1(E) + \mu_e(E_{\sigma,\varepsilon} - E) \le \mu_1(E) + \varepsilon.$$

Therefore, $\mu_1(E) = \mu_e(E)$ on sets $E \in Q_o$ of finite outer measure. Interchanging the role of μ_1 and μ_2, we conclude that

$$\mu_1(E) = \mu_2(E) = \mu_e(E)$$

for every $E \in Q_o$ of finite outer measure.

Now fix $E \in Q_o$ not necessarily of finite outer measure. Since λ is σ-finite on Q, there exists a sequence of sets $Q_n \in Q$ such that $E = \bigcup Q_n \cap E$ and each of the intersections $Q_n \cap E$ is in Q_o and is of finite outer measure. Operating as in (10.1)–(10.3) if necessary, we may assume that the Q_n are mutually disjoint. Then

$$\mu_1(E) = \sum \mu_1 \left(Q_n \cap E \right) = \sum \mu_2 \left(Q_n \cap E \right) = \mu_2(E). \qquad \square$$

---

[16]The requirement that λ be σ-finite is essential to insure a unique extension (see Section 11.1 of the Problems and Complements).

12   The Lebesgue measure of sets in $\mathbb{R}^N$

The collection Q of the diadic cubes, including the empty set, is a semialgebra of subsets of $\mathbb{R}^N$.[17]

The Euclidean measure λ of cubes in $\mathbb{R}^N$ provides a nonnegative, finitely additive set function defined on Q from which one may construct the outer measure μ_e as indicated in (4.1).

Proposition 12.1. *Let $\mathcal{M}$ be the σ-algebra generated by the Euclidean measure of cubes in $\mathbb{R}^N$.*

   (i) *$\mathcal{M}$ contains the diadic cubes Q.*

  (ii) *$\mathcal{M}$ contains all the open sets in $\mathbb{R}^N$.*

 (iii) *$\mathcal{M}$ contains all the closed sets in $\mathbb{R}^N$.*

 (iv) *$\mathcal{M}$ contains all the sets in $\mathbb{R}^N$ of the type $\mathcal{F}_\sigma, \mathcal{G}_\delta, \mathcal{F}_{\sigma\delta}, \mathcal{G}_{\delta\sigma}, \dots$.*

Proof. The collection of cubes and the Euclidean measure on them satisfies the assumption of Proposition 9.1. This proves (i). Open sets are the countable union of closed diadic cubes. Therefore, they are in $\mathcal{M}$.

The remaining statements follow from this and Proposition 6.1.   □

The restriction of μ_e to $\mathcal{M}$ is the Lebesgue measure in $\mathbb{R}^N$, and sets in $\mathcal{M}$ are said to be Lebesgue measurable.

The outer measure of a singleton $\{y\} \in \mathbb{R}^N$ is zero since $\{y\}$ may be included into cubes or arbitrarily small measure. From this and the countable subadditivity of μ_e it follows that any countable set in $\mathbb{R}^N$ has outer measure zero. Therefore, by Proposition 6.1(iii), any countable set in $\mathbb{R}^N$ is Lebesgue measurable and has Lebesgue measure zero. In particular, the set $\mathbf{Q}$ of the rational numbers is measurable and has measure zero. Analogously, the set of points in $\mathbb{R}^N$ of rational coordinates is measurable and has measure zero.

Every Borel set is Lebesgue measurable. Indeed, the σ-algebra $\mathcal{M}$ contains the open sets and the Borel sets $\mathcal{B}$ form the smallest σ-algebra containing the open sets. The converse is false as the inclusion $\mathcal{B} \subset \mathcal{M}$ is strict.[18]

12.1   A necessary and sufficient condition of measurability. Let μ be the Lebesgue measure in $\mathbb{R}^N$. For a subset E of $\mathbb{R}^N$, define

$$\mu_e'(E) = \inf\{\mu(\mathcal{O}) | \text{where } \mathcal{O} \text{ is open and } E \subset \mathcal{O}\}. \qquad (12.1)$$

The definition is analogous to that of the outer measure (4.1) except for the class where the infimum is taken.

---

[17]These cubes might be open or closed or neither. It is required that they have the structure (1.1) with some, none, or all the inequalities defining them being strict.

[18]In Section 14, we exhibit a measurable subset of [0, 1] that is not a Borel set.

Since every open set is the countable union of closed cubes with pairwise-disjoint interior, the class of the open sets containing E is contained in the class of the countable unions of cubes, containing E. Thus $\mu_e(E) \leq \mu'_e(E)$.

Proposition 12.2. *Let E be a subset of $\mathbb{R}^N$ of finite outer measure. Then $\mu_e(E) = \mu'_e(E)$.*

Proof. Having fixed $\varepsilon > 0$, let $\{Q_{\varepsilon,n}\}$ be a countable collection of cubes whose union contains E and satisfying (4.2). For each n, there exists a cube $Q'_{\varepsilon,n}$ congruent to $Q_{\varepsilon,n}$ such that

$$Q_{\varepsilon,n} \subset \mathring{Q}'_{\varepsilon,n}4 \quad \text{and} \quad \mu(\mathring{Q}'_{\varepsilon,n}4 - Q_{\varepsilon,n}) \leq \frac{1}{2^n}\varepsilon.$$

The union of the $\mathring{Q}'_{\varepsilon,n}4$ is open and contains E. Therefore,

$$\mu'_e(E) \leq \mu\left(\bigcup \mathring{Q}'_{\varepsilon,n}4\right) \leq \sum \mu(\mathring{Q}'_{\varepsilon,n}4)$$
$$\leq \sum \mu(Q_{\varepsilon,n}) + \sum \frac{1}{2^n}\varepsilon$$
$$\leq \mu_e(E) + 2\varepsilon. \qquad \square$$

Proposition 12.3. *A set $E \subset \mathbb{R}^N$ of finite outer measure is Lebesgue measurable if and only if for every $\varepsilon > 0$ there exists an open set $E_{o,\varepsilon}$ such that*

$$E \subset E_{o,\varepsilon} \quad \text{and} \quad \mu_e(E_{o,\varepsilon} - E) \leq \varepsilon. \tag{12.2}$$

Equivalently, a set $E \subset \mathbb{R}^N$ of finite outer measure E is Lebesgue measurable if and only if there exists a set E_δ of the type $\mathcal{G}_\delta$ such that

$$E \subset E_\delta \quad \text{and} \quad \mu_e(E_\delta - E) = 0. \tag{12.2}'$$

Proof. The Lebesgue measure can be regarded as the extension of its restriction to the Borel sets. Thus (12.2)–(12.2)' follow from Proposition 10.2. $\qquad \square$

Proposition 12.4. *A bounded set $E \subset \mathbb{R}^N$ is Lebesgue measurable if and only if for every $\varepsilon > 0$, there exists a closed set $E_{c,\varepsilon}$ such that*

$$E_{c,\varepsilon} \subset E \quad \text{and} \quad \mu_e(E - E_{c,\varepsilon}) \leq \varepsilon. \tag{12.3}$$

Equivalently, a bounded set $E \subset \mathbb{R}^N$ is Lebesgue measurable if and only if there exists a set E_σ of the type $\mathcal{F}_\sigma$ such that

$$E_\sigma \subset E \quad \text{and} \quad \mu_e(E - E_\sigma) = 0. \tag{12.3}'$$

Proof. Let Q be a cube containing E. The set $(Q - E)$ is of finite outer measure. Thus the characteristic conditions (12.3)–(12.3)' follow from the characteristic conditions (12.2)–(12.2)' $\qquad \square$

13   A nonmeasurable set

The following construction, due to Vitali, exhibits a subset of $[0, 1]$ that is not Lebesgue measurable.[19]

Let $\bullet : [0, 1) \times [0, 1) \rightarrow [0, 1)$ be the addition modulo 1 acting on pairs $x, y \in [0, 1)$; i.e.,

$$x \bullet y = \begin{cases} x + y & \text{if } x + y < 1, \\ x + y - 1 & \text{if } x + y \geq 1. \end{cases}$$

If E is a Lebesgue-measurable subset of $[0, 1)$, then for every fixed $y \in [0, 1)$, the set

$$E \bullet y \equiv \{x \bullet y \mid x \in E\}$$

is Lebesgue measurable and

$$\mu(E \bullet y) = \mu(E).$$

Next, we introduce an equivalence relation $\sim$ in $[0, 1)$ by

$$x \sim y \quad \text{if } x - y \text{ is rational.}$$

Such a relation identifies equivalence classes in $[0, 1)$. If $\mathcal{E}$ is one such class, then two elements of $\mathcal{E}$ differ by a rational number. In particular, the rational numbers in $[0, 1)$ all belong to one such equivalence class.

We select one and only one element out of each class to form a set E that by this procedure contains one and only one element from each of these equivalence classes. In particular, any two distinct elements $x, y \in E$ are not equivalent. Such a selection is possible by the axiom of choice.

Now let $r_o = 0$, let $\{r_n\}$ denote the sequence of rational numbers in $(0, 1)$, and set

$$E_n = E \bullet r_n, \quad n = 0, 1, 2, \ldots.$$

The sets E_n are pairwise disjoint. Indeed, if $x \in E_n \cap E_m$, there exist two elements $x_n, x_m \in E$ and two rational numbers r_n and r_m such that

$$x_n \bullet r_n = x_m \bullet r_m.$$

This and the definition of the operation $\bullet$ implies that $x_n - x_m$ is a rational number. Thus $x_n \sim x_m$. This, however, contradicts the definition of E unless $m = n$.

Next, we observe that each element of $[0, 1)$ belongs to some E_n. Indeed, every $x \in [0, 1)$ must belong to some equivalence class, and therefore there must exist some $y \in E$ such that $x - y$ is rational.

---

[19]G. Vitali, Sul problema della misura dei gruppi di punti di una retta, *Mem. Accad. Sci. Bologna*, 1905.

If $x - y \geq 0$, then $x = y + r_n$ for some r_n and hence $x \in E_n$. If $x - y < 0$, then $x = y - r_m$ for some r_m. This can be rewritten as

$$x = y + (1 - r_m) - 1 \quad \text{or, equivalently, as} \quad x = y \bullet (1 - r_m).$$

Thus in either case, $x \in E_n$ for some n. We conclude that

$$[0, 1) = \bigcup E_n. \tag{13.1}$$

If E were Lebesgue measurable, E_n would also be Lebesgue measurable, and it would have the same measure. Since the sets E_n are mutually disjoint,

$$\mu([0, 1)) = \sum \mu(E_n). \tag{13.2}$$

This, however, is a contradiction since the right-hand side is either zero or infinity.

14 Borel sets, measurable sets, and incomplete measures

Proposition 14.1. *There exists a subset E of $[0, 1]$ that is Lebesgue measurable but is not a Borel set.*

Proposition 14.2. *The restriction of the Lebesgue measure on $\mathbb{R}$ to the σ-algebra of the Borel sets in $\mathbb{R}$ is not a complete measure.*

The next sections prepare for the proof of these propositions, which is given in Section 14.3.

14.1   A continuous increasing function $f : [0, 1] \to [0, 1]$. Inductively construct a nonincreasing sequence of functions $\{f_n\}$ defined in $[0, 1]$ by setting $f_o(x) = x$,

$$f_1(x) = \begin{cases} \dfrac{1}{2}x & \text{if } 0 \leq x \leq \dfrac{1}{3}, \\[2mm] 2x - \dfrac{1}{2} & \text{if } \dfrac{1}{3} \leq x \leq \dfrac{1}{2}, \\[2mm] \dfrac{1}{2}x + \dfrac{1}{4} & \text{if } \dfrac{1}{2} \leq x \leq \dfrac{5}{6}, \\[2mm] 2x - 1 & \text{if } \dfrac{5}{6} \leq x \leq 1. \end{cases}$$

The function f_1 has been constructed by first dividing $[0, 1]$ into the two subintervals $[0, \frac{1}{2}]$ and $[\frac{1}{2}, 1]$. The first subinterval is subdivided, in turn, into the two subintervals $[0, \frac{1}{3}]$ and $[\frac{1}{3}, \frac{1}{2}]$. In the first of these, f_1 is affine and has derivative $\frac{1}{2}$, whereas in the second, f_1 is affine and has derivative 2. The second subinterval $[\frac{1}{2}, 1]$ is divided into two intervals $[\frac{1}{2}, \frac{5}{6}]$ and $[\frac{5}{6}, 1]$ in such a way that f_1 is affine on each of them and has derivative $\frac{1}{2}$ in the first and 2 in the second. The resulting function f_1 is continuous and increasing in $[0, 1]$. Moreover, $f_1 \leq f_o$ and $f_1 = f_o$ at each of the endpoints of the subdivision. The functions f_n for $n \geq 2$ are constructed inductively to satisfy the following:

(i) Each f_n is continuous and increasing in $[0, 1]$. Moreover, $[0, 1]$ is subdivided into 4^n subintervals in such a way that f_n is affine on each of them and has derivative either 2^{-n} or 2^n.

(ii) $f_{n+1}(x) \leq f_n(x)$ for all $n \in \mathbb{N}$ and all $x \in [0, 1]$.

(iii) If α is any one of the endpoints of the 4^n intervals into which $[0, 1]$ has been subdivided, then $f_m(\alpha) = f_n(\alpha)$ for all $m \geq n$.

(iv) If $[\alpha, \beta]$ is an interval where f_n is affine, then $(\beta - \alpha) \leq 2^{-n}$ and $f_n(\beta) - f_n(\alpha) \leq 2^{-n}$.

Let f_n be constructed and let $[\alpha, \beta]$ be one of the intervals where f_n is affine. Then f_{n+1} restricted to $[\alpha, \beta]$ can be constructed by the following graphical procedure. Set

$$A = (\alpha, f_n(\alpha)), \qquad B = (\beta, f_n(\beta))$$

and let C be the midpoint of the segment $\overline{AB}$. Next, let D be the unique point below the segment $\overline{AC}$ such that the slope of $\overline{AD}$ is 2^{-n} and the slope of $\overline{DC}$ is 2^n. Likewise, let E be the unique point below the segment $\overline{CB}$ such that the slope of $\overline{CE}$ is 2^{-n} and the slope of $\overline{CB}$ is 2^n. Then the polygonal $ADCEB$ is the graph of f_{n+1} within $[\alpha, \beta]$.

Since the f_n are continuous and strictly increasing in $[0, 1]$, their inverses f_n^{-1} are also continuous and strictly increasing in $[0, 1]$. Moreover, by construction, such inverses satisfy properties (i)–(iv) except (ii), where the inequality is reversed.

For a fixed $n \in \mathbb{N}$, any fixed $x \in [0, 1]$ belongs to at least one of the 4^n closed subintervals where f_n is affine. If, for example, $x \in [\alpha, \beta]$, then for all $m \geq n$,

$$0 \leq f_n(x) - f_m(x) \leq f_n(\beta) - f_n(\alpha) \leq 2^{-n}.$$

Since $x \in [0, 1]$ is arbitrary, the sequence $\{f_n\}$ converges uniformly in $[0, 1]$ to a nondecreasing, uniformly continuous function f in $[0, 1]$.

Having fixed $x < y$ in $[0, 1]$, there exists n so large that one of the intervals $[\alpha, \beta]$ where f_n is affine is contained in $[x, y]$. Therefore,

$$f(x) \leq f(\alpha) = f_n(\alpha) < f_n(\beta) = f(\beta) \leq f(y).$$

Thus f is strictly increasing in $[0, 1]$ and has a continuous, strictly increasing inverse f^{-1} in $[0, 1]$. Set

$$A_n = \bigcup \{\text{intervals } [\alpha, \beta], \text{ where } f_n \text{ is affine and } f_n' = 2^n\},$$

$$B_n = \bigcup \{\text{intervals } [\alpha, \beta], \text{ where } f_n \text{ is affine and } f_n' = 2^{-n}\}.$$

From the definition,

$$[\alpha, \beta] \in A_n \implies f(\beta) - f(\alpha) = 2^n(\beta - \alpha),$$
$$[\alpha, \beta] \in B_n \implies f(\beta) - f(\alpha) = 2^{-n}(\beta - \alpha).$$

Therefore, adding over all such intervals,

$$\mu(f(A_n)) = 2^n \mu(A_n), \qquad\qquad \mu(A_n) + \mu(B_n) = 1,$$
$$\mu(f(B_n)) = 2^{-n} \mu(B_n), \qquad\qquad \mu(f(A_n)) + \mu(f(B_n)) = 1.$$

From this, we compute

$$\mu(A_n) = \frac{2^n - 1}{2^{2n} - 1}, \qquad\qquad \mu(B_n) = 2^n \frac{2^n - 1}{2^{2n} - 1},$$
$$\mu(f(A_n)) = 2^n \frac{2^n - 1}{2^{2n} - 1}, \qquad\qquad \mu(f(B_n)) = \frac{2^n - 1}{2^{2n} - 1}.$$

Set

$$S_n = \bigcup_{j=n}^{\infty} A_j, \qquad S = \bigcap_{n=1}^{\infty} S_n.$$

The set S is measurable, and we compute

$$0 \le \mu(S) = \lim \mu(S_n) \le \lim \sum_{j=n}^{\infty} \mu(A_j) = 0.$$

The sets $f(A_n)$ are measurable, being the finite union of intervals. Therefore, the sets

$$f(S_n) = \bigcup_{j=n}^{\infty} f(A_j), \qquad f(S) = \bigcap_{n=1}^{\infty} f(S_n)$$

are also measurable. Then we compute

$$1 \ge \mu(f(S)) = \lim \mu(f(S_n)) \ge \lim \mu(f(A_n)) = 1.$$

Therefore, f maps the set $S \subset [0, 1]$ of measure zero onto $f(S) \subset [0, 1]$ of measure 1. Likewise, f maps the set $[0, 1] - S$ of measure 1 onto $[0, 1] - f(S)$ of measure zero.

14.2 On the preimage of a measurable set. Since f is continuous, the preimage of an open or closed subset of $[0, 1]$ is open or closed and hence Lebesgue measurable.[20]

---

[20]Open and closed here are meant in the relative topology of $[0, 1]$.

More generally, one might consider the family

$$\mathcal{F} = \left\{ \begin{array}{l} \text{the collection of the subsets } E \text{ of } [0, 1] \\ \text{such that } f^{-1}(E) \text{ is Lebesgue measurable} \end{array} \right\}. \tag{14.1}$$

Since f is strictly increasing, the complement of any set in $\mathcal{F}$ is in $\mathcal{F}$.
If $\{E_n\}$ is a countable collection of elements of $\mathcal{F}$, then

$$f^{-1}\left(\bigcup E_n \right) = \bigcup f^{-1}(E_n)$$

and

$$f^{-1}\left(\bigcap E_n \right) = \bigcap f^{-1}(E_n).$$

Therefore, the countable union or intersection of elements in $\mathcal{F}$ remains in $\mathcal{F}$.
Thus $\mathcal{F}$ is a σ-algebra of subsets of $[0, 1]$. It follows that $\mathcal{F}$ must contain the Borel
sets $\mathcal{B}$ of $[0, 1]$ since they form the smallest σ-algebra containing the open sets. In
particular, the preimage of a Borel set is measurable.

Since f is continuous and increasing, the same argument shows that the preimage of a Borel set is a Borel set.

14.3   Proof of Propositions 14.1 and 14.2. Since the Lebesgue measure is complete, every subset of $\mathcal{S}$ is measurable and has measure zero. Likewise, every subset of $[0, 1] - f(\mathcal{S})$ is measurable and has measure zero.

Let E be the Vitali nonmeasurable subset of $[0, 1]$. Then $(E - \mathcal{S})$ is also nonmeasurable. Indeed, if it were measurable, E would be the union of the measurable sets $(E - \mathcal{S})$ and $E \bigcap \mathcal{S}$.

The set $\mathcal{D} = f(E - \mathcal{S})$ is contained in $[0, 1] - f(\mathcal{S})$. Therefore, $\mathcal{D}$ is measurable and it has measure zero. The preimage of $\mathcal{D}$ is not measurable.

The measurable set $\mathcal{D}$ is not a Borel set, for otherwise $f^{-1}(\mathcal{D})$ would be measurable.

Since $\mathcal{D}$ is measurable and has measure zero, by $(12.2)'$ of Proposition 12.3, there exists a set $\mathcal{D}_\delta$ of the type $\mathcal{G}_\delta$ such that $\mathcal{D} \subset \mathcal{D}_\delta$ and $\mu(\mathcal{D}_\delta) = 0$.

Since $\mathcal{D}$ is not a Borel set, the restriction of the Lebesgue measure to the σ-algebra of the Borel sets is not complete. $\qquad\square$

Remark 14.1. Returning to the family $\mathcal{F}$ defined in (14.1), the same example shows that $\mathcal{F}$ does not contain the σ-algebra $\mathcal{M}$ of the measurable subsets of $[0, 1]$, as there exist measurable sets whose preimage is not measurable. Interchanging the roles of f and f^{-1} shows that, in general, $\mathcal{F}$ is not contained in $\mathcal{M}$.

15   More on Borel measures

A feature of the Lebesgue measure in $\mathbb{R}^N$ is that the measure of a measurable
set $E \subset \mathbb{R}^N$ of finite measure can be approximated by the measure of open sets

containing E or closed sets contained in E. This is the content of Propositions 12.3 and 12.4.

A Borel measure μ in $\mathbb{R}^N$ is a measure defined on a σ-algebra containing the Borel sets $\mathcal{B}$.[21] Such a requirement alone does not guarantee that the μ-measure of a Borel set $E \subset \mathbb{R}^N$ of finite μ-measure can be approximated by the μ-measure of open sets containing E and closed sets contained in E.

As an example, consider the counting measure on $\mathbb{R}$. Since such a measure is defined on all subsets of $\mathbb{R}$, it is defined on the Borel sets of $\mathbb{R}$ and hence is a Borel measure. A single point is a Borel set with counting measure 1, and every open set that contains it has counting measure infinity.

However, if μ is a finite Borel measure in $\mathbb{R}^N$ and E is a Borel set, then $\mu(E)$ can be approximated by the μ-measure of closed sets included in E or by the μ-measure of open sets containing E.

Proposition 15.1. *Let μ be a finite Borel measure in $\mathbb{R}^N$ and let E be a Borel set. For every $\varepsilon > 0$, there exists a closed set $E_{c,\varepsilon} \subset E$ such that*

$$E_{c,\varepsilon} \subset E \quad \text{and} \quad \mu(E - E_{c,\varepsilon}) \leq \varepsilon. \tag{15.1}$$

Moreover, for every $\varepsilon > 0$, there exists an open set $E_{o,\varepsilon}$ such that

$$E \subset E_{o,\varepsilon} \quad \text{and} \quad \mu(E_{o,\varepsilon} - E) \leq \varepsilon. \tag{15.2}$$

Proof of (15.1). Let $\mathcal{A}$ be the σ-algebra where μ is defined and set

$$\mathcal{C}_o = \left\{ \begin{array}{l} \text{the collection of sets } E \in \mathcal{A} \text{ such that for every } \varepsilon > 0 \\ \text{there exists a closed set } C \subset E \text{ such that } \mu(E - C) \leq \varepsilon \end{array} \right\}.$$

Such a collection is nonempty since the closed sets are in $\mathcal{C}_o$. Let $\{E_n\}$ be a countable collection of sets in $\mathcal{C}_o$ and, having fixed $\varepsilon > 0$, select closed sets $C_n \subset E_n$ such that $\mu(E_n - C_n) \leq 2^{-n}\varepsilon$ for all $n \in \mathbb{N}$. Then

$$\mu\left(\bigcap E_n - \bigcap C_n\right) \leq \mu\left(\bigcup (E_n - C_n)\right)$$
$$\leq \sum \mu(E_n - C_n) \leq \varepsilon.$$

Since $\bigcap C_n$ is closed, the intersection $\bigcap E_n$ belongs to $\mathcal{C}_o$. Next, by (3.3)–(3.4) of Proposition 3.1,[22]

$$\lim_{m \to \infty} \mu\left(\bigcup E_n - \bigcup_{n=1}^{m} C_n\right) = \mu\left(\bigcup E_n - \bigcup C_n\right)$$
$$\leq \mu\left(\bigcup (E_n - C_n)\right) \leq \varepsilon.$$

---

[21] Some authors define it as a measure μ whose domain of definition is *exactly* $\mathcal{B}$.

[22] The requirement that μ be finite enters at this stage.

Therefore, there exists a positive integer m_ε such that

$$\mu\left(\bigcup E_n - \bigcup_{n=1}^{m_\varepsilon} C_n\right) \le 2\varepsilon.$$

Since the union $\bigcup_{n=1}^{m_\varepsilon} C_n$ is closed, the union $\bigcup E_n$ belong to $\mathcal{C}_o$.

The collection $\mathcal{C}_o$ contains trivially the closed sets and, in particular, the closed diadic cubes in $\mathbb{R}^N$. Since every open set is the countable union of such cubes, $\mathcal{C}_o$ also contains the open sets. Set

$$\mathcal{C} = \{\text{the collection of sets } E \in \mathcal{C}_o \text{ such that } (\mathbb{R}^N - E) \in \mathcal{C}_o\}.$$

If $E \in \mathcal{C}$, then the definition implies that $(\mathbb{R}^N - E)$ belongs to $\mathcal{C}$. Thus $\mathcal{C}$ is closed for the operation of taking complements. In particular, $\mathcal{C}$ contains all the open sets and all the closed sets.

Let $\{E_n\}$ be a countable collections of sets in $\mathcal{C}$, i.e., both E_n and $(\mathbb{R}^N - E_n)$ belong to $\mathcal{C}_o$ for all n. Then $\bigcup E_n \in \mathcal{C}_o$ and

$$\mathbb{R}^N - \bigcup E_n = \bigcap (\mathbb{R}^N - E_n) \in \mathcal{C}_o.$$

Analogously, $\bigcap E_n \in \mathcal{C}_o$ and

$$\mathbb{R}^N - \bigcap E_n = \bigcup (\mathbb{R}^N - E_n) \in \mathcal{C}_o.$$

Thus $\mathcal{C}$ is a σ-algebra. Since $\mathcal{C}$ contains the open sets, it contains the σ-algebra of the Borel sets. $\square$

Proof of (15.2). Let $E \in \mathcal{B}$. Then $(\mathbb{R}^N - E)$ is a Borel set, and by (15.1), having fixed $\varepsilon > 0$, there exists a closed set $C \subset (\mathbb{R}^N - E)$ such that

$$\mu((\mathbb{R}^N - E) - C) = \mu((\mathbb{R}^N - C) - E) \le \varepsilon.$$

Since $\mathbb{R}^N - C$ is open, (15.2) follows. $\square$

Corollary 15.2. *Let μ be a finite Borel measure in $\mathbb{R}^N$ and let E be a Borel set. There exists a set E_σ of the type $\mathcal{F}_\sigma$ such that*

$$E_\sigma \subset E \quad \text{and} \quad \mu(E - E_\sigma) = 0. \tag{15.1$'$}$$

Moreover, there exists a set E_δ of the type $\mathcal{G}_\delta$ such that

$$E \subset E_\delta \quad \text{and} \quad \mu(E_\delta - E) = 0. \tag{15.2$'$}$$

15.1   Some extensions to general Borel measures. The approximation with closed sets contained in E continues to hold for Borel measures that are not necessarily finite, provided E is of finite measure.

Proposition 15.3. *Let μ be a Borel measure in $\mathbb{R}^N$ and let E be a Borel set of finite measure. For every $\varepsilon > 0$, there exists a closed set $E_{c,\varepsilon} \subset E$ such that* (15.1) *holds.*

Proof. Let $\mathcal{A}$ be the σ-algebra where μ is defined. The set E being fixed, set

$$\mu_E(A) = \mu\left(E \bigcap A\right) \quad \text{for all } A \in \mathcal{A}.$$

Then μ_E is a finite Borel measure in $\mathbb{R}^N$. $\qquad\qquad\square$

A statement of the type of (15.2) is false for general Borel measures even if $\mu(E) < \infty$, as indicated by the counting measure on $\mathbb{R}$. However, a statement of the type of (15.2) continues to hold for Borel measures that are finite on bounded sets.

Proposition 15.4. *Let μ be a Borel measure in $\mathbb{R}^N$ that is finite on bounded sets, and let E be a Borel set of finite measure. For every $\varepsilon > 0$, there exists an open set $E_{o,\varepsilon} \supset E$ such that* (15.2) *holds.*

Proof. Let Q_n be the open cube centered at the origin, of edge n and with faces parallel to the coordinate planes. Since E is a Borel set, $(Q_n - E)$ is also a Borel set and $\mu(Q_n - E) < \infty$.

By Proposition 15.3, having fixed $\varepsilon > 0$, there exists a closed set $C_n \subset (Q_n - E)$ such that

$$\mu((Q_n - E) - C_n) = \mu((Q_n - C_n) - E) \leq \frac{1}{2^n}\varepsilon.$$

The set $(Q_n - C_n)$ is open and contains $Q_n \bigcap E$. Therefore,

$$E = \bigcup Q_n \bigcap E \subset \bigcup (Q_n - C_n) = E_{o,\varepsilon}.$$

The set $E_{o,\varepsilon}$ is open and contains E, and

$$\mu(E_{o,\varepsilon} - E) = \mu\left(\bigcup(Q_n - C_n) - E\right) \leq \sum \frac{1}{2^n}\varepsilon. \qquad\square$$

15.2   Regular Borel measures and Radon measures. A Borel measure μ in $\mathbb{R}^N$ is *regular* if for every Borel set E,

$$\mu(E) = \inf\{\mu(\mathcal{O}), \quad \text{where } \mathcal{O} \text{ is open and } E \subset \mathcal{O}\}. \tag{15.3}$$

The Lebesgue measure in $\mathbb{R}^N$ is regular, whereas the counting measure in $\mathbb{R}$ is not regular.

A Radon measure in $\mathbb{R}^N$ is a Borel measure that is finite on compact subsets of $\mathbb{R}^N$. The Lebesgue measure in $\mathbb{R}^N$ is a Radon measure. The Lebesgue–Stieltjes measure on $\mathbb{R}$ is a Radon measure. The Dirac measure δ_x with mass concentrated at x is a Radon measure.

The counting measure on $\mathbb{R}$ is a Borel measure but not a Radon measure. The Hausdorff measure μ_α is a Borel measure in $\mathbb{R}^N$ but not a Radon measure for all $\alpha \in [0, N)$.

It follows from Proposition 15.4 that Radon measures are regular.

16   Regular outer measures and Radon measures

Let μ be a Borel measure in $\mathbb{R}^N$, and let μ_e be an outer measure in $\mathbb{R}^N$ that coincides with μ on the Borel sets; i.e.,

$$\mu(E) = \mu_e(E) \quad \text{for all Borel sets } E. \tag{16.1}$$

The outer measure μ_e is *regular* with respect to μ if for every set $E \subset \mathbb{R}^N$ of finite outer measure, there exists a set E_δ of the type $\mathcal{G}_\delta$ such that

$$E \subset E_\delta \quad \text{and} \quad \mu_e(E) = \mu(E_\delta).$$

Proposition 16.1. *Let $\mathcal{Q}$ be the collection of all open sets in $\mathbb{R}^N$. If μ_e is generated by a nonnegative set function λ on $\mathcal{Q}$ and (16.1) holds, then μ_e is regular with respect to μ.*

Proof. It follows from the process (4.1) of generating an outer measure μ_e from a set function λ on a sequential covering $\mathcal{Q}$ combined with Proposition 10.1.   □

Corollary 16.2. *The Lebesgue–Stieltjes outer measure $\mu_{f,e}$ is regular with respect to the Lebesgue–Stieltjes measure μ_f.*

Proposition 16.3. *The Hausdorff outer measure $\mathcal{H}_\alpha$ is regular with respect to the Hausdorff measure μ_α.*

Proof. Having fixed a set $E \subset \mathbb{R}^N$ of finite $\mathcal{H}_\alpha$ outer measure and $m \in \mathbb{N}$, there exists a countable collection of sets $\{E_{n,m}\}$, each of diameter less than $\frac{1}{m}$, whose union contains E and satisfying

$$\mathcal{H}_{\alpha,\frac{1}{m}}(E) \geq \sum_n \text{diam}(E_{n,m})^\alpha - \frac{1}{m}.$$

The sets

$$\mathcal{O}_{n,m} = \bigcup \left\{ x \in \mathbb{R}^N \mid \text{dist}\{x; E_{n,m}\} < \frac{1}{m} \text{diam}(E_{n,m}) \right\}$$

are open and satisfy

$$\text{diam}(\mathcal{O}_{n,m}) \leq \left(1 + \frac{1}{m}\right) \text{diam}(E_{n,m}).$$

From this,

$$\left(1 + \frac{1}{m}\right)^{\alpha} \mathcal{H}_{\alpha}(E) \geq \sum_{n} \mathrm{diam}(\mathcal{O}_{n,m})^{\alpha} - \frac{2^{\alpha}}{m}$$

$$\geq \mathcal{H}_{\alpha,\varepsilon_m}(E_{\delta}) - \frac{2^{\alpha}}{m},$$

where

$$E_{\delta} = \bigcap_{m}\bigcup_{n} \mathcal{O}_{n,m} \quad \text{and} \quad \varepsilon_m = \left(1 + \frac{1}{m}\right)\frac{1}{m}.$$

The set E_{δ} is of the type $\mathcal{G}_{\delta}$ and $E \subset E_{\delta}$.   □

16.1 More on Radon measures. Let μ be a Radon measure in $\mathbb{R}^N$, and for every set $E \subset \mathbb{R}^N$, set

$$\mu_e(E) = \inf\{\mu(\mathcal{O}), \text{ where } \mathcal{O} \text{ is open and } E \subset \mathcal{O}\}. \qquad (16.2)$$

This is an outer measure that coincides with μ on the Borel sets, since μ is regular. By Proposition 16.1, such an outer measure μ_e is regular with respect to μ. Such a construction fails for general Borel measures, as indicated by the counting measure on $\mathbb{R}$.

Proposition 16.4. *By (16.2), a Radon measure μ in $\mathbb{R}^N$ generates an outer measure μ_e that coincides with μ on the Borel sets and is regular with respect to μ.*

17 Vitali coverings

Let $\{X, \mathcal{A}, \mu\}$ be $\mathbb{R}^N$ endowed with the Lebesgue measure, and let $\mathcal{F}$ denote a family of closed, nontrivial cubes in $\mathbb{R}^N$.

We say that $\mathcal{F}$ is a *fine* Vitali covering for a set $E \subset \mathbb{R}^N$ if for every $x \in E$ and every $\varepsilon > 0$, there exists a cube $Q \in \mathcal{F}$ such that $x \in Q$ and $\mathrm{diam}\{Q\} < \varepsilon$.

The collection of N-dimensional closed diadic cubes of diameter not exceeding some given positive number is an example of a fine Vitali covering for any set $E \subset \mathbb{R}^N$.

Theorem 17.1 (Vitali[23]). *Let E be a bounded, Lebesgue-measurable set in $\mathbb{R}^N$ and let $\mathcal{F}$ be a fine Vitali covering for E. There exists a countable collection $\{Q_n\}$ of cubes $Q_n \in \mathcal{F}$ with pairwise-disjoint interior such that*

$$\mu\left(E - \bigcup Q_n\right) = 0. \qquad (17.1)$$

---

[23]G. Vitali, Sui gruppi di punti e sulle funzioni di variabili reali, *Atti Accad. Sci. Torino*, **43** (1908), 75–92. The proof presented here is due to S. Banach, Sur un théoréme de Vitali, *Fund. Math.*, **5** (1924), 130–136. The theorem is more general in that E need not be Lebesgue measurable. See Sections 17.1–17.3 of the Problems and Complements.

Remark 17.1. The theorem does not claim that $\bigcup Q_n$ covers E. Rather, $\bigcup Q_n$ covers E in a measure-theoretical sense. However, the proof shows that $E \subset \bigcup Q_n''$, where Q_n'' are the cubes congruent to Q_n and with double the edge.

Remark 17.2. The proof relies on the structure of the Lebesgue measure in $\mathbb{R}^N$ and would not hold for a general Radon measure in $\mathbb{R}^N$.

Remark 17.3. If $\mathcal{F}$ is a covering of E, but not necessarily a fine Vitali covering, a similar statement holds in a weaker form (see Section 1 of Chapter VIII).

Proof of the Vitali covering theorem. Without loss of generality, we may assume that E and the cubes making up the family $\mathcal{F}$ are all included in some larger cube Q. Label by $\mathcal{F}_o$ the family $\mathcal{F}$, and out of $\mathcal{F}_o$ select a cube Q_o. If Q_o covers E then the theorem is proven. Otherwise, introduce the family of cubes

$$\mathcal{F}_1 \equiv \left\{ \begin{array}{l} \text{the collection of cubes } Q \in \mathcal{F}_o \text{ whose interior is} \\ \text{disjoint from the interior of } Q_o, \text{ i.e., } \overset{\circ}{Q} \cap \overset{\circ}{Q}_o = \emptyset \end{array} \right\}.$$

If Q_o does not cover E, such a family is nonempty. Also introduce the number

$$d_1 \equiv \{\text{the supremum of the diameters of the cubes } Q \in \mathcal{F}_1\}.$$

Then out of $\mathcal{F}_1$ select a cube Q_1 whose diameter is larger than $\frac{1}{2}d_1$. If $Q_o \bigcup Q_1$ covers E, then the theorem is proven. Otherwise, introduce the family of cubes

$$\mathcal{F}_2 \equiv \left\{ \begin{array}{l} \text{the collection of cubes } Q \in \mathcal{F}_1 \text{ whose interior is} \\ \text{disjoint from the interior of } Q_1, \text{ i.e., } \overset{\circ}{Q} \cap \overset{\circ}{Q}_1 = \emptyset \end{array} \right\}$$

and the number

$$d_2 \equiv \{\text{the supremum of the diameters of the cubes } Q \in \mathcal{F}_2\}.$$

Then out of $\mathcal{F}_2$ select a cube Q_2 whose diameter is larger than $\frac{1}{2}d_2$. Proceeding in this fashion, we inductively define families $\{\mathcal{F}_n\}$, positive numbers $\{d_n\}$, and cubes $\{Q_n\}$ by the recursive procedure

$$\mathcal{F}_n \equiv \left\{ \begin{array}{l} \text{the collection of cubes } Q \in \mathcal{F}_{n-1} \text{ whose interior is} \\ \text{disjoint from the interior of } Q_{n-1}, \text{ i.e., } \overset{\circ}{Q} \cap \overset{\circ}{Q}_{n-1} = \emptyset \end{array} \right\}.$$

$$d_n \equiv \{\text{the supremum of the diameters of the cubes } Q \in \mathcal{F}_n\}.$$

$$Q_n \equiv \left\{ a \text{ cube selected out of } \mathcal{F}_n \text{ such that } \text{diam}\{Q_n\} > \frac{1}{2}d_n \right\}.$$

The cubes $\{Q_n\}$ have pairwise-disjoint interior, and they are all included in some larger cube Q. Therefore,

$$\sum \left(\frac{\text{diam}\{Q_n\}}{\sqrt{N}} \right)^N = \sum \mu(Q_n) < \infty. \tag{17.2}$$

The convergence of the series implies that lim diam$\{Q_n\} = 0$. To prove (17.1), we proceed by contradiction; i.e., we assume that

$$\mu \left(E - \bigcup Q_n \right) \geq 2\varepsilon \quad \text{for some } \varepsilon > 0. \tag{17.3}$$

First, for each Q_n, we construct a larger cube Q'_n of diameter

$$\text{diam}\{Q'_n\} = (4\sqrt{N} + 1) \, \text{diam}\{Q_n\} \tag{17.4}$$

with the same center as Q_n and faces parallel to the faces of Q_n. By the convergence of the series in (17.2), there exists some $n_\varepsilon \in \mathbb{N}$ such that

$$\mu \left(\bigcup_{n=n_\varepsilon+1}^{\infty} Q'_n \right) \leq \sum_{n=n_\varepsilon+1}^{\infty} \mu(Q'_n) \leq \varepsilon. \tag{17.5}$$

Using this inequality and (17.3), we estimate

$$\mu \left(\left(E - \bigcup_{n=1}^{n_\varepsilon} Q_n \right) - \bigcup_{n=n_\varepsilon+1}^{\infty} Q'_n \right) \geq \mu \left(E - \bigcup Q_n \right) - \mu \left(\bigcup_{n=n_\varepsilon+1}^{\infty} Q'_n \right) \geq \varepsilon. \tag{17.6}$$

This implies that there exists an element

$$x \in \left(E - \bigcup_{n=1}^{n_\varepsilon} Q_n \right) - \bigcup_{n=n_\varepsilon+1}^{\infty} Q'_n. \tag{17.7}$$

Such an element must have positive distance 2σ from the union of the first n_ε cubes. Indeed, such a finite union is closed and x does not belong to any of the cubes $Q_n, n = 1, 2, \ldots, n_\varepsilon$.

By the definition of a fine Vitali covering, given such a σ, there exists a cube $Q_\delta \in \mathcal{F}$ of positive diameter $0 < \delta \leq \sigma$ that covers x. By construction, Q_δ does

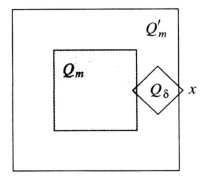

Figure 17.1.

not intersect the interior of any of the first n_ε cubes Q_n; i.e.,

$$Q_\delta \cap \mathring{Q}_n = \emptyset, \quad n = 1, 2, \ldots, n_\varepsilon.$$

It follows that Q_δ belongs to the family $\mathcal{F}_{n_\varepsilon+1}$. Next, we claim that

$$Q_\delta \cap \mathring{Q}_n \neq \emptyset \quad \text{for some } n = n_\varepsilon + 1, n_\varepsilon + 2, \ldots.$$

Indeed, if Q_δ did not intersect the interior of any such cubes, it would belong to all the families $\mathcal{F}_n$. This, however, would imply that

$$0 < \delta = \text{diam}\{Q_\delta\} \le d_n \to 0 \quad \text{as } n \to \infty.$$

Then let $m \ge (n_\varepsilon + 1)$ be the smallest positive integer for which $Q_\delta \cap \mathring{Q}_m \neq \emptyset$. In particular,

$$Q_\delta \notin \mathcal{F}_{m+1}, \quad Q_\delta \in \mathcal{F}_m, \quad \text{and} \quad \delta \le d_m.$$

By the selection (17.7), the element x does not belong to Q'_m. Therefore, the intersection $Q_\delta \cap \mathring{Q}_m$ can be nonempty only if

$$\delta = \text{diam}\{Q_\delta\} > \frac{1}{2\sqrt{N}}(\text{diam}\{Q'_m\} - \text{diam}\{Q_m\}).$$

From this and (17.4), we find the contradiction $d_m \ge \delta > d_m$. □

Corollary 17.2. *Let E be a bounded, Lebesgue-measurable set in $\mathbb{R}^N$, and let $\mathcal{F}$ be a fine Vitali covering for E. For every $\varepsilon > 0$, there exists a finite collection of cubes*

$$\mathcal{F}_\varepsilon \equiv \{Q_1, Q_2, \ldots, Q_{n_\varepsilon}\} \quad (Q_n \in \mathcal{F})$$

with pairwise-disjoint interior such that

$$\sum \mu(Q_n) - \varepsilon \le \mu(E) \le \mu\left(\bigcup_{n=1}^{n_\varepsilon} E \cap Q_n\right) + \varepsilon. \tag{17.8}$$

Proof. Having fixed $\varepsilon > 0$, let $E_{o,\varepsilon}$ be an open set containing E and satisfying (13.2). Also introduce the subfamily

$$\mathcal{F}_\varepsilon \equiv \left\{ \begin{array}{c} \text{the collection of the cubes out of } \mathcal{F} \\ \text{that are contained in } E_{o,\varepsilon} \end{array} \right\},$$

and out of $\mathcal{F}_\varepsilon$ select a countable collection of closed cubes $\{Q_n\}$ with pairwise-disjoint interior satisfying (17.1). By construction,

$$\sum \mu(Q_n) \le \mu(E_{o,\varepsilon}) \le \mu(E) + \varepsilon. \tag{17.9}$$

This, in turn, implies that there exists a positive integer n_ε such that

$$\sum_{n_\varepsilon+1}^{\infty} \mu(Q_n) \le \varepsilon.$$

From this and (17.1),

$$\mu(E) = \mu\left(\bigcup E \cap Q_n\right) \le \mu\left(\bigcup_{n=1}^{n_\varepsilon} E \cap Q_n\right) + \varepsilon. \qquad (17.10)$$

The corollary follows from this and (17.9).   □

18   The Besicovitch covering theorem

Let E be a subset of $\mathbb{R}^N$. A collection $\mathcal{F}$ of nontrivial closed balls in $\mathbb{R}^N$ is a Besicovitch covering for E if each $x \in E$ is the center of a nontrivial ball $B(x)$ belonging to $\mathcal{F}$.

Theorem 18.1 (Besicovitch[24]). *Let E be a bounded subset of $\mathbb{R}^N$ and let $\mathcal{F}$ be a Besicovitch covering for E. There exist a countable collection $\{x_n\}$ of points in E and a corresponding collection of balls $\{B_n\}$ in $\mathcal{F}$,*

$$B_n = B_{\rho_n}(x_n) \text{ balls centered at } x_n \text{ and radius } \rho_n,$$

such that $E \subset \bigcup B_n$. Moreover, there exists a positive integer c_N depending only upon the dimension N and independent of E and the covering $\mathcal{F}$ such that the balls $\{B_n\}$ can be organized into at most c_N subcollections

$$\mathcal{B}_1 = \{B_{n_1}\},$$
$$\mathcal{B}_2 = \{B_{n_2}\},$$
$$\cdots,$$
$$\mathcal{B}_{c_N} = \{B_{n_{c_N}}\}$$

in such a way that the balls $\{B_{n_j}\}$ of each subcollection $\mathcal{B}_j$ are disjoint.

Remark 18.1. The theorem continues to hold, by essentially the same proof, if the balls making up the Besicovitch covering $\mathcal{F}$ are replaced by cubes with faces parallel to the coordinate planes.[25]

Proof of Theorem 18.1. Since E is bounded, we may assume that E and the balls making up the family $\mathcal{F}$ are all included in some large ball B_o centered at the

---

[24]A. S. Besicovitch, A general form of the covering principle and relative differentiation of additive functions, *Proc. Cambridge Philos. Soc.*, **I**-41 (1945), 103–110; **II**-42 (1946), 1–10.
[25]A. P. Morse, A theory of covering and differentiation, *Trans. Amer. Math. Soc.*, **55** (1944), 205–235.

origin. Set $E_1 = E$ and

$$\mathcal{F}_1 = \{\text{the collection of balls } B(x) \in \mathcal{F} \text{ whose center is in } E_1\},$$
$$r_1 = \{\text{the supremum of the radii of the balls in } \mathcal{F}_1\}.$$

Select $x_1 \in E_1$ and a ball

$$B_1 = B_{\rho_1}(x_1) \in \mathcal{F}_1 \quad \text{of radius } \rho_1 > \frac{3}{4}r_1.$$

If $E_1 \subset B_1$, the process terminates. Otherwise, set $E_2 = E_1 - B_1$ and

$$\mathcal{F}_2 = \{\text{the collection of balls } B(x) \in \mathcal{F} \text{ whose center is in } E_2\},$$
$$r_2 = \{\text{the supremum of the radii of the balls in } \mathcal{F}_2\}.$$

Then select $x_2 \in E_2$ and a ball

$$B_2 = B_{\rho_2}(x_2) \in \mathcal{F}_2 \quad \text{of radius } \rho_2 > \frac{3}{4}r_2.$$

Proceeding recursively, define countable collections of sets E_n, balls B_n, families $\mathcal{F}_n$, and positive numbers r_n by

$$E_n = E - \bigcup_{j=1}^{n-1} B_j, \quad x_n \in E_n.$$

$$\mathcal{F}_n = \{\text{the collection of balls } B(x) \in \mathcal{F} \text{ whose center is in } E_n\},$$
$$r_n = \{\text{the supremum of the radii of the balls in } \mathcal{F}_n\}.$$
$$B_n = B_{\rho_n}(x_n) \in \mathcal{F}_n \quad \text{of radius } \rho_n > \frac{3}{4}r_n.$$

By construction, if $m > n$,

$$\rho_n > \frac{3}{4}r_n \geq \frac{3}{4}r_m \geq \frac{3}{4}\rho_m. \tag{18.1}$$

This implies the balls $B_{\frac{1}{3}\rho_n}(x_n)$ are disjoint. Indeed, since $x_m \notin B_n$,

$$|x_n - x_m| > \rho_n = \frac{1}{3}\rho_n + \frac{2}{3}\rho_n \geq \frac{1}{3}\rho_n + \frac{1}{3}\rho_m. \tag{18.2}$$

The balls $B_{\frac{1}{3}\rho_n}(x_n)$ are all contained in B_o and are disjoint. Therefore, $\{\rho_n\} \to 0$ as $n \to \infty$.

The union of the balls $\{B_n\}$ covers E. If not, select $x \in E - \bigcup B_n$ and a nontrivial ball $B_\rho(x)$ centered at x and radius $\rho > 0$. Such a ball exists since $\mathcal{F}$ is a Besicovitch covering. By construction, $B_\rho(x)$ must belong to all the families $\mathcal{F}_n$. Therefore, $0 < \rho \leq r_n \to 0$. The contradiction implies that $E \subset \bigcup B_n$. $\square$

The proof of the last statement regarding the subcollections $\mathcal{B}_j$ is based on the following geometrical fact.

Proposition 18.2. *There exists a positive integer c_N depending only on N such that for every index k, at most c_N balls out of $\{B_1, B_2, \ldots, B_{k-1}, B_k\}$ intersect B_k.*

The collections $\mathcal{B}_j$ are constructed by regarding them initially as empty boxes to be filled with disjoint balls taken out of $\{B_n\}$. Each element of $\{B_n\}$ is allocated to some of the boxes $\mathcal{B}_j$ as follows.

First, for $j = 1, 2, \ldots, c_N$, put B_j into $\mathcal{B}_j$. Next, consider the ball B_{c_N+1}. By Proposition 18.2, at least one of the first c_N balls does not intersect B_{c_N+1}, say, for example, B_1. Then allocate B_{c_N+1} to $\mathcal{B}_1$.

Consider the subsequent ball B_{c_N+2}. At least two of the first $(c_N + 1)$ balls do not intersect B_{c_N+2}. If one of the B_j, $j = 2, \ldots, c_N$, say, for example, B_2, does not intersect B_{c_N+2}, allocate B_{c_N+2} to $\mathcal{B}_2$. If all the balls B_j, $j = 2, \ldots, c_N$, intersect B_{c_N+2}, then B_1 and B_{c_N+1} do not intersect B_{c_N+2} since at least two of the first $(c_N + 1)$ balls do not intersect B_{c_N+2}. Then allocate B_{c_N+2} to $\mathcal{B}_1$, which now would contain three disjoint balls.

Proceeding recursively, assume that all the balls

$$B_1, \ldots, B_{c_N}, \ldots, B_{c_N+n-1} \quad \text{for some } n \in \mathbb{N}$$

have been allocated so that at the $(n-1)$th step of the process, each of the $\mathcal{B}_j$ contains at most n disjoint balls. To allocate B_{c_N+n} observe that by Proposition 18.2, at least n of the first $(c_N + n - 1)$ balls must be disjoint from B_{c_N+n}. This implies that the elements of at least one of the boxes $\mathcal{B}_j$, $j = 1, 2, \ldots, c_N$, are all disjoint from B_{c_N+n}. Allocate B_{c_N+n} to one such a box and proceed inductively. $\square$

19 Proof of Proposition 18.2

Fix some positive integer k, consider those balls B_j for $j = 1, 2, \ldots, k$ that intersect $B_k = B_{\rho_k}(x_k)$, and divide them into two sets:

$$\mathcal{G}_1 = \left\{ \begin{array}{c} \text{the collection of balls } B_j = B_{\rho_j}(x_j) \text{ for } j = 1, \ldots, k \\ \text{that intersect } B_k \text{ and such that } \rho_j \leq \frac{3}{4}M\rho_k \end{array} \right\},$$

$$\mathcal{G}_2 = \left\{ \begin{array}{c} \text{the collection of balls } B_j = B_{\rho_j}(x_j) \text{ for } j = 1, \ldots, k \\ \text{that intersect } B_k \text{ and such that } \rho_j > \frac{3}{4}M\rho_k \end{array} \right\}.$$

where $M > 3$ is a positive integer to be chosen.

Lemma 19.1. *The number of balls in $\mathcal{G}_1$ does not exceed $4^N(M+1)^N$.*

Proof. Let $\{B_{\rho_j}(x_j)\}$ be the collection of balls in $\mathcal{G}_1$ and let $\#\{\mathcal{G}_1\}$ denote their number. The balls $\{B_{\frac{1}{3}\rho_j}(x_j)\}$ are disjoint and are contained in $B_{(M+1)\rho_k}(x_k)$. Indeed, since $B_j \cap B_k \neq \emptyset$,

$$|x_j - x_k| \leq \rho_j + \rho_k \leq \left(\frac{3}{4}M + 1\right)\rho_k.$$

Moreover, for any $x \in B_{\frac{1}{3}\rho_j}(x_j)$,

$$|x - x_k| \leq |x - x_j| + |x_j - x_k|$$

$$\leq \frac{1}{3}\rho_j + \left(\frac{3}{4}M + 1\right)\rho_k \leq (M + 1)\rho_k.$$

From this, denoting by κ_N the volume of the unit ball in $\mathbb{R}^N$,

$$\sum_{j:B_j \in \mathcal{G}_1} \kappa_N \left(\frac{1}{3}\rho_j\right)^N \leq \kappa_N(M + 1)^N \rho_k^N.$$

Since $j < k$, it follows from (18.1) that $\frac{1}{3}\rho_j > \frac{1}{4}\rho_k$. Therefore,

$$\#\{\mathcal{G}_1\}\kappa_N \left(\frac{1}{4}\rho_k\right)^N \leq \kappa_N(M + 1)^N \rho_k^N. \qquad \square$$

An upper estimate of the number of balls in $\mathcal{G}_2$ is derived by counting the number of rays originating from the center x_k of B_k to each of the centers x_j of $B_j \in \mathcal{G}_2$. We first establish that the angle between any two such rays is not less than an absolute angle θ_o. Then we estimate the number of rays originating from x_k and mutually forming an angle of at least θ_o.

Let $B_{\rho_n}(x_n)$ and $B_{\rho_m}(x_m)$ be any two balls in $\mathcal{G}_2$ and set

$$\theta = \{\text{angle between the rays from } x_k \text{ to } x_n \text{ and } x_m\}.$$

Lemma 19.2. *The number M can chosen so that $\theta > \theta_o = \arccos\{\frac{5}{6}\}$.*

Proof. Assume $n < m < k$. By construction, $x_m \notin B_{\rho_n}(x_n)$; i.e.,

$$|x_n - x_m| > \rho_n. \qquad (19.1)$$

Also $x_k \notin B_{\rho_n}(x_n) \bigcup B_{\rho_m}(x_m)$; i.e.,

$$\rho_n < |x_n - x_k| \quad \text{and} \quad \rho_m < |x_m - x_k|.$$

Since both $B_{\rho_n}(x_n)$ and $B_{\rho_m}(x_m)$ intersect B_k and are in $\mathcal{G}_2$,

$$\frac{3}{4}M\rho_k < \rho_n \leq |x_n - x_k| \leq \rho_n + \rho_k,$$
$$\frac{3}{4}M\rho_k < \rho_m \leq |x_m - x_k| \leq \rho_m + \rho_k. \qquad (19.2)$$

By elementary trigonometry,[26]

$$\cos\theta = \frac{|x_n - x_k|^2 + |x_m - x_k|^2 - |x_n - x_m|^2}{2|x_n - x_k||x_m - x_k|}.$$

---

[26]The Carnot formula applied to the triangle of vertices x_k, x_n, x_m.

Assuming $\cos\theta > 0$ and using (19.1) and (19.2), estimate

$$\cos\theta \leq \frac{(\rho_n + \rho_k)^2 + (\rho_m + \rho_k)^2 - \rho_n^2}{2\rho_n\rho_m}$$

$$\leq \frac{\rho_m^2 + 2\rho_k^2 + 2\rho_k(\rho_n + \rho_m)}{2\rho_n\rho_m}$$

$$\leq \frac{1}{2}\frac{\rho_m}{\rho_n} + \frac{\rho_k}{\rho_n}\frac{\rho_k}{\rho_m} + \frac{\rho_k}{\rho_m} + \frac{\rho_k}{\rho_n}$$

$$\leq \frac{1}{2}\frac{\rho_m}{\rho_n} + \left(\frac{4}{3}\right)^2 \frac{1}{M^2} + 2\frac{4}{3}\frac{1}{M}.$$

Since $m > n$, from (18.1) it follows that $\rho_n > \frac{3}{4}\rho_m$. Therefore,

$$\cos\theta \leq \frac{2}{3} + \frac{4}{3}\frac{1}{M}\left(\frac{4}{3}\frac{1}{M} + 2\right).$$

Now choose M so large that the $\cos\theta \leq \frac{5}{6}$. ☐

If $N = 2$, the number of rays originating from the origin and mutually forming an angle $\theta > \theta_o$ is at most $2\pi/\theta_o$.

If $N \geq 3$, let $C(\theta_o)$ be a circular cone in $\mathbb{R}^N$ with vertex at the origin whose axial cross-section with a two-dimensional hyperplane forms an angle $\frac{1}{2}\theta_o$. Denote by $\sigma_N(\theta_o)$ the solid angle corresponding to $C(\theta_o)$.[27] Then the number of rays originating from the origin and mutually forming an angle $\theta > \theta_o$ is at most $\omega_N/\sigma_N(\theta_o)$.

The number c_N claimed by Proposition 18.2 is estimated by

$$c_N = \#\{\mathcal{G}_1\} + \#\{\mathcal{G}_2\} \leq 4^N(M+1)^N + \frac{\omega_N}{\sigma_N(\theta_o)}.$$

20   The Besicovitch measure-theoretical covering theorem

Let $\mathcal{F}$ denote a family of nontrivial closed balls in $\mathbb{R}^N$. We say that $\mathcal{F}$ is a *fine* Besicovitch covering for a set $E \subset \mathbb{R}^N$ if for every $x \in E$ and every $\varepsilon > 0$, there exists a ball $B_\rho(x) \in \mathcal{F}$ centered at x and of radius $\rho < \varepsilon$.

A fine Besicovitch covering of a set $E \subset \mathbb{R}^N$ differs from a fine Vitali covering in that each $x \in E$ is required to be the center of a ball of arbitrarily small radius.

The next measure-theoretical covering, called the Besicovitch covering theorem, holds for any Radon measure μ and its associated outer measure μ_e (Section 16.1). The set E to be covered in a measure-theoretical sense is not required to be μ-measurable.

---

[27] That is, the area of the intersection of $C(\theta_o)$ with the unit sphere in $\mathbb{R}^N$. The area of the unit sphere in $\mathbb{R}^N$ is denoted by ω_N. Accordingly, the solid angle of the unit sphere is ω_N.

Theorem 20.1 (Besicovitch). *Let E be a bounded set in $\mathbb{R}^N$ and let $\mathcal{F}$ be a fine Besicovitch covering for E. Let μ be a Radon measure in $\mathbb{R}^N$ and let μ_e be the outer measure associated with it.*

There exists a countable collection $\{B_n\}$ of disjoint balls $B_n \in \mathcal{F}$ such that

$$\mu_e\left(E - \bigcup B_n\right) = 0. \tag{20.1}$$

Remark 20.1. It is not claimed here that $E \subset \bigcup B_n$. The collection $\{B_n\}$ forms a measure-theoretical covering of E in the sense of (20.1).

Proof of the Besicovitch covering theorem. We may assume that $\mu_e(E) > 0$; otherwise, the statement is trivial. Since E is bounded, we may assume that both E and all the balls making up the covering $\mathcal{F}$ are contained in some larger ball B_o.

Let $\mathcal{B}_j$, $j = 1, 2, \ldots, c_N$, be the subcollections of disjoint balls claimed by Theorem 18.1. Since

$$E \subset \bigcup_{j=1}^{c_N} \bigcup_{n_j=1}^{\infty} B_{n_j},$$

it holds that

$$\mu_e\left(E \cap \bigcup_{j=1}^{c_N} \bigcup_{n_j=1}^{\infty} B_{n_j}\right) = \mu_e(E) > 0.$$

Therefore, there exists some index $j \in \{1, 2, \ldots, c_N\}$ for which

$$\mu_e\left(E \cap \bigcup_{n_j=1}^{\infty} B_{n_j}\right) \geq \frac{1}{c_N}\mu_e(E).$$

Since all the balls B_{n_j} are disjoint and are all included in B_o,

$$\mu_e\left(E \cap \bigcup_{n_j=1}^{\infty} B_{n_j}\right) \leq \sum_{n_j=1}^{\infty} \mu(B_{n_j}) \leq \mu(B_o) < \infty.$$

Therefore, there exists some index m_1 such that

$$\mu_e\left(E \cap \bigcup_{n_j=1}^{m_1} B_{n_j}\right) \geq \frac{1}{2c_N}\mu_e(E). \tag{20.2}$$

The finite union of balls is μ-measurable. Therefore, by the Carathéodory criterion of measurability (6.2) and the lower estimate in (20.2),

$$\mu_e(E) = \mu_e\left(E \cap \bigcup_{n_j=1}^{m_1} B_{n_j}\right) + \mu_e\left(E - \bigcup_{n_j=1}^{m_1} B_{n_j}\right)$$

$$\geq \frac{1}{2c_N}\mu_e(E) + \mu_e\left(E - \bigcup_{n_j=1}^{m_1} B_{n_j}\right).$$

Therefore,

$$\mu_e\left(E - \bigcup_{n_j=1}^{m_1} B_{n_j}\right) \leq \eta\mu_e(E), \quad \eta = 1 - \frac{1}{2c_N} \in (0, 1). \tag{20.3}$$

Now set

$$E_1 = E - \bigcup_{n_j=1}^{m_1} B_{n_j}.$$

If $\mu_e(E_1) = 0$, the the process terminates and the theorem is proven. Otherwise, let $\mathcal{F}_1$ denote the collection of balls in $\mathcal{F}$ that do not intersect any of the balls B_{n_j} for $n_j = 1, 2, \ldots, m_1$. Since $\mathcal{F}$ is a fine Besicovitch covering for E, the family $\mathcal{F}_1$ is nonempty, and it is a fine Besicovitch covering for E_1.

Repeating the previous selection process for the set E_1 and the Besicovitch covering $\mathcal{F}_1$ yields a finite number m_2 of closed disjoint balls B_{n_ℓ} in $\mathcal{F}_1$ such that

$$\mu_e\left(E_1 - \bigcup_{n_\ell=1}^{m_2} B_{n_\ell}\right) \leq \eta\mu_e(E_1) = \eta\mu_e\left(E - \bigcup_{n_j=1}^{m_1} B_{n_j}\right) \leq \eta^2\mu_e(E).$$

Relabelling the balls B_{n_j} and B_{n_ℓ} yields a finite number s_2 of closed, disjoint balls B_n in $\mathcal{F}$ such that

$$\mu_e\left(E - \bigcup_{n=1}^{s_2} B_n\right) \leq \eta^2\mu_e(E).$$

Repeating the process k times gives a collection of s_k closed disjoint balls in $\mathcal{F}$ such that

$$\mu_e\left(E - \bigcup_{n=1}^{s_k} B_n\right) \leq \eta^k\mu_e(E). \tag{20.4}$$

If for some $k \in \mathbb{N}$

$$\mu_e\left(E - \bigcup_{n=1}^{s_k} B_n\right) = 0,$$

the process terminated and the theorem is proven. Otherwise, (20.4) holds for all $k \in \mathbb{N}$. Letting $k \to \infty$ proves (20.1). $\qquad\square$

PROBLEMS AND COMPLEMENTS

1 PARTITIONING OPEN SUBSETS OF $\mathbb{R}^N$

1.1   The diadic cubes covering an open set E can be chosen so that their diameter is proportional to their distance from the boundary of E.

Proposition 1.1c (Whitney[28]). *Every open set $E \subset \mathbb{R}^N$ can be partitioned into the countable union of closed diadic cubes $\{Q_n\}$ with pairwise-disjoint interior and satisfying*

$$\text{diam}\{Q_j\} \le \text{dist}\{Q_j; \partial E\} \le 4\,\text{diam}\{Q_j\} \quad \textit{for all } j \in \mathbb{N}.$$

2 LIMITS OF SETS, CHARACTERISTIC FUNCTIONS, AND σ-ALGEBRAS

2.1   From the definition (2.1), it follows that $E' \subset E''$. There are sequences of sets $\{E_n\}$ for which the inclusion is strict.

2.2   Prove that

$$\limsup \chi_{E_n} = \chi_{E''}, \qquad\qquad \liminf \chi_{E_n} = \chi_{E'},$$
$$(\limsup E_n)^c = \liminf E_n^c, \qquad\qquad (\liminf E_n)^c = \limsup E_n^c.$$

2.3   Let $\mathcal{A}$ be the collection of subsets $E \subset X$ such that either E or E^c is finite. Then if X is not finite $\mathcal{A}$ is an algebra but not a σ-algebra.

2.4   Construct the smallest σ-algebra generated by two elements of X.

2.5   Construct the smallest σ-algebra generated by the collection of all finite subsets of X.

3 MEASURES

3.1   COMPLETION OF A MEASURE SPACE $\{X, \mathcal{A}, \mu\}$.   If $\{X, \mathcal{A}, \mu\}$ is not complete, it can be completed as follows. First, set

$$\overline{\mathcal{A}} = \left\{ \begin{array}{c} \text{the collection of sets of the type } E \bigcup N, \text{ where } E \in \mathcal{A} \\ \text{and } N \text{ is a subset of a set in } \mathcal{A} \text{ of measure zero} \end{array} \right\}.$$

Then set

$$\overline{\mu}\left(E \bigcup N\right) = \mu(E) \quad \text{for all } E \bigcup N \in \overline{\mathcal{A}}.$$

---

[28] H. Whitney, Analytic extensions of functions defined in closed sets, *Trans. Amer. Math. Soc.*, **36** (1934), 63–89.

Proposition 3.1c. *The definition does not depend on the choices of E and N identifying the same element $E \bigcup N \in \overline{\mathcal{A}}$; i.e., if*

$$E_1 \bigcup N_1 = E_2 \bigcup N_2, \qquad E_i \bigcup N_i \in \overline{\mathcal{A}}, \quad i = 1, 2, \tag{3.1c}$$

then

$$\overline{\mu}\left(E_1 \bigcup N_1\right) = \overline{\mu}\left(E_2 \bigcup N_2\right).$$

Moreover, $\overline{\mathcal{A}}$ is a σ-algebra and $\overline{\mu}$ is a complete measure defined in $\overline{\mathcal{A}}$.

Proof. From (3.1c), it follows that

$$E_1 \subset E_2 \bigcup N_2' \quad \text{and} \quad E_2 \subset E_1 \bigcup N_1',$$

where N_i' are sets of measure zero.   □

3.2 Let X be an infinite set and let $\mathcal{A}$ be the collection of subsets of X that are either finite or have a finite complement. Also let $\mu : \mathcal{A} \to \mathbb{R}^*$ be a set function defined by $\mu(E) = 0$ if E is finite and $\mu(E) = \infty$ if E has finite complement. The collection $\mathcal{A}$ is not a σ-algebra and μ is not countably additive.

3.3 Let X be an uncountable set and let $\mathcal{A}$ be the collection of subsets of X that are either countable or have a countable complement. Also let $\mu : \mathcal{A} \to \mathbb{R}^*$ be a set function defined by $\mu(E) = 0$ if E is countable and $\mu(E) = \infty$ if E has countable complement. The collection $\mathcal{A}$ is a σ-algebra and μ is a measure.

4 OUTER MEASURES

4.1 If $\mu_e(N) = 0$, then $\mu_e(E) = \mu_e(E \bigcup N)$ for every $E \subset X$.

4.2 A finitely additive outer measure is a measure.

4.3 The countable sum of outer measures is an outer measure.

4.4 Construct the Lebesgue–Stieltjes outer measure corresponding to the functions

$$f(x) = e^x, \quad f(x) = \begin{cases} [\![x]\!] & \text{for } x \geq 0, \\ 0 & \text{for } x < 0, \end{cases} \quad f(x) = \begin{cases} 1 & \text{for } x \geq 0, \\ -1 & \text{for } x < 0. \end{cases}$$

4.5 Let $\mathcal{Q}$ consist of $X, \emptyset$, and all the singletons in X. Define the set functions λ_1 and λ_2 from $\mathcal{Q}$ into $\mathbb{R}^+$ by

$$\lambda_1(X) = \infty, \quad \lambda_1(\emptyset) = 0, \quad \lambda_1(E) = 1 \quad \text{for all } E \in \mathcal{Q}, E \neq X, \emptyset,$$
$$\lambda_2(X) = 1, \quad \lambda_2(E) = 0 \quad \text{for all } E \in \mathcal{Q} \text{ such that } E \neq X.$$

Each of these is a set function on a sequential covering of X. Describe the outer measures generated by these.

5   The Hausdorff outer measure in $\mathbb{R}^N$

5.1   The Hausdorff dimension of a set $E \subset \mathbb{R}^N$. From (5.3), it follows that if $\mu_e(E)$ is finite, then $\mathcal{H}_{N+\eta}(E) = 0$ for all $\eta > 0$. If $E \subset \mathbb{R}^2$ is a segment, then $\mu_e(E) = \mathcal{H}_2(E) = 0$. Moreover,

$$\mathcal{H}_{1+\eta}(E) = 0 \quad \text{for all } \eta > 0 \quad \text{and} \quad \mathcal{H}_1(E) = \{\text{length of } E\}.$$

The Hausdorff dimension of a set $E \subset \mathbb{R}^N$ is defined by

$$\dim_{\mathcal{H}}(E) = \inf\{\alpha \mid \mathcal{H}_\alpha(E) = 0\}. \tag{5.1c}$$

5.2   Let $\{E_n\}$ be a countable collection of sets in $\mathbb{R}^N$ with the same Hausdorff dimension d. Then their union has the same Hausdorff dimension d.

5.3   The Hausdorff dimension of a point in $\mathbb{R}^N$ is zero. The Hausdorff dimension of a countable set in $\mathbb{R}^N$ is zero.

5.4   The Hausdorff dimension of the Cantor set is $\ln 2 / \ln 3$.

For $\delta \in (0, 1)$, let $\mathcal{C}_\delta$ be the generalized Cantor set introduced in (2.6c) of the Problems and Complements of the Preliminaries. For $n \in \mathbb{N}$, consider the intervals $\{J_{n,j}\}$, $j = 1, 2 \ldots, 2^n$, introduced in the same section. For each $n \in \mathbb{N}$, they form a finite sequence of disjoint closed intervals covering $\mathcal{C}_\delta$ and each of length δ^n. Therefore, by the definition of the outer measure $\mathcal{H}_\alpha$, for $\alpha > 0$,

$$\mathcal{H}_\alpha(\mathcal{C}_\delta) \le \lim_{n \to \infty} \sum_{j=1}^{2^n} (\operatorname{diam}\{J_{n,j}\})^\alpha = \lim_{n \to \infty} 2^n \delta^{\alpha n}. \tag{5.2c}$$

Therefore, if $\alpha > \log_{1/\delta} 2$, then $\mathcal{H}_a(\mathcal{C}_\delta) = 0$.

It follows from (5.1c) that the Hausdorff dimension of $\mathcal{C}_\delta$ does not exceed $\log_{1/\delta} 2$. If $\delta = \frac{1}{3}$ the $\mathcal{C}_\delta$ coincides with the standard Cantor set $\mathcal{C}$. Thus

$$\dim_{\mathcal{H}}(\mathcal{C}) \le \frac{\ln 2}{\ln 3}.$$

To prove the converse inequality, it suffices to establish a version of (5.2c) with the reverse inequality. For this, if $\mathcal{H}_\alpha(\mathcal{C}_\delta)$ is finite, given any $\varepsilon > 0$, there exists a countable collection of open intervals $\{\mathcal{I}_m\}$, each of diameter less than ε, whose union covers $\mathcal{C}_\delta$ and such that

$$\sum_{m=1}^\infty (\operatorname{diam}\{\mathcal{I}_m\})^\alpha \le \mathcal{H}_\alpha(\mathcal{C}_\delta) + \varepsilon. \tag{5.3c}$$

Since $\mathcal{C}_\delta$ is compact, we may assume that the collection $\{\mathcal{I}_m\}$ is finite. The proof is then concluded by establishing that the intervals $\mathcal{I}_m$ have essentially the same structure as the $J_{n,j}$ and that, as a consequence, (5.3c) is a reverse version of (5.2c). While the idea of the proof is simple, its technical implementation is rather involved.[29]

---

[29] F. Hausdorff, Dimension und äusseres Mass, *Math. Ann.*, **79** (1919), 157–179.

5.5 METRIC OUTER MEASURES. An outer measure μ_e in $\mathbb{R}^N$ is a *metric* outer measure if for any two sets E and F at positive mutual distance,

$$\mu\left(E\bigcup F\right) = \mu_e(E) + \mu_e(F).$$

The Lebesgue–Stieltjes outer measure on $\mathbb{R}$ is a metric outer measure. The Hausdorff outer measure $\mathcal{H}_\alpha$ is metric.

11 MORE ON EXTENSIONS FROM SEMIALGEBRAS TO σ-ALGEBRAS

11.1 Let X be the rationals in $[0, 1]$ and let $\mathcal{Q}$ be the algebra of the finite unions of sets of the type $(a, b] \cap X$, where a, b are real numbers and $0 \leq a < b \leq 1$.

The set function $\lambda\{(a, b] \cap X\} = \infty$ and $\lambda(\emptyset) = 0$ is a measure on $\mathcal{Q}$ that is not σ-finite. Then λ has at least two extensions μ_1 and μ_2 on the smallest σ-algebra $\mathcal{Q}_o$ generated by $\mathcal{Q}$; i.e.,

$$\mu_1 = \{\text{the counting measure on } \mathcal{Q}_o\},$$
$$\mu_2 = \{\mu_2(E) = \infty \text{ for every } E \in \mathcal{Q}_o \text{ and } \mu(\emptyset) = 0\}.$$

11.2 Let μ_e be an outer measure on X and let $\{X, \mathcal{A}, \mu\}$ be the measure space induced by μ_e. Using $\mathcal{A}$ as a sequential covering of X and $\mu : \mathcal{A} \to \mathbb{R}^*$ as an outer measure, we may construct another outer measure μ'. The new outer measure satisfies $\mu' \geq \mu_e$. Moreover, $\mu'(E) = \mu_e(E)$ if and only if there exists a μ-measurable set $A \supset E$ such that $\mu(A) = \mu_e(E)$.

12 THE LEBESGUE MEASURE OF SETS IN $\mathbb{R}^N$

In this section, *measure* means the Lebesgue measure in $\mathbb{R}^N$ and μ_e is the outer measure from which the Lebesgue measure is constructed.

12.1 The Lebesgue measure of a polyhedron coincides with its Euclidean measure.

12.2 The Lebesgue outer measure is translation invariant; i.e.,

$$\mu_e(y + E) = \mu_e(E) \quad \text{for all } E \subset \mathbb{R}^N.$$

In particular, the Lebesgue measure is translation invariant.

12.3 The interval $[0, 1]$ is not countable, for otherwise it would have measure zero. The Cantor set provides an example of a measurable uncountable set of measure zero.

12.4 There exist unbounded sets with finite measure.

12.5 An hyperplane in $\mathbb{R}^N$ has measure zero (the dimension of the hyperplane is less than N).

12.6 The boundary of a ball in $\mathbb{R}^N$ has measure zero. However, there exist open sets in $\mathbb{R}^N$ whose boundary has positive measure, for example, the complement of the generalized Cantor set.[30]

Such a set also provides an example of an open set $E \subset [0, 1]$ that is dense in $[0, 1]$. Its measure is less than 1, and for every interval $I \subset [0, 1]$, the measure of $I \bigcap E$ is positive.

12.7 The set of the rational numbers has Lebesgue measure zero, and its boundary has infinite measure.

12.8 Let E be a bounded measurable set in $\mathbb{R}^N$. There exists a set E_δ of the type $\mathcal{G}_\delta$ such that

$$\mu_e \left(A \bigcap E \right) = \mu_e \left(A \bigcap E_\delta \right)$$

for all $A \subset \mathbb{R}^N$. This is false if E is not measurable.

12.9 **INNER MEASURE AND MEASURABILITY.** Define the inner measure $\mu_i(E)$ of a bounded set $E \subset \mathbb{R}^N$ as

$$\mu_i(E) = \sup\{\mu(C) | \text{where } C \text{ is closed and } C \subset E\}. \tag{12.1c}$$

It follows from the definition that for every $\varepsilon > 0$, there exists a closed set $E_{c,\varepsilon}$ such that

$$E_{c,\varepsilon} \subset E \quad \text{and} \quad \mu_i(E) \leq \mu(E_{c,\varepsilon}) + \varepsilon.$$

Proposition 12.1c. *A bounded set $E \subset \mathbb{R}^N$ is Lebesgue measurable if and only if $\mu_i(E) = \mu_e(E)$.*

12.10 **THE PEANO–JORDAN MEASURE OF BOUNDED SETS IN $\mathbb{R}^N$.** Let E be a bounded set in $\mathbb{R}^N$ and construct the two classes of sets

$$\mathcal{O}_E \equiv \left\{ \begin{array}{c} \text{the sets that are the finite union of open} \\ \text{cubes in } \mathbb{R}^N \text{ and that contain } E \end{array} \right\},$$

$$\mathcal{I}_E \equiv \left\{ \begin{array}{c} \text{the sets that are the finite union of closed} \\ \text{cubes in } \mathbb{R}^N \text{ and that are contained in } E \end{array} \right\}.$$

The Peano–Jordan outer and inner measure of E are defined as

$$\mu^e_{\mathcal{P}-\mathcal{J}}(E) = \inf_{O \in \mathcal{O}_E} \mu(O), \qquad \mu^i_{\mathcal{P}-\mathcal{J}}(E) = \sup_{I \in \mathcal{I}_E} \mu(I). \tag{12.2c}$$

A bounded set $E \subset \mathbb{R}^N$ is Peano–Jordan measurable if its Peano–Jordan outer and inner measures coincide. From the definitions of Lebesgue outer and inner measures, it follows that

$$\mu^i_{\mathcal{P}-\mathcal{J}}(E) \leq \mu_i(E) \leq \mu_e(E) \leq \mu^e_{\mathcal{P}-\mathcal{J}}(E).$$

---

[30]See Section 2.3 of the Problems and Complements of the Preliminaries.

Thus a Peano–Jordan measurable set is Lebesgue measurable. The converse is false. The set $\mathbf{Q} \cap [0, 1]$ is Lebesgue measurable and its measure is zero. However,

$$\mu^e_{\mathcal{P}-\mathcal{J}}\left(\mathbf{Q} \cap [0, 1]\right) = 1, \qquad \mu^i_{\mathcal{P}-\mathcal{J}}\left(\mathbf{Q} \cap [0, 1]\right) = 0.$$

Thus $\mathbf{Q} \cap [0, 1]$ is not Peano–Jordan measurable. This last example shows that the Peano–Jordan measure is not a measure in the sense of SS3, since its domain is not a σ-algebra.

12.11   Since a measure is generated from a given outer measure, the Peano–Jordan measure can be constructed from $\mu^e_{\mathcal{P}-\mathcal{J}}$ the same way as the Lebesgue measure was generated by the outer measure μ_e. Thus the difference between these two measures stems from the difference of the generating outer measures. In the Peano–Jordan outer measure (12.2c), the infimum is taken over the family of sets that are a *finite* union of cubes, whereas in the Lebesgue outer measure the infimum is taken over the family of sets that are a *countable* union of cubes.

13   A NONMEASURABLE SET

13.1   Every measurable subset A of the nonmeasurable set E has measure zero. Indeed, the sets $A_n = A \bullet r_n \subset E_n$ for $r_n \in \mathbf{Q} \cap [0, 1]$ are disjoint, have each measure equal to the measure of A, and their union is contained in $[0, 1]$. Thus

$$\sum \mu(A_n) = \mu\left(\bigcup A_n\right) \le 1.$$

13.2   Every set $A \subset [0, 1]$ of positive outer measure, contains a nonmeasurable set. Indeed, at least one of the intersections $A \cap E_n$ is nonmeasurable. If all such intersections were measurable, then by the subadditive property of the outer measure,

$$0 < \mu_e(A) \le \sum \mu\left(A \cap E_n\right) = 0$$

since all the $A \cap E_n$ are measurable subsets of E.

13.3   Let $E \subset [0, 1]$ be the Vitali nonmeasurable set. Then $\mu_e(E) > 0$.

13.4   The Lebesgue measure on $[0, 1]$ is a set function defined of the Lebesgue-measurable subset of $[0, 1]$ satisfying the following:

  (i) μ is nonnegative.

  (ii) μ is countably additive.

  (iii) μ is translation invariant.

  (iv) If $I \subset [0, 1]$ is an interval, then $\mu(I)$ coincides with the Euclidean measure of I.

It is impossible to define a set function μ satisfying (i)–(iv) and defined on all the subsets of $[0, 1]$.

The construction of the nonmeasurable set $E \subset [0, 1]$ uses only properties (i)–(iii) of the function μ, and it is independent of the particular construction of the Lebesgue measure. The final contradiction argument uses property (iv).

If a function μ satisfying (i)–(iv) and defined in all the subsets of $[0, 1]$ were to exist, the same construction would imply that $\mu([0, 1])$ is either zero or infinity.

The requirement (iv) cannot be removed from these remarks. Indeed, the counting measure or the identically zero measure would satisfy (i)–(iii) but not (iv).

14 BOREL SETS, MEASURABLE SETS, AND INCOMPLETE MEASURES

The strict inclusion $\mathcal{B} \subset \mathcal{M}$ can be established by an indirect cardinality argument.[31]

14.1 Let $\mathcal{B}_o$ denote the collection of all open intervals of $[0, 1]$. The cardinality of $\mathcal{B}_o$ does not exceed the cardinality of $\mathbb{R}$. Indeed, the topology inherited by $[0, 1]$ from the Euclidean topology in $\mathbb{R}$ has a countable base. Therefore, every open set in $[0, 1]$ is the countable union of elements of the base. It follows that the cardinality of $\mathcal{B}_o$ does not exceed the cardinality of the set of sequences of countable elements; i.e.,

$$\text{card}(\mathcal{B}_o) \leq \text{card}(\mathbb{N}^{\mathbb{N}}) = \text{card}(\mathbb{R}).$$

14.2 Let $\mathcal{B}_1$ be the collection of all sets that can be obtained by taking countable unions, countable intersections, and complements of elements of $\mathcal{B}_o$. Then

$$\text{card}(\mathcal{B}_1) = \text{card}(\mathcal{B}_o^{\mathbb{N}}) \leq \text{card}(\mathbb{R}).$$

14.3 Define inductively $\mathcal{B}_n$ for all $n \in \mathbb{N}$; i.e., $\mathcal{B}_n$ is the collection of all sets that can be obtained by taking countable unions, countable intersections, and complements of elements of $\mathcal{B}_{n-1}$. Then for all $n \in \mathbb{N}$,

$$\text{card}(\mathcal{B}_n) \leq \text{card}(\mathbb{R}^{\mathbb{N}}) = \text{card}(\mathbb{R}).$$

14.4 Let Ω be the first uncountable. For each $\alpha < \Omega$ define $\mathcal{B}_\alpha$ as the collection of sets that can be obtained by taking countable unions, countable intersections, and complements of elements of $\mathcal{B}_\beta$ for all ordinal numbers $\beta < \alpha$. By the definition of the first uncountable, the cardinality of $\mathcal{B}_\alpha$ does not exceed $\text{card}(\mathbb{R}^{\mathbb{N}}) = \text{card}(\mathbb{R})$.

14.5 The smallest σ-algebra containing the open sets of $[0, 1]$ can be constructed by this procedure by setting

$$\mathcal{B} = \bigcup_{\alpha < \Omega} \mathcal{B}_\alpha.$$

By transfinite induction, the cardinality of $\mathcal{B}$ does not exceed $\text{card}(\mathbb{R})$.

---

[31] See the cardinality statements of Sections 3 and 4 of the Preliminaries.

14.6   Since the Lebesgue measure is complete, every subset of the Cantor set is measurable and has measure zero. Since the cardinality of the Cantor set is the cardinality of $\mathbb{R}$, the cardinality of all the subsets of the Cantor sets is $\text{card}(2^{\mathbb{R}})$.

14.7   The cardinality of the Lebesgue-measurable subsets of $[0, 1]$ is not less than $\text{card}(2^{\mathbb{R}})$. Thus the cardinality of the Borel subsets of $[0, 1]$ does not exceed $\text{card}(\mathbb{R})$ and the cardinality of the Lebesgue-measurable subsets of $[0, 1]$ is not less than $\text{card}(2^{\mathbb{R}})$.

17   VITALI COVERINGS

The next proposition can be regarded as a measure-theoretical separation criterion for disjoint measurable sets.

Proposition 17.1c.   *Let $\{E_1, E_2, \ldots, E_n\}$ be a finite collection of bounded, Lebesgue-measurable, disjoint sets in $\mathbb{R}^N$, and let $\mathcal{F}$ be a fine Vitali covering for each of them. Then for each E_j, there exists a finite collection of cubes*

$$Q_{j,1}, Q_{j,2}, \ldots, Q_{j,n_j}, \quad n_j \in \mathbb{N},$$

with pairwise-disjoint interior such that

$$\mu\left(\bigcup_{i=1}^{n_j} E_j \cap Q_{j,i}\right) \geq \frac{1}{2}\mu(E_j).$$

Moreover, the collections of cubes

$$\{Q_{h,\ell}\}_{\ell=1}^{n_h}, \quad \{Q_{k,s}\}_{s=1}^{n_k}, \quad h \neq k, \quad h, k = 1, 2, \ldots, n,$$

have pairwise-disjoint interior; i.e., any two cubes, the first out of $\{Q_{h,\ell}\}_{\ell=1}^{n_h}$ and the second out of $\{Q_{k,s}\}_{s=1}^{n_k}$, have disjoint interior.

17.1   The notion of Vitali's covering of a set $E \subset \mathbb{R}^N$ is independent of the measurability of E. Let μ_e be the outer measure associated to the Lebesgue measure in $\mathbb{R}^N$ and generated by (16.2). Extend the Vitali covering theorem to the case of a set E of finite outer measure $\mu_e(E)$.

17.2   Let E be a bounded subset of $\mathbb{R}^N$ of finite outer measure that admits a fine Vitali covering with nontrivial cubes contained in E. Then E is Lebesgue measurable. The requirement that the cubes making up the Vitali covering be contained in E is essential. Give a counterexample.

17.3   Let $\{Q_\alpha\}$ be an uncountable family of nontrivial, closed cubes in $\mathbb{R}^N$. Then $\bigcup Q_\alpha$ is Lebesgue measurable.

18   THE BESICOVITCH COVERING THEOREM

18.1   A SIMPLER FORM OF THE BESICOVITCH THEOREM. The next is a covering statement, based only on the geometry of cubes in $\mathbb{R}^N$ and independent of measures.

Theorem 18.1c (Besicovitch). *Let E be a bounded subset of $\mathbb{R}^N$ and let $\mathcal{F}$ be a collection of cubes in $\mathbb{R}^N$ with faces parallel to the coordinate planes and such that each $x \in E$ is the center of a nontrivial cube $Q(x)$ belonging to $\mathcal{F}$.*

There exist a countable collection $\{x_n\}$ of points $x_n \in E$ and a corresponding collection of cubes $\{Q(x_n)\}$ in $\mathcal{F}$ such that

$$E \subset \bigcup Q(x_n) \quad and \quad \sum \chi_{Q(x_n)} \leq 4^N. \tag{18.1c}$$

Remark 18.1c. The second part of (18.1c) asserts that each point $x \in \mathbb{R}^N$ is covered by at most 4^N cubes out of $\{Q(x_n)\}$. Equivalently, at most 4^N cubes overlap at each given point.

It is remarkable that the largest number of possible overlaps of the cubes $Q(x_n)$ at each given point is independent of the set E and the covering $\mathcal{F}$, and depends only on the geometry of the cubes in $\mathbb{R}^N$.

Lemma 18.2c. *Let $\{Q(x_n)\}$ be a countable collection of cubes in $\mathbb{R}^N$ with centers at x_n and satisfying the following:*

$$If\ n < m,\ then\ x_m \notin Q(x_n)\ and\ \mu(Q(x_m)) \leq 2\mu(Q(x_n)). \tag{18.2c}$$

Then each point $x \in \mathbb{R}^N$ is covered by at most 4^N cubes out of $\{Q(x_n)\}$.

Proof. Assume first $N = 2$. Having fixed $x \in \mathbb{R}^2$, we may assume up to a translation that x is the origin. Denote by $2\rho_n$ the edge of the cube $Q(x_n)$, and let $\{Q_j\}$ be the collection of squares containing the origin whose center x_j is in the first quadrant and of edge $2\rho_j$.

Starting from Q_1 and the corresponding edge $2\rho_1$, consider the four closed squares

$$S_1 = [0, \rho_1] \times [0, \rho_1], \qquad S_2 = [0, \rho_1] \times [\rho_1, 2\rho_1],$$
$$S_3 = [\rho_1, 2\rho_1] \times [\rho_1, 2\rho_1], \qquad S_4 = [\rho_1, 2\rho_1] \times [0, \rho_1].$$

Also let

$$S_o = [0, 2\rho_1] \times [0, 2\rho_1] = S_1 \bigcup S_2 \bigcup S_3 \bigcup S_4.$$

By construction, Q_1 covers S_1, and by the second part of (18.2c), the center x_j of each of the Q_j (for $j = 2, 3, \dots$) cannot lie outside S_o. Indeed, if it did, since Q_j contains the origin, $\rho_j > 2\rho_1$ and $\mu(Q_j) > 4\mu(Q_1)$.

Thus the centers x_j of the Q_j for $j = 2, 3, \dots$ must belong to some of the squares S_1, S_2, S_3, S_4. Now x_j cannot belong to S_1 because otherwise the first part of (18.2c) would be violated, since $S_1 \subset Q_1$.

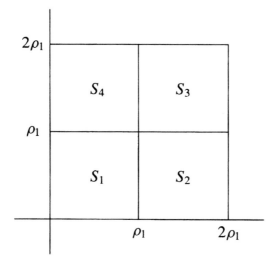

Figure 18.1c.

If some $x_j \in S_2$, then since Q_j contains the origin, $S_2 \subset Q_j$. Therefore, by the first part of (18.2c), no other center x_ℓ for $\ell > j$ belongs to S_2. This implies that S_2 contains at most one center x_j.

By a similar argument, S_3 and S_4 contain each at most one center x_j of a cube Q_j. Thus the collection $\{Q_j\}$ contains at most four cubes.

Defining analogously the collections of cubes containing the origin and whose centers are, respectively, in the second, third, and fourth quadrants, we conclude that each of these collections contains at most four cubes. Thus the origin is covered by at most 16 cubes out of the collection $\{Q_{(x_n)}\}$.

A similar argument for $N = 3$ gives that each point is covered by at most 64 cubes. The general case follows by induction.   □

Proof of Theorem 18.1c. Set

$$E_1 = E \quad \text{and} \quad \lambda_1 = \sup\{\mu(Q(x)) | x \in E\}.$$

If $\lambda_1 = \infty$ there exists cubes $Q(x) \in \mathcal{F}$ of arbitrarily large edge and centered at some $x \in E$. Since E is bounded, we select one such cube.

If $\lambda_1 < \infty$ select $x_1 \in E_1$ and a cube $Q(x_1)$ such that $\mu(Q(x_1)) > \frac{1}{2}\lambda_1$. Then set

$$E_2 = E_1 - Q(x_1) \quad \text{and} \quad \lambda_2 = \sup\{\mu(Q(x)) | x \in E_2\}.$$

If $E_1 \subset Q(x_1)$ the process terminates. Otherwise, $\lambda_2 > 0$ and we select $x_2 \in E_2$ and a cube $Q(x_2)$ such that $\mu(Q(x_2)) > \frac{1}{2}\lambda_2$.

Proceeding recursively, define countable collections of sets E_n, points $x_n \in E_n$, corresponding cubes $Q(x_n)$, and positive numbers λ_n by

$$E_n = E - \bigcup_{j=1}^{n-1} Q(x_j), \qquad \lambda_n = \sup\{\mu(Q(x))|x \in E_n\},$$

where the points $x_n \in E_n$ and the corresponding cubes $Q(x_n)$ are selected to satisfy

$$\mu(Q(x_n)) \geq \frac{1}{2}\lambda_n.$$

By construction, $\{\lambda_n\}$ is a decreasing sequence and

$$\mu(Q(x_m)) \leq \lambda_m \leq \lambda_n \leq 2\mu(Q(x_n)) \quad \text{for } n < m$$

as long as $\lambda_m > 0$. Therefore, by the previous lemma, at most 4^N of the cubes $\{Q(x_n)\}$ overlap at each $x \in \mathbb{R}^N$. It remains to prove that $E \subset \bigcup Q(x_n)$.

If $E_n = \emptyset$ for some n, then $E \subset \bigcup_{j=1}^{n-1} Q(x_j)$ and the process terminates. If $E_n \neq \emptyset$ for all n, we claim that $\lim \lambda_n = 0$. To this end, compute

$$\limsup \mu(Q(x_n)) = \delta.$$

Let x_o be a cluster point of $\{x_n\}$. If $\delta > 0$, then x_o would be covered by infinitely many cubes $Q(x_n)$. Therefore, $\delta = 0$. The relations

$$\frac{1}{2}\lambda_n \leq \mu(Q(x_n)) \leq \lambda_n, \qquad \lambda_{n+1} \leq \lambda_n,$$

now imply that $\lim \lambda_n = 0$.

If $x \in E - \bigcup Q(x_n)$, there exists a nontrivial cube $Q(x) \in \mathcal{F}$ such that $\mu(Q(x)) < \lambda_n$ for all n. Therefore, $\mu(Q(x)) = 0$ and $Q(x)$ is a trivial cube. $\square$

18.2   ANOTHER BESICOVITCH-TYPE COVERING.

Proposition 18.3c (Giusti[32]). *Let E be a bounded subset of $\mathbb{R}^N$ and let $x \to \rho(x)$ be a function from E into $(0, 1)$.*[33]

There exists a countable collection $\{x_n\}$ of points in E such that the open balls $B(x_n, \rho(x_n))$ with center at x_n and radius $\rho(x_n)$ are pairwise disjoint and

$$E \subset \bigcup B(x_n, 3\rho(x_n)). \tag{18.3c}$$

---

[32]E. Giusti, Precisazione delle funzioni $H^{1,p}$ e singolarità delle soluzioni deboli di sistemi ellittici non lineari, *Boll. Univ. Mat. Ital.*, 1968, 71–76.

[33]Neither E nor $x \to \rho(x)$ is required to be measurable.

Proof. Let $\mathcal{F}_1$ be the collection of pairwise-disjoint balls $B(x, \rho(x))$ such that $\frac{1}{2} \le \rho(x) < 1$. Since E is bounded, $\mathcal{F}_1$ contains at most a finite number of such balls, say, for example,

$$B(x_1, \rho(x_1)), B(x_2, \rho(x_2)), \ldots, B(x_{n_1}, \rho(x_{n_1})) \qquad (18.4c)_1$$

for some $n_1 \in \mathbb{N}$. If the union of these balls covers E, the construction terminates. If not, let $\mathcal{F}_2$ be the collection of pairwise-disjoint balls $B(x, \rho(x))$ not intersecting any of the balls selected in $(18.4c)_1$ and such that $2^{-2} \le \rho(x) < 2^{-1}$.

Since E is bounded, $\mathcal{F}_2$, if nonempty, contains at most a finite number of such balls, say, for example,

$$B(x_{n_1+1}, \rho(x_{n_1+1})), B(x_{n_1+2}, \rho(x_{n_1+2})), \ldots, B(x_{n_2}, \rho(x_{n_2}))$$

for some $n_2 \in \mathbb{N}$. If the union of the balls

$$B(x_i, \rho(x_i)) \quad \text{for } i = 1, 2, \ldots, n_1, n_1 + 1, \ldots, n_2 \qquad (18.4c)_2$$

covers E, the construction terminates. If $\mathcal{F}_2$ is empty or if the union of these balls does not cover E, construct the collection $\mathcal{F}_3$ of pairwise-disjoint balls $B(x, \rho(x))$ not intersecting any of the balls selected in $(18.4c)_2$ and such that $2^{-3} \le \rho(x) < 2^{-2}$.

Proceeding recursively, assume that we have selected a finite collection of pairwise-disjoint balls

$$B(x_1, \rho(x_1)), B(x_2, \rho(x_2)), \ldots, B(x_{n_j}, \rho(x_{n_j})) \qquad (18.4c)_j$$

for some $n_j \in \mathbb{N}$. If their union does not cover E, construct the family $\mathcal{F}_{j+1}$ of pairwise-disjoint balls $B(x, \rho(x))$ not intersecting any of the balls selected in $(18.4c)_j$ and such that $2^{-(j+1)} \le \rho(x) < 2^{-j}$.

Since E is bounded, $\mathcal{F}_{j+1}$, if nonempty, contains at most a finite number of such balls, which we add to the collection in $(18.4c)_j$.

If for some $j \in \mathbb{N}$ the union of the balls in $(18.4c)_j$ covers E, the process terminates. Otherwise, this recursive procedure generates a countable collection of pairwise-disjoint balls $\{B(x_n, \rho(x_n))\}$.

It remains to prove that such a collection satisfies (18.3c). Let $x \in E$. By construction,

$$B(x, \rho(x)) \bigcap B(x_j, \rho(x_j)) \ne \emptyset \quad \text{for some } j.$$

Therefore, $\rho(x) \le 2\rho(x_j)$. For one such x_j fixed,

$$|x - x_j| \le \rho(x) + \rho(x_j) \le 3\rho(x_j).$$

Thus $x \in B(x_j, 3\rho(x_j))$. $\qquad\qquad\qquad\qquad\qquad\qquad\qquad$ $\square$

III

The Lebesgue Integral

1 Measurable functions

Let $\{X, \mathcal{A}, \mu\}$ be a measure space and $E \in \mathcal{A}$. For a function $f : E \to \mathbb{R}^*$ and $c \in \mathbb{R}$, set

$$[f > c] = \{x \in E | f(x) > c\}. \tag{1.1}$$

The sets $[f \geq c]$, $[f < c]$, $[f \leq c]$ are defined similarly. Alternatively, we may define them starting from (1.1) as

$$[f \geq c] = \bigcap \left[f > c - \frac{1}{n} \right],$$
$$[f \leq c] = E - [f > c], \tag{1.2}$$
$$[f < c] = E - [f \geq c].$$

Therefore, if $[f > c]$ is measurable for all $c \in \mathbb{R}$, the sets in (1.2) also are measurable for all $c \in \mathbb{R}$. Now let $[f \geq c]$ be measurable for all $c \in \mathbb{R}$. Then from

$$[f > c] = \bigcup \left[f \geq c + \frac{1}{n} \right],$$

it follows that $[f > c]$ is measurable for all $c \in \mathbb{R}$. Similarly, if any one of the four sets

$$[f > c], \qquad [f \geq c], \qquad [f < c], \qquad [f \leq c] \tag{1.3}$$

is measurable for all $c \in \mathbb{R}$, the remaining three also are measurable for all $c \in \mathbb{R}$.

A function $f : E \to \mathbb{R}^*$ is measurable if at least one of the sets in (1.3) is measurable for all $c \in \mathbb{R}$.

Remark 1.1. The notion of measurable function depends only on the σ-algebra $\mathcal{A}$ and is independent of the measure μ defined on $\mathcal{A}$.

Proposition 1.1. *A function $f : E \to \mathbb{R}^*$ is measurable if and only if at least one of the sets in* (1.3) *is measurable for all rational numbers c.*

Proof. Assume, for example, that the second of the sets in (1.3) is measurable for all $c \in \mathbf{Q}$. Having fixed some $c \in \mathbb{R} - \mathbf{Q}$, let $\{q_n\}$ be a sequence of rational numbers decreasing to c. Then the set

$$[f > c] = \bigcup [f \geq q_n]$$

is measurable. $\square$

Proposition 1.2. *Let $f : E \to \mathbb{R}^*$ be measurable and let $\alpha \in \mathbb{R} - \{0\}$.*

(i) *The functions $|f|, \alpha \cdot f, \alpha + f, f^2$ are measurable.*

(ii) *If $f \neq 0$, then $\frac{1}{f}$ also is measurable.*

(iii) *For any measurable subset $E' \subset E$, the restriction $f|_{E'}$ is measurable.*

Proof. The statements in (i) and (ii) follow from the set identities

$$[|f| > c] = \begin{cases} [f > c] \bigcup [f < -c] & \text{if } c \geq 0, \\ E & \text{if } c < 0, \end{cases}$$

$$[\alpha + f > c] = [f > c - \alpha],$$

$$[\alpha \cdot f > c] = \begin{cases} \left[f > \dfrac{c}{\alpha} \right] & \text{if } \alpha > 0, \\ \left[f < \dfrac{c}{\alpha} \right] & \text{if } \alpha < 0, \end{cases}$$

$$[f^2 > c] = \begin{cases} [f > \sqrt{c}] \bigcup [f < -\sqrt{c}] & \text{if } c \geq 0, \\ E & \text{if } c < 0. \end{cases}$$

$$\left[\frac{1}{f} > c \right] = \begin{cases} [f > 0] \bigcap \left[f < \dfrac{1}{c} \right] & \text{if } c > 0, \\ [f > 0] & \text{if } c = 0, \\ [f > 0] \bigcup \left[f < \dfrac{1}{c} \right] & \text{if } c < 0. \end{cases}$$

To prove (iii), it suffices to observe that $[f|_{E'} > c] = [f > c] \bigcap E'$. $\square$

Proposition 1.3. *Let $f : E \to \mathbb{R}^*$ and $g : E \to \mathbb{R}$ be measurable. Then*

(i) *The set $[f > g]$ is measurable.*

(ii) *The functions $f \pm g$ are measurable.*

(iii) *The function $f \cdot g$ is measurable.*

(iv) *If $g \neq 0$, the function $\frac{f}{g}$ is measurable.*

Proof. Let $\{q_n\}$ denote the sequence of the rational numbers. Then

$$[f > g] = \bigcup [f \geq q_n] \bigcap [g < q_n].$$

This proves (i). To prove (ii), we observe that for all $c \in \mathbb{R}$,

$$[f \pm g > c] = [f > \mp g + c],$$

and the latter is measurable in view of (i). By Proposition 1.2, $(f \pm g)^2$ also are measurable, and this implies (iii) in view of the identity

$$fg = \frac{1}{4}(f + g)^2 - \frac{1}{4}(f - g)^2.$$

Finally, (iv) follows from parts (iii) and (ii) of Proposition 1.2 $\square$

Proposition 1.4. *Let $\{f_n\}$ be a sequence of measurable functions in E. Then the functions*

$$\varphi = \sup f_n, \qquad \psi = \inf f_n,$$
$$f'' = \limsup f_n, \qquad f' = \liminf f_n,$$

are measurable.

Proof. From the definitions, it follows that for every $c \in \mathbb{R}$

$$[\varphi > c] = \bigcup [f_n > c], \qquad [\psi \geq c] = \bigcap [f_n \geq c]. \qquad (1.4)$$

Thus φ and ψ are measurable. Set

$$\varphi_n = \sup_{j \geq n} f_j \quad \text{and} \quad \psi_n = \inf_{j \geq n} f_j.$$

By (1.4), the functions φ_n and ψ_n are measurable for all $n \in \mathbb{N}$. Moreover, also by (1.4),

$$f'' = \inf \varphi_n \quad \text{and} \quad f' = \sup \psi_n$$

are measurable. $\square$

Let f and g be two functions defined in E. We say that $f = g$ almost everywhere (a.e.) in E if there exists a set $\mathcal{E} \subset E$ of measure zero such that

$$f(x) = g(x) \quad \text{for all } x \in E - \mathcal{E} \quad \text{and} \quad \mu(\mathcal{E}) = 0.$$

More generally, a property of real valued functions defined on a measure space $\{X, \mathcal{A}, \mu\}$ is said to hold *almost everywhere (a.e.)* if it does hold for all $x \in X$ except possibly for a measurable set $\mathcal{E} \subset X$ of measure zero.

Lemma 1.5. *Let* $\{X, \mathcal{A}, \mu\}$ *be complete. If* f *is measurable and* $f = g$ *a.e. in* E, *then* g *also is measurable.*

Proof. Let $\mathcal{E} = [f \neq g]$. Since every subset of $\mathcal{E}$ is measurable, the set

$$[g > c] = \left\{ [f > c] \bigcap (E - \mathcal{E}) \right\} \bigcup \left\{ [g > c] \bigcap \mathcal{E} \right\}$$

is measurable. □

Corollary 1.6. *Let* $\{X, \mathcal{A}, \mu\}$ *be complete and let* $\{f_n\}$ *be a sequence of measurable functions defined in* $E \in \mathcal{A}$ *and taking values in* $\mathbb{R}^*$. *Assume that*

$$f(x) = \lim f_n(x) \quad \text{exists a.e. in } E.$$

Then f *is measurable.*

2 The Egorov theorem

Let $\{f_n\}$ be a sequence of functions defined in a measurable set E with values in $\mathbb{R}^*$ and set

$$f''(x) = \limsup f_n(x), \qquad f'(x) = \liminf f_n(x), \quad x \in E. \tag{2.1}$$

The functions f'' and f' are defined in E and take values in $\mathbb{R}^*$. We will assume throughout that they are a.e. finite in E, i.e., that there exist measurable sets $\mathcal{E}''$ and $\mathcal{E}'$ contained in E such that

$$\begin{aligned} f''(x) \in \mathbb{R} \quad \text{for all } x \in E - \mathcal{E}'' \quad &\text{and} \quad \mu(\mathcal{E}'') = 0, \\ f'(x) \in \mathbb{R} \quad \text{for all } x \in E - \mathcal{E}' \quad &\text{and} \quad \mu(\mathcal{E}') = 0. \end{aligned} \tag{2.2}$$

The upper limit in (2.1) is uniform if for every $\varepsilon > 0$, there exists an index n_ε such that

$$f_n(x) \leq f''(x) + \varepsilon \quad \text{for all } n \geq n_\varepsilon \quad \text{and} \quad \text{for all } x \in E - \mathcal{E}''. \tag{2.2}''$$

Similarly, the lower limit in (2.1) is uniform if for every $\varepsilon > 0$, there exists an index n_ε such that

$$f_n(x) \geq f'(x) - \varepsilon \quad \text{for all } n \geq n_\varepsilon \quad \text{and} \quad \text{for all } x \in E - \mathcal{E}'. \tag{2.1}'$$

Proposition 2.1. *Let* $\{f_n\}$ *be a sequence of measurable functions defined on a measurable set* $E \in \mathcal{A}$ *of finite measure and with values in* $\mathbb{R}^*$.

Assume that, for example, the second (first) part of (2.2) holds. Then for every $\eta > 0$, *there exists a measurable set* $E_\eta \subset E$ *such that* $\mu(E - E_\eta) \leq \eta$ *and the lower (upper) limit in (2.1) is uniform in* E_η.

Proof. The statement is only proved for the lower limit, the arguments for the upper limit being analogous.

Fix two positive integers m and n and introduce the sets

$$E_{m,n} = \bigcap_{j=n}^{\infty} \left\{ x \in (E - \mathcal{E}') | f_j(x) \geq f'(x) - \frac{1}{m} \right\}.$$

For $m \in \mathbb{N}$ fixed, the sets $E_{m,n}$ are measurable and expanding. By the definitions of the function f' and of the set $\mathcal{E}'$,

$$E - \mathcal{E}' = \bigcup_{n=1}^{\infty} E_{m,n} \quad \text{and} \quad \mu(E) = \lim_{n \to \infty} \mu(E_{m,n}).$$

Therefore, having fixed $\eta > 0$, there exists an index $n(m, \eta)$ such that

$$\mu(E - E_{m,n(m,\eta)}) \leq \frac{1}{2^m} \eta.$$

The set claimed by the proposition is

$$E_\eta = \bigcap_{m=1}^{\infty} E_{m,n(m,\eta)}. \tag{2.3}$$

Indeed, E_η is measurable, and by construction,

$$\mu(E - E_\eta) = \mu \left(E - \bigcap_{m=1}^{\infty} E_{m,n(m,\eta)} \right) = \mu \left(\bigcup_{m=1}^{\infty} (E - E_{m,n(m,\eta)}) \right)$$
$$\leq \sum_{m=1}^{\infty} \mu(E - E_{m,n(m,\eta)}) \leq \eta. \tag{2.4}$$

Fix an arbitrary $\varepsilon > 0$ and let m_ε be the smallest positive integer such that $\varepsilon m_\varepsilon \geq 1$. From the inclusion

$$E_\eta \subset E_{m_\varepsilon,n(m_\varepsilon,\eta)}$$

and the definition of the sets $E_{m,n}$, it follows that

$$f_n(x) \geq f'(x) - \varepsilon \quad \text{for all } n \geq n_\varepsilon \equiv n(m_\varepsilon, \eta) \quad \text{and} \quad \text{for all } x \in E_\eta.$$

Thus, in particular, the lower limit in (2.1) is uniform within E_η. $\qquad \square$

Theorem 2.2 (Egorov[1]). *Let $\{f_n\}$ be a sequence of measurable functions defined in a measurable set E of finite measure and with values in $\mathbb{R}^*$. Assume that the sequence converges a.e. in E to a function $f : E \to \mathbb{R}^*$, which is finite a.e. in E. Then for every $\eta > 0$, there exists a measurable set $E_\eta \subset E$ such that $\mu(E - E_\eta) \leq \eta$ and the limit in (2.1) is uniform in E_η.*

Remark 2.1. Egorov's theorem is, in general, false if E is not of finite measure.

---

[1] D. Th. Egorov, Sur les suites des fonctions mesurables, *C. R. Acad. Sci. Paris*, **152** (1911), 244–246.

2.1   The Egorov theorem in $\mathbb{R}^N$. If the measure space $\{X, \mathcal{A}, \mu\}$ is $\mathbb{R}^N$ endowed with the Lebesgue measure, the Egorov theorem can be given a stronger form.

Proposition 2.3. *Let $\{f_n\}$ be a sequence of Lebesgue-measurable functions defined on a bounded, measurable set $E \subset \mathbb{R}^N$ and with values in $\mathbb{R}^*$. Assume that, for example, the second part of (2.2) holds. Then for every $\eta > 0$, there exists a closed set $E_{c,\eta} \subset E$ such that $\mu(E - E_{c,\eta}) \leq \eta$ and the lower limit in (2.1) is uniform in $E_{c,\eta}$.*

Proof. The set E_η in (2.3) is a bounded and Lebesgue-measurable subset of $\mathbb{R}^N$. Therefore, by Proposition 12.4 of Chapter II, there exists a closed set

$$E_{c,\eta} \subset E_\eta \quad \text{such that} \quad \mu(E_\eta - E_{c,\eta}) \leq \eta. \qquad \square$$

2.2   More on Egorov's theorem. The proof of Proposition 2.1 for $f' = \liminf f_n$ does not directly use the fact that f' is the lower limit of $\{f_n\}$, and it is based only on the pointwise inequalities generated by the notion of lower limit; i.e.,

> for all fixed $x \in E$ and $\varepsilon > 0$, there exists an index $n_\varepsilon(x)$
> depending upon ε and x such that

$$f_n(x) \geq f'(x) - \varepsilon \quad \text{for all } n \geq n_\varepsilon(x). \qquad (2.5)$$

Starting from such a pointwise inequality, the proposition concludes that (2.5) actually holds uniformly within a measurable set $E_\eta \subset E$ arbitrarily close to E in a measure-theoretical sense.

These remarks imply that the conclusion of the proposition continues to hold if f' is replaced by any measurable function $\zeta : E \to \mathbb{R}^*$ that is a.e. finite in E and satisfies (2.5).

Proposition 2.4. *Let $\{f_n\}$ be a sequence of measurable defined on a measurable set E of finite measure and with values in $\mathbb{R}^*$. Assume that there exists a measurable function $\zeta : E \to \mathbb{R}$ that is a.e. finite in E and satisfies the following property:*

For all fixed $x \in E$ and $\varepsilon > 0$, there exists an index $n_\varepsilon(x)$ depending upon ε and x such that

$$f_n(x) \geq \zeta(x) - \varepsilon \quad \text{for all } n \geq n_\varepsilon. \qquad (2.5)_\zeta$$

Then for all $\eta > 0$, there exists a measurable set $E_\eta \subset E$ such that $\mu(E - E_\eta) \leq \eta$ and such that $(2.5)_\zeta$ holds uniformly in E_η; i.e., for all $\varepsilon > 0$, there exists an index n_ε depending only upon ε such that

$$f_n(x) \geq \zeta(x) - \varepsilon \quad \text{for all } n \geq n_\varepsilon \quad \text{and} \quad \text{for all } x \in E_\eta. \qquad (2.5)_\eta$$

3   Approximating measurable functions by simple functions

A function f from a measurable set E with values in $\mathbb{R}$ is *simple* if it is measurable and if it takes a finite number of values. The characteristic function of a measurable set is simple.

Let f be simple in E, let $\{f_1, f_2, \ldots, f_n\}$ be the distinct values taken by f in E, and set

$$E_i = \{x \in E \mid f(x) = f_i\}. \tag{3.1}$$

The sets E_i are measurable and disjoint, and f can be written in its *canonical form*,

$$f = \sum_{i=1}^{n} f_i \chi_{E_i}. \tag{3.2}$$

Given measurable sets $E_1, E_2, \ldots, E_n$ and real numbers $f_1, f_2, \ldots, f_n$, the expression in (3.2) would give a simple function but not, in general, its canonical representation. This occurs only if the E_i are disjoint and the f_i are distinct.

The sum and the product of simple functions are simple functions.

If f and g are simple functions written in their canonical form (3.2), then $(f+g)$ and (fg) are still simple but not necessarily in their canonical form.

Proposition 3.1. *Let $f : E \to \mathbb{R}^*$ be a nonnegative measurable function. There exists a sequence of simple functions $\{f_n\}$ such that $f_n \leq f_{n+1}$ and*

$$f(x) = \lim f_n(x) \quad \text{for all } x \in E. \tag{3.3}$$

Proof. For a fixed $n \in \mathbb{N}$, set

$$f_n(x) = \begin{cases} n & \text{if } f(x) \geq n, \\ \dfrac{j}{2^n} & \text{if } \dfrac{j}{2^n} \leq f(x) < \dfrac{j+1}{2^n}, \\ & j = 0, 1, \ldots, n2^n - 1. \end{cases} \tag{3.4}$$

By construction, $f_n \leq f_{n+1}$. Since f is measurable, the sets

$$\left[f \geq \frac{j}{2^n} \right] - \left[f \geq \frac{j+1}{2^n} \right], \quad j = 0, 1, 2, \ldots, n2^n - 1,$$

and the set $[f \geq n]$ are measurable and disjoint. Thus the f_n are simple.

Fix some $x \in E$. If $f(x) \in \mathbb{R}$, by choosing some positive integer $n_o \geq f(x)$, the construction of f_n implies that

$$0 \leq f(x) - f_n(x) \leq \frac{1}{2^n} \quad \text{for all } n \geq n_o. \tag{3.5}$$

If $f(x) = \infty$, then $f_n(x) = n$ for all positive integers n. Thus (3.3) holds in either case. $\qquad \square$

A function f from a set E into $\mathbb{R}^*$ can be decomposed as

$$f = f^+ - f^-, \tag{3.6}$$

where

$$f^+ = \frac{1}{2}(|f| + f) \quad \text{and} \quad f^- = \frac{1}{2}(|f| - f). \tag{3.6}'$$

Corollary 3.2. *Let $f : E \to \mathbb{R}^*$ be a measurable function. There exists a sequence of simple functions $\{f_n\}$ such that $f(x) = \lim f_n(x)$ for all $x \in E$.*

4   Convergence in measure

Let $\{f_n\}$ be a sequence of measurable functions from a measurable set E of finite measure into $\mathbb{R}^*$, and let $f : E \to \mathbb{R}^*$ be measurable and a.e. finite in E. The sequence $\{f_n\}$ converges *in measure* to f if for any $\eta > 0$,[2]

$$\lim \mu\{x \in E \mid |f_n(x) - f(x)| > \eta\} = 0.$$

Proposition 4.1. *The convergence in measure identifies the limit uniquely up to a set of measure zero; i.e., if $\{f_n\}$ converges in measure to f and g, then $f = g$ a.e. in E.*

Proof. For $n \in \mathbb{N}$ and a.e. $x \in E$,

$$|f(x) - g(x)| \le |f(x) - f_n(x)| + |f_n(x) - g(x)|.$$

Therefore, for all $\varepsilon > 0$,

$$\{x \in E \mid |f(x) - g(x)| > \varepsilon\} \subset \left\{x \in E \mid |f(x) - f_n(x)| > \frac{1}{2}\varepsilon\right\}$$

$$\bigcup \left\{x \in E \mid |f_n(x) - g(x)| > \frac{1}{2}\varepsilon\right\}.$$

Taking the measure of both sides and letting $n \to \infty$ gives

$$\mu(\{x \in E \mid |f(x) - g(x)| > \varepsilon\}) = 0$$

for all $\varepsilon > 0$. $\qquad\square$

Proposition 4.2. *Let $\{X, \mathcal{A}, \mu\}$ be a complete measure space and let $E \in \mathcal{A}$ be of finite measure. If $\{f_n\}$ converges a.e. in E to a function $f : E \to \mathbb{R}^*$ that is finite a.e. in E, then $\{f_n\}$ converges to f in measure.*

Proof. Having fixed an arbitrary $\varepsilon > 0$, by the Egorov theorem, there exists a measurable set $E_\varepsilon \subset E$ such that $\mu(E - E_\varepsilon) \le \varepsilon$ and $\{f_n\}$ converges to f uniformly in E_ε. Therefore, for any $\eta > 0$,

$$\limsup \mu\{x \in E \mid |f_n(x) - f(x)| > \eta\} \le \varepsilon. \qquad\square$$

Remark 4.1. Proposition 4.2 is, in general, false if E is not of finite measure.

Remark 4.2. Convergence in measure does not imply a.e. convergence, as shown by the following example. For $m, n \in \mathbb{N}$, let

$$\varphi_{nm}(x) \equiv \begin{cases} 1 & \text{for } x \in \left[\dfrac{m-1}{n}, \dfrac{m}{n}\right], \\[2mm] 0 & \text{for } x \in [0, 1] - \left[\dfrac{m-1}{n}, \dfrac{m}{n}\right]. \end{cases} \qquad (4.1)$$

---

[2]Notion introduced by F. Riesz, Sur les systémes ortogonaux de fonctions, *C. R. Acad. Sci. Paris*, **144** (1907), 615–619 and F. Fischer, Sur la convergence en moyenne, *C. R. Acad. Sci. Paris*, **144** (1907), 1148–1150.

Then construct a sequence of functions $f_n : [0, 1] \to \mathbb{R}$ by setting

$$f_1 = \varphi_{11}, \quad f_2 = \varphi_{21}, \quad f_3 = \varphi_{22}, \quad f_4 = \varphi_{31}, \quad f_5 = \varphi_{32},$$
$$f_6 = \varphi_{33}, \quad f_7 = \varphi_{41}, \quad f_8 = \varphi_{42}, \quad f_9 = \varphi_{43}, \quad \dots$$

The sequence $\{f_n\}$ converges in measure to zero in $[0, 1]$. However, $\{f_n\}$ does not converge to zero anywhere in $[0, 1]$. Indeed, for any fixed $x \in [0, 1]$, there exist infinitely many indices $m, n \in \mathbb{N}$ such that

$$\frac{m-1}{n} \le x \le \frac{m}{n} \quad \text{and hence} \quad \varphi_{nm}(x) = 1.$$

Even though the sequence $\{f_n\}$ does not converge to zero anywhere in $[0, 1]$, it contains a subsequence $\{f_{n'}\} \subset \{f_n\}$ converging to zero a.e. in $[0, 1]$. For example, one might select $\{f_{n'}\} = \{\varphi_{n1}\}$.

Proposition 4.3 (F. Riesz[3]). *Let $\{X, \mathcal{A}, \mu\}$ be a complete measure space. Let $\{f_n\}$ be a sequence of measurable functions from a measurable set E of finite measure, into $\mathbb{R}^*$, and let $f : E \to \mathbb{R}^*$ be measurable and a.e. finite in E. If $\{f_n\}$ converges in measure to f, there exists a subsequence $\{f_{n'}\} \subset \{f_n\}$ converging to f a.e. in E.*

Proof. For $m, n \in \mathbb{N}$, by the triangle inequality,

$$|f_n(x) - f_m(x)| \le |f_n(x) - f(x)| + |f_m(x) - f(x)| \quad \text{for a.e. } x \in E.$$

Therefore, for all $\eta > 0$,

$$\mu([|f_n - f_m| > \eta]) \le \mu\left(\left[|f_n - f| > \frac{1}{2}\eta\right]\right) + \mu\left(\left[|f_m - f| > \frac{1}{2}\eta\right]\right).$$

From this and the definition of convergence in measure, it follows that

$$\lim_{n,m \to \infty} \mu\{x \in E \mid |f_n(x) - f_m(x)| > \eta\} = 0. \tag{4.2}$$

We will establish the proposition under assumption (4.2).

From (4.2), it follows that for every $j \in \mathbb{N}$, there exists a positive integer n_j such that

$$\mu\left\{x \in E \mid |f_n(x) - f_m(x)| > \frac{1}{2^j}\right\} \le \frac{1}{2^{j+1}} \quad \text{for all } m, n \ge n_j.$$

The numbers n_j may be chosen so that $n_j < n_{j+1}$. Setting

$$E_{n_j} = \left\{x \in E \mid |f_{n_j}(x) - f_{n_{j+1}}(x)| \le \frac{1}{2^j}\right\},$$

---

[3]F. Riesz, Sur les suites des fonctions mesurables, *C. R. Acad. Sci. Paris*, **148** (1909), 1303–1305.

it follows that

$$\mu(E - E_{n_j}) \le \frac{1}{2^{j+1}}.$$

This construction implies that for any fixed $m \in \mathbb{N}$,

$$\mu\left(E - \bigcap_{j=m}^{\infty} E_{n_j}\right) \le \frac{1}{2^m}.$$

The subsequence $\{f_{n_j}\}$ selected out of $\{f_n\}$ is convergent for all $x \in \bigcap_{j=m}^{\infty} E_{n_j}$. Indeed, for any such x and any pair of indices

$$m \le n_j < n_{j+\ell}, \quad \ell \in \mathbb{N},$$

it holds that

$$|f_{n_j}(x) - f_{n_{j+\ell}}(x)| \le \sum_{i=j}^{j+\ell-1} |f_{n_i}(x) - f_{n_{i+1}}(x)| \le \frac{1}{2^{j-1}}.$$

Since $m \in \mathbb{N}$ is arbitrary, $\{f_{n_j}\}$ converges a.e. in E. $\square$

The next proposition can be regarded as a Cauchy-type criterion for a sequence $\{f_n\}$ to converge in measure.

Proposition 4.4. *Let $\{X, \mathcal{A}, \mu\}$ be a complete measure space, and let $\{f_n\}$ be a sequence of measurable functions from a measurable set E of finite measure into $\mathbb{R}^*$. The sequence $\{f_n\}$ converges in measure if and only if (4.2) holds for all $\eta > 0$.*

Proof. The necessary condition has been established in the first part of the proof of Proposition 4.3, leading to (4.2). To prove its sufficiency, let (4.2) hold for all $\eta > 0$ and let $\{f_{n'}\}$ be a subsequence selected out of $\{f_n\}$ and convergent a.e. in E. The limit

$$f(x) = \lim f_{n'}(x) \quad \text{for a.e. } x \in E$$

defines a measurable function $f : E \to \mathbb{R}^*$, which is finite a.e. in E.

We regard f as being defined everywhere in E by setting it to be equal to zero on the set where such a limit does not exist. Having fixed positive numbers η and ε, by virtue of (4.2), there exists an index n_ε such that

$$\mu\left\{x \in E \mid |f_n(x) - f_m(x)| > \frac{1}{2}\eta\right\} \le \frac{1}{2}\varepsilon \quad \text{for all } n, m \ge n_\varepsilon.$$

Since $\{f_{n'}\} \to f$ a.e. in E, by the Egorov theorem, there exist a measurable set $E_\varepsilon \subset E$ and an index n'_ε such that $\mu(E - E_\varepsilon) \le \frac{1}{2}\varepsilon$ and

$$|f_{n'}(x) - f(x)| \le \frac{1}{2}\eta \quad \text{for all } x \in E_\varepsilon \quad \text{and for all } n' \ge n'_\varepsilon.$$

From the inequality

$$|f_n(x) - f(x)| \leq |f_n(x) - f_{n'}(x)| + |f_{n'}(x) - f(x)|,$$

it follows that for all indices $n \geq n_\varepsilon$ and $n' \geq \max\{n_\varepsilon; n'_\varepsilon\}$,

$$\{x \in E \mid |f_n(x) - f(x)| > \eta\} \subset \left\{x \in E \mid |f_n(x) - f_{n'}(x)| > \frac{1}{2}\eta\right\}$$
$$\bigcup \left\{x \in E \mid |f_{n'}(x) - f(x)| > \frac{1}{2}\eta\right\}.$$

From this,

$$\mu\{x \in E \mid |f_n(x) - f(x)| > \eta\} \leq \varepsilon \quad \text{for all } n \geq n_\varepsilon. \qquad \square$$

5   Quasi-continuous functions and Lusin's theorem

Let $\{X, \mathcal{A}, \mu\}$ be $\mathbb{R}^N$ endowed with the Lebesgue measure and let $E \subset \mathbb{R}^N$ be measurable. A measurable function $f : E \to \mathbb{R}^*$ is *quasi-continuous* if for every $\varepsilon > 0$, there exists a closed set $E_{c,\varepsilon} \subset E$ such that

$$\mu(E - E_{c,\varepsilon}) \leq \varepsilon \quad \text{and} \quad \text{the restriction of } f \text{ to } E_{c,\varepsilon} \text{ is continuous.} \qquad (5.1)$$

Proposition 5.1. *A simple function defined in a bounded measurable set $E \subset \mathbb{R}^N$ is quasi-continuous.*

Proof. Let $f : E \to \mathbb{R}$ be simple and let $\{f_1, f_2, \ldots, f_n\}$ be its range. Since the sets E_i, defined in (3.1) are measurable, having fixed $\varepsilon > 0$, there exist closed sets $E_{c,i} \subset E_i$ such that[4]

$$\mu(E_i - E_{c,i}) \leq \frac{\varepsilon}{n}, \quad i = 1, 2, \ldots, n.$$

Setting

$$E_{c,\varepsilon} = \bigcup_{i=1}^{n} E_{c,i}, \quad \text{it holds that } \mu(E - E_{c,\varepsilon}) \leq \varepsilon.$$

The sets $E_{c,i}$, being closed, bounded, and disjoint, are at positive mutual distance. Since f is constant on each of them, it is continuous in $E_{c,\varepsilon}$. $\qquad \square$

Theorem 5.2 (Lusin[5]). *Let E be a bounded measurable set in $\mathbb{R}^N$. A function $f : E \to \mathbb{R}$ is measurable if and only if it is quasi-continuous.*

---

[4] See Proposition 12.4 of Chapter II.
[5] N. Lusin, Sur les propriété des fonctions mesurables, *C. R. Acad. Sci. Paris*, **154** (1912), 1688–1690.

Proof (necessity). Assume first that $f \geq 0$. By Proposition 3.1, there exist a sequence of simple functions $\{f_n\}$ that converges to f pointwise in E. Since each of the f_n is quasi-continuous, having fixed $\varepsilon > 0$, there exist closed sets $E_{c,n} \subset E$ such that

$$\mu(E - E_{c,n}) \leq \frac{1}{2^{n+1}}\varepsilon,$$

and the restriction of f_n to $E_{c,n}$ is continuous. By the Egorov theorem, set in $\mathbb{R}^N$, there exists a closed set $E_{c,o} \subset E$ such that

$$\mu(E - E_{c,o}) \leq \frac{1}{2}\varepsilon \quad \text{and} \quad f_n \text{ converges uniformly to } f \text{ in } E_{c,o}.$$

The set

$$E_{c,\varepsilon} = \bigcap_{n=0}^{\infty} E_{c,n}$$

is closed, and

$$\mu(E - E_{c,\varepsilon}) = \mu\left(E - \bigcap_{n=0}^{\infty} E_{c,n}\right)$$

$$= \mu\left(\bigcup_{n=0}^{\infty}(E - E_{c,n})\right)$$

$$\leq \sum_{n=0}^{\infty} \mu(E - E_{c,n}) \leq \varepsilon.$$

Since the functions f_n are continuous in $E_{c,\varepsilon}$ and converge to f uniformly in $E_{c,\varepsilon}$, f also is continuous in E_ε. This establishes the necessary part of the theorem if f is nonnegative. If f is of variable sign, by the decomposition (3.6)–(3.6)′, it can be written as the difference of two quasi-continuous functions. Thus f is quasi-continuous. □

Proof (sufficiency). Let f be quasi-continuous in E. Having fixed $\varepsilon > 0$, let $E_{c,\varepsilon}$ be the closed set satisfying (5.1). To show that $[f \geq c]$ is measurable, write

$$[f \geq c] = \left([f \geq c] \bigcap E_{c,\varepsilon}\right) \bigcup \left([f \geq c] \bigcap (E - E_{c,\varepsilon})\right).$$

Since the restriction of f to $E_{c,\varepsilon}$ is continuous, $[f \geq c] \bigcap E_{c,\varepsilon}$ is closed. Moreover,

$$\mu_e\left([f \geq c] \bigcap (E - E_{c,\varepsilon})\right) \leq \mu(E - E_{c,\varepsilon}) \leq \varepsilon.$$

Therefore, $[f \geq c]$ is measurable by Proposition 12.4 of Chapter II. □

6   Integral of simple functions

Let $\{X, \mathcal{A}, \mu\}$ be a measure space, and let $E \in \mathcal{A}$. For a measurable set A and $\alpha \in \mathbb{R}$, define

$$\int_E \alpha \chi_A d\mu = \begin{cases} \alpha\mu\left(E \bigcap A\right) & \text{if } \alpha \neq 0, \\ 0 & \text{if } \alpha = 0. \end{cases}$$

Since $\mu(E \bigcap A) \in \mathbb{R}^*$, the first of these is well defined as an element of $\mathbb{R}^*$ for all $\alpha \in \mathbb{R} - \{0\}$.

Let $f : E \to \mathbb{R}$ be a nonnegative simple function, with canonical representation

$$f = \sum_{i=1}^{n} f_i \chi_{E_i}, \tag{6.1}$$

where $\{E_1, E_2, \ldots, E_n\}$ is a finite collection of mutually disjoint measurable sets exhausting E and $\{f_1, f_2, \ldots, f_n\}$ is a finite collection of mutually distinct, nonnegative numbers. The Lebesgue integral of a nonnegative simple function f is defined by

$$\int_E f(x)d\mu = \sum_{i=1}^{n} \int_E f_i \chi_{E_i} d\mu = \sum_{i=1}^{n} f_i \mu(E_i). \tag{6.2}$$

This could be finite or infinite. If it is finite, then f is said to be *integrable* in E.

Remark 6.1. If $f : E \to \mathbb{R}^*$ is nonnegative, simple, and integrable, the set $[f > 0]$ has finite measure.

Now let $\{E_1, E_2, \ldots, E_m\}$ be a finite collection of measurable disjoint sets exhausting E, and consider the nonnegative, simple function

$$f = \sum_{j=1}^{m} f_j \chi_{E_j}, \tag{6.1}'$$

where f_j are nonnegative numbers. This is not, in general, in canonical form. Since the sets E_j are mutually disjoint and exhaust E, this would occur if the numbers f_j are mutually distinct. We put it into its canonical form by setting

$$E_i = \bigcup\{E_j | f_j = f_i\}, \quad i = 1, 2, \ldots n, \tag{6.3}$$

and then by writing the representation (6.1) by means of the sets E_i.

From the definitions (6.2) and (6.3), it follows that

$$\int_E f d\mu = \sum_{i=1}^{n} f_i \sum_{\{j : f_j = f_i\}} \mu(E_j) = \sum_{j=1}^{m} f_j \mu(E_j).$$

Thus the integral of a nonnegative simple function is independent of the representation of f.

Let $f, g : E \to \mathbb{R}$ be nonnegative simple functions. If $f \geq g$ a.e. in E, then

$$\int_E f(x)d\mu \geq \int_E g(x)d\mu. \tag{6.4}$$

If both f and g are nonnegative, simple, and integrable,

$$\int_E (\alpha f + \beta g)d\mu = \alpha \int_E f d\mu + \beta \int_E g d\mu \tag{6.5}$$

for all $\alpha, \beta \in \mathbb{R}$.

7   The Lebesgue integral of nonnegative functions

Let $f : E \to \mathbb{R}^*$ be measurable and nonnegative, and let $\mathcal{S}_f$ denote the collection of all nonnegative simple functions $\zeta : E \to \mathbb{R}$ such that $\zeta \leq f$. Since $\zeta \equiv 0$ is one such function, the class $\mathcal{S}_f$ is nonempty.

The Lebesgue integral of f over E is defined by

$$\int_E f d\mu = \sup_{\zeta \in \mathcal{S}_f} \int_E \zeta d\mu. \tag{7.1}$$

This could be finite or infinite. The elements $\zeta \in \mathcal{S}_f$ are *not* required to vanish outside a set of finite measure. For example, if f is a positive constant on a measurable set of infinite measure, its integral is well defined by (7.1) and is infinity.

The key new idea of this notion of integral is that the *range* of a nonnegative function f is partioned, as opposed to its *domain*, as in the notion of the Riemann integral.[6]

If $f : E \to \mathbb{R}^*$ is measurable and nonpositive, we define

$$\int_E f d\mu = - \int_E (-f)d\mu \quad (f \leq 0). \tag{7.1$_-$}$$

A nonnegative measurable function $f : E \to \mathbb{R}^*$ is said to be *integrable* if the number defined by (7.1) is finite. For example, if μ is the counting measure on $\mathbb{N}$, a nonnegative function $f : \mathbb{N} \to \mathbb{R}$ is integrable if and only if $\sum f(n) < \infty$.

Remark 7.1. If $f, g : E \to \mathbb{R}^*$ are measurable and nonnegative and $f \leq g$ a.e. in E, then $\mathcal{S}_f \subset \mathcal{S}_g$. Thus

$$\int_E f d\mu \leq \int_E g d\mu.$$

---

[6]H. Lebesgue, Sur une généralisation de l'intégrale définite, *C. R. Acad. Sci. Paris*, **132** (1901), 1025–1028. See also Section 7.7 of the Problems and Complements.

A measurable function $f : E \to \mathbb{R}^*$ is said to be integrable if $|f|$ is integrable. From the decomposition (3.6)–(3.6)', it follows that $f^\pm \le |f|$. Therefore, if f is integrable, $f^\pm$ also are integrable. If f is integrable, its integral is defined by

$$\int_E f d\mu = \int_E f^+ d\mu - \int_E f^- d\mu. \qquad (7.2)$$

If $E' \subset E$ is measurable and $f : E \to \mathbb{R}^*$ is integrable, then $f \chi_{E'}$ also is integrable and

$$\int_{E'} f d\mu = \int_E f \chi_{E'} d\mu. \qquad (7.3)$$

The integral of a nonnegative function $f : E \to \mathbb{R}^*$ is always defined, finite or infinite, by (7.1). More generally, we set

$$\int_E f d\mu \equiv \begin{cases} +\infty & \text{if } \int_E f^+ d\mu = +\infty \text{ and } \int_E f^- dx < \infty, \\ -\infty & \text{if } \int_E f^+ d\mu < \infty \text{ and } \int_E f^- dx = \infty. \end{cases} \qquad (7.4)$$

8 Fatou's lemma and the monotone convergence theorem

Given a measure space $\{X, \mathcal{A}, \mu\}$ and a measurable set E, we let $\{f_n\}$ denote a sequence of measurable functions from E with values in $\mathbb{R}^*$.

Lemma 8.1 (Fatou[7]). *Let $\{f_n\}$ be a sequence of measurable and a.e. nonnegative functions in E. Then*

$$\int_E \liminf f_n d\mu \le \liminf \int_E f_n d\mu. \qquad (8.1)$$

Proof. Set $f = \liminf f_n$ and select a nonnegative simple function $\zeta \in \mathcal{S}_f$. Assume first that ζ is integrable so that it vanishes outside a set $F \subset E$ of finite measure. For all fixed $x \in F$ and $\varepsilon > 0$, there exists an index $n_\varepsilon(x)$ such that

$$f_n(x) \ge \zeta(x) - \varepsilon \quad \text{for all } n \ge n_\varepsilon(x).$$

By the version of Egorov's theorem as stated in Proposition 2.4, having fixed $\eta > 0$, there exists a set $F_\eta \subset F$ such that $\mu(F - F_\eta) \le \eta$ and this inequality holds uniformly in F_η; i.e., for every fixed $\varepsilon > 0$, there exists n_ε such that

$$f_n(x) \ge \zeta(x) - \varepsilon \quad \text{for all } n \ge n_\varepsilon \quad \text{and} \quad \text{for all } x \in F_\eta.$$

[7]P. J. Fatou, Séries trigonometriques et séries de Taylor, *Acta Math.*, **30** (1906), 335–400.

From this, for $n \geq n_\varepsilon$,

$$\int_E f_n d\mu \geq \int_{F_\eta} f_n d\mu \geq \int_{F_\eta} (\zeta - \varepsilon) d\mu$$

$$\geq \int_F \zeta d\mu - \int_{F - F_\eta} \zeta d\mu - \varepsilon\mu(F)$$

$$\geq \int_E \zeta d\mu - \eta \sup \zeta - \varepsilon\mu(F).$$

Since $\mu(F)$ is finite, this implies

$$\liminf \int_E f_n d\mu \geq \int_E \zeta d\mu \quad \text{for all integrable } \zeta \in \mathcal{S}_f. \tag{8.2}$$

If ζ is not integrable, it equals some positive number δ on a measurable set $F \subset E$ of infinite measure. Having fixed $\varepsilon \in (0, \delta)$, set

$$F_n = \{x \in E \mid f_j(x) \geq \delta - \varepsilon \text{ for all } j \geq n\}.$$

From the definition of lower limit $F_n \subset F_{n+1}$ and $F \subset \bigcup F_n$. Therefore,[8]

$$\liminf \mu(F_n) \geq \mu(\liminf F_n) = \infty.$$

From this,

$$\liminf \int_E f_n d\mu \geq \liminf \int_{F_n} f_n d\mu \geq (\delta - \varepsilon) \liminf \mu(F_n).$$

Thus in either case, (8.2) holds for all $\zeta \in \mathcal{S}_f$. □

In the conclusion (8.1) of Fatou's lemma, equality does not hold in general. For example, in $\mathbb{R}$ with the Lebesgue measure, the sequence

$$f_n(x) = \begin{cases} 1 & \text{for } x \in [n, (n+1)], \\ 0 & \text{otherwise} \end{cases}$$

satisfies (8.1) with strict inequality. This raises the issue of when (8.1) holds with equality or, equivalently, when one can pass to the limit under the integral.

Theorem 8.2 (monotone convergence). *Let $\{f_n\}$ be a monotone increasing sequence of measurable, nonnegative functions in E; i.e.,*

$$0 \leq f_n(x) \leq f_{n+1}(x) \quad \text{for all } x \in E \quad \text{and} \quad \text{for all } n \in \mathbb{N}.$$

Then

$$\lim \int_E f_n d\mu = \int_E \lim f_n d\mu.$$

---

[8] See (3.3) of Proposition 3.1 of Chapter II.

Remark 8.1. The integrals are meant in the sense of (7.1). In particular, both sides could be infinite.

Proof of Theorem 8.2. The sequence $\{f_n\}$ converges for all $x \in E$ to a measurable function $f : E \to \mathbb{R}^*$. Therefore, by Fatou's lemma,

$$\int_E f d\mu = \int_E \lim f_n d\mu \le \lim \int_E f_n d\mu \le \int_E f d\mu. \qquad \square$$

9 Basic properties of the Lebesgue integral

Proposition 9.1. *Let* $f, g : E \to \mathbb{R}$ *be integrable. Then for all* $\alpha, \beta \in \mathbb{R}$,

$$\int_E (\alpha f + \beta g) d\mu = \alpha \int_E f d\mu + \beta \int_E g d\mu. \qquad (9.1)$$

If $f \ge g$ *a.e. in* E, *then*

$$\int_E f d\mu \ge \int_E g d\mu. \qquad (9.2)$$

$$\left| \int_E f d\mu \right| \le \int_E |f| d\mu. \qquad (9.3)$$

If E' *is a measurable subset of* E, *then*

$$\int_E f d\mu = \int_{E-E'} f d\mu + \int_{E'} f d\mu. \qquad (9.4)$$

Proof. For $\alpha \ge 0$, denote by αS_f the collection of functions of the form $\alpha \zeta$, where $\zeta \in S_f$. If $\alpha \ge 0$ and $f \ge 0$, then $\alpha S_f = S_{\alpha f}$. Therefore,

$$\int_E \alpha f d\mu = \sup_{\eta \in S_{\alpha f}} \int_E \eta d\mu = \alpha \sup_{\zeta \in S_f} \int_E \zeta d\mu = \alpha \int_E f d\mu.$$

Similarly, if $\alpha < 0$, we use (7.1)_ and conclude that for every nonnegative measurable function $f : E \to \mathbb{R}^*$,

$$\int_E \alpha f d\mu = \alpha \int_E f d\mu. \qquad (9.5)$$

If $\alpha > 0$ and f is integrable and of variable sign, then (9.5) continues to hold in view of (7.2) and the decomposition

$$\alpha f = (\alpha f)^+ - (\alpha f)^-.$$

A similar argument applies if $\alpha < 0$, and we conclude that (9.5) holds true for every integrable function and every $\alpha \in \mathbb{R}$. Therefore, it suffices to prove (9.1) for $\alpha = \beta = 1$.

Assume first that both f and g are nonnegative. There exist monotone increasing sequences of simple functions $\{\zeta_n\}$ and $\{\xi_n\}$ converging pointwise in E to f and g, respectively. By the monotone convergence theorem,

$$\int_E (f+g)d\mu = \lim \int_E (\zeta_n + \xi_n)d\mu$$

$$= \lim \int_E \zeta_n d\mu + \lim \int_E \xi_n d\mu$$

$$= \int_E f d\mu + \int_E g d\mu.$$

Next, we assume that $f \geq 0$ and $g \leq 0$. First, we observe that $(f+g)$ is integrable since $|f+g| \leq |f| + |g|$. From the decomposition

$$(f+g)^+ - g = (f+g)^- + f$$

and (9.1) proven for the sum of two nonnegative functions,

$$\int_E (f+g)^+ d\mu + \int_E -g d\mu = \int_E (f+g)^- d\mu + \int_E f d\mu.$$

This and the definition (7.2) proves (9.1) for $f \geq 0$ and $g \leq 0$.

If f and g are integrable with no further sign restriction,

$$\int_E (f+g)d\mu = \int_E \{(f^+ + g^+) - (f^- + g^-)\}d\mu = \int_E f d\mu + \int_E g d\mu.$$

To prove (9.2), observe that from $(f-g) \geq 0$ and (9.1),

$$0 \leq \int_E (f-g)d\mu = \int_E f d\mu - \int_E g d\mu.$$

Inequality (9.3) follows from (9.2) and

$$-|f| \leq f \leq |f|.$$

Finally, (9.4) follows from (9.1) upon writing

$$f = f \chi_{E'} + f \chi_{E-E'}. \qquad \square$$

Corollary 9.2. *Let* $f : E \to \mathbb{R}^*$ *be integrable and let* E *be of finite measure. Then*

$$\mu(E) \inf_{x \in E} f(x) \leq \int_E f d\mu \leq \mu(E) \sup_{x \in E} f(x).$$

10   Convergence theorems

The properties of the Lebesgue integral permit one to formulate various versions of Fatou's lemma and of the monotone convergence theorem.

For example, the conclusion of Fatou's lemma continues to hold if the functions f_n are of variable sign, provided they are uniformly bounded below by some integrable function g.

Proposition 10.1. *Let $g : E \to \mathbb{R}^*$ be integrable and assume that $f_n \geq g$ a.e. in E for all $n \in \mathbb{N}$. Then*

$$\liminf \int_E f_n d\mu \geq \int_E \liminf f_n d\mu.$$

Proof. Since $(f_n - g) \geq 0$, the sequence $\{f_n - g\}$ satisfies the assumptions of Fatou's lemma. Thus

$$\liminf \int_E f_n d\mu \geq \int_E g d\mu + \int_E (\liminf f_n - g) d\mu. \qquad \square$$

Proposition 10.2. *Let $\{f_n\}$ be a sequence of nonnegative, measurable functions on E. Then*

$$\sum \int_E f_n d\mu = \int_E \sum f_n d\mu.$$

Proof. The sequence $\{\sum_{i=1}^{n} f_i\}$ is a monotone sequence of nonnegative, measurable functions. $\qquad \square$

Remark 10.1. It is not required that the f_n be integrable or that $\sum f_n$ be integrable. The integral of measurable, nonnegative functions, finite or infinite, is well defined by (7.1).

Theorem 10.3 (dominated convergence). *Let $\{f_n\}$ be a dominated and convergent sequence of integrable functions in E, i.e.,*

$$\lim f_n(x) = f(x) \quad \text{for all } x \in E,$$

and there exists an integrable function $g : E \to \mathbb{R}^*$ such that

$$|f_n| \leq g \quad \text{a.e. in } E \quad \text{for all } n \in \mathbb{N}.$$

Then the limit function $f : E \to \mathbb{R}^*$ is integrable and

$$\lim \int_E f_n d\mu = \int_E \lim f_n d\mu.$$

Proof. The limit function f is measurable, and by Fatou's lemma,

$$\int_E |f| d\mu \leq \lim \int_E |f_n| d\mu \leq \int_E g d\mu < \infty.$$

Thus f is integrable. Next,

$$(g - f_n) \geq 0 \quad \text{and} \quad (f_n + g) \geq 0 \quad \text{for all } n \in \mathbb{N}.$$

Therefore, by Fatou's lemma,

$$\int_E f d\mu \leq \liminf \int_E f_n \leq \limsup \int_E f_n d\mu \leq \int_E f d\mu. \qquad \square$$

11   Absolute continuity of the integral

Theorem 11.1 (Vitali[9]). *Let E be measurable, and let $f : E \to \mathbb{R}^*$ be integrable. For every $\varepsilon > 0$, there exists $\delta > 0$ such that for every measurable subset $\mathcal{E} \subset E$ of measure less than δ,*

$$\int_{\mathcal{E}} |f| d\mu < \varepsilon.$$

Proof. We may assume that $f \geq 0$. For $n = 1, 2, \ldots$, consider the functions

$$f_n = \begin{cases} f(x) & \text{if } f(x) < n, \\ n & \text{if } f(x) \geq n. \end{cases}$$

Since $\{f_n\}$ is increasing,

$$\int_E f_n d\mu \leq \int_E f d\mu \quad \text{and} \quad \lim \int_E f_n d\mu = \int_E f d\mu.$$

Having fixed $\varepsilon > 0$, there exists some index n_ε such that

$$\int_E f_{n_\varepsilon} d\mu > \int_E f d\mu - \frac{1}{2}\varepsilon.$$

Choose $\delta = \frac{\varepsilon}{2n_\varepsilon}$. Then for every measurable set $\mathcal{E} \subset E$ of measure less than δ,

$$\int_{\mathcal{E}} f d\mu \leq \int_{\mathcal{E}} f_{n_\varepsilon} d\mu + \int_{E \setminus \mathcal{E}} (f_{n_\varepsilon} - f) d\mu + \frac{1}{2}\varepsilon$$

$$\leq n_\varepsilon \mu(\mathcal{E}) + \frac{1}{2}\varepsilon < \varepsilon. \qquad \square$$

12   Product of measures

Let $\{X, \mathcal{A}, \mu\}$ and $\{Y, \mathcal{B}, \nu\}$ be two measure spaces. Any pair of sets $A \subset X$ and $B \subset Y$ generates a subset $A \times B$ of the Cartesian product $(X \times Y)$ called a generalized rectangle.

---

[9]G. Vitali, Sulle funzioni integrali, *Atti Rend. Accad. Sci. Torino*, **40** (1905), 1021–1034.

There are subsets of $(X \times Y)$ that are not rectangles. The intersection of any two rectangles is a rectangle by the formula

$$(A_1 \times B_1) \bigcap (A_2 \times B_2) = \left(A_1 \bigcap A_2 \right) \times \left(B_1 \bigcap B_2 \right).$$

The mutual complement of any two rectangles, while not a rectangle, can be written as the disjoint union of two rectangles by the decomposition

$$(A_2 \times B_2) - (A_1 \times B_1) = \{(A_2 - A_1) \times B_2\}$$
$$\bigcup \left\{ (A_1 \bigcap A_2) \times (B_2 - B_1) \right\}.$$

Thus the collection $\mathcal{R}$ of all rectangles is a semialgebra.[10]

If $A \in \mathcal{A}$ and $B \in \mathcal{B}$, the rectangle $A \times B$ is called a *measurable rectangle*. The collection of all measurable rectangles is denoted by $\mathcal{R}_o$. By the previous remarks, $\mathcal{R}_o$ is a semialgebra.

Since $X \times Y \in \mathcal{R}_o$, such a collection forms a sequential covering of $(X \times Y)$.[11] The semialgebra $\mathcal{R}_o$ can be endowed with a nonnegative set function by setting

$$\lambda(A \times B) = \mu(A)\nu(B) \tag{12.1}$$

for all measurable rectangles $A \times B$.

Proposition 12.1. *Let $\{A_n \times B_n\}$ be a countable collection of disjoint, measurable rectangles whose union is a measurable rectangle $A \times B$. Then*

$$\lambda(A \times B) = \sum \lambda(A_n \times B_n).$$

Proof. For each $x \in A$,

$$B = \bigcup \{B_j | (x, y) \in A_j \times B_j; \, y \in B\}.$$

Since, for each $x \in A$ fixed, this is a disjoint union,

$$\nu(B)\chi_A(x) = \sum \nu(B_j)\chi_{A_j}(x).$$

Integrating in $d\mu$ over A and using Proposition 10.2 now gives

$$\mu(A)\nu(B) = \sum \mu(A_n)\nu(B_n). \qquad \square$$

Thus λ is unambiguously defined since the measure of a measurable rectangle does not depend on its partitions into countably many pairwise-disjoint measurable rectangles.

Proposition 12.1 also implies that λ is a measure on the semialgebra $\mathcal{R}_o$. Therefore, λ can be extended to a complete measure $(\mu \times \nu)$ on $(X \times Y)$, which coincides with λ on $\mathcal{R}_o$.[12]

---

[10]See Section 9 of Chapter II.

[11]See Section 4 of Chapter II.

[12]See Theorem 11.1 of Chapter II.

Theorem 12.2. *Every pair $\{X, \mathcal{A}, \mu\}$ and $\{Y, \mathcal{B}, \nu\}$ of measure spaces generates a complete product measure space*

$$\{(X \times Y), (\mathcal{A} \times \mathcal{B}), (\mu \times \nu)\},$$

where $(\mathcal{A} \times \mathcal{B})$ is a σ-algebra containing $\mathcal{R}_o$ and $(\mu \times \nu)$ is a measure on $(\mathcal{A} \times \mathcal{B})$ that coincides with (12.1) *on measurable rectangles.*

13   On the structure of $(\mathcal{A} \times \mathcal{B})$

Denote by $(\mathcal{A} \times \mathcal{B})_o$ the smallest σ-algebra generated by the collection of all measurable rectangles. Also set

$$\mathcal{R}_\sigma = \{\text{countable unions of elements of } \mathcal{R}_o\},$$
$$\mathcal{R}_{\sigma\delta} = \{\text{countable intersections of elements of } \mathcal{R}_\sigma\}.$$

By construction,

$$\mathcal{R}_o \subset \mathcal{R}_\sigma \subset \mathcal{R}_{\sigma\delta} \subset (\mathcal{A} \times \mathcal{B})_o \subset (\mathcal{A} \times \mathcal{B}).$$

For each $E \subset (X \times Y)$, the two sets

$$Y \supset E_x = \{y | (x, y) \in E\} \quad \text{for a fixed } x \in X,$$
$$X \supset E_y = \{x | (x, y) \in E\} \quad \text{for a fixed } y \in Y$$

are, respectively, the X-section and the Y-section of E.

Proposition 13.1. *Let $E \in (\mathcal{A} \times \mathcal{B})_o$. Then for every $y \in Y$, the Y-section E_y is in $\mathcal{A}$, and for every $x \in X$, the X-section E_x is in $\mathcal{B}$.*

Proof. The collection $\mathcal{F}$ of all sets $E \in (\mathcal{A} \times \mathcal{B})$ such that $E_x \in \mathcal{B}$ for all $x \in X$ is a σ-algebra. Since $\mathcal{F}$ contains all the measurable rectangles, it must contain the smallest σ-algebra generated by the measurable rectangles. □

Remark 13.1. The converse is false as there exist nonmeasurable sets $E \subset (X \times Y)$ such that all the x and y sections are measurable. An example is in Section 13.5 of the Problems and Complements.

Remark 13.2. There exist $(\mu \times \nu)$-measurable rectangles $A \times B$ that are not measurable rectangles. To construct an example, let $A_o \subset X$ not be μ-measurable but be included into a measurable set of finite μ-measure. Also let $B_o \in \mathcal{B}$ be of zero ν-measure. The rectangle $A_o \times B_o$ is $(\mu \times \nu)$-measurable and has measure zero. For each $\varepsilon > 0$, there exists a measurable rectangle R_ε containing $A_o \times B_o$ and of measure less than ε. Therefore, $A_o \times B_o$ is $(\mu \times \nu)$-measurable by the criterion of measurability of Proposition 10.2 of Chapter II.

Remark 13.3. This example implies that Proposition 13.1 does not hold if $(\mathcal{A} \times \mathcal{B})_o$ is replaced by $(\mathcal{A} \times \mathcal{B})$. In particular, the inclusion $(\mathcal{A} \times \mathcal{B})_o \subset (\mathcal{A} \times \mathcal{B})$ is strict.

Proposition 13.2. *Let $E \in \mathcal{R}_{\sigma\delta}$ be of finite measure. Then the function $x \to \nu(E_x)$ is μ-measurable and the function $y \to \mu(E_y)$ is ν-measurable. Moreover,*

$$\int_{(X \times Y)} \chi_E d(\mu \times \nu) = \int_X \nu(E_x) d\mu = \int_Y \mu(E_y) d\nu.$$

Proof. The statement is obvious if E is a measurable rectangle. If $E \in \mathcal{R}_\sigma$, it can be decomposed into the countable union of disjoint measurable rectangles E_n. The functions

$$x \to \nu(E_x) = \sum \nu(E_{n,x})$$

and

$$y \to \mu(E_y) = \sum \mu(E_{n,y})$$

are measurable, and by monotone convergence,

$$\int_{(X \times Y)} \chi_E d(\mu \times \nu) = \sum \int_{(X \times Y)} \chi_{E_n} d(\mu \times \nu)$$

$$= \sum \int_X \nu(E_{n,x}) d\mu = \sum \int_Y \mu(E_{n,y}) d\nu$$

$$= \int_X \sum \nu(E_{n,x}) d\mu = \int_Y \sum \mu(E_{n,y}) d\nu$$

$$= \int_X \nu(E_x) d\mu = \int_Y \mu(E_y) d\nu.$$

If $E \in \mathcal{R}_{\sigma\delta}$, there exists a countable collection $\{E_n\}$ of elements of $\mathcal{R}_\sigma$ such that $E_{n+1} \subset E_n$ and $E = \bigcap E_n$. The functions

$$x \to \nu(E_x) = \lim \nu(E_{n,x})$$

and

$$y \to \mu(E_y) = \lim \mu(E_{n,y}),$$

are measurable. Since $(\mu \times \nu)(E) < \infty$, we may assume that

$$(\mu \times \nu)(E_1) < \infty.$$

Then by dominated convergence,

$$\int_{(X \times Y)} \chi_E d(\mu \times \nu) = \lim \int_{(X \times Y)} \chi_{E_n} d(\mu \times \nu)$$

$$= \lim \int_X \nu(E_{n,x}) d\mu = \lim \int_Y \mu(E_{n,y}) d\nu$$

$$= \int_X \lim \nu(E_{n,x}) d\mu = \int_Y \lim \mu(E_{n,y}) d\nu$$

$$= \int_X \nu(E_x) d\mu = \int_Y \mu(E_y) d\nu. \qquad \square$$

Proposition 13.3. *Assume that* $\{X, \mathcal{A}, \mu\}$ *and* $\{Y, \mathcal{B}, \nu\}$ *are complete measure spaces. Let* $E \in (\mathcal{A} \times \mathcal{B})$ *be of measure zero. Then*

$$E_x \text{ are } \nu\text{-measurable} \quad and \quad \nu(E_x) = 0, \quad \mu\text{--a.e. in } X,$$
$$E_y \text{ are } \mu\text{-measurable} \quad and \quad \mu(E_y) = 0, \quad \nu\text{--a.e. in } Y.$$

Proof. If $E \in \mathcal{R}_{\sigma\delta}$ the conclusion follows from the previous proposition. If E is $(\mu \times \nu)$-measurable, there exists a set $E_\delta \in \mathcal{R}_{\sigma\delta}$ such that[13]

$$E \subset E_\delta \quad and \quad (\mu \times \nu)(E_\delta - E) = 0.$$

Then

$$E_x \subset E_{\delta,x} \quad and \quad E_y \subset E_{\delta,y},$$

and the conclusion follows since μ and ν are complete measures. $\qquad\square$

Proposition 13.4. *Assume that* $\{X, \mathcal{A}, \mu\}$ *and* $\{Y, \mathcal{B}, \nu\}$ *are complete measure spaces and let* $E \in (\mathcal{A} \times \mathcal{B})$ *be of finite measure. Then*

$$E_x \text{ are } \nu\text{-measurable for } \mu\text{--a.e. } x \in X \quad and \quad x \to \nu(E_x) \text{ is integrable,}$$
$$E_y \text{ are } \mu\text{-measurable for } \nu\text{--a.e. } y \in Y \quad and \quad y \to \mu(E_y) \text{ is integrable.}$$

Moreover,

$$\int_{(X \times Y)} \chi_E d(\mu \times \nu) = \int_X \nu(E_x) d\mu = \int_Y \mu(E_y) d\nu.$$

Proof. If $E \in \mathcal{R}_{\sigma\delta}$, this is the content of Proposition 13.2. If E is measurable and $(\mu \times \nu)(E) < \infty$, there exists $E_\delta \in \mathcal{R}_{\sigma\delta}$ such that $E \subset E_\delta$ and the difference $E_\delta - E = \mathcal{E}$ has $(\mu \times \nu)$-measure zero. Therefore, by Proposition 13.3, the sets $E_x = E_{\delta,x} - \mathcal{E}_x$ are ν-measurable for μ-almost all $x \in X$, and

$$\nu(E_x) = \nu(E_{\delta,x}) \quad \text{for } \mu\text{-almost all } x \in X.$$

A similar statement holds for almost all E_y. Since E and $\mathcal{E}$ are disjoint,

$$\int_{(X \times Y)} \chi_E d(\mu \times \nu) = \int_{(X \times Y)} \chi_{E_\delta} d(\mu \times \nu)$$
$$= \int_X \nu(E_{\delta,x}) d\mu = \int_Y \mu(E_{\delta,y}) d\nu$$
$$= \int_X \nu(E_x) d\mu = \int_Y \mu(E_y) d\nu. \qquad\square$$

---

[13] See Proposition 10.3 of Chapter II.

14 The Fubini–Tonelli theorem

Theorem 14.1 (Fubini[14]). *Let $\{X, \mathcal{A}, \mu\}$ and $\{Y, \mathcal{B}, \nu\}$ be two complete measure spaces, and let*

$$(X \times Y) \ni (x, y) \longrightarrow f(x, y) \text{ be integrable in } (X \times Y).$$

Then

$$X \ni x \longrightarrow f(x, y) \text{ is } \mu\text{-integrable in } X \text{ for } \nu\text{-almost all } y \in Y,$$
$$Y \ni y \longrightarrow f(x, y) \text{ is } \nu\text{-integrable in } Y \text{ for } \mu\text{-almost all } x \in X.$$

Moreover,

$$X \ni x \longrightarrow \int_Y f(x, y) d\nu \text{ is } \mu\text{-integrable in } X,$$
$$Y \ni y \longrightarrow \int_X f(x, y) d\mu \text{ is } \nu\text{-integrable in } Y, \tag{14.1}$$

and

$$\int_{(X \times Y)} f(x, y) d(\mu \times \nu) = \int_X \left(\int_Y f(x, y) d\nu \right) d\mu$$
$$= \int_Y \left(\int_X f(x, y) d\mu \right) d\nu. \tag{14.2}$$

Proof. By the decomposition $f = f^+ - f^-$, we may assume that $f \geq 0$. By Proposition 13.4, the statement holds if f is the characteristic function of a measurable set E of finite measure. If f is nonnegative and integrable, there exists a sequence $\{f_n\}$ of nonnegative integrable simple functions such that $f_n \nearrow f$ a.e. in $(X \times Y)$. Since each of the f_n is integrable, it vanishes outside a set of finite measure. Therefore, Proposition 13.4 holds for each of such f_n. Then by monotone convergence,

$$\int_{(X \times Y)} f(x, y) d(\mu \times \nu) = \lim \int_{(X \times Y)} f_n(x, y) d(\mu \times \nu)$$
$$= \lim \int_X \left(\int_Y f_n(x, y) d\nu \right) d\mu$$
$$= \int_X \lim \left(\int_Y f_n(x, y) d\nu \right) d\mu$$
$$= \int_X \left(\int_Y f(x, y) d\nu \right) d\mu. \qquad \square$$

---

[14]G. Fubini, Sugli integrali multipli, *Rend. Accad. Lincei Roma*, **16** (1907), 608–614.

14.1   The Tonelli version of the Fubini theorem. The double-integral formula (14.2) requires that f be integrable in the product measure $(\mu \times \nu)$. Tonelli observed that if f is nonnegative, the integrability requirement can be relaxed, provided $(\mu \times \nu)$ is σ-finite.

Theorem 14.2 (Tonelli[15]). *Let $\{X, \mathcal{A}, \mu\}$ and $\{Y, \mathcal{B}, \nu\}$ be complete and σ-finite, and let $f : (X \times Y) \to \mathbb{R}^*$ be measurable and nonnegative. Then the measurability statements in* (14.1) *and the double-integral formula* (14.2) *hold. The integrals in* (14.2) *could be either finite or infinite.*

Proof. The integrability requirement in the Fubini theorem was used to insure the existence of a sequence $\{f_n\}$ of integrable functions each vanishing outside a set of finite measure and converging to f. The positivity of f and the σ-finiteness in the Tonelli theorem provide similar information.   □

If f is integrable in $(\mu \times \nu)$, then Fubini's theorem holds and equality occurs in (14.2). If f is not integrable, then the left-hand side of (14.2) is infinite. Tonelli's theorem asserts that in such a case, the right-hand side also is well defined and is infinity, provided $(\mu \times \nu)$ is σ-finite.

In particular, Tonelli's theorem could be used to establish whether a nonnegative, measurable function $f : (X \times Y) \to \mathbb{R}^*$ is integrable through the equality of the two right-hand sides of (4.2).

The requirement that $(\mu \times \nu)$ be σ-finite cannot be removed, as shown by the example in 14.5 of the Problems and Complements.

15   Some applications of the Fubini–Tonelli theorem

15.1   Integrals in terms of distribution functions. Let $f : E \to \mathbb{R}^*$ be measurable and nonnegative. The *distribution function* of f relative to E is defined as

$$\mathbb{R}^+ \ni t \longrightarrow \mu([f > t]). \tag{15.1}$$

This is a nonincreasing function of t, and if f is finite a.e. in E, then

$$\lim_{t \to \infty} \mu([f > t]) = 0 \quad \text{unless } \mu([f > t]) \equiv \infty. \tag{15.2}$$

If f is integrable, such a limit can be given a quantitative form. Indeed,

$$t\mu([f > t]) = \int_E t \chi_{[f > t]} d\mu \le \int_E f d\mu < \infty. \tag{15.3}$$

---

[15]L. Tonelli, Sull'integrazione per parti, *Atti Accad. Naz. Lincei* (5), **18**-2 (1909), 246–253.

Proposition 15.1. *Let $\{X, \mathcal{A}, \mu\}$ be σ-finite and let $f : E \to \mathbb{R}^*$ be measurable and nonnegative. Also let ν be a σ-finite measure on $\mathbb{R}^+$ such that $\nu([0, t)) = \nu([0, t])$. Then*

$$\int_E \nu([0, f])d\mu = \int_0^\infty \mu([f > t])d\nu. \tag{15.4}$$

In particular, if $\nu([0, t]) = t^p$ for some $p > 0$, then

$$\int_E f^p d\mu = p\int_0^\infty t^{p-1}\mu([f > t])dt, \tag{15.4$_p$}$$

where dt is the Lebesgue measure on $\mathbb{R}^+$.

Proof. The function $f : E \to \mathbb{R}^*$, when regarded as a function from $E \times \mathbb{R}^+$ into $\mathbb{R}^*$, is measurable in the product measure $(\mu \times \nu)$. Likewise, the function $g(t) = t$ from $\mathbb{R}^+$ into $\mathbb{R}^*$, when regarded as a function from $E \times \mathbb{R}$ into $\mathbb{R}^*$, is measurable in the product measure $(\mu \times \nu)$. Therefore, the difference $f - t$ is measurable in the product measure $(\mu \times \nu)$. This implies that the set

$$[f - t > 0] = [f > t]$$

is measurable in the product measure $(\mu \times \nu)$. Therefore, by the Tonelli theorem,

$$\int_0^\infty \mu([f > t])d\nu = \int_0^\infty \left(\int_E \chi_{[f>t]}d\mu\right)d\nu$$

$$= \int_E \left(\int_0^\infty \chi_{[f>t]}d\nu\right)d\mu = \int_E \left(\int_0^f d\nu\right)d\mu$$

$$= \int_E \nu([0, f])d\mu. \qquad \square$$

Both sides of (15.4) could be infinity and the formula could be used to verify whether $\nu([0, f])$ is μ-integrable over E.

In the next two applications in Sections 15.2 and 15.3, the measure space $\{X, \mathcal{A}, \mu\}$ is $\mathbb{R}^N$ with the Lebesgue measure.

15.2 Convolution integrals. A measurable function f from $\mathbb{R}^N$ into $\mathbb{R}^*$, when regarded as a function defined on $\mathbb{R}^{2N}$, is measurable. Indeed, for every $c \in \mathbb{R}$, the set $[f > c] \times \mathbb{R}^N$ is a measurable rectangle in the product space. This implies that the function

$$\mathbb{R}^{2N} \ni (x, y) \longrightarrow f(x - y)$$

is also measurable with respect to the product measure. Indeed, the set

$$\{(x, y) \in \mathbb{R}^{2N} | f(x - y) > c\}$$

coincides with the measurable rectangle $[f > c] \times \mathbb{R}^N$ in the rotated coordinates $\xi = x - y$.

Given any two nonnegative measurable functions $f, g : \mathbb{R}^N \to \mathbb{R}^*$, their *convolution* is defined as

$$x \longrightarrow (f * g)(x) = \int_{\mathbb{R}^N} g(y) f(x - y) dy. \tag{15.5}$$

Since f and g are both nonnegative, the right-hand side, finite or infinite, is well defined for all $x \in \mathbb{R}^N$.

Proposition 15.2. *Let f and g be nonnegative and integrable in $\mathbb{R}^N$. Then $(f * g)(x)$ is finite for a.e. $x \in \mathbb{R}^N$, the function $(f * g)$ is integrable in $\mathbb{R}^N$ and*

$$\int_{\mathbb{R}^N} (f * g) dx \le \left(\int_{\mathbb{R}^N} f dx \right) \left(\int_{\mathbb{R}^N} g dx \right). \tag{15.6}$$

Proof. The function $(x, y) \to g(y) f(x - y)$ is nonnegative and measurable with respect to the product measure. Therefore, by the Tonelli theorem,

$$\iint_{\mathbb{R}^{2N}} g(y) f(x - y) dx dy = \int_{\mathbb{R}^N} \left(\int_{\mathbb{R}^N} g(y) f(x - y) dx \right) dy$$

$$= \left(\int_{\mathbb{R}^N} f dx \right) \left(\int_{\mathbb{R}^N} g dx \right). \qquad \square$$

The convolution of any two integrable functions f and g is defined as in (15.5). Since

$$|g(y) f(x - y)| \le |g(y)| |f(x - y)|,$$

the convolution $(f * g)$ is well defined as an integrable function over $\mathbb{R}^{2N}$.

15.3 The Marcinkiewicz integral. Let E be a nonvoid set in $\mathbb{R}^N$ and let $\delta(x)$ denote the distance from x to E; i.e.,

$$\delta_E(x) = \inf_{z \in E} |x - z|.$$

By definition, $\delta_E(x) = 0$ for all $x \in E$.

Lemma 15.3. *Let E be a nonempty set in $\mathbb{R}^N$. Then the distance function $x \to \delta(x)$ is Lipschitz continuous with Lipschitz constant 1.*

Proof. Fix x and y in $\mathbb{R}^N$ and assume that $\delta_E(x) \ge \delta_E(y)$. From the definition of $\delta_E(y)$, having fixed $\varepsilon > 0$, there exists $z' \in E$ such that

$$\delta_E(y) \ge |y - z'| - \varepsilon.$$

Then estimate

$$0 \le \delta_E(x) - \delta_E(y) \le \inf_{z \in E} |x - z| - |y - z'| + \varepsilon$$
$$\le |x - z'| - |y - z'| + \varepsilon \le |x - y| + \varepsilon. \tag{15.7}$$

The conclusion follows since $\varepsilon > 0$ is arbitrary. $\square$

Let E be a bounded, closed set in $\mathbb{R}^N$. Fix a positive number λ and a cube Q containing E. The Marcinkiewicz integral relative to E and λ is the function

$$\mathbb{R}^N \ni x \longrightarrow M_{E,\lambda}(x) = \int_Q \frac{\delta_E^\lambda(y)}{|x-y|^{N+\lambda}} dy. \tag{15.8}$$

The right-hand side is well defined as the integral of a measurable, nonnegative function.

Proposition 15.4 (Marcinkiewicz[16]). *The Marcinkiewicz integral $M_{E,\lambda}(x)$ is finite for a.e. $x \in E$. Moreover, the function $x \to M_{E,\lambda}(x)$ is integrable in E and*

$$\int_E M_{E,\lambda}dx \leq \frac{\omega_N}{\lambda}\mu(Q-E), \tag{15.9}$$

where ω_N is the measure of the unit sphere in $\mathbb{R}^N$.

Proof. Since $\delta_E(y) = 0$ for all $y \in E$, by the Tonelli theorem,

$$\begin{aligned}
\int_E M_{E,\lambda}(x)dx &= \int_Q \delta_E^\lambda(y) \left(\int_E \frac{dx}{|x-y|^{N+\lambda}} \right) dy \\
&= \int_{Q-E} \delta_E^\lambda(y) \left(\int_E \frac{dx}{|x-y|^{N+\lambda}} \right) dy.
\end{aligned} \tag{15.10}$$

For each fixed $y \in (Q-E)$ and $x \in E$, since E is closed, we estimate

$$|x-y| \geq \delta_E(y) > 0.$$

Therefore, for each fixed $y \in (Q-E)$,

$$\begin{aligned}
\int_E \frac{dx}{|x-y|^{N+\lambda}} &\leq \int_{|x-y|\geq\delta_E(y)} \frac{d(x-y)}{|x-y|^{N+\lambda}} \\
&\leq \omega_N \int_{\delta_E(y)}^\infty \frac{ds}{s^{1+\lambda}} = \frac{\omega_N}{\lambda\delta_E^\lambda(y)}.
\end{aligned}$$

Using this estimate in (15.10) yields

$$\int_E M_{E,\lambda}(x)dx \leq \frac{\omega_N}{\lambda} \int_{Q-E} dy. \qquad \square$$

16 Signed measures and the Hahn decomposition

Let μ_1 and μ_2 be two measures defined on the same σ-algebra $\mathcal{A}$. If one of them is finite, the set function

$$\mathcal{A} \ni E \longrightarrow \mu(E) = \mu_1(E) - \mu_2(E) \tag{16.1}$$

---

[16]J. Marcinkiewicz, Sur les series de Fourier, *Fund. Math.*, **27** (1936), 38–69; Sur quelques integrales du type de Dini, *Ann. Soc. Polon. Math.* **17** (1938), 42–50.

is well defined and countably additive on $\mathcal{A}$. However, since it is not necessarily nonnegative, it is called a *signed measure*.

Signed measures are also generated by an integrable function f on a measure space $\{X, \mathcal{A}, \mu\}$ by the formula

$$\mathcal{A} \ni E \longrightarrow \int_E f d\mu = \int_E f^+ d\mu - \int_E f^- d\mu. \tag{16.2}$$

More generally, a *signed measure* on X is a set function μ satisfying the following:

(i) The domain of μ is a σ-algebra $\mathcal{A}$ of subsets of X.

(ii) $\mu(\emptyset) = 0$.

(iii) μ takes at most one of the values $\pm\infty$.

(iv) μ is countably additive; i.e., if $\{E_n\}$ is a countable collection of disjoint elements in $\mathcal{A}$, then

$$\mu\left(\bigcup E_n\right) = \sum \mu(E_n).$$

This is meant in the sense that if the left-hand side is finite, the series on the right-hand side is convergent, and if the left-hand side is $\pm\infty$, then the series on the right-hand side diverges accordingly.

Any linear real combination of measures defined on the same σ-algebra is a signed measure, provided all but one are finite.

Let $\{X, \mathcal{A}, \mu\}$ be a measure space for a signed measure μ. A measurable set $E \subset \mathcal{A}$ is said to be *positive* (*negative*) if $\mu(A) \geq (\leq)0$ for all measurable subsets $A \subset E$. The difference and the union of two positive (negative) sets is positive (negative). Since μ is countably additive, the countable, disjoint union of positive (negative) sets is positive (negative). From this, it follows that any countable union of positive (negative) sets is positive (negative).

Lemma 16.1. *Let $\{X, \mathcal{A}, \mu\}$ be a measure space for a signed measure μ. Let $E \subset X$ be measurable and such that $|\mu(E)| < \infty$. Then every measurable set $A \subset E$ satisfies $|\mu(A)| < \infty$.*

Proof. Assume, for example, that μ does not take the value $+\infty$. Let $A \subset E$. If $\mu(A) > 0$, then $\mu(A) < \infty$. If $\mu(A) < 0$, taking the measure of the disjoint union $E = (E - A) \bigcup A$ gives

$$\mu(E) = \mu(E - A) + \mu(A),$$

which, in turn, implies

$$0 < -\mu(A) = \mu(E - A) - \mu(E) < \infty. \qquad \square$$

Proposition 16.2. *Let $\{X, \mathcal{A}, \mu\}$ be a measure space for a signed measure μ. Every measurable set E of positive, finite measure contains a positive subset A of positive measure.*

Proof. If E is positive we take $A = E$. Otherwise, E contains a set of negative measure. Let n_1 be the smallest positive integer for which there exists $B_1 \subset E$ such that

$$\mu(B_1) \leq -\frac{1}{n_1}.$$

If $A_1 = (E - B_1)$ is positive, then we take $A = A_1$. Otherwise, there exists a subset $B_2 \subset (E - B_1)$ of negative measure. We then let n_2 be the smallest positive integer such that

$$\mu(B_2) \leq -\frac{1}{n_2}.$$

Proceeding in this fashion, if for some finite m the set

$$A_m = E - \bigcup_{j=1}^{m} B_j$$

is positive, the process terminates. Otherwise, the indicated procedure generates the sequences of sets $\{B_j\}$ and $\{A_m\}$. We establish that the set

$$A = \bigcap A_m = E - \bigcup B_j$$

is positive by showing that every measurable subset $C \subset A$ has nonnegative measure.

Since $A \subset E$ and E is of finite measure, by Lemma 16.1 $|\mu(A)| < \infty$. By construction, the sets B_j and A are measurable and disjoint. Since μ is countably additive,

$$0 < \mu(E) = \mu(A) + \sum \mu(B_j) \leq \mu(A) - \sum \frac{1}{n_j}.$$

This implies that the series $\sum n_j^{-1}$ is convergent, and therefore $n_j \to \infty$ as $j \to \infty$. It also implies that $\mu(A) > 0$.

Let C be a measurable subset of A. Since A belongs to all A_j, by construction,

$$\mu(C) \geq -\frac{1}{n_j - 1} \longrightarrow 0 \quad \text{as } j \to \infty. \qquad \square$$

Theorem 16.3 (Hahn decomposition[17]). *Let $\{X, \mathcal{A}, \mu\}$ be a measure space for a signed measure μ. Then X can be decomposed into a positive set X^+ and a negative set X^-.*

---

[17]See [21, Volume 1].

Proof. Assume, for example, that μ does not take the value $+\infty$, and set

$$M = \{\sup \mu(A), \text{ where } A \in \mathcal{A} \text{ is positive}\}.$$

Let $\{A_n\}$ be a sequence of positive sets such that $\mu(A_n)$ increases to M, and set

$$A = \bigcup A_n.$$

The set A is positive and, by construction $\mu(A) \leq M$. On the other hand, for all n,

$$A = (A - A_n) \bigcup A_n.$$

Since this is a disjoint union and A is positive,

$$\mu(A) = \mu(A - A_n) + \mu(A_n) \geq \mu(A_n) \quad \text{for all } n.$$

Thus $\mu(A) = M$ and $M < \infty$.

The complement $(X - A)$ is a negative set, for otherwise it would contain a set E of positive measure, which in turn would contain a positive set A_o. Then A and A_o are disjoint and $A \bigcup A_o$ is a positive set. Therefore,

$$\mu\left(A \bigcup A_o\right) = \mu(A) + \mu(A_o) > M,$$

contradicting the definition of M.

The Hahn decomposition is realized by taking

$$X^+ = A \quad \text{and} \quad X^- = X - X^+. \qquad \square$$

A set $E \in \mathcal{A}$ is a null set if every subset of E is measurable and has measure zero. There exist measurable sets of zero measure that are not null sets.

By removing out of X^+ a null set $\mathcal{E}$ and adding it to X^-, the set $(X^+ - \mathcal{E})$ remains a positive set and $(X^- \bigcup \mathcal{E})$ remains negative. Moreover,

$$X = (X^+ - \mathcal{E}) \bigcup \left(X^- \bigcup \mathcal{E}\right).$$

Thus the Hahn decomposition is not unique. However, it is unique up to null sets.

17 The Radon–Nikodým theorem

Let μ and ν be two measures defined on the same σ-algebra $\mathcal{A}$. The measure ν is *absolutely continuous* with respect to μ if $\mu(E) = 0$ implies $\nu(E) = 0$.

Let $\{X, \mathcal{A}, \mu\}$ be a measure space and let $f : X \to \mathbb{R}^*$ be measurable and nonnegative. The set function

$$\mathcal{A} \ni E \longrightarrow \nu(E) = \int_E f \, d\mu \qquad (17.1)$$

is a measure defined on $\mathcal{A}$ and absolutely continuous with respect to μ.

Theorem 17.1 (Radon–Nikodým[18]). *Let* $\{X, \mathcal{A}, \mu\}$ *and* $\{X, \mathcal{A}, \nu\}$ *be two* σ*-finite measure spaces on the same* σ*-algebra* $\mathcal{A}$, *and let* ν *be absolutely continuous with respect to* μ. *There exists a nonnegative* μ*-measurable function* $f : X \to \mathbb{R}^*$ *such that* ν *has the representation* (17.1). *Such a* f *is unique up to a set of* μ*-measure zero.*

Remark 17.1. The function f that appears in representation (17.1) is called the Radon–Nikodým derivative of ν with respect to μ since formally $d\nu = f d\mu$.

Remark 17.2. It is not asserted that f is μ-integrable. This would occur if and only if ν is finite.

Remark 17.3. The assumption that both μ and ν be σ-finite cannot be removed, as shown by counterexamples in Sections 17.1 and 17.2 of the Problems and Complements.

Proof of Theorem 17.1. Assume first that both μ and ν are finite. Let Φ be the family of all measurable nonnegative functions $\varphi : X \to \mathbb{R}^*$ such that

$$\int_E \varphi d\mu \le \nu(E) \quad \text{for all } E \in \mathcal{A}.$$

Since $0 \in \Phi$, such a class is nonempty. For two given functions φ_1 and φ_2 in Φ, the function $\max\{\varphi_1; \varphi_2\}$ is in Φ. Indeed, for any $E \in \mathcal{A}$,

$$\int_E \max\{\varphi_1; \varphi_2\} d\mu = \int_{E \bigcap [\varphi_1 \ge \varphi_2]} \varphi_1 d\mu + \int_{E \bigcap [\varphi_2 > \varphi_1]} \varphi_2 d\mu$$
$$\le \nu\left(E \bigcap [\varphi_1 \ge \varphi_2]\right) + \nu\left(E \bigcap [\varphi_2 > \varphi_1]\right) = \nu(E).$$

Since ν is finite, the number

$$M = \sup_{\varphi \in \Phi} \int_X \varphi d\mu \le \nu(X) < \infty$$

is finite. Let $\{\varphi_n\}$ be a sequence of functions in Φ such that

$$\lim \int_X \varphi_n d\mu = M,$$

and set

$$f_n = \max\{\varphi_1, \varphi_2, \dots, \varphi_n\}.$$

---

[18] J. Radon, Theorie und Anwendungen der absolut additiven Mengenfunktionen, *Sitzungsber. Akad. Wiss. Wien,* **122** (1913), 1295–1438; O. M. Nikodým, Sur les functions d'ensembles, in *Comptes Rendues du 1ére Congrès de Mathématiques des Pays Slaves,* Warsaw, 1929, 304–313; O. M. Nikodým, Sur une généralisation des intégrales de M. J. Radon, *Fund. Math.,* **15** (1930), 131–179. Although commonly referred to as the Radon–Nikodým theorem, the first version of this theorem, in the context of a measure in $\mathbb{R}^N$ absolutely continuous with respect to the Lebesgue measure, is due to H. Lebesgue, Sur l'intégration des fonctions discontinues, *Ann. Sci. École Norm. Sup.,* **27** (1910), 361–450. Radon extended it to Radon measures and Nikodým to general measures.

The sequence $\{f_n\}$ is nondecreasing, and μ is a.e. convergent to a function f that belongs to Φ. Indeed, by the monotone convergence theorem, for every measurable set E,

$$\int_E f d\mu = \lim \int_E f_n d\mu \leq v(E).$$

Such a limiting function is the f claimed by the theorem. For this, it suffices to establish that the measure

$$\mathcal{A} \ni E \longrightarrow \eta(E) = v(E) - \int_E f d\mu$$

is identically zero. If not, there exists a set $A \in \mathcal{A}$ such that $\eta(A) > 0$.

Since both v and η are absolutely continuous with respect to μ, for such a set $\mu(A) > 0$. Also, since μ is finite, there exists $\varepsilon > 0$ such that

$$\xi(A) = \eta(A) - \varepsilon\mu(A) > 0.$$

The set function

$$\mathcal{A} \ni E \to \xi(E) = \eta(E) - \varepsilon\mu(E)$$

is a signed measure on $\mathcal{A}$. Therefore, by Proposition 16.2, the set A contains a positive subset A_o of positive measure. In particular,

$$\eta\left(E \bigcap A_o\right) - \varepsilon\mu\left(E \bigcap A_o\right) \geq 0 \quad \text{for all } E \in \mathcal{A}.$$

From this and the definition of η,

$$\varepsilon\mu\left(E \bigcap A_o\right) \leq v\left(E \bigcap A_o\right) - \int_{E \bigcap A_o} f d\mu \quad \text{for all } E \in \mathcal{A}.$$

The function $(f + \varepsilon\chi_{A_o})$ belongs to Φ. Indeed, for every measurable set E,

$$\int_E (f + \varepsilon\chi_{A_o})dx = \int_{E-A_o} f d\mu + \int_{E \cap A_o} (f + \varepsilon)d\mu$$

$$\leq v(E - A_o) + v\left(E \bigcap A_o\right) = v(E).$$

This, however, contradicts the definition of M since

$$\int_X (f + \chi_{A_o})dx = M + \mu(A_o) > M.$$

If $g : X \to \mathbb{R}^*$ is another nonnegative measurable function by which the measure v can be represented, let

$$A_n = \left\{x \in X | f(x) - g(x) \geq \frac{1}{n}\right\}.$$

Then for all positive integers n,

$$\frac{1}{n}\mu(A_n) \le \int_{A_n} (f - g)d\mu = \nu(A_n) - \nu(A_n) = 0.$$

Thus $f = g$, μ—a.e. in X.

Assume next that μ is σ-finite and ν is finite. Let E_n be a sequence of expanding sets such that

$$\mu(E_n) \le \mu(E_{n+1}) < \infty \quad \text{and} \quad X = \bigcup E_n,$$

and denote by μ_n the restriction of μ to E_n. Let f_n be the unique function claimed by the Radon–Nikodým theorem for the pair of finite measures, $\{\mu_n; \nu\}$. By construction,

$$f_{n+1}|_{E_n} = f_n \quad \text{for all } n \in \mathbb{N}.$$

The function f claimed by the theorem is

$$f(x) = \sup f_n(x).$$

Indeed, if $E \in \mathcal{A}$,

$$\nu(E) = \lim \nu \left(E \bigcap E_n \right) = \lim \int_E f_n d\mu = \int_E f d\mu.$$

The uniqueness of such a f is proved as in the case of μ finite. Finally, a similar argument establishes the theorem when ν also is σ-finite. $\quad\square$

18   Decomposing measures

Two measures μ and ν on the same space $\{X, \mathcal{A}\}$ are *mutually singular* if X can be decomposed into two measurable, disjoint sets X_μ and X_ν such that for every $E \in \mathcal{A}$,

$$\mu \left(E \bigcap X_\nu \right) = 0 \quad \text{and} \quad \nu \left(E \bigcap X_\mu \right) = 0.$$

An example of mutually singular measures is given by (16.2).

If μ and ν are mutually singular, we write $\nu \perp \mu$. If $\nu \ll \mu$ and $\nu \perp \mu$, then $\nu \equiv 0$.

18.1   The Jordan decomposition. Given a measure space $\{X, \mathcal{A}, \mu\}$ for a signed measure μ, let

$$X = X^+ \bigcup X^-$$

be the corresponding Hahn decomposition of X. For every $E \in \mathcal{A}$, set

$$\mu^+(E) = \mu \left(E \bigcap X^+ \right) \quad \text{and} \quad \mu^-(E) = -\mu \left(E \bigcap X^- \right).$$

The set functions $\mu^\pm$ are measures on $\mathcal{A}$, and

- at least one of them is finite;

- they are independent of the particular Hahn decomposition;

- by construction, they are mutually singular.

For every set $E \in \mathcal{A}$,

$$\mu(E) = \mu^+(E) - \mu^-(E). \tag{18.1}$$

Theorem 18.1 (Jordan[19]). *Let $\{X, \mathcal{A}, \mu\}$ be a measure space for a signed measure μ. There exists a unique pair (μ^+, μ^-) of mutually singular measures, one of which is finite, such that $\mu = \mu^+ - \mu^-$.*

Proof. Let $X = X^+ \bigcup X^-$ be the Hahn decomposition of X relative to μ and determined up to a null set. The existence of the two mutually singular measures μ^+ and μ^- decomposing μ follows from the remarks leading to (18.1). It remains to show that such a decomposition is unique.

If $\mu = \alpha - \beta$ is another such decomposition into mutually singular measures α and β, there exist sets X_α and X_β such that

$$X = X_\alpha \bigcup X_\beta, \qquad X_\alpha \bigcap X_\beta = \emptyset,$$

and for all $E \in \mathcal{A}$,

$$\alpha \left(E \bigcap X_\beta \right) = \beta \left(E \bigcap X_\alpha \right) = 0.$$

By the definition, X_α is a positive set and X_β is a negative set. Thus $X^+ = X_\alpha$ and $X^- = X_\beta$ up to null sets. It follows that for every $E \in \mathcal{A}$,

$$\mu^+(E) = \mu \left(E \bigcap X^+ \bigcap X_\alpha \right) = \alpha(E).$$

Analogously, $\mu^-(E) = \beta(E)$ for all $E \in \mathcal{A}$. $\qquad\qquad\square$

The two measures $\mu^\pm$ are the *upper* and *lower* variations of μ. The measure

$$|\mu| = \mu^+ + \mu^-$$

is the *total variation* of μ. Both measures $\mu^\pm$ are absolutely continuous with respect to $|\mu|$. Moreover, for every $E \in \mathcal{A}$,

$$-\mu^-(E) \le \mu(E) \le \mu^+(E) \quad \text{and} \quad |\mu(E)| \le |\mu|(E).$$

---

[19]See [26, Volume 1].

18.2   The Lebesgue decomposition. Two signed measures μ and ν on the same space $\{X, \mathcal{A}\}$ are mutually singular, and we write $\nu \perp \mu$ if the measures $|\mu|$ and $|\nu|$ are mutually singular. The signed measure ν is absolutely continuous with respect to μ, and we write $\nu \ll \mu$ if

$$|\nu(E)| = 0 \quad \text{whenever } |\mu|(E) = 0.$$

Theorem 18.2 (Lebesgue). *Let* $\{X, \mathcal{A}, \mu\}$ *be a σ-finite measure space for a signed measure μ. and let ν be a σ-finite signed measure defined on $\mathcal{A}$. There exists a unique pair* (ν_o, ν_1) *of σ-finite, signed measures defined on the same σ-algebra $\mathcal{A}$ such that*

$$\nu = \nu_o + \nu_1 \quad \text{and} \quad \nu_o \perp \mu, \quad \nu_1 \ll \mu.$$

Proof. If $\nu = \nu'_o + \nu'_1$ is another such decomposition, then

$$\nu_o - \nu'_o = \nu_1 - \nu'_1.$$

This implies that $(\nu_o - \nu'_o)$ is both singular and absolutely continuous with respect to μ and therefore identically zero. Analogously, $(\nu_1 - \nu'_1) \equiv 0$.

The notions of mutually singular signed measures and absolute continuity of a signed measure ν with respect to a signed measure μ are set in terms of the same notions for their total variations $|\nu|$ and $|\mu|$. Therefore, we may assume that μ is a measure.

If $\nu^\pm$ is the Jordan decomposition of ν, by treating ν^+ and ν^- separately, we may assume that also ν is a measure.

Assume first that both μ and ν are finite measures and set $\lambda = \mu + \nu$. Both μ and ν are absolutely continuous with respect to λ. Therefore, by the Radon–Nikodým theorem, there exist measurable, nonnegative functions f and g such that

$$\mu(E) = \int_E f\, d\lambda \quad \text{and} \quad \nu(E) = \int_E g\, d\lambda \quad \text{for all } E \in \mathcal{A}.$$

Define ν_o and ν_1 by

$$\nu_o(E) = \nu\left(E \bigcap [f = 0]\right), \qquad \nu_1(E) = \nu\left(E \bigcap [f > 0]\right).$$

By construction $\nu = \nu_o + \nu_1$. The measure ν_o is singular with respect to μ since

$$X = [f > 0] \bigcup [f = 0],$$

and for every measurable set E,

$$\nu_o\left(E \bigcap [f > 0]\right) = \mu\left(E \bigcap [f = 0]\right) = 0.$$

The measure ν_1 is absolutely continuous with respect to μ. Indeed, $\mu(E) = 0$ implies that $f = 0$, λ—a.e. on E. Therefore,

$$0 = \lambda\left(E \bigcap [f > 0]\right) \geq \nu\left(E \bigcap [f > 0]\right) = \nu_1(E).$$

If μ and ν are both σ-finite measures, there exists a countable collection $\{X_n\}$ of disjoint measurable sets whose union is X and such that $\mu(X_n)$ and $\nu(X_n)$ are both finite. Let μ_n and ν_n denote the restrictions of μ and ν to X_n and let

$$\nu_n = \nu_{o,n} + \nu_{1,n}, \quad n = 1, 2, \ldots,$$

be the corresponding Lebesgue decompositions of ν_n. The Lebesgue decomposition of ν is obtained by setting

$$\begin{aligned} \nu_o(E) &= \sum \nu_{o,n}\left(E \bigcap X_n\right), \\ \nu_1(E) &= \sum \nu_{1,n}\left(E \bigcap X_n\right) \end{aligned} \quad \text{for all } E \in \mathcal{A}. \qquad \square$$

18.3 A general version of the Radon–Nikodým theorem.

Theorem 18.3. *Let $\{X, \mathcal{A}, \mu\}$ be a σ-finite measure space and let ν be a σ-finite, signed measure on the same space $\{X, \mathcal{A}\}$. If $\nu \ll \mu$, there exists a measurable function $f : X \to \mathbb{R}^*$ such that*

$$\nu(E) = \int_E f \, d\mu \quad \text{for all } E \in \mathcal{A}. \qquad (18.2)$$

The function f need not be integrable; however, at least one of f^+ or f^- must be integrable, and the integral in (18.2) is well defined in the sense of (7.4). Precisely, if the signed measure ν does not take the value $+\infty$ ($-\infty$), then the upper (lower) variation ν^+ (ν^-) of ν is finite and f^+ (f^-) is integrable.

Such a function f is unique up to a set of μ-measure zero.

Proof. Determine the Hahn decomposition $X = X^+ \bigcup X^-$, up to a null set, and the corresponding Jordan decomposition $\nu = \nu^+ - \nu^-$.

The upper and lower variations $\nu^\pm$ are absolutely continuous with respect to μ. Applying the Radon–Nikodým theorem to the pairs $(\nu_\pm, \mu)$ determines nonnegative μ-measurable functions $f_\pm$ such that

$$\nu_\pm(E) = \nu\left(E \bigcap X^\pm\right) = \int_E f_\pm d\mu \quad \text{for all } E \in \mathcal{A}.$$

One verifies that $f_\pm$ vanish μ—a.e. in $X^\pm$ and that the function claimed by the theorem is $f = f_+ - f_-$. $\qquad \square$

PROBLEMS AND COMPLEMENTS

1 MEASURABLE FUNCTIONS

$\{X, \mathcal{A}, \mu\}$ IS A MEASURE SPACE AND $E \in \mathcal{A}$

1.1 The characteristic function of a set E is measurable if and only if E is measurable.

1.2 A function f is measurable if and only if its restriction to any measurable subset of its domain is measurable.

1.3 A function f is measurable if and only if $f^+ = \max\{f; 0\}$ and $f^- = \max\{-f; 0\}$ are both measurable.

1.4 Let $\{X, \mathcal{A}, \mu\}$ be complete. A function defined on a set of measure zero is measurable.

1.5 Let $\{X, \mathcal{A}, \mu\}$ not be complete. Then there exists a measurable set A of measure zero that contains a nonmeasurable set B. The two functions $f = \chi_A$ and $g = \chi_{A-B}$ differ on a set of outer measure zero. However, f is measurable and g is not.

1.6 The sum $f + g$ and product $f \cdot g$ of two measurable functions $f, g : E \to \mathbb{R}^*$ are measurable. (Both f and g are allowed to take values in $\mathbb{R}^*$.)

1.7 If f is measurable then $[f = c]$ is measurable for all c in the range of f. The converse, however, is false.

1.8 Let f be measurable. Then $|f|^{p-1} f$ also is measurable for all $p > 0$.

1.9 $|f|$ measurable does not imply that f is measurable. Likewise, f^2 measurable does not imply that f is measurable.

$\{\mathbb{R}^N, \mathcal{M}, \mu\}$ IS $\mathbb{R}^N$ WITH THE LEBESGUE MEASURE AND $E \in \mathcal{M}$

1.10 A continuous function from a closed set $E \subset \mathbb{R}^N$ into $\mathbb{R}$ is measurable.

1.11 A function $f : E \to \mathbb{R}^*$ is upper (lower) semicontinuous if for each $x \in E$,

$$\limsup_{y \to x} f(y) \leq f(x) \qquad (\liminf_{y \to x} f(y) \geq f(x)).$$

A function $f : E \to \mathbb{R}^*$ is upper (lower) semicontinuous if and only if

$$\{x \in E | f(x) < (>) c\} \text{ is open for all } c \in \mathbb{R}.$$

Upper-(lower-)semicontinuous functions are measurable.

1.12 A monotone function f in some interval $(a, b) \subset \mathbb{R}$ is measurable.

1.13 Let $f : \mathbb{R} \to \mathbb{R}$ be measurable and let $g : \mathbb{R} \to \mathbb{R}$ be continuous. The composition $g(f) : E \to \mathbb{R}$ is measurable. However, the composition $f(g) : E \to \mathbb{R}$ is, in general, not measurable.

1.14 Let $f : [0, 1] \to \mathbb{R}^*$ be measurable. Then $\chi_{\mathbb{Q} \cap [0,1]}(f)$ is measurable.

2 THE EGOROV THEOREM

$\{X, \mathcal{A}, \mu\}$ IS A MEASURE SPACE AND $E \in \mathcal{A}$ IS OF FINITE MEASURE

2.1 State and prove a version of the Egorov theorem in the case when $f_n \to \infty$ in a measurable subset of E of positive measure.

2.2 Let $\{f_n\}$ be a sequence of measurable functions from E into $\mathbb{R}^*$. Assume that for a.e. $x \in E$, the set $\{f_n(x)\}$ is bounded. For every $\varepsilon > 0$, there exist a measurable set $E_\varepsilon \subset E$ and a positive number k_ε such that

$$\mu(E - E_\varepsilon) \le \varepsilon \quad \text{and} \quad |f_n| \le k_\varepsilon \quad \text{on } E_\varepsilon \quad \text{for all } n \in \mathbb{N}.$$

Proposition 2.1c. *Let $\{f_n\}$ and f be measurable functions from E into $\mathbb{R}^*$. Assume that f is finite a.e. in E. Then $\{f_n\} \to f$ a.e. in E if and only if for every $\eta > 0$,*

$$\lim \mu \left(\bigcup_{j=n}^{\infty} \{x \in E \,\|\, |f_n(x) - f(x)| \ge \eta\} \right) = 0. \qquad (2.1c)$$

Hint. Denoting by A the set where $\{f_n\}$ is not convergent,

$$A = \bigcup_{m=1}^{\infty} \limsup \left\{ x \in E \,\|\, |f_n(x) - f(x)| \ge \frac{1}{m} \right\}.$$

Then $\mu(A) = 0$ if and only if (2.1c) holds.

4 CONVERGENCE IN MEASURE

$\{X, \mathcal{A}, \mu\}$ IS COMPLETE AND $E \in \mathcal{A}$ IS A SET OF FINITE MEASURE

4.1 Let $f : E \to \mathbb{R}^*$ be measurable and assume that $|f| > 0$ a.e. on E. For every $\varepsilon > 0$, there exist a measurable set $E_\varepsilon \subset E$ and a positive number δ_ε such that $\mu(E - E_\varepsilon) \le \varepsilon$ and $|f| > \delta_\varepsilon$ on E_ε.

4.2 Let $\{f_n\} : E \to \mathbb{R}^*$ be a sequence of measurable functions converging to f a.e. in E. Assume that $|f| > 0$ and $|f_n| > 0$ a.e. on E for all $n \in \mathbb{N}$. For every $\varepsilon > 0$, there exist a measurable set $E_\varepsilon \subset E$ and a positive number δ_ε such that

$$\mu(E - E_\varepsilon) \le \varepsilon \quad \text{and} \quad |f_n| > \delta_\varepsilon \quad \text{on } E_\varepsilon \quad \text{for all } n \in \mathbb{N}.$$

4.3 Let μ be the counting measure on the rationals of $[0, 1]$. Then convergence in measure is equivalent to uniform convergence.

4.4 Let $\{f_n\} : E \to \mathbb{R}^*$ be a sequence of measurable functions and assume that $f_n(x)$ is finite for a.e. $x \in E$. There exists a sequence of positive numbers $\{k_n\}$ such that $f_n k_n^{-1} \to 0$ a.e. in E.

4.5   Let $\{f_n\}, \{g_n\} : E \to \mathbb{R}^*$ be sequences of measurable functions converging in measure to f and g respectively and let $\alpha, \beta \in \mathbb{R}$. Then

$$\{\alpha f_n + \beta g_n\}, \{|f_n|\}, \{f_n g_n\} \quad \longrightarrow \quad \alpha f + \beta g, |f|, fg$$

in measure. Moreover, if $f \neq 0$ a.e. on E and $f_n \neq 0$ a.e. on E for all n, then $\frac{1}{f_n}$ converges to $\frac{1}{f}$ in measure. (*Hint*: Use Sections 4.1 and 4.2 of the Problems and Complements.)

7   THE LEBESGUE INTEGRAL OF NONNEGATIVE MEASURABLE FUNCTIONS

7.1   COMPARING THE LEBESGUE INTEGRAL WITH THE PEANO–JORDAN INTEGRAL. Let E be a bounded, Peano–Jordan-measurable set in $\mathbb{R}^N$. Denote by $\mathcal{P} = \{E_n\}$ a finite partition of E into pairwise-disjoint Peano–Jordan-measurable sets. For a bounded function $f : E \to \mathbb{R}$, set

$$h_n = \inf_{x \in E_n} f(x), \qquad \mathcal{F}_{\mathcal{P}}^- = \sum h_n \mu_{\mathcal{P}-\mathcal{J}}(E_n),$$

$$k_n = \sup_{x \in E_n} f(x), \qquad \mathcal{F}_{\mathcal{P}}^+ = \sum k_n \mu_{\mathcal{P}-\mathcal{J}}(E_n).$$

A bounded function $f : E \to \mathbb{R}$ is Peano–Jordan integrable if for every $\varepsilon > 0$, there exists a partition $\mathcal{P}_\varepsilon$ of E into Peano–Jordan-measurable sets E_n such that

$$\mathcal{F}_{\mathcal{P}}^+ - \mathcal{F}_{\mathcal{P}}^- \leq \varepsilon.$$

If the sets E_n making up the partition $\{E_n\}$ are Peano–Jordan measurable, they are also Lebesgue measurable. Therefore,

$$\sum h_n \mu_{\mathcal{P}-\mathcal{J}}(E_n) \leq \int_E f d\mu \leq \sum k_n \mu_{\mathcal{P}-\mathcal{J}}(E_n).$$

Thus if a bounded function f is Peano–Jordan integrable, it is also Lebesgue integrable. The converse is false. Indeed, the characteristic function of the rationals $\mathbf{Q}$ of $[0, 1]$ is Lebesgue integrable and not Peano–Jordan integrable.

This is not longer the case, however, if f is not bounded. Following Riemann's notion of improper integral, the function

$$f(x) = \frac{1}{x} \sin \frac{1}{x} \quad \text{for } x \in (0, 1]$$

is Riemann integrable in $(0, 1)$ but not Lebesgue integrable.

7.2   Let $f : E \to \mathbb{R}^*$ be measurable and nonnegative. If $f = \infty$ in a set $\mathcal{E} \subset E$ of measure zero,

$$\int_E f d\mu = \int_{E-\mathcal{E}} f d\mu.$$

7.3   Let $f : X \to \mathbb{R}^*$ be measurable and nonnegative. Then

$$\int_E f d\mu = 0 \quad \text{for all } E \in \mathcal{A} \quad \text{implies} \quad f = 0 \quad \text{a.e. in } X.$$

7.4   Let $f : E \to \mathbb{R}^*$ be integrable. If the integral of f over every measurable subset $A \subset E$ is nonnegative, then $f \geq 0$ a.e. on E.

7.5   Let $f : \mathbb{R} \to \mathbb{R}$ be Lebesgue integrable. Then for every $h \in \mathbb{R}$ and every interval $[a, b] \subset \mathbb{R}$,

$$\int_{[a,b]} f d\mu = \int_{[a+h,b+h]} f(x - h) d\mu.$$

7.6   Construct the Lebesgue integral of a nonnegative μ-measurable function $f : \mathbb{R}^N \to \mathbb{R}$ when μ is the Dirac delta-measure δ_x concentrated at some $x \in \mathbb{R}^N$.[20]

7.7   ON THE DEFINITION OF THE LEBESGUE INTEGRAL.   The original definition of Lebesgue was based only on the Lebesgue measure in $\mathbb{R}^N$ and was in two stages.

First, f was assumed to be measurable, nonnegative, and bounded. Then the integral of such an f was defined as in (7.1), where the supremum was taken over the class of simple functions $\zeta \leq f$ and vanishing outside a set of finite measure.

Second, the integral of a measurable, nonnegative function f, defined in E, was defined by first setting

$$\Phi_f = \left\{ \begin{array}{c} \text{the collection of all nonnegative, measurable, bounded} \\ \text{functions } \varphi \text{ defined in } E \text{ and such that } \varphi \leq f \text{ a.e. in } E \\ \text{and vanishing outside a set of finite measure} \end{array} \right\}$$

and then by defining

$$\int_E f d\mu = \sup_{\varphi \in \Phi_f} \int_E \varphi d\mu.$$

Such a definition, while adequate for the Lebesgue measure in $\mathbb{R}^N$, is not adequate for a general measure space $\{X, \mathcal{A}, \mu\}$. For example, let $\{X, \mathcal{A}, \mu\}$ be the measure space of Section 3.3 of Chapter II. Then

$$\infty = \int_X 1 d\mu = \sup_{\varphi \in \Phi_1} \int_X \varphi d\mu = 0.$$

10   CONVERGENCE THEOREMS

$\{X, \mathcal{A}, \mu\}$ is a measure space and $E \in \mathcal{A}$.

---

[20] See (3.5)–(3.5)′ of Section 3.2 of Chapter II.

10.1   Let $f : E \to \mathbb{R}^*$ be integrable and let $\{E_n\}$ be a countable collection of measurable, disjoint subsets of E such that $E = \bigcup E_n$. Then

$$\int_E f d\mu = \sum \int_{E_n} f d\mu. \tag{7.1c}$$

10.2   Let μ be the counting measure on the positive rationals $\{r_1, r_2, \dots\}$, and let

$$f_n(r_j) = \begin{cases} \dfrac{1}{j} & \text{if } j < n, \\ 0 & \text{if } j \geq n. \end{cases}$$

Then $\{f_n\}$ is a sequence of integrable functions, uniformly convergent to a nonintegrable function.

10.3   Let μ be a finite measure and let $\{f_n\}$ be a sequence of integrable functions converging uniformly in X. The limiting function f is integrable, and

$$\lim \int f_n d\mu = \int \lim f_n d\mu.$$

10.4   Assume $\{X, \mathcal{A}, \mu\}$ is complete and $E \in \mathcal{A}$ is of finite measure. Let $\{f_n\}$ be a sequence of measurable functions in E dominated a.e. by some integrable function g. If $f_n \to f$ in measure, then

$$\lim \int_E f_n d\mu = \int_E f d\mu.$$

10.5   Let f be integrable in E. For every $\varepsilon > 0$, there exists a simple function φ such that

$$\int_E |f - \varphi| d\mu \leq \varepsilon.$$

10.6   **ANOTHER VERSION OF DOMINATED CONVERGENCE.**

Theorem 10.1c.   *Let $\{f_n\}$ be a sequence of integrable functions in E converging a.e. in E to some f. Assume that there exists a sequence of integrable functions $\{g_n\}$ converging a.e. in E to an integrable function g and such that*

$$\lim \int_E g_n d\mu = \int_E g d\mu$$

and, moreover,

$$|f_n(x)| \leq g_n(x) \quad \text{for a.e. } x \in E \text{ and for all } n \in \mathbb{N}.$$

Then f is integrable and

$$\lim \int_E f_n d\mu = \int_E f d\mu.$$

In Sections 10.7–10.10, μ is the Lebesgue measure in $\mathbb{R}^N$.

10.7 For $n \in \mathbb{N}$, let

$$f_n(x) = \begin{cases} n & \text{if } x \in \left[0, \dfrac{1}{n}\right], \\ 0 & \text{otherwise.} \end{cases}$$

Then

$$\int_0^1 f_n dx = 1 \neq \int_0^1 \lim f_n dx = 0.$$

10.8 Let $f_n : \mathbb{R}^+ \to \mathbb{R}$ be defined by

$$f_n = \begin{cases} \dfrac{1}{n} & \text{for } 0 \le x \le n, \\ 0 & \text{for } x > n. \end{cases}$$

Then

$$\lim \int_{\mathbb{R}^+} f_n dx = 1 \neq \int_{\mathbb{R}^+} \lim f_n dx = 0.$$

10.9 Let $f : \mathbb{R} \to \mathbb{R}$ be Lebesgue measurable and locally bounded. Assume that f is Riemann integrable on $\mathbb{R}$. Then f is Lebesgue integrable on $\mathbb{R}$ and

$$\int_{\mathbb{R}} f d\mu = \lim \int_{-n}^{n} f d\mu.$$

10.10 Let $\{f_n\}$ be the sequence of nonnegative integrable functions defined on $\mathbb{R}^N$ by

$$f_n(x) = \begin{cases} n^N \exp\left\{\dfrac{-1}{1 - n^2|x|^2}\right\} & \text{if } |x| < \dfrac{1}{n}, \\ 0 & \text{if } |x| \ge \dfrac{1}{n}. \end{cases}$$

The sequence $\{f_n\}$ converges to zero a.e. in $\mathbb{R}^N$ and each f_n is integrable with uniformly bounded integral. However,

$$\lim \int_{\mathbb{R}^N} f_n dx \neq \int_{\mathbb{R}^N} \lim f_n dx.$$

10.11 Let $\{f_n\}$ be the sequence defined in Section 7.7 of the Problems and Complements of Chapter I. The assumptions of the dominated convergence theorem fail, and

$$\lim_{n \to \infty} \int_0^1 f_n(x) dx = \frac{1}{2} \neq \int_0^1 \lim_{n \to \infty} f_n(x) dx = 0.$$

13 ON THE STRUCTURE OF $(\mathcal{A} \times \mathcal{B})$

13.1 Let $\{X, \mathcal{A}, \mu\}$ and $\{Y, \mathcal{B}, \nu\}$ be complete measure spaces. If $A \subset X$ is not μ-measurable, $B \subset Y$ is ν-measurable, $\nu(B) > 0$, then $A \times B$ is not $(\mu \times \nu)$-measurable.

13.2 Let $\{X, \mathcal{A}, \mu\}$ be noncomplete and let $A \subset X$ not be μ-measurable but included in a μ-measurable set A' of measure zero. For every ν-measurable set $B \subset Y$, the rectangle $A \times B$ is $(\mu \times \nu)$-measurable. As a consequence, the assumption that both $\{X, \mathcal{A}, \mu\}$ and $\{Y, \mathcal{B}, \nu\}$ be complete cannot be removed from Proposition 13.4.

13.3 The Lebesgue measure in $\mathbb{R}^N$ coincides with the product of N copies of the Lebesgue measure in $\mathbb{R}$.

13.4 Let $E \subset [0, 1]$ be the Vitali nonmeasurable set. The diagonal set $\mathcal{E} = \{(x, x) | x \in E\}$ is measurable in $\mathbb{R}^2$ and has measure zero. The rectangle $E \times E$ is not Lebesgue measurable in $\mathbb{R}^2$.

13.5 Let $X = \mathbb{N} \bigcup \Omega$, where Ω is the first uncountable. Let $\mathcal{A}$ be the σ-algebra of the subsets of X that are either countable or their complement is countable. For $E \in \mathcal{A}$, let $\mu(E) = 0$ if E is countable and $\mu(E) = 1$ otherwise. The set $E = \{(x, y) \in X \times X | x < y\}$ is not $(\mu \times \mu)$-measurable.

If E were measurable, it would have finite measure since

$$E \subset (X \times X) \quad \text{and} \quad (\mu \times \mu)(E) \le \mu(X)\mu(X) = 1.$$

Therefore, it would have to satisfy Proposition 13.4. However,

$$\int_X \mu(E_y)d\nu = 0 \quad \text{and} \quad \int_Y \nu(E_x)d\mu = 1.$$

Note that all the x and y sections of E are measurable.

14 THE FUBINI–TONELLI THEOREM

14.1 Let ω_N be the measure of the unit sphere in $\mathbb{R}^N$. By the Fubini–Tonelli theorem,

$$\omega_{N+1} = 2\omega_N \int_0^{\frac{\pi}{2}} (\sin t)^{N-1} dt.$$

14.2 By the Fubini–Tonelli theorem,

$$\int_{\mathbb{R}^N} e^{-|x|^2} dx = \prod_{i=1}^N \int_{\mathbb{R}} e^{-x_i^2} dx_i = \pi^{\frac{N}{2}}.$$

14.3 Let $\{X, \mathcal{A}, \mu\}$ be $[0, 1]$ with the Lebesgue measure, and let $\{Y, \mathcal{B}, \nu\}$ be the rationals in $[0, 1]$ with the counting measure. The function $f(x, y) = x$ is integrable on $\{X, \mathcal{A}, \mu\}$ and not integrable on the product space.

14.4 The function $f(x, y) = (x^2 - y^2)(x^2 + y^2)^{-2}$ is not Lebesgue integrable on a neighborhood of the origin of $\mathbb{R}^2$.

14.5 Let $[0, 1]$ be equipped with the Lebesgue measure μ and the counting measure ν, both acting on the same σ-algebra of the Borel subsets of $[0, 1]$. The corresponding product measure space is not σ-finite since ν is not σ-finite. The diagonal set $E = \{x = y\}$ is $(\mu \times \nu)$-measurable, and

$$\nu(E_x) = 1 \quad \text{for all } x \in [0, 1] \quad \text{and} \quad \mu(E_y) = 0 \quad \text{for all } y \in [0, 1].$$

Therefore,

$$\int_{[0,1]} \nu(E_x)d\mu = 1 \quad \text{and} \quad \int_{[0,1]} \mu(E_y)d\nu = 0.$$

Moreover,

$$\iint_{[0,1] \times [0,1]} \chi_E d(\mu \times \nu) = \infty.$$

14.6 Let f and g be integrable functions on complete measure spaces $\{X, \mathcal{A}, \mu\}$ and $\{Y, \mathcal{B}, \nu\}$, respectively. Then $F(x, y) = f(x)g(y)$ is integrable on the product space $(X \times Y)$.

Each of the two functions f and g, when regarded as functions of two sets of variables, are measurable on the product space $(X \times Y)$. Therefore, the product fg is measurable on the product space.

To establish that fg is integrable, we may assume that both are nonnegative. If they are both bounded and vanish outside sets of finite measure, the conclusion follows from Fubini's theorem. Otherwise, they can be approximated by monotone increasing sequences of bounded functions on their respective measure spaces, each vanishing outside a set of finite measure.

15   SOME APPLICATIONS OF THE FUBINI–TONELLI THEOREM

15.1   INTEGRALS IN TERMS OF DISTRIBUTION FUNCTIONS.

Proposition 15.1c. *Assume that $\mu([f > t]) \neq \infty$. Then*

$$\lim_{\varepsilon \to 0} \mu([f > t + \varepsilon]) = \mu([f > t]),$$

$$\lim_{\varepsilon \to 0} \mu([f > t - \varepsilon]) = \mu([f \geq t]).$$

Therefore, the distribution function $t \to \mu([f > t])$ is right continuous, and it is continuous at a point t if and only if $\mu([f = t]) = 0$.

Proposition 15.2c. *Let $\{f_n\} \to f$ in measure. Then for every $\varepsilon > 0$,*

$$\limsup \mu([f_n > t]) \leq \mu([f > t - \varepsilon]),$$

$$\liminf \mu([f_n > t]) \geq \mu([f > t + \varepsilon]).$$

Therefore, $\mu([f_n > t]) \to \mu([f > t])$ only at those t where the distribution function of f is continuous.

15.2 Let $f : E \to \mathbb{R}^*$ be measurable, and let μ be σ-finite. Then

$$\int_E |f|^p d\mu = p \int_0^\infty t^{p-1} \mu([|f| > t]) dt. \tag{15.1c}$$

15.3 The Marcinkiewicz integral $M_{E,\lambda}(x)$ is infinity for all $x \in (Q - E)$.

17 THE RADON–NIKODÝM THEOREM

17.1 Let ν be the Lebesgue measure in $[0, 1]$, and let μ be the counting measure on the same σ-algebra of the Lebesgue-measurable subsets of $[0, 1]$. Then ν is absolutely continuous with respect to μ, but there does not exist a nonnegative, μ-measurable function $f : [0, 1] \to \mathbb{R}^*$ for which ν can be represented as in (17.1).

17.2 In $[0, 1]$, let $\mathcal{A}$ be the σ-algebra of the sets that either are countable or have countable complement. Let μ be the counting measure on $\mathcal{A}$, and let $\nu : \mathcal{A} \to \mathbb{R}^*$ be defined by $\nu(E) = 0$ if E is countable and $\nu(E) = 1$ otherwise. Then ν is absolutely continuous with respect to μ, but it does not have a Radon–Nikodým derivative.

17.3 Under the assumptions of the Radon–Nikodým theorem,

$$\nu \left\{ x \in X \middle| \frac{d\nu}{d\mu} = 0 \right\} = 0.$$

17.4 Let the assumptions of the Radon–Nikodým theorem hold. If g is ν-integrable, $g(\frac{d\nu}{d\mu})$ is μ-integrable, and for every measurable set E,

$$\int_E g \, d\nu = \int_E g \frac{d\nu}{d\mu} d\mu.$$

17.5 Let μ, ν_1, and ν_2 be σ-finite measures defined on the same σ-algebra $\mathcal{A}$. If ν_1 and ν_2 are absolutely continuous with respect to μ,

$$\frac{d(\nu_1 + \nu_2)}{d\mu} = \frac{d\nu_1}{d\mu} + \frac{d\nu_2}{d\mu}, \quad \mu\text{—a.e.}$$

17.6 Let μ, ν, and η be σ-finite measures on X defined on the same σ-algebra $\mathcal{A}$. Assume that μ is absolutely continuous with respect to η and that ν is absolutely continuous with respect to μ. Then

$$\frac{d\nu}{d\eta} = \frac{d\nu}{d\mu} \frac{d\mu}{d\eta} \quad \text{a.e. with respect to } \eta.$$

17.7 Let μ and ν be two σ-finite measures, not identically zero, defined on the same σ-algebra $\mathcal{A}$ and mutually absolutely continuous. Then

$$\frac{d\nu}{d\mu} \neq 0 \quad \text{and} \quad \frac{d\mu}{d\nu} = \left(\frac{d\nu}{d\mu} \right)^{-1}.$$

17.8   Let $\{X, \mathcal{A}, \mu\}$ be $[0, 1]$ with the Lebesgue measure. Let $\{E_n\}$ be a measurable partition of $[0, 1]$ and let $\{\alpha_n\}$ be a sequence of positive numbers such that $\sum \alpha_n < \infty$. Find the Radon–Nikodým derivative of the measure

$$\mathcal{A} \ni E \longrightarrow \nu(E) = \sum \alpha_n \mu \left(E \bigcap E_n \right)$$

with respect to the Lebesgue measure.

Proposition 17.1c. *Let μ and ν be two measures defined on the same σ-algebra $\mathcal{A}$. Assume that ν is finite. Then ν is absolutely continuous with respect to μ if and only if for every $\varepsilon > 0$, there exists $\delta > 0$ such that $\nu(E) < \varepsilon$ for every set $E \in \mathcal{A}$ such that $\mu(E) < \delta$.*

Proof (necessity).   If not, there exist $\varepsilon > 0$ and a sequence of measurable sets $\{E_n\}$ such that

$$\nu(E_n) \geq \varepsilon \quad \text{and} \quad \mu(E_n) \leq \frac{1}{2^n}.$$

Let $E = \limsup E_n$ and compute

$$\nu(E) = \nu(\limsup E_n) \geq \limsup \nu(E_n) \geq \varepsilon.$$

On the other hand, for all $n \in \mathbb{N}$,

$$\mu(E) \leq \mu \left(\bigcup_{j=n}^{\infty} E_j \right) \leq \sum_{j=n}^{\infty} \mu(E_j) \leq \frac{1}{2^{n-1}}. \qquad \square$$

17.9   The proposition might fail if ν is not finite. Let $X = \mathbb{N}$, and for every subset $E \subset \mathbb{N}$, set

$$\mu(E) = \sum_{n \in E} \frac{1}{2^n}, \qquad \nu(E) = \sum_{n \in E} 2^n.$$

IV

Topics on Measurable Functions of Real Variables

1 Functions of bounded variations

Let f be a real-valued function defined and bounded in some interval $[a, b] \subset \mathbb{R}$. Denote by

$$\mathcal{P} \equiv \{a = x_o < x_1 < \cdots < x_n = b\}$$

a partition of $[a, b]$ and set

$$\mathcal{V}_f[a, b] = \sup_{\mathcal{P}} \sum_{i=1}^{n} |f(x_i) - f(x_{i-1})|.$$

This number, finite or infinite, is called the *total variation* of f in $[a, b]$. If $\mathcal{V}_f[a, b]$ is finite, the function f is said to be of *bounded variation* in $[a, b]$.[1]

If f is monotone on $[a, b]$, it is of bounded variation on $[a, b]$ and

$$\mathcal{V}_f[a, b] = |f(b) - f(a)|.$$

More generally, if f is the difference of two monotone functions on $[a, b]$, then it is of bounded variation on $[a, b]$.

If f is Lipschitz continuous on $[a, b]$ with Lipschitz constant L, then f is of bounded variation on $[a, b]$, and

$$\mathcal{V}_f[a, b] \leq L(b - a).$$

---

[1]C. Jordan, Sur la Série de Fourier, *C. R. Acad. Sci. Paris*, **92** (1881), 228–230.

Continuity does not imply bounded variation. The function

$$f(x) = \begin{cases} x \cos \dfrac{\pi}{x} & \text{for } x \in (0, 1], \\ 0 & \text{for } x = 0 \end{cases} \tag{1.1}$$

is continuous in $[0, 1]$ and not of bounded variation on $[0, 1]$. Consider the partition of $[0, 1]$,

$$\mathcal{P}_n = \left\{ 0 < \frac{1}{n} < \frac{1}{n-1} < \cdots < \frac{1}{n - (n-1)} = 1 \right\}.$$

Then by direct computation,

$$\mathcal{V}_f[0, 1] \geq \lim_{n \to \infty} \left\{ \left| \frac{\cos n\pi}{n} \right| + \sum_{j=1}^{n-1} \left| \frac{\cos \pi (n-j)}{n-j} - \frac{\cos \pi (n-j+1)}{n-j+1} \right| \right\}$$

$$= \lim_{n \to \infty} \frac{1}{n} + \sum_{j=1}^{n-1} \left(\frac{1}{n-j} + \frac{1}{n-j+1} \right)$$

$$\geq \lim_{n \to \infty} \sum_{i=1}^{n} \frac{1}{i}.$$

Bounded variation does not imply continuity. The function

$$f(x) = \begin{cases} 0 & \text{for } x \in [-1, 1] - \{0\}, \\ 1 & \text{for } x = 0 \end{cases}$$

is discontinuous and of bounded variation. Indeed, $\mathcal{V}_f[-1, 1] = 2$.

Proposition 1.1. *Let f and g be of bounded variation in $[a, b]$ and let α and β be real numbers. Then $(\alpha f + \beta g)$ and (fg) are of bounded variation in $[a, b]$. If $|g| \geq \varepsilon$ for some $\varepsilon > 0$, then (f/g) is of bounded variation on $[a, b]$.*

Proposition 1.2. *Let f be of bounded variation on $[a, b]$. Then f is of bounded variation in every closed subinterval of $[a, b]$. Moreover, for every $c \in [a, b]$,*

$$\mathcal{V}_f[a, b] = \mathcal{V}_f[a, c] + \mathcal{V}_f[c, b]. \tag{1.2}$$

The *positive* and *negative* variations of f in $[a, b]$ are defined by

$$\mathcal{V}_f^+[a, b] = \sup_{\mathcal{P}} \sum_{j=1}^{n} [f(x_j) - f(x_{j-1})]^+,$$

$$\mathcal{V}_f^-[a, b] = \sup_{\mathcal{P}} \sum_{j=1}^{n} [f(x_j) - f(x_{j-1})]^-.$$

Proposition 1.3. *Let f be of bounded variation in $[a, b]$. Then*

$$\mathcal{V}_f[a, b] = \mathcal{V}_f^+[a, b] + \mathcal{V}_f^-[a, b],$$
$$f(b) - f(a) = \mathcal{V}_f^+[a, b] - \mathcal{V}_f^-[a, b].$$

In particular, for every $x \in [a, b]$, there holds the Jordan decomposition

$$f(x) = f(a) + \mathcal{V}_f^+[a, x] - \mathcal{V}_f^-[a, x]. \tag{1.3}$$

Since $x \to \mathcal{V}_f^\pm[a, x]$ are both nondecreasing, a function f of bounded variation can be written as the difference of two nondecreasing functions. We have already observed that the difference of two monotone functions in $[a, b]$ is of bounded variation in $[a, b]$. Thus a function f is of bounded variation in $[a, b]$ if and only if it is the difference of two monotone functions in $[a, b]$.

Proposition 1.4. *A function f of bounded variation on $[a, b]$ has at most countably many jump discontinuities in $[a, b]$.*

Proof. We may assume that f is monotone increasing. Then for every $c \in (a, b)$, the limits

$$\lim_{x \to c^+} f(x) = f(c^+), \qquad \lim_{x \to c^-} f(x) = f(c^-)$$

exist and are finite. If $f(c^+) > f(c^-)$, we select one and only one rational number out of the interval $(f(c^-), f(c^+))$. This way, the set of jump discontinuities of f is put in one-to-one correspondence with a subset of $\mathbb{N}$.   □

2 Dini derivatives

Let f be a real-valued function defined in $[a, b]$. For a fixed $x \in [a, b]$, set

$$D_\pm f(x) = \liminf_{h \to 0^\pm} \frac{f(x + h) - f(x)}{h},$$
$$D^\pm f(x) = \limsup_{h \to 0^\pm} \frac{f(x + h) - f(x)}{h}.$$

These are the four *Dini numbers*, or the four *Dini derivatives*, of f at x.[2] If f is differentiable at x, these four numbers all coincide with $f'(x)$.

Proposition 2.1. *Let f be real valued and nondecreasing in $[a, b]$. Then the functions $x \to D_\pm f(x)$, $D^\pm f(x)$ are measurable.*[3]

---

[2] See [13].

[3] W. Sierpiński, Sur les fonctions derivées des fonctions discontinues, *Fund. Math.*, **3** (1922), 123–127; S. Banach, Sur les fonctions dérivée des fonctions measurable, *Fund. Math.*, **3** (1922), 128–132.

Proof. We prove that $D^+ f$ is measurable, the arguments for the remaining ones being similar. For $n \in \mathbb{N}$, set

$$u_n(x) = \sup_{0 < h < \frac{1}{n}} \frac{f(x+h) - f(x)}{h}.$$

Since

$$D^+ f(x) = \lim u_n(x),$$

it suffices to prove that the u_n are measurable. By monotonicity, f is measurable and for h fixed, the difference quotients in the definition of u_n are measurable.

We prove that the supremum of such difference quotients for $h \in (0, \frac{1}{n}]$ can be realized for h ranging over a countable subset of $(0, \frac{1}{n}]$. In such a way, u_n would be the supremum of a countable collection of measurable functions.

Let $\mathbb{Q}_n$ be the set of rational numbers in $(0, \frac{1}{n}]$, and set

$$v_n(x) = \sup_{h \in \mathbb{Q}_n} \frac{f(x+h) - f(x)}{h}.$$

By construction, $v_n(x) \le u_n(x)$. To establish the reverse inequality, having fixed $\varepsilon > 0$, there exists $\tau \in (0, \frac{1}{n}]$ such that

$$\frac{f(x+\tau) - f(x)}{\tau} > u_n(x) - \varepsilon.$$

Having fixed such a τ, there exist $h \in \mathbb{Q}_n$ and $h \ge \tau$ such that

$$\frac{1}{\tau} < \frac{1}{h} + \frac{\varepsilon}{|f(x+\tau)| + |f(x)| + 1}.$$

Therefore, since $f(x+\tau) \le f(x+h)$,

$$\frac{f(x+h) - f(x)}{h} + \varepsilon \ge \frac{f(x+\tau) - f(x)}{\tau}$$

$$> u_n(x) - \varepsilon.$$

Thus

$$v_n(x) \ge u_n(x) - 2\varepsilon \quad \text{for all } \varepsilon > 0. \qquad \square$$

For a function f defined in $[a, b]$, set

$$D'' f(x) = \max\{D^- f(x); D^+ f(x)\},$$
$$D' f(x) = \min\{D_- f(x); D_+ f(x)\}.$$

If f is nondecreasing, the two functions $D'' f$ and $D' f$ are both measurable.

Proposition 2.2. *Let f be a real-valued, nondecreasing function in $[a, b]$. Then*

$$f(b) - f(a) \geq k\mu([D''f > k]) \quad \text{for all } k \in \mathbb{R}.$$

Proof. The assertion is trivial if $\mu([D''f > k]) = 0$ or if $k \leq 0$. Assuming that

$$\mu([D''f > k]) > 0 \quad \text{and} \quad k > 0,$$

let $\mathcal{F}$ denote the family of all closed intervals $[\alpha, \beta] \subset [a, b]$ such that at least one of the extremes α or β is in $[D''f > k]$ and such that

$$\frac{f(\beta) - f(\alpha)}{\beta - \alpha} > k.$$

By the definition of $D''f$, having fixed $x \in [D''f > k]$ and $\delta > 0$, there exists some interval $[\alpha, \beta] \in \mathcal{F}$ of length less than δ and such that $x \in [\alpha, \beta]$. Therefore, $\mathcal{F}$ is a Vitali covering for $[D''f > k]$.

By Corollary 17.2 of Chapter II, for any fixed $\varepsilon > 0$, there exists a finite collection of intervals

$$[\alpha_i, \beta_i] \in \mathcal{F} \quad \text{for } i = 1, 2, \ldots, n$$

with pairwise-disjoint interior such that

$$\sum_{i=1}^{n} (\beta_i - \alpha_i) > \mu([D''f > k]) - \varepsilon.$$

From this and the definition of $\mathcal{F}$,

$$f(b) - f(a) \geq \sum_{i=1}^{n} [f(\beta_i) - f(\alpha_i)]$$

$$> \sum_{i=1}^{n} k(\beta_i - \alpha_i)$$

$$> k\mu([D''f > k]) - k\varepsilon. \qquad \square$$

Corollary 2.3. *Let f be a real-valued, nondecreasing function defined in $[a, b]$. Then $D''f$ and $D'f$ are a.e. finite in $[a, b]$.*

Proof. For all $k > 0$,

$$\mu([D'f = \infty]) \leq \mu([D''f = \infty])$$

$$\leq \mu([D''f > k]) \leq \frac{f(b) - f(a)}{k}.$$

The corollary follows by letting $k \to \infty$. $\qquad \square$

3   Differentiating functions of bounded variation

A real-valued function f defined in $[a, b]$ is differentiable at some $x \in (a, b)$ if and only if $D'' f$ and $D' f$ are finite at x and

$$D'' f(x) = D' f(x).$$

A real-valued function f defined in $[a, b]$ is a.e. differentiable in $[a, b]$ if and only if $D'' f$ and $D' f$ are a.e. finite in $[a, b]$ and

$$\mu([D'' f > D' f]) = 0.$$

Theorem 3.1 (Lebesgue[4]). *A real-valued, nondecreasing function f in $[a, b]$ is a.e. differentiable in $[a, b]$.*

Proof. By Corollary 2.3, $D'' f$ and $D' f$ are a.e. finite in $[a, b]$. Assume that $\mu([D'' f > D' f]) > 0$, and for $p, q \in \mathbb{N}$, set

$$\mathcal{I}_{p,q} = \left\{ x \in [D'' f > D' f] \,|\, D' f(x) < \frac{p}{q} < \frac{p+1}{q} < D'' f(x) \right\}.$$

Since

$$[D'' f > D' f] = \bigcup_p \bigcup_q \mathcal{I}_{p,q},$$

there exists a pair p, q of positive integers such that $\mu(\mathcal{I}_{p,q}) > 0$.

Let $\mathcal{F}$ be the family of closed intervals $[\alpha, \beta] \subset [a, b]$ such that at least one of the extremes α and β belongs to $\mathcal{I}_{p,q}$ and such that

$$\frac{f(\beta) - f(\alpha)}{\beta - \alpha} < \frac{p}{q}.$$

Having fixed $x \in \mathcal{I}_{p,q}$ and some $\delta > 0$, there exists an interval $[\alpha, \beta] \in \mathcal{F}$ of length less than δ and such that $x \in [\alpha, \beta]$. Therefore, $\mathcal{F}$ is a Vitali covering of $\mathcal{I}_{p,q}$. By Corollary 17.2 of Chapter II, having fixed an arbitrary $\varepsilon > 0$, we may extract out of $\mathcal{F}$ a finite collection of intervals

$$[\alpha_i, \beta_i] \in \mathcal{F} \quad \text{for } i = 1, 2, \ldots, ni$$

with pairwise-disjoint interior such that

$$\sum_{i=1}^{n} (\beta_i - \alpha_i) - \varepsilon < \mu(\mathcal{I}_{p,q}) < \mu \left(\mathcal{I}_{p,q} \bigcap \bigcup_{i=1}^{n} [\alpha_i, \beta_i] \right) + \varepsilon.$$

---

[4]See [33]; also in A. Rajchman and S. Saks, Sur la derivabilité des fonctions monotones, *Fund. Math.*, 4 (1923), 204–213. A proof independent of measure theory appears in [44, pp. 5–9]. In [44], this proof is taken as a starting point to introduce a new notion of measure.

Therefore, by the construction of the family $\mathcal{F}$,

$$\sum_{i=1}^{n}[f(\beta_i) - f(\alpha_i)] < \frac{p}{q}\sum_{i=1}^{n}(\beta_i - \alpha_i)$$

$$< \frac{p}{q}\mu(\mathcal{I}_{p,q}) + \frac{p}{q}\varepsilon.$$

By Proposition 2.2 applied to f restricted to the interval $[\alpha_i, \beta_i]$, we derive

$$f(\beta_i) - f(\alpha_i) > \frac{p+1}{q}\mu\left(\mathcal{I}_{p,q}\bigcap[\alpha_i, \beta_i]\right).$$

Adding these inequalities for $i = 1, 2, \ldots, n$ gives

$$\sum_{i=1}^{n}[f(\beta_i) - f(\alpha_i)] > \frac{p+1}{q}\sum_{i=1}^{n}\mu\left(\mathcal{I}_{p,q}\bigcap[\alpha_i, \beta_i]\right)$$

$$\geq \frac{p+1}{q}\mu\left(\mathcal{I}_{p,q}\bigcap\bigcup_{i=1}^{n}[\alpha_i, \beta_i]\right)$$

$$> \frac{p+1}{q}\mu(\mathcal{I}_{p,q}) - \frac{p+1}{q}\varepsilon.$$

Combining the inequalities involving $\mu(\mathcal{I}_{p,q})$,

$$\frac{p}{q}\mu(\mathcal{I}_{p,q}) + \frac{p}{q}\varepsilon > \frac{p+1}{q}\mu(\mathcal{I}_{p,q}) - \frac{p+1}{q}\varepsilon.$$

From this,

$$\mu(\mathcal{I}_{p,q}) < (2p+1)\varepsilon \quad \text{for all } \varepsilon > 0. \qquad \square$$

Corollary 3.2. *A real-valued function f of bounded variation in $[a, b]$ is a.e. differentiable in $[a, b]$.*

4   Differentiating series of monotone functions

Theorem 4.1 (Fubini[5]). *Let $\{f_n\}$ be a sequence of real-valued, nondecreasing functions in $[a, b]$, and assume that the series $\sum f_n$ is convergent in $[a, b]$ to a real-valued function f defined in $[a, b]$. Then f is a.e. differentiable in $[a, b]$, and*

$$f'(x) = \sum f_n'(x) \quad \text{for a.e. } x \in [a, b].$$

---

[5]G. Fubini, Sulla derivazione per serie, *Atti Accad. Naz. Lincei Rend.*, **16** (1907), 608–614; also in L. Tonelli, Successioni di curve e derivazione per serie, *Atti Accad. Lincei*, **25** (1916), 85–91.

Proof. By possibly replacing f_n with $(f_n - f_n(a))$, we may assume that $f_n(a) = 0$ and $f_n \geq 0$. For $n \in \mathbb{N}$, write

$$f = \sum_{i=1}^{n} f_i + R_n, \quad \text{where } R_n = \sum_{j=n+1}^{\infty} f_j.$$

The functions R_n are nondecreasing and hence a.e. differentiable in $[a, b]$. The difference

$$R_n - R_{n+1} = f_{n+1},$$

is also nondecreasing and a.e. differentiable in $[a, b]$. Therefore,

$$R_n' - R_{n+1}' = f_{n+1}' \geq 0 \quad \text{a.e. in } [a, b].$$

It follows that the sequence $\{R_n'\}$ is a.e. nonincreasing with respect to n in $[a, b]$ and therefore has a limit

$$\lim R_n'(x) = g(x) \quad \text{for a.e. } x \in [a, b].$$

The sum f is nondecreasing and hence a.e. differentiable in $[a, b]$. Therefore,

$$f'(x) = \sum_{i=1}^{n} f_i'(x) + R_n'(x) \quad \text{for a.e. } x \in [a, b].$$

To prove the theorem, it suffices to show that $g = 0$ a.e. in $[a, b]$. For $m \in \mathbb{N}$, set

$$\mathcal{I}_m = \left\{ x \in [a, b] \,|\, g(x) > \frac{1}{m} \right\}$$

and observe that

$$\{x \in [a, b] \,|\, g(x) > 0\} = \bigcup \mathcal{I}_m.$$

For each $n \in \mathbb{N}$, there exists a set $\mathcal{E}_n \subset [a, b]$ of measure zero such that

$$R_n'(x) \geq g(x) \quad \text{for all } x \in \mathcal{I}_m - \mathcal{E}_n.$$

Therefore, by Proposition 2.2 applied to the function R_n,

$$R_n(b) \geq \frac{1}{m} \mu \left(\left[R_n' > \frac{1}{m} \right] \right)$$

$$\geq \frac{1}{m} \mu(\mathcal{I}_m) \quad \text{for all } n \in \mathbb{N}.$$

Since the series is convergent everywhere in $[a, b]$, the left-hand side goes to zero as $n \to \infty$. Thus $\mu(\mathcal{I}_m) = 0$ for all $m \in \mathbb{N}$. $\square$

5   Absolutely continuous functions

A real-valued function f defined in $[a, b]$ is *absolutely continuous* in $[a, b]$ if for every $\varepsilon > 0$, there exists a positive number δ such that for every finite collection of disjoint intervals $(a_j, b_j) \subset [a, b]$, $j = 1, 2, \ldots, n$, of total length not exceeding δ,[6]

$$\sum_{j=1}^{n} |f(b_j) - f(a_j)| < \varepsilon \quad \left(\sum_{j=1}^{n} (b_j - a_j) < \delta \right). \tag{5.1}$$

If g is integrable in $[a, b]$, then the function

$$x \longrightarrow \int_a^x g(t)dt$$

is absolutely continuous in $[a, b]$. This follows from the absolute continuity of the integral.

If f is Lipschitz continuous in $[a, b]$, it is absolutely continuous in $[a, b]$. The converse is false. A counterexample can be constructed using Section 5.1 of the Problems and Complements.

Absolute continuity implies continuity, but the converse is false. The function in (1.1) is continuous and not absolutely continuous.

The linear combination of two absolutely continuous functions f and g in $[a, b]$ as well as their product (fg) are absolutely continuous. Their quotient f/g is absolutely continuous if $|g| \geq c_o > 0$ in $[a, b]$.

Proposition 5.1. *Let f be absolutely continuous on $[a, b]$. Then f is of bounded variation on $[a, b]$.*

Proof. Having fixed $\varepsilon > 0$ and the corresponding δ, partition $[a, b]$ by points $a = x_o < x_1 < \cdots < x_n = b$ such that

$$\frac{1}{2}\delta < (x_i - x_{i-1}) < \delta, \quad i = 1, 2, \ldots, n.$$

The number n of the intervals making up the partition does not exceed $2(b-a)/\delta$. In each of them, the variation of f is less than ε. Then by (1.2) of Proposition 1.2,

$$\mathcal{V}_f[a, b] = \sum_{i=1}^{n} \mathcal{V}_f[x_{i-1}, x_i] \leq 2(b-a)\frac{\varepsilon}{\delta}. \qquad \square$$

Corollary 5.2. *Let f be absolutely continuous on $[a, b]$. Then f is a.e. differentiable on $[a, b]$.*

Proposition 5.3. *Let f be absolutely continuous in $[a, b]$. If $f' \geq 0$ a.e. in $[a, b]$, then f is nondecreasing in $[a, b]$.*

---

[6]See [33]; also in G. Vitali, Sulle funzioni integrali, *Atti Accad. Sci. Torino*, **40** (1905), 753–766.

Proof. For an interval $[\alpha, \beta] \subset [a, b]$, we will show that $f(\beta) \geq f(\alpha)$. Having fixed the interval $[\alpha, \beta]$, set

$$\mathcal{I} = \{x \in [\alpha, \beta] \text{ such that } f'(x) \geq 0\}$$

by the assumption $\mu(\mathcal{I}) = (\beta - \alpha)$. Having fixed $\varepsilon > 0$, let δ be a corresponding positive number claimed by the absolute continuity of f. For every $x \in \mathcal{I}$, there exist some $\sigma_x > 0$ such that

$$f(x + h) - f(x) > -\varepsilon h \quad \text{for all } h \in (0, \sigma_x).$$

The collection of all the intervals $[x, x + h]$ for x ranging over $\mathcal{I}$ and $h \in (0, \sigma_x)$, is a Vitali covering for $\mathcal{I}$. Therefore, in correspondence of the previously fixed $\delta > 0$, we may extract a finite collection of intervals $[\alpha_i, \beta_i], i = 1, 2, \ldots, n$, with pairwise-disjoint interior such that

$$\mu(\mathcal{I}) - \delta \leq \mu \left(\mathcal{I} \cap \bigcup_{i=1}^{n} (\alpha_i - \beta_i) \right)$$

and

$$f(\beta_i) - f(\alpha_i) > -\varepsilon (\beta_i - \alpha_i).$$

The complement

$$[\alpha, \beta] - \bigcup_{i=1}^{n} (\alpha_i, \beta_i),$$

consists of finitely many disjoint intervals

$$[a_j, b_j], \quad j = 1, 2, \ldots, m \quad \text{for some } m \in \mathbb{N},$$

of total length not exceeding δ. Therefore, (5.1) holds for such a finite collection. We next compute

$$f(\beta) - f(\alpha) = \sum_{i=1}^{n} [f(\beta_i) - f(\alpha_i)] + \sum_{j=1}^{m} [f(b_j) - f(a_j)]$$

$$\geq -\varepsilon \sum_{i=1}^{n} (\beta_i - \alpha_i) - \varepsilon$$

$$\geq -\varepsilon[1 + (\beta - \alpha)] \quad \text{for all } \varepsilon > 0.$$

Corollary 5.4. *Let f be absolutely continuous in $[a, b]$. If $f' = 0$ a.e. on $[a, b]$, then f is constant in $[a, b]$.*

Remark 5.1. The conclusions of Proposition 5.3 and Corollary 5.4 are false if f is of bounded variation and not absolutely continuous. A counterexample is the function of the jumps introduced in Sections 1.6 and 2.4 of the Problems and Complements.

The assumption of absolute continuity cannot be relaxed to the mere continuity, as shown by the Cantor ternary function and its variants (see Sections 5.3–5.6 of the Problems and Complements). The same examples also show that bounded variation and continuity do not imply absolute continuity.

6 Density of a measurable set

Let $E \subset [a, b]$ be Lebesgue measurable. The density function of E,

$$x \longrightarrow d_E(x) = \int_a^x \chi_E(t)dt,$$

is absolutely continuous and nondecreasing in $[a, b]$. Since the complement $[a, b] - E$ is measurable, the function $d_{[a,b]-E}$ is well defined and

$$d_E(x) + d_{[a,b]-E}(x) = x - a.$$

Therefore,

$$d'_E(x) + d'_{[a,b]-E}(x) = 1 \quad \text{a.e. in } [a, b].$$

Proposition 6.1 (Lebesgue[7]). *Let $E \subset [a, b]$ be Lebesgue measurable. Then*

$$d'_E(x) = 1 \quad \text{a.e. in } E \quad \text{and} \quad d'_E(x) = 0 \quad \text{a.e. in } [a, b] - E.$$

Proof. It suffices to prove the first of these. If E is open, then $d'_E = 1$ in E. Now assume that E is of the type $\mathcal{G}_\delta$, i.e., that there exists a countable collection of open sets $\{E_n\}$ such that

$$E_{n+1} \subset E_n \quad \text{and} \quad E = \bigcap E_n.$$

By dominated convergence,

$$d_E(x) = \int_a^x \chi_E(t)dt = \lim \int_a^x \chi_{E_n}(t)dt = \lim d_{E_n}(x).$$

Therefore,

$$d_E = d_{E_1} + \sum (d_{E_{n+1}} - d_{E_n}).$$

Each of the terms of the series is nonincreasing since $d'_{E_{n+1}} \le d'_{E_n}$. Therefore, by the Fubini theorem,

$$d'_E = d'_{E_1} + \sum (d_{E_{n+1}} - d_{E_n})' \quad \text{a.e. in } [a, b].$$

If $x \in E$, then $d'_{E_n} = 1$ for all $n \in \mathbb{N}$, and the assertion follows.

---

[7] See [33, pp. 185–187].

If E is a measurable subset of $[a, b]$, there exists a set E_δ of the type $\mathcal{G}_\delta$ such that

$$E \subset E_\delta \quad \text{and} \quad \mu(E_\delta - E) = 0.$$

This implies that $d_E = d_{E_\delta}$ and thus

$$d'_E(x) = d'_{E_\delta}(x) = 1 \quad \text{for a.e. } x \in E. \qquad \square$$

7 Derivatives of integrals

We have observed that if f is Lebesgue integrable in $[a, b]$, then the function

$$x \longrightarrow F(x) = \int_a^x f(t)dt$$

is absolutely continuous and hence a.e. differentiable in $[a, b]$.

Proposition 7.1. *Let f be Lebesgue integrable in $[a, b]$. Then $F' = f$ a.e. in $[a, b]$.*

Proof. Assume first that f is simple. Then if $\{\lambda_1, \lambda_2, \ldots, \lambda_n\}$ are the distinct values taken by f, there exist disjoint, measurable sets

$$E_i \subset [a, b], \quad i = 1, 2, \ldots, n, \qquad E_i \bigcap E_j \quad \text{for } i \neq j$$

such that

$$f = \sum_{i=1}^n \lambda_i \chi_{E_i} \quad \text{and} \quad F = \sum_{i=1}^n \lambda_i d_{E_i}.$$

Therefore, the assertion follows from Proposition 6.1.

Next, assume that f is integrable and nonnegative in $[a, b]$. There exists a sequence of simple functions $\{f_n\}$ such that

$$f_n \leq f_{n+1} \quad \text{and} \quad \lim f_n(x) = f(x) \quad \text{for all } x \in [a, b].$$

By dominated convergence,

$$F(x) = \lim F_n(x), \quad \text{where } F_n(x) = \int_a^x f_n(t)dt.$$

Therefore,

$$F = F_1 + \sum (F_{n+1} - F_n).$$

The terms of the series are nondecreasing since

$$(F_{n+1} - F_n)' = f_{n+1} - f_n \geq 0 \quad \text{a.e. in } [a, b].$$

Therefore, by the Fubini theorem,

$$F' = \lim F'_n = \lim f_n = f$$

a.e. in $[a, b]$. A general integrable f is the difference of two nonnegative integrable functions. $\qquad \square$

Proposition 7.2 (Lebesgue[8]). *Let f be absolutely continuous in $[a, b]$. Then f' is integrable and*

$$f(x) = f(a) + \int_a^x f'(t)dt. \tag{7.1}$$

Proof. Assume first that f is nondecreasing so that $f' \geq 0$ a.e. in $[a, b]$. Defining $f(x) = f(b)$ for $x \geq b$, the limit

$$f'(x) = \lim \frac{f\left(x + \frac{1}{n}\right) - f(x)}{\frac{1}{n}}$$

exists a.e. in $[a, b]$. Then by Fatou's lemma,

$$\int_a^b f'dx \leq \liminf n \int_a^b \left[f\left(x + \frac{1}{n}\right) - f(x) \right] dx$$

$$= \liminf n \left\{ \int_b^{b+\frac{1}{n}} f(x)dx - \int_a^{a+\frac{1}{n}} f(x)dx \right\}$$

$$\leq \liminf n \frac{f(b) - f(a)}{n}$$

$$= f(b) - f(a).$$

Since f is absolutely continuous, it is of bounded variation and, by the Jordan decomposition, is the difference of two nondecreasing functions. Thus f' is integrable on $[a, b]$. The function

$$g(x) = f(a) + \int_a^x f'(t)dt$$

is absolutely continuous, and $g' = f'$ a.e. in $[a, b]$. Thus $(g - f)' = 0$ a.e. in $[a, b]$, and by Corollary 5.4, $g = f + \text{const}$ on $[a, b]$. Since $g(a) = f(a)$, the conclusion follows. $\qquad\square$

Remark 7.1. The proof of Proposition 7.2 contains the following.

Corollary 7.3. *Let f be of bounded variation in $[a, b]$. Then f' is integrable in $[a, b]$.*

Remark 7.2. Proposition 7.2 is false if f is only of bounded variation on $[a, b]$. A counterexample is given by a nonconstant, nondecreasing simple function. For such a function, the representation (7.1) does not hold.

---

[8]See [33, p. 188].

The proposition continues to be false even by requiring that f be continuous. The Cantor ternary function (see Section 5.3 of the Problems and Complements) is of bounded variation and continuous in $[0, 1]$, and its derivative is integrable. However, it is not absolutely continuous, and (7.1) does not hold. A similar conclusion holds for the function in Section 5.5 of the Problems and Complements.

8   Differentiating Radon measures

Let f be a nonnegative, Lebesgue-measurable, real-valued function defined in $\mathbb{R}^N$ and integrable on compact subsets of $\mathbb{R}^N$, and let μ denote the Lebesgue measure in $\mathbb{R}^N$. The notion of differentiating the integral of f at some point $x \in \mathbb{R}^N$ is replaced by

$$\lim_{\rho \to 0} \frac{1}{\mu(B_\rho(x))} \int_{B_\rho(x)} f \, d\mu = \lim_{\rho \to 0} \frac{\nu(B_\rho(x))}{\mu(B_\rho(x))}, \quad d\nu = f \, d\mu,$$

where $B_\rho(x)$ denotes the closed ball in $\mathbb{R}^N$ centered at x and radius ρ and provided the limit exists. More generally, given any two Radon measures μ and ν in $\mathbb{R}^N$, set

$$D_\mu^+ \nu(x) = \limsup_{\rho \to 0} \frac{\nu(B_\rho(x))}{\mu(B_\rho(x))}, \tag{8.1}$$

$$D_\mu^- \nu(x) = \liminf_{\rho \to 0} \frac{\nu(B_\rho(x))}{\mu(B_\rho(x))}, \tag{8.2}$$

provided $\mu(B_\rho(x)) > 0$ for all $\rho > 0$, and

$$D_\mu^+ \nu(x) = D_\mu^- \nu(x) = \infty \tag{8.3}$$

if $\mu(B_\rho(x)) = 0$ for some $\rho > 0$.

If for some $x \in \mathbb{R}^N$ the upper limit in (8.1) and the lower limit in (8.2) are equal and finite, we set

$$D_\mu^+ \nu(x) = D_\mu^- \nu(x) = D_\mu \nu(x)$$

and say that the Radon measure ν is differentiable at x with respect to the Radon measure μ.

Proposition 8.1. *Let μ and ν be two Radon measures in $\mathbb{R}^N$, and let μ_e and ν_e be their associated outer measures.[9] For every $t > 0$ and every set*

$$E \subset \{x \in \mathbb{R}^N | D_\mu^+ \nu(x) \geq t\}, \tag{8.4}_+$$

it holds that

$$\mu_e(E) \leq \frac{1}{t} \nu_e(E). \tag{8.5}_+$$

---

[9]See Section 16.1 of Chapter II.

Analogously, for every $t > 0$ and every set

$$E \subset \{x \in \mathbb{R}^N | D_\mu^- \nu(x) \leq t\}, \tag{8.4}_-$$

it holds that

$$\mu_e(E) \geq \frac{1}{t} \nu_e(E). \tag{8.5}_-$$

Proof. Fix $t > 0$. In proving $(8.5)_+$, assume first that the set E satisfying $(8.4)_+$ is bounded. Such a set being fixed, let $\mathcal{O}$ be an open set that contains E. Having fixed $\varepsilon \in (0, t)$, by the definition of $D_\mu^+ \nu(x)$, for every $x \in E$, there exists a ball $B_\rho(x)$ centered at x and of arbitrarily small radius ρ such that

$$(t - \varepsilon)\mu(B_\rho(x)) < \nu(B_\rho(x)). \tag{8.6}$$

Set

$$\mathcal{F} = \left\{ \begin{array}{l} \text{collection of balls } B_\rho(x) \text{ for } x \in E \\ \text{satisfying (8.6) and contained in } \mathcal{O} \end{array} \right\}.$$

Since $\mathcal{O}$ is open and ρ is arbitrarily small, such a collection is not empty and forms a fine Besicovitch covering for E. By the Besicovitch measure-theoretical covering theorem, there exists a countable collection $\{B(x_n)\}$ of disjoint, closed balls in $\mathcal{F}$ such that

$$\mu_e \left(E - \bigcup B_n \right) = 0.$$

From this and (8.6),

$$\mu_e(E) \leq \sum \mu(B_n) \leq \frac{1}{t - \varepsilon} \sum \nu(B_n) \leq \frac{1}{t - \varepsilon} \nu(\mathcal{O}).$$

Since ν is regular, there exists a set E_δ of the type $\mathcal{G}_\delta$ and containing E such that $\nu_e(E) = \nu(E_\delta)$. Therefore,[10]

$$\nu_e(E) = \nu(E_\delta) = \inf\{\nu(\mathcal{O}), \text{ where } \mathcal{O} \text{ is open and contains } E_\delta\}.$$

Thus

$$\mu_e(E) \leq \frac{1}{t - \varepsilon} \nu_e(E) \quad \text{for all } \varepsilon \in (0, t).$$

This proves $(8.5)_+$ if E is bounded. If not, construct a countable collection $\{E_n\}$ of bounded sets such that $E_n \subset E_{n+1}$ whose union if E. Then apply $(8.5)_+$ to each of the E_n to obtain

$$\mu_e(E_n) \leq \frac{1}{t} \nu_e(E) \quad \text{for all } n \in \mathbb{N}. \tag{8.7}$$

---

[10]See Sections 15.2 and 16 of Chapter II.

For each n, let $E_{n,\delta}$ be a set of the type $\mathcal{G}_\delta$ such that

$$E_n \subset E_{n,\delta} \quad \text{and} \quad \mu_e(E_n) = \mu(E_{n,\delta}).$$

By construction, $E \subset \liminf E_{n,\delta}$. Therefore,[11]

$$\mu_e(E) \le \mu_e(\liminf E_{n,\delta}) \le \liminf \mu(E_{n,\delta}) = \liminf \mu(E_n).$$

Letting $n \to \infty$ in (8.7) proves (8.5)$_+$ for any set E satisfying (8.4)$_+$.
The statement of (8.4)$_-$–(8.5)$_-$ is proved similarly. $\square$

9   Existence and measurability of $D_\mu \nu$

The next Proposition asserts that ν is differentiable with respect to μ for μ-almost all $x \in \mathbb{R}^N$. Equivalently $D_\mu \nu(x)$ exists μ-a.e. in $\mathbb{R}^N$.

Proposition 9.1. *There exists a Borel set $\mathcal{E} \subset \mathbb{R}^N$ such that*

$$\mu(\mathcal{E}) = 0 \quad \text{and} \quad D_\mu \nu = D_\mu^\pm \nu \quad \text{in } \mathbb{R}^N - \mathcal{E}.$$

Moreover, $D_\mu^\pm \nu$ is finite in $\mathbb{R}^N - \mathcal{E}$.

Proof. Assume first that both μ and ν are finite and set

$$\mathcal{E}_\infty^\pm = [D_\mu^\pm \nu = \infty].$$

By (8.5)$_+$,

$$\mu_e([D_\mu^+ \nu > t]) \le \frac{1}{t} \nu(\mathbb{R}^N) \quad \text{for all } t > 0.$$

Since

$$\mathcal{E}_\infty^- \subset \mathcal{E}_\infty^+ \subset [D_\mu^+ \nu > t] \quad \text{for all } t > 0,$$

letting $t \to \infty$ implies $\mu_e(\mathcal{E}_\infty^\pm) = 0$. Therefore, there exist Borel sets $\mathcal{E}_{\infty,\delta}^\pm$ of the type $\mathcal{G}_\delta$ such that

$$\mathcal{E}_\infty^\pm \subset \mathcal{E}_{\infty,\delta}^\pm \quad \text{and} \quad \mu_e(\mathcal{E}_\infty^\pm) = \mu(\mathcal{E}_{\infty,\delta}^\pm) = 0.$$

Next, for positive rational numbers $\alpha < \beta$, set

$$\mathcal{E}_{\alpha,\beta} = \{x \in \mathbb{R}^N - (\mathcal{E}_{\infty,\delta}^+ \cup \mathcal{E}_{\infty,\delta}^-) \mid D_\mu^- \nu(x) < \alpha < \beta < D_\mu^+ \nu(x)\}.$$

By (8.5)$_+$ and (8.5)$_-$,

$$\beta \mu_e(\mathcal{E}_{\alpha,\beta}) \le \nu_e(\mathcal{E}_{\alpha,\beta}) \le \alpha \mu_e(\mathcal{E}_{\alpha,\beta}).$$

---

[11] By (3.3) of Chapter II. The latter can be applied since μ_e restricted to the Borel sets is a measure.

Therefore, $\mu_e(\mathcal{E}_{\alpha,\beta}) = 0$. From this,

$$\mu_e([D_\mu^- v < D_\mu^+ v]) = \mu_e \left(\bigcup_{\substack{0<\alpha<\beta \\ \alpha,\beta \text{ rationals}}} \mathcal{E}_{\alpha,\beta} \right)$$

$$\leq \sum_{\substack{0<\alpha<\beta \\ \alpha,\beta \text{ rationals}}} \mu_e(\mathcal{E}_{\alpha,\beta}) = 0.$$

There exists a Borel set $[D_\mu^- v < D_\mu^+ v]_\delta$ of the type $\mathcal{G}_\delta$ and containing $[D_\mu^- v < D_\mu^+ v]$ such that

$$\mu_e([D_\mu^- v < D_\mu^+ v]) = \mu([D_\mu^- v < D_\mu^+ v]_\delta) = 0.$$

Setting

$$\mathcal{E} = \mathcal{E}_{\infty,\delta}^+ \bigcup \mathcal{E}_{\infty,\delta}^- \bigcup [D_\mu^- v < D_\mu^- v]_\delta$$

proves the proposition if μ and v are finite. For the general case, one first proves the proposition for the restrictions μ_n and v_n of μ and v to the ball B_n centered at the origin and radius n and then lets $n \to \infty$. □

Henceforth, we regard $D_\mu v$ as defined in the whole $\mathbb{R}^N$ by setting it to be zero on the Borel set $\mathcal{E}$ claimed by Proposition 9.1. Set

$$f_\rho(x) = \begin{cases} \dfrac{v(B_\rho(x))}{\mu(B_\rho(x))} & \text{for } x \in \mathbb{R}^N - \mathcal{E}, \\ 0 & \text{for } x \in \mathcal{E}. \end{cases} \tag{9.1}$$

The limit as $\rho \to 0$ exists everywhere in $\mathbb{R}^N$ and equals $D_\mu v$. Taking such a limit along a countable collection $\{\rho_n\} \to 0$,

$$D_\mu v = \lim f_{\rho_n} \quad \text{for all } x \in \mathbb{R}^N. \tag{9.2}$$

Proposition 9.2. *For all $t \in \mathbb{R}$, the sets $[D_\mu v \geq t]$ are Borel sets. In particular, $D_\mu v$ is Borel measurable.*

The proof hinges on the following lemma.

Lemma 9.3. *For all $\rho > 0$, the two functions*

$$x \longrightarrow \mu(B_\rho(x)), \quad v(B_\rho(x))$$

are upper semicontinuous.

Proof. The statement for $x \to \mu(B_\rho(x))$ reduces to[12]

$$\limsup_{y \to x} \mu(B_\rho(y)) \le \mu(B_\rho(x)) \quad \text{for all } x \in \mathbb{R}^N.$$

Let $\{x_n\}$ be a sequence of points in $\mathbb{R}^N$ converging to x. Since the balls $B_\rho(x_n)$ are closed,

$$\limsup \chi_{B_\rho(x_n)} \le \chi_{B_\rho(x)} \quad \text{pointwise in } \mathbb{R}^N.$$

Equivalently,

$$\liminf(1 - \chi_{B_\rho(x_n)}) \ge (1 - \chi_{B_\rho(x)}) \quad \text{pointwise in } \mathbb{R}^N.$$

By Fatou's lemma,

$$\begin{aligned}
\mu(B_{2\rho}(x)) - \mu(B_\rho(x)) &= \int_{B_{2\rho}} (1 - \chi_{B_\rho(x)}) d\mu \\
&\le \int_{B_{2\rho}} \liminf(1 - \chi_{B_\rho(x_n)}) d\mu \\
&\le \liminf \int_{B_{2\rho}} (1 - \chi_{B_\rho(x_n)}) d\mu \\
&= \liminf(\mu(B_{2\rho}(x)) - \mu(B_\rho(x_n))) \\
&= \mu(B_{2\rho}(x)) - \limsup \mu(B_\rho(x_n)).
\end{aligned}$$

The statement for $x \to \nu(B_\rho(x))$ is proved analogously. $\qquad \square$

9.1   Proof of Proposition 9.2. If $t \le 0$, then $[D_\mu\nu \ge t] = \mathbb{R}^N$. Therefore, it suffices to consider $t > 0$. From (9.1)–(9.2),

$$D_\mu\nu = \inf \varphi_n, \quad \text{where } \varphi_n = \sup_{j \ge n} f_{\rho_j}.$$

From the second part of (1.4) of Chapter III,

$$[D_\mu\nu \ge t] = \bigcap_n [\varphi_n \ge t].$$

Now using the first parts of (1.2) and (1.4) of Chapter III,

$$\begin{aligned}
[\varphi_n \ge t] &= \bigcap_k \left[\varphi_n > t - \frac{1}{k} \right] \\
&= \bigcap_k \bigcup_{j \ge n} \left[f_{\rho_j} > t - \frac{1}{k} \right].
\end{aligned}$$

---

[12] See Section 7 of Chapter I and Section 1.11 of the Problems and Complements of Chapter III.

Thus it suffices to show that the sets $[f_\rho > \tau]$ are Borel sets for all $\rho, \tau > 0$. From (9.1),

$$[f_\rho > \tau] = \{x \in \mathbb{R}^N | [\nu(B_\rho(x)) > \tau \mu(B_\rho(x))]\} - \mathcal{E}.$$

Since $\mathcal{E}$ is a Borel set, it suffices to show that $[\nu(B_\rho) > \tau \mu(B_\rho)]$ is a Borel set. Let $\{q_n\}$ denote the rational numbers. For a fixed $\tau > 0$,[13]

$$[\nu(B_\rho) > \tau \mu(B_\rho)] = \bigcup_n [\nu(B_\rho) \geq q_n] \bigcap [\tau \mu(B_\rho) < q_n].$$

Since the two functions $x \to \mu(B_\rho(x)), \nu(B_\rho(x))$ are upper semicontinuous, the sets $[\tau \mu(B_\rho) < q_n]$ are open and the sets $[\nu(B_\rho) \geq q_n]$ are closed for all $q_n \in \{q_n\}$. $\qquad\square$

10   Representing $D_\mu \nu$

In representing $D_\mu \nu$, assume that the two Radon measures μ and ν are defined on the same σ-algebra $\mathcal{A}$. The measurable function $D_\mu \nu$ can be identified by considering separately the cases when ν is absolutely continuous or singular with respect to μ.

For general Radon measures, $D_\mu \nu$ is identified by combining these two cases and applying the Lebesgue decomposition theorem of ν into two measures ν_o and ν_1, where the first is absolutely continuous and the second is singular with respect to μ.[14]

10.1   Representing $D_\mu \nu$ for $\nu \ll \mu$.

Lemma 10.1. *Assume that ν is absolutely continuous with respect to μ. Then*

$$\nu([D_\mu \nu = 0]) = 0.$$

Proof. Let $\mathcal{E}$ be the Borel set claimed by Proposition 9.1 and appearing in (9.1). Then for all $t > 0$,

$$[D_\mu \nu = 0] \subset \mathcal{E} \bigcup [D_\mu^- \nu < t].$$

From this and (8.5)$_-$, since $\nu \ll \mu$,

$$\nu([D_\mu \nu = 0]) \leq \nu([D_\mu^- \nu < t]) \leq t \mu([D_\mu \nu = 0]).$$

If μ is finite, the conclusion follows by letting $t \to 0$. In general, first restrict μ and ν to a ball B_n centered at the origin and radius n and then let $n \to \infty$. $\qquad\square$

---

[13]Compare with Proposition 1.3(i) of Chapter III.
[14]See Theorem 18.2 of Chapter III.

It follows from Lemma 10.1 that

$$\nu([D_\mu\nu = 0]) = \int_{[D_\mu\nu=0]} D_\mu\nu\, d\mu = 0.$$

The next proposition asserts that such a formula actually holds with the set $[D_\mu\nu = 0]$ replaced by any μ-measurable set E.

Proposition 10.2. *Assume that ν is absolutely continuous with respect to μ. Then for every μ-measurable set E,*

$$\nu(E) = \int_E D_\mu\nu\, d\mu. \tag{10.1}$$

Proof. Let $E \subset \mathbb{R}^N$ be μ-measurable, and for $t > 1$ and $n \in \mathbb{Z}$, set

$$E_n = \{x \in E - [D_\mu\nu = 0] \mid t^n \le D_\mu\nu < t^{n+1}\}.$$

By construction,

$$E - \bigcup_{n\in\mathbb{Z}} E_n \subset [D_\mu\nu = 0].$$

Therefore,

$$\nu\left(E - \bigcup_{n\in\mathbb{Z}} E_n\right) = 0.$$

From this and $(8.5)_-$,

$$\nu(E) = \sum_{n\in\mathbb{Z}} \nu(E_n) \le \sum_{n\in\mathbb{Z}} t^{n+1}\mu(E_n)$$

$$= t\sum_{n\in\mathbb{Z}} t^n\mu(E_n) \le t\sum_{n\in\mathbb{Z}} \int_{E_n} D_\mu\nu\, d\mu$$

$$= t\int_E D_\mu\nu\, d\mu.$$

Similarly, using $(8.5)_+$,

$$\nu(E) = \sum_{n\in\mathbb{Z}} \nu(E_n) \ge \sum_{n\in\mathbb{Z}} t^n\mu(E_n)$$

$$\ge \frac{1}{t}\sum_{n\in\mathbb{Z}} t^{n+1}\mu(E_n) \ge \frac{1}{t}\sum_{n\in\mathbb{Z}} \int_{E_n} D_\mu\nu\, d\mu$$

$$= \frac{1}{t}\int_E D_\mu\nu\, d\mu.$$

From this,

$$\frac{1}{t}\int_E D_\mu\nu\, d\mu \le \nu(E) \le t\int_E D_\mu\nu\, d\mu$$

for all $t > 1$. Letting $t \to 1$ proves (10.1). $\qquad\square$

10.2   Representing $D_\mu v$ for $v \perp \mu$. Continue to assume that μ and v are two Radon measures in $\mathbb{R}^N$ defined on the same σ-algebra $\mathcal{A}$.

Proposition 10.3. *Assume that v is singular with respect to μ. There exists a Borel set $\mathcal{E}_\perp$ of μ-measure zero such that*

$$D_\mu v(x) = 0 \quad \text{for all } x \in \mathbb{R}^N - \mathcal{E}_\perp. \tag{10.2}$$

Proof. Since μ and v are singular, $\mathbb{R}^N$ can be partitioned into two disjoints sets $\mathbb{R}^N_\mu$ and $\mathbb{R}^N_v$ such that for every $E \in \mathcal{A}$,[15]

$$\mu\left(E \bigcap \mathbb{R}^N_v\right) = 0 \quad \text{and} \quad v\left(E \bigcap \mathbb{R}^N_\mu\right) = 0.$$

Let $\{t_n\}$ be the sequence of the positive rational numbers, and set

$$E_n = [D_\mu v > t_n] \quad \text{and} \quad E = [D_\mu v > 0] = \cup E_n.$$

The sets E_n are Borel sets and

$$E_n \cap \mathbb{R}^N_\mu \subset \{x \in \mathbb{R}^N \mid D_\mu v(x) > t_n\}.$$

Therefore, by $(8.5)_+$ of Proposition 8.1,

$$\mu(E_n) = \mu(E_n \cap \mathbb{R}^N_\mu) \leq \frac{1}{t_n} v(E_n \cap \mathbb{R}^N_\mu) = 0.$$

From this, $\mu(E) \leq \sum \mu(E_n) = 0$. $\square$

11   The Lebesgue differentiation theorem

Let μ be a Radon measure in $\mathbb{R}^N$ defined on a σ-algebra $\mathcal{A}$. A function $f : \mathbb{R}^N \to \mathbb{R}^*$ measurable with respect to μ is said to be locally μ-integrable in $\mathbb{R}^N$ if

$$\int_E |f| d\mu < \infty \quad \text{for every bounded set } E \in \mathcal{A}.$$

If f is nonnegative, the formula

$$\mathcal{A} \ni E \longrightarrow v(E) = \int_E f d\mu$$

defines a Radon measure v in $\mathbb{R}^N$, absolutely continuous with respect to μ, whose Radon–Nikodým derivative with respect to μ is f. Moreover, such an f is unique, up to a set of μ-measure zero. Therefore, by Proposition 10.2,

$$D_\mu v = \frac{dv}{d\mu} = f \quad \mu\text{—a.e. in } \mathbb{R}^N.$$

---

[15] See Section 18 of Chapter III.

Now let f be locally μ-integrable in $\mathbb{R}^N$ and of variable sign. Writing $f = f^+ - f^-$ and applying the same reasoning separately to $f^\pm$, an N-dimensional version of the Lebesgue differentiation theorem for general Radon measures is proved.

Theorem 11.1 (Lebesgue[16]). *Let μ be a Radon measure in $\mathbb{R}^N$ and let $f : \mathbb{R}^N \to \mathbb{R}^*$ be locally μ-integrable. Then*

$$\lim_{\rho \to 0} \frac{1}{\mu(B_\rho(x))} \int_{B_\rho(x)} f d\mu = f(x) \quad for \; \mu\text{---a.e. } x \in \mathbb{R}^N \tag{11.1}$$

If μ is the Lebesgue measure in $\mathbb{R}$ and f is locally Lebesgue integrable, the limit in (11.1) takes the form

$$\lim_{h \to 0} \frac{1}{2h} \int_{x-h}^{x+h} f(y) dy = f(x) \quad \text{for a.e. } x \in \mathbb{R}. \tag{11.1$_{N=1}$}$$

In this sense, (11.1) can be regarded as an N-dimensional notion of taking the derivative of an integral at a fixed point $x \in \mathbb{R}^N$.

11.1   Points of density. Let $E \subset \mathbb{R}^N$ be μ measurable. Applying (11.1) with $f = \chi_E$ gives

$$\lim_{\rho \to 0} \frac{\mu\left(E \cap B_\rho(x)\right)}{\mu(B_\rho(x))} = \chi_E(x), \quad \mu\text{---a.e. in } E.$$

A point $x \in E$ for which such a limit is 1 is a *point of density* of E.

Corollary 11.2. *Almost every point of a μ-measurable set $E \subset \mathbb{R}^N$ is a point of density for E.*

11.2   Lebesgue points of an integrable function. Let μ be a Radon measure in $\mathbb{R}^N$ and let f be locally μ-integrable. The points $x \in \mathbb{R}^N$ where (11.1) holds form a set called the *set of differentiability* of f.

A point x is a *Lebesgue point* for f if

$$\lim_{\rho \to 0} \frac{1}{\mu(B_\rho(x))} \int_{B_\rho(x)} |f(y) - f(x)| d\mu = 0. \tag{11.2}$$

A Lebesgue point is a differentiability point for f. The converse is false.

Theorem 11.3. *Let μ be a Radon measure in $\mathbb{R}^N$, and let f be locally μ-integrable. There exists a μ-measurable set $\mathcal{E} \subset \mathbb{R}^N$ of μ-measure zero such that (11.2) holds for all $x \in \mathbb{R}^N - \mathcal{E}$.*

---

[16]A. S. Besicovitch, A general form of the covering principle and relative differentiation of additive functions, *Proc. Cambridge Philos. Soc.*, **I-41** (1945), 103–110, **II-42** (1946), 1–10.

Proof. Let r_n be a rational number. The function $|f - r_n|$ is locally μ-integrable. Therefore, there exists a Borel set $\mathcal{E}_n \subset \mathbb{R}^N$ of μ-measure zero such that

$$\lim_{\rho \to 0} \frac{1}{\mu(B_\rho(x))} \int_{B_\rho(x)} |f - r_n| d\mu = |f(x) - r_n|.$$

Since f is locally μ-integrable in $\mathbb{R}^N$, there exists a μ-measurable set $\mathcal{E}_o \subset \mathbb{R}^N$ of μ-measure zero such that $f(x)$ is finite for all $x \in \mathbb{R}^N - \mathcal{E}_o$. The set

$$\mathcal{E} = \bigcup_{n=0}^{\infty} \mathcal{E}_n$$

is of μ-measure zero, and for all $x \in \mathbb{R}^N - \mathcal{E}$,

$$\lim_{\rho \to 0} \frac{1}{\mu(B_\rho(x))} \int_{B_\rho(x)} |f - f(x)| d\mu$$

$$\leq \lim_{\rho \to 0} \frac{1}{\mu(B_\rho(x))} \int_{B_\rho(x)} |f - r_n| d\mu$$

$$+ \lim_{\rho \to 0} \frac{1}{\mu(B_\rho(x))} \int_{B_\rho(x)} |r_n - f(x)| d\mu$$

$$\leq 2|f(x) - r_n|$$

for all rational numbers $\{r_n\}$. Since $f(x)$ is finite, there exists a sequence of rational numbers $\{r_{n'}\}$ converging to $f(x)$. Thus (11.2) holds for all $x \in (\mathbb{R}^N - \mathcal{E})$. $\quad\square$

Corollary 11.4. *Let μ be a Radon measure in $\mathbb{R}^N$, and let f be locally μ-integrable. Almost every point $x \in \mathbb{R}^N$ is a Lebesgue point for f.*

12 Regular families

Let μ be a Radon measure in $\mathbb{R}^N$. For a fixed $x \in \mathbb{R}^N$, a family $\mathcal{F}_x$ of μ-measurable subsets of $\mathbb{R}^N$ is said to be *regular* at x if the following hold:

(i) For every $\varepsilon > 0$, there exists $S \in \mathcal{F}_x$ such that diam$\{S\} \leq \varepsilon$.

(ii) There exists a constant $c \geq 1$ such that for each $S \in \mathcal{F}_x$,

$$\mu(B(x)) \leq c\mu(S), \tag{12.1}$$

where $B(x)$ is the smallest ball in $\mathbb{R}^N$ centered at x and containing S.

Condition (i) asserts, roughly speaking, that the sets $S \in \mathcal{F}_x$ shrink to x although x is not required to be in any of the sets $S \in \mathcal{F}_x$. Condition (ii) says that each S is, roughly speaking, comparable to a ball centered at x.

If μ is the Lebesgue measure in $\mathbb{R}^N$, examples of regular families $\mathcal{F}_o$ at the origin include the collection of cubes, ellipsoids, or regular polygons centered at the origin.

An example of a regular family $\mathcal{F}_o$ whose sets S do not contain the origin is the collection of spherical annuli of the form $\{\frac{1}{2}\rho < |x| < \rho\}$. The sets in $\mathcal{F}_x$ have no symmetry restrictions.

The sets shrinking to a point x in Theorem 11.1 need not be balls, provided they shrink to x along a regular family $\mathcal{F}_x$.

Proposition 12.1. *Let f be locally μ-integrable in $\mathbb{R}^N$. Then if x is a Lebesgue point for f and $\mathcal{F}_x$ is a regular family at x,*

$$\lim_{\substack{\operatorname{diam}\{S\}\to 0 \\ S\in\mathcal{F}_x}} \frac{1}{\mu(S)} \int_S |f(y) - f(x)| d\mu = 0. \qquad (12.2)$$

In particular,

$$f(x) = \lim_{\substack{\operatorname{diam}\{S\}\to 0 \\ S\in\mathcal{F}_x}} \frac{1}{\mu(S)} \int_S f(y) d\mu. \qquad (12.3)$$

Proof. Having fixed $S \in \mathcal{F}_x$, let $B(x)$ be the ball satisfying (12.1). Then

$$\frac{1}{\mu(S)} \int_S |f(y) - f(x)| d\mu \leq \frac{c}{\mu(B(x))} \int_{B(x)} |f(y) - f(x)| d\mu. \qquad \square$$

Referring back to $(11.1)_{N=1}$, these remarks imply that for locally Lebesgue integrable functions of one variable,

$$\lim_{h\to 0} \frac{1}{h} \int_x^{x+h} f(y) = f(x) \quad \text{for a.e. } x \in \mathbb{R}. \qquad (11.1)'_{N=1}$$

13   Convex functions

A function f from an open interval (a, b) into $\mathbb{R}^*$ is convex if for every pair $x, y \in (a, b)$ and every $t \in [0, 1]$,

$$f(tx + (1 - t)y) \leq tf(x) + (1 - t)f(y).$$

A function f is concave if $-f$ is convex. The two-dimensional set

$$\mathcal{G}_f = \{(x, y) \in \mathbb{R}^2 | x \in (a, b), \ y \geq f(x)\}$$

is the *epigraph* of f. The function f is convex if and only if its epigraph is convex.

The positive linear combination of convex functions is convex and the limit of a sequence of convex functions is convex.

Proposition 13.1. *Let $\{f_\alpha\}$ be a family of convex functions defined in (a, b). Then the function $f = \sup f_\alpha$ is convex on (a, b).*

Proof. Fix $x, y \in (a, b)$ and $t \in [0, 1]$, and assume first that $f(tx + (1 - t)y)$ is finite. Having fixed an arbitrary $\varepsilon > 0$, there exists α such that

$$f(tx + (1 - t)y) \le f_\alpha(tx + (1 - t)y) + \varepsilon$$
$$\le t f_\alpha(x) + (1 - t) f_\alpha(y) + \varepsilon$$
$$\le t f(x) + (1 - t) f(y) + \varepsilon.$$

If $f(tx + (1 - t)y) = \infty$, having fixed an arbitrarily large number k, there exists α such that

$$k \le t f_\alpha(x) + (1 - t) f_\alpha(y). \qquad \square$$

Proposition 13.2. *Let f be a real-valued convex function in some interval $(a, b) \subset \mathbb{R}$. Then the function*

$$y \longrightarrow \mathcal{F}(x; y) = \frac{f(x) - f(y)}{x - y}, \quad x, y \in (a, b), \quad x \ne y,$$

is nondecreasing.

Proof. Assume that $y > x$. It suffices to show that $\mathcal{F}(x; z) \le \mathcal{F}(x; y)$ for $z = tx + (1 - t)y$ for all $t \in (0, 1)$. By the convexity of f,

$$\mathcal{F}(x; z) = \frac{f(tx + (1 - t)y) - f(x)}{(1 - t)(y - x)}$$
$$\le \frac{(1 - t)[f(y) - f(x)]}{(1 - t)(y - x)} = \mathcal{F}(x; y).$$

The proof for $y < x$ is analogous. $\qquad \square$

By symmetry, the function $x \to \mathcal{F}(x; y)$ is also nondecreasing.

Proposition 13.3. *Let f be a real-valued convex function in some interval $[a, b] \subset \mathbb{R}$. Then f is locally Lipschitz continuous in (a, b).*

Proof. Fix a subinterval $[c, d] \subset (a, b)$. Then

$$\mathcal{F}(c; a) = \frac{f(c) - f(a)}{c - a}$$
$$\le \frac{f(x) - f(y)}{x - y}$$
$$\le \frac{f(b) - f(d)}{b - d} = \mathcal{F}(b; d)$$

for all $x, y \in [c, d]$. If the difference quotient $\mathcal{F}(x; y)$ is nonnegative, $\mathcal{F}(b; d)$ also is nonnegative. Therefore,

$$|f(x) - f(y)| \le \mathcal{F}(b; d)|x - y|.$$

If the difference quotient $\mathcal{F}(x; y)$ is negative, then

$$|f(x) - f(y)| \le -\mathcal{F}(a; c)|x - y|. \qquad \square$$

Proposition 13.4. *Let f be a real-valued convex function in (a, b). Then f is differentiable a.e. in (a, b). Moreover, the right and left derivatives $D_{\pm} f(x)$ exist and are finite at each $x \in (a, b)$ and are both monotone nondecreasing functions. Also, $D_- f(x) \leq D_+ f(x)$ for all $x \in (a, b)$.*

Proof. For each $x \in (a, b)$ fixed, the function

$$(h, k) \longrightarrow \mathcal{F}(x + h; x + k)$$

is nondecreasing in both variables. Therefore, the limits

$$D_- f(x) = \lim_{h \nearrow 0} \mathcal{F}(x + h; x) \leq \lim_{k \searrow 0} \mathcal{F}(x; x + k) = D_+ f(x).$$

exist and are finite. Since f is absolutely continuous in every closed subinterval of (a, b), it is a.e. differentiable in (a, b), and $D_- f(x) = D_+ f(x)$ for a.e. $x \in (a, b)$. If $x < y$,

$$D_+ f(x) = \lim_{h \searrow 0} \mathcal{F}(x + h; x) \leq \lim_{h \searrow 0} \mathcal{F}(y + h; y) = D_+ f(y).$$

Thus $D_{\pm} f$ are both nondecreasing.   □

14 Jensen's inequality

Let φ be a real-valued, convex function in some interval (a, b). For a fixed $x \in (a, b)$, consider the set

$$\mathcal{D}_x \varphi = [D_- \varphi(x), D_+ \varphi(x)].$$

If φ is differentiable at x, then $\mathcal{D}_x \varphi = \varphi'(x)$. Otherwise, $\mathcal{D}_x \varphi$ is an interval. Fix $\alpha \in (a, b)$ and $m \in \mathcal{D}_\alpha \varphi$. Since φ is convex, the line through $(\alpha, \varphi(\alpha))$ and slope m lies below the epigraph of φ. In particular,

$$m(\eta - \alpha) + \varphi(\alpha) \leq \varphi(\eta) \quad \text{for all } \eta \in (a, b). \tag{14.1}$$

Proposition 14.1 (Jensen[17]). *Let E be a measurable set of finite measure, and let $f : E \to \mathbb{R}$ be integrable in E. Then for every real-valued, convex function φ defined in $\mathbb{R}$,*

$$\varphi \left(\frac{1}{\mu(E)} \int_E f \, d\mu \right) \leq \frac{1}{\mu(E)} \int_E \varphi(f) \, d\mu. \tag{14.2}$$

Proof. Applying (14.1) for the choices

$$\alpha = \frac{1}{\mu(E)} \int_E f \, d\mu, \qquad \eta = f(x) \quad \text{for a.e. } x \in E$$

---

[17]J. Jensen, Sur les fonctions convexes et les inégalités entre les valeurs moyennes, *Acta Math.*, **30** (1906), 175–193. No restriction is placed on the underlying measure space $\{X, \mathcal{A}, \mu\}$.

yields

$$\varphi\left(\frac{1}{\mu(E)}\int_E f\,dx\right) + m\left(f(x) - \frac{1}{\mu(E)}\int_E f\,dx\right) \le \varphi(f(x)).$$

Integrate over E and divide by the measure of E. $\square$

15   Extending continuous functions

Let f be a continuous function defined on a set $E \subset \mathbb{R}^N$ with values in $\mathbb{R}$ and with modulus of continuity

$$\omega_f(s) = \sup_{\substack{|x-y|<s \\ x,y\in E}} |f(x) - f(y)|, \quad s > 0.$$

The function $s \to \omega_f(s)$ is nonnegative and nondecreasing in $[0, \infty)$. We also assume that $\omega_f(\cdot)$ is dominated in $[0, \infty)$ by some increasing, affine function $\ell(\cdot)$; i.e.,

$$\omega_f(s) \le as + b \quad \text{for all } s \in [0, \infty) \quad \text{for some } a, b \in \mathbb{R}^+. \tag{15.1}$$

The function f is uniformly continuous in E if and only if $\omega_f(s) \to 0$ as $s \to 0$. We denote by $s \to c_f(s)$ the concave modulus of continuity of f, i.e., the smallest concave function in $[0, \infty)$ whose graph lies above the graph of $s \to \omega_f(s)$. Such a function can be constructed as

$$c_f(s) = \inf\{\ell(s)|\ell \text{ is affine and } \ell \ge \omega_f \text{ in } [0, \infty)\}.$$

It follows from the definitions that

$$c_f(|x - y|) - |f(x) - f(y)| \ge 0 \quad \text{for all } x, y \in E. \tag{15.2}$$

Theorem 15.1 (Kirzbraun–Pucci[18]). *Let f be a real-valued, uniformly continuous function on a set $E \subset \mathbb{R}^N$ with modulus of continuity ω_f satisfying (15.1). There exists a continuous function $\tilde{f}$ defined on $\mathbb{R}^N$ that coincides with f on E. Moreover, f and $\tilde{f}$ have the same concave modulus of continuity c_f, and*

$$\sup_{\mathbb{R}^N} \tilde{f} = \sup_E f, \quad \inf_{\mathbb{R}^N} \tilde{f} = \inf_E f.$$

---

[18]If the modulus of continuity is of Lipschitz type; i.e., if $\omega_f(s) = Ls$ for some positive constant L, then $c_f(s) = Ls$. In such a case, the theorem is in M. D. Kirzbraun, Über die zusammenziehenden und Lipschitzschen Transformationen, *Fund. Math.*, **22** (1934), 77–108. For a general modulus of continuity, the extension has been taken from the 1974 lectures on real analysis by C. Pucci at the University of Florence, Italy.

Proof. For each $x \in \mathbb{R}^N$, set

$$g(x) = \inf_{y \in E}\{f(y) + c_f(|x - y|)\}.$$

The required extension is

$$\tilde{f}(x) = \min\{g(x); \sup_E f\}.$$

If $x \in E$, by (15.2),

$$f(y) + c_f(|x - y|) \geq f(x) + c_f(|x - y|) - |f(x) - f(y)| \geq f(x)$$

for all $y \in E$. Therefore, $g = f$ within E. Next, for all $x \in \mathbb{R}^N$ and all $y \in E$,

$$\inf_E f + \inf_{y \in E} c_f(|x - y|) \leq g(x) \leq f(y) + c_f(|x - y|).$$

Therefore,

$$\inf_{\mathbb{R}^N} g = \inf_E f \quad \text{and} \quad \sup_{\mathbb{R}^N} \tilde{f} = \sup_E f.$$

To prove that f and $\tilde{f}$ have the same concave modulus of continuity, it suffices to prove that g has the same concave modulus of continuity as f. Fix $x_1, x_2 \in \mathbb{R}^N$ and $\varepsilon > 0$. There exists $y \in E$ such that

$$g(x_1) \geq f(y) + c_f(|x_1 - y|) - \varepsilon.$$

Therefore, for such $y \in E$,

$$g(x_1) - g(x_2) \geq c_f(|x_1 - y|) - c_f(|x_2 - y|) - \varepsilon.$$

If $|x_2 - y| \leq |x_1 - x_2|$,

$$g(x_1) - g(x_2) \geq -c_f(|x_1 - x_2|) - \varepsilon.$$

Otherwise,

$$|x_1 - y| > |x_2 - y| - |x_1 - x_2| > 0.$$

Since $s \to c_f(s)$ is concave, $-c_f(\cdot)$ is convex, and by Proposition 13.2,

$$\frac{c_f(|x_1 - y|) - c_f(0)}{|x_1 - y|} \geq \frac{c_f(|x_1 - y| + |x_2 - x_1|) - c_f(|x_1 - x_2|)}{|x_1 - y| + |x_2 - x_1| - |x_1 - x_2|}$$
$$\geq \frac{c_f(|x_2 - y|) - c_f(|x_1 - x_2|)}{|x_1 - y|}.$$

From this, taking into account that $c_f(0) = 0$, we deduce

$$c_f(|x_1 - y|) - c_f(|x_2 - y|) \geq -c_f(|x_1 - x_2|).$$

Thus in either case,

$$g(x_1) - g(x_2) \geq -c_f(|x_1 - x_2|) - \varepsilon.$$

Interchanging the roles of x_1 and x_2 and taking into account that $\varepsilon > 0$ is arbitrary gives

$$|g(x_1) - g(x_2)| \leq c_f(|x_1 - x_2|). \qquad \square$$

16   The Weierstrass approximation theorem

Theorem 16.1 (Weierstrass[19]). *Let f be a real-valued, uniformly continuous function defined on a bounded set $E \subset \mathbb{R}^N$. There exists a sequence of polynomials $\{P_j\}$ such that*

$$\sup_E |f - P_j| \longrightarrow 0 \quad \text{as } j \longrightarrow \infty.$$

Proof. By Theorem 15.1, we may regard f as defined in the whole $\mathbb{R}^N$ with modulus of continuity ω_f. After a translation and dilation, we may assume that E is contained in the interior of the unit cube Q centered at the origin of $\mathbb{R}^N$ and with faces parallel to the coordinate planes.

For $x \in \mathbb{R}^N$ and $\delta > 0$, we let $Q_\delta(x)$ denote the cube of edge 2δ centered at x and congruent to Q. For $j \in \mathbb{N}$, set

$$p_j(x) = \frac{1}{\alpha_j^N} \prod_{i=1}^{N} (1 - x_i^2)^j, \quad \alpha_j = \int_{-1}^{1} (1 - t^2)^j dt.$$

These are polynomials of degree $2jN$ satisfying

$$\int_{Q_1(x)} p_j(x - y)dy = \int_Q p_j(y)dy = 1$$

for all $j \in \mathbb{N}$ and all $x \in \mathbb{R}^N$. The approximating polynomials claimed by the theorem are

$$P_j(x) = \int_Q f(y)p_j(x - y)dy. \tag{16.1}$$

These are called the Stieltjes polynomials relative to f. For $x \in E$, compute

$$P_j(x) - f(x) = \int_Q f(y)p_j(x - y)dy - \int_{Q_1(x)} f(x)p_j(x - y)dy.$$

Let $\delta > 0$ be so small that $Q_\delta(x) \subset Q$. Then

$$|P_j(x) - f(x)| \leq \int_{Q_\delta(x)} |f(x) - f(y)|p_j(x - y)dy$$

$$+ \left| \int_{Q - Q_\delta(x)} f(y)p_j(x - y)dy \right|$$

$$+ \left| \int_{Q_1(x) - Q_\delta(x)} f(x)p_j(x - y)dy \right|$$

$$\leq \omega_f(\sqrt{N}\delta) + \sup_E |f| \int_{Q - Q_\delta(x)} p_j(x - y)dy$$

[19]K. Weierstrass, Über die analytische Darstellbarkeit sogenannter willkürlicher Funktionen einer reellen Veränderlichen, *Könress Preussichen Akad. Wiss.*, 1885.

$$+ \sup_{E} |f| \int_{Q_1(x)-Q_\delta(x)} p_j(x - y)dy.$$

To estimate the last two integrals, we observe that for $y \notin Q_\delta(x)$,

$$p_j(x - y) \le \alpha_j^{-N} (1 - \delta^2)^{jN}.$$

Moreover, from the definition of α_j,

$$\alpha_j \ge 2 \int_0^1 (1 - t)^j dt = \frac{2}{j + 1}.$$

Combining these calculations, we estimate

$$\sup_{E} |f - P_j| \le \omega_f\left(\sqrt{N}\delta\right) + 2 \sup_{E} |f| \left(\frac{j + 1}{2}\right)^N (1 - \delta^2)^{jN}. \qquad \square$$

Corollary 16.2. *Let E be a compact subset of $\mathbb{R}^N$. Then $C(E)$ endowed with the topology of the uniform convergence is separable.*

Proof. The collection of polynomials in the real variables $x_1, x_2, \ldots, x_N$ with rational coefficients is a countable, dense subset of $C(E)$. $\qquad \square$

17 The Stone–Weierstrass theorem

Let $\{X; \mathcal{U}\}$ be a compact Hausdorff space, and denote by $C(X)$ the collection of all real-valued, continuous functions defined in X. Setting

$$d(f, g) = \sup_{x \in X} |f(x) - g(x)|, \quad f, g \in C(X), \tag{17.1}$$

defines a complete metric in $C(X)$. We continue to denote by $C(X)$ the resulting metric space.

The sum of two functions in $C(X)$ is in $C(X)$, and the product of a function in $C(X)$ by a real number is an element of $C(X)$. Thus $C(X)$ is a vector space. One verifies that the operations of the sum and product by real numbers,

$$+ : C(X) \times C(X) \to C(X), \qquad \bullet : \mathbb{R} \times C(X) \to C(X),$$

are continuous with respect to the corresponding product topologies. Thus $C(X)$ is a topological vector space.

The space $C(X)$ is also an *algebra* in the sense that the product of any two functions in $C(X)$ remains in $C(X)$. More generally, a subset $\mathcal{F} \subset C(X)$ is an algebra if it is closed under the operations of sum, product, and product by real numbers. For example, the collection of functions $f \in C(X)$ that vanish at some fixed point $x_o \in X$ is an algebra. The intersection of all algebras containing a given subset of $C(X)$ is an algebra.

Since $\{X; \mathcal{U}\}$ is Hausdorff, its points are closed. Therefore, having fixed $x \neq y$ in X, there exists a continuous function $f : X \to [0, 1]$ such that $f(x) = 0$ and $f(y) = 1$.[20] Thus there exists an element of $C(X)$ that distinguishes any two fixed, distinct points in X.

More generally an algebra $\mathcal{F} \subset C(X)$ *separates* points of X if for any pair of distinct points $x, y \in X$, there exists a function $f \in \mathcal{F}$ such that $f(x) \neq f(y)$.

For example, if E is a bounded, open subset of $\mathbb{R}^N$, the collection of all polynomials in the coordinate variables forms an algebra $\mathcal{P}$ of functions in $C(\overline{E})$. Such an algebra trivially separates points.

The classical Weierstrass theorem asserts that every $f \in C(\overline{E})$ can be approximated by elements of $\mathcal{P}$ in the metric of (17.1). Equivalently, $C(\overline{E})$ is the closure of $\mathcal{P}$ in the metric (17.1). The proof was based on constructing explicitly the approximating polynomials to a given $f \in C(\overline{E})$. Stone's theorem identifies the structure that a subset of $C(X)$ must possess to be dense in $C(X)$.

Theorem 17.1 (Stone[21]). *Let $\{X; \mathcal{U}\}$ be a compact Hausdorff space, and let $\mathcal{F} \subset C(X)$ be an algebra that separates points and that contains the constant functions. Then $\overline{\mathcal{F}} = C(X)$.*

18   Proof of the Stone–Weierstrass theorem

Proposition 18.1. *Let $\{X; \mathcal{U}\}$ be a compact Hausdorff space and let $\mathcal{F} \subset C(X)$ be an algebra. Then*

(i) *the closure $\overline{\mathcal{F}}$ in $C(X)$ is an algebra;*

(ii) *if $f \in \mathcal{F}$, then $|f| \in \overline{\mathcal{F}}$;*

(iii) *if f and g are in $\mathcal{F}$, then $\max\{f; g\}$ and $\min\{f; g\}$ are in $\overline{\mathcal{F}}$.*

Proof. The first statement follows from the structure of an algebra and the notion of closure in the metric (17.1). To prove (ii), we may assume without loss of generality that $|f| \leq 1$.

Regard f as a variable ranging over $[-1, 1]$. By the classical Weierstrass approximation theorem applied to the function

$$[-1, 1] \ni f \longrightarrow |f|,$$

having fixed $\varepsilon > 0$, there exists a polynomial $\mathcal{P}_\varepsilon(f)$ in the variable f such that

$$\sup_{f \in [-1,1]} ||f| - \mathcal{P}_\varepsilon(f)| \leq \varepsilon.$$

---

[20]Urysohn's lemma; see Section 2 of Chapter I.

[21]M. H. Stone, Generalized Weierstrass approximation theorem, *Math. Magazine*, **21** (1947/1948), 167–184, 237–254.

This, in turn, implies that

$$\sup_{x \in X} \left| |f(x)| - \mathcal{P}_\varepsilon(f(x)) \right| \leq \varepsilon.$$

Since $\mathcal{F}$ is an algebra, $\mathcal{P}_\varepsilon(f) \in \mathcal{F}$. Thus $|f|$ is in the closure of $\mathcal{F}$.

The last statements follow from (ii) and the identities

$$\max\{f; g\} = \frac{1}{2}(f + g) + \frac{1}{2}|f - g|,$$

$$\min\{f; g\} = \frac{1}{2}(f + g) - \frac{1}{2}|f - g|. \qquad \square$$

18.1 Proof of Stone's theorem. Having fixed $f \in C(X)$ and $\varepsilon > 0$, we exhibit a function $\varphi \in \overline{\mathcal{F}}$ such that $d(f, \varphi) \leq \varepsilon$. Since $\mathcal{F}$ separates points of X, for any two distinct points $\xi, \eta \in X$, there exists $h \in \mathcal{F}$ such that $h(\xi) \neq h(\eta)$. Since $\mathcal{F}$ contains the constants, there exist numbers λ and μ such that the function

$$\varphi_{\xi\eta} = \lambda h + 1 \cdot \mu,$$

is in $\mathcal{F}$ and satisfies

$$\varphi_{\xi\eta}(\xi) = f(\xi) \quad \text{and} \quad \varphi_{\xi\eta}(\eta) = f(\eta).$$

By keeping ξ fixed, regard $\varphi_{\xi\eta}$ as a family of continuous functions, parameterized with $\eta \in X$. Since $\varphi_{\xi\eta}$ and f coincide at η and are both continuous, for each $\eta \in X$, there exists an open set $\mathcal{O}_\eta$ containing η and such that

$$\varphi_{\xi\eta}(x) < f(x) + \varepsilon \quad \text{for all } x \in \mathcal{O}_\eta.$$

The collection of open sets $\mathcal{O}_\eta$ as η ranges over X is an open covering for X, from which we extract a finite one, for example, $\{\mathcal{O}_{\eta_1}, \mathcal{O}_{\eta_2}, \ldots, \mathcal{O}_{\eta_n}\}$, for some finite n. Set

$$\varphi_\xi = \min\{\varphi_{\xi\eta_1}, \varphi_{\xi\eta_2}, \ldots, \varphi_{\xi\eta_n}\}.$$

By Proposition 18.1(iii), $\varphi_\xi \in \overline{\mathcal{F}}$. Moreover, by construction,

$$\varphi_\xi(x) \leq f(x) + \varepsilon \quad \text{for all } x \in X$$

and

$$\varphi_\xi(\xi) = f(\xi) \quad \text{for all} \quad \xi \in X.$$

Since φ_ξ and f coincide at ξ and they are both continuous, for each $\xi \in X$, there exists an open set $\mathcal{O}_\xi$ containing ξ and such that

$$\varphi_\xi(x) > f(x) - \varepsilon \quad \text{for all } x \in \mathcal{O}_\xi.$$

The collection of open sets $\mathcal{O}_\xi$ as ξ ranges over X is an open covering for X, from which we extract a finite one, for example, $\{\mathcal{O}_\xi, \mathcal{O}_\xi, \dots, \mathcal{O}_{\xi_m}\}$, for some finite m. Set

$$\varphi = \max\{\varphi_{\xi_1}, \varphi_{\xi_2}, \dots, \varphi_{\xi_m}\}.$$

By Proposition 18.1(iii), $\varphi \in \overline{\mathcal{F}}$. Moreover, by construction,

$$|f(x) - \varphi(x)| \le \varepsilon \quad \text{for all } x \in X. \qquad \square$$

19   The Ascoli–Arzelà theorem

Let $E \subset \mathbb{R}^N$ be open. A sequence of functions $\{f_n\}$ from E into $\mathbb{R}$ is equibounded if there exists $M > 0$ such that

$$\sup_E |f_n| \le M \quad \text{for all } n \in \mathbb{N}.$$

The sequence $\{f_n\}$ is equicontinuous in E if there exists a continuous increasing function $\omega : \mathbb{R}^+ \to \mathbb{R}^+$ such that $\omega(0) = 0$ and for all $x, y \in E$,

$$|f_n(x) - f_n(y)| \le \omega(|x - y|) \quad \text{for all } n \in \mathbb{N}.$$

Theorem 19.1 (Ascoli[22]). *Let $\{f_n\}$ be a sequence of equibounded and equicontinuous functions in E. There exist a subsequence $\{f_{n'}\}$ and a continuous function $f : E \to \mathbb{R}$ such that*

(i) $f_{n'}(x) \longrightarrow f(x)$ *for all $x \in E$;*

(ii) $|f(x) - f(y)| \le \omega(|x - y|)$ *for every $x, y \in E$;*

(iii) $f_{n'} \longrightarrow f$ *uniformly on compact sets $\mathcal{K} \subset E$.*

Proof. Let $\mathbf{Q}$ denote the set of points of $\mathbb{R}^N$ whose coordinates are rational. Such a set is countable and dense in E. Let $x_1 \in \mathbf{Q} \cap E$. Since the sequence of numbers $\{f_n(x_1)\}$ is bounded, we may select a subsequence $\{f_{n_1}(x_1)\}$ convergent to some real number that we denote with $f(x_1)$; i.e.,

$$f_{n_1}(x_1) \longrightarrow f(x_1).$$

If $x_2 \in \mathbf{Q} \cap E$, the sequence of numbers $\{f_{n_1}(x_2)\}$ is bounded, and we may select a convergent subsequence

$$f_{n_{1,2}}(x_2) \longrightarrow f(x_2).$$

---

[22]G. Ascoli, Le curve limiti di una varietà data di curve, *Rend. Accad. Lincei*, **18** (1884), 521–586; C. Arzelà, Sulle funzioni di linee, *Mem. Accad. Sci. Bologna*, **5-5** (1894/1895), 225–244. First proved by Ascoli for equi-Lipschitz functions and extended by Arzelà to a general family of equicontinuous functions.

Proceeding in this fashion, we may, by a diagonalization process, select out of $\{f_n\}$ a subsequence $\{f_{n'}\}$ such that

$$f_{n'}(x) \longrightarrow f(x) \quad \text{for all } x \in \mathbf{Q} \cap E.$$

Next, fix $x \in (E - \mathbf{Q})$. Since $\mathbf{Q}$ is dense in E, for each $\varepsilon > 0$, there exists $x_\ell \in \mathbf{Q} \cap E$ such that $|x - x_\ell| < \varepsilon$. Therefore, by the assumption of equicontinuity,

$$\begin{aligned}
|f_{n'}(x) - f_{m'}(x)| &\leq |f_{n'}(x) - f_{n'}(x_\ell)| \\
&\quad + |f_{m'}(x) - f_{m'}(x_\ell)| \\
&\quad + |f_{n'}(x_\ell) - f_{m'}(x_\ell)| \\
&\leq 2\omega(\varepsilon) + |f_{n'}(x_\ell) - f_{m'}(x_\ell)|.
\end{aligned}$$

Since $\{f_{n'}(x_\ell)\}$ is convergent, there exists a positive integer $m(x_\ell)$ large enough that

$$|f_{n'}(x_\ell) - f_{m'}(x_\ell)| \leq \varepsilon \quad \text{for all } n', m' > m(x_\ell).$$

Therefore, for all such n' and m',

$$|f_{n'}(x) - f_{m'}(x)| \leq \varepsilon + 2\omega(\varepsilon).$$

This implies that $\{f_{n'}(x)\}$ is a Cauchy sequence, and we denote its limit by $f(x)$. To prove (iii), fix $x, y \in E$ and write

$$|f(x) - f(y)| = \lim |f_{n'}(x) - f_{n'}(y)| \leq \omega(|x - y|).$$

Let $\mathcal{K}$ be a compact subset of E and fix $\varepsilon > 0$. The collection of balls $B_\varepsilon(x)$ of radius ε centered at points $x \in E$ covers $\mathcal{K}$, and we may select a finite subcover, say,

$$B_\varepsilon(x_\ell), \quad \ell = 1, 2, \ldots k \quad \text{for some } k \in \mathbb{N}.$$

We may also select a positive integer $m(k)$ large enough that

$$|f_{n'}(x_\ell) - f(x_\ell)| \leq \varepsilon \quad \text{for all } n' > m(k) \quad \text{for all } \ell = 1, 2, \ldots, k.$$

Each $x \in \mathcal{K}$ is contained in some ball $B_\varepsilon(x_\ell)$. Therefore, for all $n' \geq m(k)$,

$$\begin{aligned}
|f_{n'}(x) - f(x)| &\leq |f_{n'}(x) - f_{n'}(x_\ell)| + |f_{n'}(x_\ell) - f(x_\ell)| + |f(x) - f(x_\ell)| \\
&\leq \varepsilon + 2\omega(\varepsilon). \qquad \square
\end{aligned}$$

19.1   Precompact subsets of $C(\overline{E})$. Let E be a bounded, open subset of $\mathbb{R}^N$, and denote by $C(\overline{E})$ the collection of all real-valued, continuous functions defined in $\overline{E}$ endowed with the metric (17.1).

Proposition 19.2. *Let $K \subset C(\overline{E})$ be a subset of equibounded and equicontinuous functions. Then K is precompact in $C(\overline{E})$.*

Proof. Let $\overline{K}$ be the closure of K in the metric (17.1). The functions in $\overline{K}$ are equibounded and equicontinuous. Therefore, every sequence $\{f_n\}$ of functions in $K(M; \omega)$ contains a subsequence convergent in the same metric to a function $f \in K$. Thus $\overline{K}$ is sequentially compact and hence compact.[23]   □

Corollary 19.3. *Let $K \subset C(\overline{E})$ be a subset of equibounded and equicontinuous functions. Then K admits a finite ε-net for all $\varepsilon > 0$.*

PROBLEMS AND COMPLEMENTS

1 FUNCTIONS OF BOUNDED VARIATIONS

1.1 The continuous function on $[0, 1]$

$$f(x) = \begin{cases} x^2 \cos \dfrac{\pi}{x} & \text{for } x \in (0, 1], \\ 0 & \text{for } x = 0 \end{cases}$$

is of bounded variation in $[0, 1]$.

1.2 Let f be the continuous function defined in $[0, 1]$ by

$$f(0) = 0, \quad f\left(\frac{1}{2n+1}\right) = 0, \quad f\left(\frac{1}{2n}\right) = \frac{1}{2n} \quad \text{for all } n \in \mathbb{N},$$

$$f \text{ is affine on the intervals } \left[\frac{1}{m+1}, \frac{1}{m}\right] \quad \text{for all } m \in \mathbb{N}.$$

Such an f is not of bounded variation in $[0, 1]$.

1.3 Prove Propositions 1.1–1.3.

1.4 Let $\{f_n\}$ be a sequence of functions on $[a, b]$ converging pointwise in $[a, b]$ to a function f. Then

$$\mathcal{V}_f[a, b] \leq \liminf \mathcal{V}_{f_n}[a, b],$$

and strict inequality may occur as shown by the sequence

$$f_n(x) = \begin{cases} 0 & \text{for } x = 0, \\ \dfrac{1}{2} & \text{for } x \in \left(0, \dfrac{1}{n}\right], \\ 0 & \text{for } x \in \left(\dfrac{1}{n}, 1\right]. \end{cases}$$

---

[23] See Proposition 17.5 of Chapter I.

1.5 Let f be continuous and of bounded variation in $[a, b]$. Then the functions

$$x \longrightarrow V_f[a, x], \quad V_f^+[a, x], \quad V_f^-[a, x]$$

are continuous in $[a, b]$.

1.6 THE FUNCTION OF THE JUMPS. Let f be of bounded variation in $[a, b]$, and let $\mathcal{D}$ denote the set of points of discontinuity of f. For $c \in \mathcal{D}$, set

$$\lambda(c) = f(c) - f(c^-), \quad \text{the left jump of } f \text{ at } c,$$
$$\rho(c) = f(c^+) - f(c), \quad \text{the right jump of } f \text{ at } c.$$

The function of the jumps of f is defined by

$$J_f(x) = \sum_{c \in \mathcal{D} \cap [a, x]} \lambda(c) + \sum_{c \in \mathcal{D} \cap [a, x)} \rho(c).$$

The difference $(f - J_f)$ is continuous in $[a, b]$. Also, J_f is of bounded variation on $[a, b]$, and

$$V_f[a, b] = V_{(f - J_f)}[a, b] + V_{J_f}[a, b].$$

Therefore, a function f of bounded variation in $[a, b]$ can be decomposed into the continuous function $(f - J_f)$ and J_f. The latter bears the possible discontinuities of f in $[a, b]$.[24]

1.7 Construct a nondecreasing function in $[0, 1]$ that is discontinuous at all the rational points of $[0, 1]$.

1.8 THE SPACE $BV[a, b]$. Let $[a, b] \subset \mathbb{R}$ be a finite interval, and denote by $BV[a, b]$ the collection of all functions $f : [a, b] \to \mathbb{R}$ of bounded variation in $[a, b]$. One verifies that $BV[a, b]$ is a linear vector space. Also, setting

$$d(f, g) = |f(a) - g(a)| + V_{f-g}[a, b] \quad \text{for } f, g \in BV[a, b] \tag{1.1c}$$

defines a distance on $BV[a, b]$ by which $\{BV[a, b]; d\}$ is a metric space.

For any two functions $f, g \in BV[a, b]$, it holds that

$$\sup_{[a, b]} |f - g| \leq |f(a) - g(a)| + V_{f-g}[a, b]. \tag{1.2c}$$

Therefore, a Cauchy sequence in $BV[a, b]$ is also Cauchy in the sup-norm. The converse is false, as illustrated in Figure 1.1c.

The sequence $\{f_n\}$ generated as in Figure 1.1c is not a Cauchy sequence in the topology of $BV[0, 1]$, while it is a Cauchy sequence in the topology of $C[0, 1]$.

---

[24]See [44, pp. 14–15].

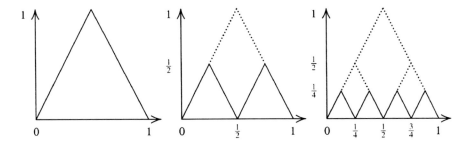

Figure 1.1c.

1.9   COMPLETENESS OF $BV[a, b]$. Let $\{f_n\}$ be a Cauchy sequence in $BV[a, b]$. There exists $f \in BV[a, b]$ such that $\{f_n\} \to f$ in the topology of $BV[a, b]$.

The proof is in two steps. First, one uses (1.2c) to identify the limit f. Then one proves that such an f is actually in $BV[a, b]$ by using that $\{f_n\}$ is Cauchy in $BV[a, b]$. As a consequence, $BV[a, b]$ is a complete metric space.

2   DINI DERIVATIVES

2.1   Compute $D^+ f(0)$ and $D_+ f(0)$ for the function in (1.1).

2.2   Let f have a maximum at some $c \in (a, b)$. Then $D^- f(c) \geq 0$.

2.3   Let f be continuous in $[a, b]$. If $D^+ f \geq 0$ in $[a, b]$, then f is nondecreasing in $[a, b]$. The assumption that f is continuous cannot be removed.

2.4   Let f be of bounded variation in $[a, b]$. Then the function of the jumps $x \to J_f(x)$ is a.e. differentiable in $[a, b]$ and $J'_f = 0$ a.e. in $[a, b]$.

2.5   A CONTINUOUS, NOWHERE-DIFFERENTIABLE FUNCTION. For a real number x, denote by $\{x\}$ the distance from x to its nearest integer and set[25]

$$f(x) = \sum_{n=0}^{\infty} \frac{\{10^n x\}}{10^n}. \tag{2.1c}$$

Each term of the series is continuous. Moreover, the series is uniformly convergent being majorized by the geometric series $\sum 10^{-n}$. Therefore, f is continuous.

Since $f(x) = f(x + j)$ for every integer j and all $x \in \mathbb{R}$, it suffices to consider $x \in [0, 1)$. Any such x has a decimal expansion of the form

$$x = 0.a_1 a_2 \ldots a_n \ldots,$$

where a_i are integers from 0 to 9. By excluding the case when $a_i = 9$ for all i larger than some m, such a representation is unique.

---

[25]B. L. Van der Waerden, Ein einfaches Beispiel einer nichtdifferenzierbaren stetigen Funktion, *Math. Z.*, **32** (1930), 474–475.

For $n \in \mathbb{N}$ fixed, compute

$$\{10^n x\} = 0.a_{n+1}a_{n+2}\cdots \qquad \text{if } 0.a_{n+1}a_{n+2}\cdots \leq \frac{1}{2},$$

$$\{10^n x\} = 1 - 0.a_{n+1}a_{n+2}\cdots \qquad \text{if } 0.a_{n+1}a_{n+2}\cdots > \frac{1}{2}.$$

Having fixed $x \in [0, 1)$, choose increments

$$h_m = \begin{cases} -10^{-m} & \text{if either } a_m = 4 \text{ or } a_m = 9, \\ +10^{-m} & \text{otherwise.} \end{cases}$$

Then form the difference quotients of f at x,

$$\frac{f(x+h_m) - f(x)}{h_m} = 10^m \sum_{n=0}^{\infty} \pm \frac{\{10^n (x \pm 10^{-m})\} - \{10^n x\}}{10^n}.$$

The numerators of the terms of this last series all vanish for $n \geq m$, whereas for $n = 0, 1, \ldots, (m-1)$, they are equal to $\pm 10^{n-m}$. Therefore, the difference quotient reduces to the sum of m terms each of the form ± 1. Such a sum is an integer, positive or negative, that has the same parity of m. Thus the limit as $h_m \to 0$ of the difference ratios does not exist.

Remark 2.1c. The function in (2.1c) is not of bounded variation in any interval $[a, b] \subset \mathbb{R}$.

2.6   AN APPLICATION OF THE BAIRE CATEGORY THEOREM. The existence of a continuous and nowhere-differentiable function can be established indirectly by a category-type argument. More generally, the Baire category theorem can be used to establish the existence of functions $f \in C[0, 1]$ with some prescribed property.

Proposition 2.1c (Banach[26]). *There exists a real-valued function in $[0, 1]$ that is continuous and such that its Dini numbers $|D_\pm f(x)|$ and $|D^\pm f(x)|$ are all infinity at every point of $[0, 1]$.*

Proof. For $n \in \mathbb{N}$, let E_n denote the collection of all functions $f \in C[0, 1]$ such that at least one of the four Dini numbers $|D_\pm f(x_o)|$ and $|D^\pm f(x_o)|$ is bounded by n for some $x_o \in [0, 1]$. Each E_n is closed and nowhere dense in $C[0, 1]$. Both statements are meant with respect to the topology of the uniform convergence in $C[0, 1]$.

To prove that E_n is nowhere dense in $C[0, 1]$, observe that any continuous function in $[0, 1]$ can be approximated in the sup-norm by continuous functions with polygonal graph of arbitrarily large Lipschitz constant. Then the complement $C[0, 1] - \bigcup E_n$ is nonempty.   □

[26]S. Banach, Über die Baire'sche Kategorie gewisser Funktionenmengen, *Stud. Math,.* **3** (1931). 174.

4   DIFFERENTIATING SERIES OF MONOTONE FUNCTIONS

4.1   Let $\{f_n\}$ be a sequence of functions of bounded variation in $[a, b]$ such that the series

$$\sum f_n(x) \quad \text{and} \quad \sum V_{f_n}[a, x]$$

are both convergent in $[a, b]$. Then the sum f of the first series is of bounded variation in $[a, b]$ and the derivative can be computed term by term a.e. in $[a, b]$.

5   ABSOLUTELY CONTINUOUS FUNCTIONS

5.1   Let f be absolutely continuous in $[a, b]$. Then f is Lipschitz continuous in $[a, b]$ if and only if f' is bounded.

5.2   The function

$$f(x) = \begin{cases} x^{1+\varepsilon} \sin \dfrac{1}{x} & \text{for } x \in (0, 1], \\ 0 & \text{for } x = 0 \end{cases}$$

is absolutely continuous in $[0, 1]$ for all $\varepsilon > 0$.

5.3   THE CANTOR TERNARY FUNCTION. Set $f(0) = 0$ and $f(1) = 1$. Divide the interval $[0, 1]$ into three equal subintervals, and on the central interval $[\frac{1}{3}, \frac{2}{3}]$, set $f = \frac{1}{2}$; i.e., f is defined to be the average of its values at the extremes of the parent interval $[0, 1]$.

Next, divide the interval $[0, \frac{1}{3}]$ into three equal subintervals, and on the central interval $[\frac{1}{3^2}, \frac{2}{3^2}]$, set $f = \frac{1}{4}$; i.e., f is defined to be the average of its values at the extremes of the parent interval $[0, \frac{1}{3}]$. Likewise, divide the interval $[\frac{2}{3}, 1]$ into three equal subintervals, and on the central interval $[\frac{7}{3^2}, \frac{8}{3^2}]$, set $f = \frac{3}{4}$; i.e., f is defined to be the average of its values at the extremes of the parent interval $[\frac{2}{3}, 1]$.

Proceeding in this fashion, we define f in $[0, 1] - C$ by successive averages and by continuity in the whole $[0, 1]$.[27]

By construction, f is nonconstant, nondecreasing, and continuous in $[0, 1]$. Since it is constant on each of the intervals making up the complement of the Cantor set C, its derivative vanishes in $[0, 1]$ except on C. Thus $f' = 0$ a.e. on $[0, 1]$.

5.4   ANOTHER CONSTRUCTION OF THE CANTOR TERNARY FUNCTION. The same function can be defined by an alternate procedure that uses the ternary expansion of the elements of the Cantor set.

---

[27]G. Cantor, Über verschiedene Theoreme aus der Theorie der Punktmengen in einem n-fach ausgedehnten stetigen Raum G_n, *Acta Math.*, **7** (1885), 105–124; H. Lebesgue, Sur les fonctions représentables analytiquement, *J. Math. Pures Appl.*, **6**-1 (1905), 139–216.

For $x \in C$, let $\{\varepsilon_n\}$ be the sequence corresponding to the ternary expansion of x, as in (2.4) of the Preliminaries. Then define

$$f(x) = f\left(2\sum_{j=1}^{\infty}\frac{1}{3^j}\varepsilon_j\right) \stackrel{\text{def}}{=} 2\sum_{j=1}^{\infty}\frac{1}{2^j}\varepsilon_j.$$

Let (α_n, β_n) be an interval removed in the nth step of the construction of the Cantor set. The extremes α_n and β_n belong to C and their ternary expansion is described in Section 2.2 of the Problems and Complements of the Preliminaries. From the form of such expansion, we compute

$$f(\alpha_n) - f(\beta_n) = 2\frac{1}{2^n} - 2\sum_{j=n+1}^{\infty}\frac{1}{2^j} = 0.$$

If (α_n, β_n) is an interval in $[0, 1] - C$, we set

$$f(x) = f(\alpha_n) \quad \text{for all } x \in [\alpha_n, \beta_n].$$

In such a way, f can be defined by continuity in the whole $[0, 1]$ and $f' = 0$ a.e. in $[0, 1]$. Moreover, such an f coincides with the one defined previously.

The Cantor ternary function is continuous of bounded variation but not absolutely continuous. This can be established indirectly by means of Corollary 5.4. Give a direct proof.

5.5   A CONTINUOUS, STRICTLY MONOTONE FUNCTION WITH A.E. ZERO DERIVATIVE.
The Cantor ternary function is piecewise constant on the complement of the Cantor set. This accounts for $f' = 0$ a.e. in $[0, 1]$. We next exhibit a continuous, strictly increasing function in $[0, 1]$, whose derivative vanishes a.e. in $[0, 1]$.

Let $t \in (0, 1)$ be fixed, define $f_o(x) = x$, and set

$$f_1(x) = \begin{cases} (1+t)x & \text{for } 0 \leq x \leq \dfrac{1}{2}, \\ (1-t)x + t & \text{for } \dfrac{1}{2} \leq x \leq 1. \end{cases}$$

The function f_1 is constructed by dividing $[0, 1]$ into two equal subintervals by setting $f_1 = f_o$ at the endpoints of $[0, 1]$ by setting

$$f_1\left(\frac{1}{2}\right) = \frac{1-t}{2}f_o(0) + \frac{1+t}{2}f_o(1) = \frac{1+t}{2}$$

and by defining f_1 to be affine in the intervals $[0, \frac{1}{2}]$ and $[\frac{1}{2}, 1]$.

This procedure permits one to construct an increasing sequence $\{f_n\}$ of strictly increasing functions in $[0, 1]$. Precisely, if f_n has been defined, it must be affine in each of the subintervals

$$\left[\frac{j}{2^n}, \frac{j+1}{2^n}\right], \quad j = 0, 1, \dots, 2^n - 1.$$

Subdivide each of these into two equal subintervals, and define f_{n+1} to be affine on each of these with values at the endpoints, given by

$$f_{n+1}\left(\frac{j}{2^n}\right) = f_n\left(\frac{j}{2^n}\right),$$

$$f_{n+1}\left(\frac{j+1}{2^n}\right) = f_n\left(\frac{j+1}{2^n}\right),$$

$$f_{n+1}\left(\frac{2j+1}{2^{n+1}}\right) = \frac{1-t}{2} f_n\left(\frac{j}{2^n}\right) + \frac{1+t}{2} f_n\left(\frac{j+1}{2^n}\right).$$

By construction, $\{f_n\}$ is increasing and

$$f_m\left(\frac{j}{2^n}\right) = f_n\left(\frac{j}{2^n}\right) \quad \text{for all } m \geq n, \quad j = 0, 1, \ldots, 2^n - 1. \qquad (5.1c)$$

The limit function f is nondecreasing. Next, we show that it is continuous and strictly increasing in $[0, 1]$.

Every fixed $x \in [0, 1]$ is included into a sequence of nested and shrinking intervals $[\alpha_n, \beta_n]$ of the type

$$\alpha_n = \frac{m_{n,x}}{2^n}, \qquad \beta_n = \frac{m_{n,x} + 1}{2^n} \quad \text{for some } m_{n,x} \in \mathbb{N}.$$

By the construction of f_{n+1}, if the parent interval of $[\alpha_n, \beta_n]$ is $[\alpha_n, \beta_{n-1}]$, then

$$f_{n+1}(\beta_n) - f_{n+1}(\alpha_n) = \frac{1+t}{2}[f_n(\beta_{n-1}) - f_n(\alpha_n)].$$

Likewise, if the parent interval of $[\alpha_n, \beta_n]$ is $[\alpha_{n-1}, \beta_n]$, then

$$f_{n+1}(\beta_n) - f_{n+1}(\alpha_n) = \frac{1-t}{2}[f_n(\beta_n) - f_n(\alpha_{n-1})].$$

Therefore, by (5.1c), either

$$f_{n+1}(\beta_n) - f_{n+1}(\alpha_n) = \frac{1+t}{2}[f_n(\beta_{n-1}) - f_n(\alpha_{n-1})] \qquad (5.2c)_+$$

or

$$f_{n+1}(\beta_n) - f_{n+1}(\alpha_n) = \frac{1-t}{2}[f_n(\beta_{n-1}) - f_n(\alpha_{n-1})]. \qquad (5.2c)_-$$

From this, by iteration,

$$f_{n+1}(\beta_n) - f_{n+1}(\alpha_n) = \prod_{i=1}^{n} \frac{1 + \varepsilon_i t}{2}, \quad \text{where } \varepsilon_i = \pm 1.$$

Since for all $m \geq n + 1$,

$$f_m(\beta_n) = f_{n+1}(\beta_n) \quad \text{and} \quad f_m(\alpha_n) = f_{n+1}(\alpha_n),$$

the previous equality implies

$$f(\beta_n) - f(\alpha_n) = \prod_{i=1}^{n} \frac{1 + \varepsilon_i t}{2}, \quad \text{where } \varepsilon_i = \pm 1.$$

For each fixed n, the right-hand side is strictly positive. Thus

$$f(\beta_n) > f(\alpha_n);$$

i.e., f is strictly monotone. On the other hand, $(5.2c)_\pm$ also imply

$$f(\beta_n) - f(\alpha_n) \le \left(\frac{1+t}{2}\right)^n \longrightarrow 0 \quad \text{for } n \to \infty.$$

Thus f is continuous in $[0, 1]$. Still using $(5.2c)_\pm$, we compute

$$\frac{f(\beta_n) - f(\alpha_n)}{\beta_n - \alpha_n} = \prod_{i=1}^{n} (1 + \varepsilon_i t).$$

As $n \to \infty$, the right-hand side either converges to zero or diverges to infinity, or the limit does not exist. However, since f is monotone, it is a.e. differentiable. Therefore, the limit must exist for a.e. $x \in [0, 1]$ and must be zero a.e. in $[0, 1]$.

By Corollary 5.4, such a function is not absolutely continuous. Give a direct proof.

5.6 The function f constructed in Section 14 of Chapter II is not absolutely continuous.

5.7 Let μ be a measure on $\mathbb{R}$ defined on the same σ-algebra of the Lebesgue-measurable sets in $\mathbb{R}$ and absolutely continuous with respect to the Lebesgue measure on $\mathbb{R}$. Then set

$$f(x) = \begin{cases} \mu([\alpha, x]) & \text{for } x \in [\alpha, \infty), \\ -\mu([x, \alpha]) & \text{for } x \in (-\infty, \alpha]. \end{cases}$$

The function f is locally absolutely continuous; i.e., its restriction to any bounded interval is absolutely continuous. The function f can be used to generate the Lebesgue–Stieltjes measure μ_f. The measure μ_f coincides with μ on the Lebesgue-measurable sets.

6 DENSITY OF A MEASURABLE SET

6.1 Construct a measurable set $E \subset (-1, 1)$ such that $d'_E(0) = \frac{1}{2}$.

7 DERIVATIVES OF INTEGRALS

7.1 Let f be absolutely continuous in $[a, b]$. Then the function $x \to V_f[a, x]$ is also absolutely continuous in $[a, b]$. Moreover,

$$V_f[a, x] = \int_a^x |f'(t)| dt \quad \text{for all } x \in [a, b].$$

7.2   Let f be of bounded variation in $[a, b]$. The singular part of f is the function[28]

$$\sigma_f(x) = f(x) - f(a) - \int_a^x f'(t)dt. \qquad (7.1c)$$

The singular part of f is of bounded variation, and $\sigma' = 0$ a.e. in $[a, b]$. It has the same singularities as f, and $(f - \sigma)$ is absolutely continuous.

Thus every function f of bounded variation on $[a, b]$ can be decomposed into the sum of an absolutely continuous function on $[a, b]$ and a singular function. Compare the σ_f with the functions of the jumps J_f given in Section 1.6 of the Problems and Complements.

7.3   Let f be Lebesgue integrable in the interval $[a, b]$, and let F denote a primitive of f. Then for every absolutely continuous function g defined in $[a, b]$,

$$\int_a^b fg\,dx = F(b)g(b) - F(a)g(a) - \int_a^b Fg'\,dx.$$

7.4   Let $f, g : [a, b] \to \mathbb{R}$ be absolutely continuous. Then

$$\int_a^b fg'\,dx + \int_a^b f'g\,dx = f(a)g(a) - f(b)g(b).$$

7.5   Let $h : [a, b] \to [c, d]$ be absolutely continuous, increasing, and such that $h(a) = c$ and $h(b) = d$. Then for every nonnegative, Lebesgue-measurable function $f : [c, d] \to \mathbb{R}$, the composition $f(h)$ is measurable and

$$\int_c^d f(s)ds = \int_a^b f(h(t))h'(t)dt.$$

This is established sequentially for f the characteristic function of an interval, the characteristic function of an open set, and the characteristic function of a measurable set for a simple function.

13   Convex functions

13.1   Give an example of a bounded, discontinuous, convex function in $[a, b]$. Give an example of a convex function unbounded in (a, b).

13.2   A continuous function f in (a, b) is convex if and only if

$$f\left(\frac{x + y}{2}\right) \le \frac{f(x) + f(y)}{2} \qquad \text{for all } x, y \in (a, b).$$

13.3   Let f be convex, nondecreasing, and nonconstant in $(0, \infty)$. Then $f(x) \to \infty$ as $x \to \infty$.

---

[28]H. Lebesgue, Sur l'intégration des fonctions discontinues, *Ann. École Norm.* (3), **27** (1910), 361–450.

13.4   Let f be convex in $[0, \infty)$. Then the limit of $x^{-1}f(x)$ as $x \to \infty$ exists, finite or infinite.

13.5   Let $\{f_n\}$ be a sequence of convex functions in (a, b) converging to some real-valued function f. The convergence is uniform within any closed subinterval of (a, b). The conclusion is false if f is permitted to take values in $\mathbb{R}^*$, as shown by the sequence $\{x^n\}$ for $x \in (0, 2)$.

13.6   Let $f \in C^2(a, b)$. Then f is convex in (a, b) if and only if $f''(x) \geq 0$ for each $x \in (a, b)$.

($\Longleftarrow$) Having fixed $x < y$ in (a, b), it suffices to prove that the function

$$[0, 1] \ni t \longrightarrow \varphi(t) = f(tx + (1 - t)y) - tf(x) - (1 - t)f(y) \qquad (13.1c)$$

is nonpositive in $[0, 1]$. Such a function vanishes at the endpoints of $[0, 1]$, and its extrema are minima since

$$\varphi''(t) = (x - y)f''(tx + (1 - t)y) \leq 0.$$

Proposition 13.1c.   *A continuous function f in (a, b) is convex if and only if either one of the two one-sided derivatives $D_{\pm}f$ is nondecreasing.*

Proof.   Assume, for example, that D_+f is nondecreasing. If the function φ in (13.1c) has a positive maximum $\varphi(t_o) > 0$ at some $t_o \in (0, 1)$, then

$$D_+\varphi(t_o) = (x - y)D_+f(t_ox + (1 - t_o)y) + f(y) - f(x) \leq 0.$$

Therefore, since D_+f is nondecreasing, $D_+\varphi(t)$ is nonpositive in $[0, t_o]$. This implies that φ is nonincreasing in $[0, t_o]$ and therefore $\varphi(t_o) \leq 0$. $\qquad\square$

13.7   The function $f(x) = |x|^p$ is convex for $p \geq 1$ and concave for $p \in (0, 1)$.

13.8   **CONVEX FUNCTIONS IN $\mathbb{R}^N$.** Let E be a convex subset of $\mathbb{R}^N$. A function $f : E \to \mathbb{R}^*$ is convex if for every pair of points x and y in E and every $t \in [0, 1]$,

$$f(tx + (1 - t)y) \leq tf(x) + (1 - t)f(y).$$

The $(N + 1)$-dimensional set

$$\mathcal{G}_f = \{(x, x_{N+1}) \in \mathbb{R}^{N+1} | x \in E, \ x_{N+1} \geq f(x)\}$$

is the epigraph of f. The function f is convex if and only if its epigraph is convex.

13.9   Let $E \subset \mathbb{R}^N$ be open and convex. A function $f \in C^2(E)$ is convex if and only if

$$\sum_{i,j=1}^{N} f_{x_ix_j}(x)\xi_i\xi_j \geq 0 \quad \text{for all } \xi \in \mathbb{R}^N.$$

Having fixed $x \in E$, let ρ denote the radius of the largest ball centered at x and contained in E. For every fixed $\xi \in \mathbb{R}^N$ such that $|\xi| = 1$, the function $\varphi(t) = f(x + t\xi)$ is convex in the interval $(-\rho, \rho)$. Therefore, $\varphi''(0) \geq 0$.

13.10 Construct a nonconvex function $f \in C^2(\mathbb{R}^2)$ such that f_{xx} and f_{yy} are both nonnegative.

13.11 Let $E \subset \mathbb{R}^N$ be convex, and let f be convex and real valued in E. Then f is continuous on $\overset{\circ}{E}$. Moreover, for every $x \in \overset{\circ}{E}$, there exist the left and right directional derivatives

$$D_{\mathbf{u}}^{\pm} f(x) = D_t^{\pm} f(x + t\mathbf{u})|_{t=0} \quad \text{for all } |\mathbf{u}| = 1.$$

Moreover, $D_{\mathbf{u}}^- f \leq D_{\mathbf{u}}^+ f$. In particular, for each $x \in \overset{\circ}{E}$, there exist the left and right derivatives $D_{x_j}^{\pm} f$ along the coordinate axes, and $D_{x_j}^- f \leq D_{x_j}^+ f$.

13.12 Let $E \subset \mathbb{R}^N$ be convex. A function f defined in E is convex if and only if

$$f(x) = \sup\{\pi(x), \text{ where } \pi \text{ is affine and } \pi \leq f \text{ in } E\}.$$

13.13 Let $f : \mathbb{R}^N \to \mathbb{R}$ be convex. There exists a positive number k such that

$$\liminf_{|x| \to \infty} \frac{f(x)}{|x|} \geq -k.$$

13.14 **THE LEGENDRE TRANSFORM.** The Legendre transform f^* of a convex function $f : \mathbb{R}^N \to \mathbb{R}^*$ is defined by[29]

$$f^*(x) = \sup_{y \in \mathbb{R}^N} \{x \cdot y - f(y)\}. \tag{13.2c}$$

Proposition 13.2c. *The Legendre transform f^* is convex in $\mathbb{R}^N$ and, moreover, $f^{**} = f$.*

Proof. The convexity of f follows from Section 13.12 of the Problems and Complements. From (13.2c),

$$f(y) + f^*(x) \geq y \cdot x \quad \text{for all } x, y \in \mathbb{R}^N.$$

Therefore,

$$f(y) \geq \sup_{x \in \mathbb{R}^N} \{y \cdot x - f^*(x)\} = f^{**}(y).$$

Also, still using (13.2c),

$$f^{**}(x) = \sup_{y \in \mathbb{R}^N} \left\{ x \cdot y - \sup_{z \in \mathbb{R}^N} \{y \cdot z - f(z)\} \right\}$$

$$= \sup_{y \in \mathbb{R}^N} \inf_{z \in \mathbb{R}^N} \{y \cdot (x - z) + f(z)\}.$$

---

[29]A. Legendre, Mémoire sur l'intégration de quelques équations aux différences partielles, *Mém. Acad. Sci.*, 1787, 309–351.

Since f is convex, for a fixed $x \in \mathbb{R}^N$, there exists a vector $\mathbf{m}$ such that

$$f(z) - f(x) \geq \mathbf{m} \cdot (z - x) \quad \text{for all } z \in \mathbb{R}^N.$$

Combining these inequalities yields

$$f^{**}(x) \geq f(x) + \sup_{y \in \mathbb{R}^N} \inf_{z \in \mathbb{R}^N} (z - x) \cdot (\mathbf{m} - y) = f(x). \qquad \square$$

13.15 FINITENESS AND COERCIVITY. The Legendre transform f^*, as defined by (13.2c), could be infinite even if f is finite in $\mathbb{R}^N$. For example, in $\mathbb{R}$,

$$|x|^* = \begin{cases} 0 & \text{if } |x| \leq 1, \\ \infty & \text{if } |x| > 1. \end{cases}$$

A convex function $f : \mathbb{R}^N \to \mathbb{R}$ is coercive at infinity if

$$\lim_{|x| \to \infty} \frac{f(x)}{|x|} = \infty.$$

Proposition 13.3c. *If f is coercive at infinity, then f^* is finite in $\mathbb{R}^N$. If f is finite, then f^* is coercive at infinity.*

Proof. Assume that f is coercive at infinity. If the sup in (13.2c) is achieved for $y = 0$, the assertion is obvious. Otherwise,

$$f^*(x) = \sup_{y \in \mathbb{R}^N - \{0\}} |y| \left\{ x \cdot \frac{y}{|y|} - \frac{f(y)}{|y|} \right\}.$$

Therefore, the supremum is achieved for some finite y, and $f^*(x)$ is finite.

To prove the converse statement, fix $\lambda > 0$ and write

$$f^*(x) = \sup_{y \in \mathbb{R}^N} \{x \cdot y - f(y)\} \geq \{x \cdot y - f(y)\}|_{y = \lambda x / |x|}$$

$$= \lambda |x| - f\left(\lambda \frac{x}{|x|}\right) \geq \lambda |x| - \sup_{|u| = \lambda} |f(\mathbf{u})|.$$

Therefore, since $x \in \mathbb{R}^N - \{0\}$ is arbitrary,

$$\lim_{|x| \to \infty} \frac{f^*(x)}{|x|} \geq \lambda \quad \text{for all } \lambda > 0. \qquad \square$$

14   DISCRETE VERSIONS OF JENSEN'S INEQUALITY

Proposition 14.1c (Hölder[30]). *Let $\{\alpha_i\}$ be a sequence of nonnegative numbers such that $\sum \alpha_i = 1$, and let $\{\xi_i\}$ be a sequence in $\mathbb{R}$. Then*

$$\exp\left(\sum \alpha_i \xi_i\right) \leq \sum \alpha_i \exp(\xi_i). \tag{14.1c}$$

---

[30]O. Hölder, *Über einen Mittelwertsatz*, *Göttinger Nachr.*, 1889, 38–47.

Proof. Apply inequality (14.2) for the function e^x with the choices $\eta = \xi_j$ and $\alpha = \sum \alpha_i \xi_i$. This gives

$$\exp\left(\sum \alpha_i \xi_i\right) + m_j \left(\xi_j - \sum \alpha_i \xi_i\right) \le \exp(\xi_j).$$

Multiplying this by α_j and adding over j proves (14.1c).   □

Corollary 14.2c. *Let $\{\alpha_i\}$ be a sequence of nonnegative numbers such that $\sum \alpha_i = 1$, and let $\{\xi_i\}$ be a sequence of positive numbers. Then*

$$\prod \xi_i^{\alpha_i} \le \sum \alpha_i \xi_i. \tag{14.2c}$$

14.1   THE INEQUALITY OF THE GEOMETRIC AND ARITHMETIC MEAN. In the case where

$$\alpha_i = 0 \quad \text{for } i > n \quad \text{and} \quad \alpha_i = \frac{1}{n} \quad \text{for } i = 1, 2, \ldots, n,$$

inequality (14.2c) reduces to the inequality between the geometric and arithmetic mean of n positive numbers; i.e.,[31]

$$(\xi_1 \xi_2 \cdots \xi_n)^{1/n} \le \frac{\xi_1 + \xi_2 + \cdots + \xi_n}{n}. \tag{14.3c}$$

15   EXTENDING CONTINUOUS FUNCTIONS

15.1   Let f be convex in a closed interval $[a, b]$, and assume that $D_+ f(a)$ and $D_- f(b)$ are both finite. Then there exists a *convex* function $\tilde{f}$ defined in $\mathbb{R}$ such that $f = \tilde{f}$ on $[a, b]$.

16   THE WEIERSTRASS APPROXIMATION THEOREM

16.1   Let $E \subset \mathbb{R}^N$ be open and bounded. Let $f \in C^1(\overline{E})$, and let P_j denote the jth Stieltjes polynomial relative to f. Then

$$\lim_{j \to \infty} \frac{\partial P_j}{\partial x_i} = \frac{\partial f}{\partial x_i}, \quad j = 1, 2, \ldots, N, \quad \text{in } E.$$

16.2   Let $E \subset \mathbb{R}^N$ be compact and let f be Lipschitz continuous in E with Lipschitz constant L. Then the Stieltjes polynomials P_j relative to f are equi-Lipschitz continuous in E with the same constant L; i.e.,

$$|P_j(x) - P_j(y)| \le L|x - y|$$

for all $x, y \in E$ and all $j \in \mathbb{N}$.

---

[31] [24, Chapter II, Section 5] contains an alternate proof of this inequality that does not use Jensen's inequality.

16.3   A continuous function $f : [0, 1] \to \mathbb{R}$ can be approximated by the Bernstein polynomials; i.e.,

$$\sup_{[0,1]} |f - B_j| \longrightarrow 0 \quad \text{as } j \longrightarrow \infty,$$

where for $j \in \mathbb{N}$,

$$B_j(x) = \sum_{i=1}^{j} \binom{j}{i} f\left(\frac{i}{j}\right) x^i (1 - x)^{j-i}.$$

The proof is analogous to the proof of Theorem 16.1. State and prove an N-dimensional version of such an approximation.[32]

16.4   Let f be uniformly continuous on a bounded set $E \subset \mathbb{R}^N$, and denote by $\mathcal{P}_n$ the set of all polynomials of degree n in the coordinate variables. Then

$$\int_E f p_n dx = 0 \quad \text{for all } p_n \in \mathcal{P}_n \text{ and all } n \in \mathbb{N}$$

implies that $f \equiv 0$.

17   THE STONE–WEIERSTRASS THEOREM

17.1   The Stone–Weierstrass theorem fails for complex-valued functions.

Let D be the closed, unit disc in the complex plane $\mathbf{C}$, and denote by $C(D; \mathbf{C})$ the linear space of all the continuous, complex-valued functions defined in D endowed with the topology generated by the metric in (17.1). Also, consider the subset $\mathcal{H}(D)$ of $C(D; \mathbf{C})$, consisting of all holomorphic functions defined in D. One verifies that $\mathcal{H}(D)$ is an algebra. Moreover, uniform limits of holomorphic functions in D are holomorphic.[33] Thus $\mathcal{H}(D)$ is closed under the metric in (17.1). The algebra $\mathcal{H}(D)$ is called the *disc algebra*.

Such an algebra separates points since it contains the holomorphic function $f(z) = z$. Moreover, $\mathcal{H}(D)$ contains the constants. However, $\mathcal{H}(D) \neq C(D; \mathbf{C})$. Indeed, the function $f(z) = \bar{z}$ is continuous but not holomorphic in D.

17.2   Let $f : \mathbb{R} \to \mathbb{R}$ be continuous and 2π-periodic. For every $\varepsilon > 0$, there exists a function of the type

$$\varphi(x) = a_o + \sum_{n=1}^{m} (b_n \cos nx + c_n \sin nx)$$

for some finite m such that

$$\sup_{x \in \mathbb{R}} |f(x) - \varphi(x)| \leq \varepsilon.$$

---

[32]G. G. Lorentz, *Bernstein Polynomials*, University of Toronto Press, Toronto, 1953.
[33]See [5, Chapter V, Théorème 1, p. 145].

19   THE ASCOLI–ARZELÀ THEOREM

19.1   A GENERAL VERSION OF THE ASCOLI–ARZELÀ THEOREM. The proof of Theorem 19.1 uses only the separability of $\mathbb{R}^N$ and the metric structure of $\mathbb{R}$. Thus it can be extended into any abstract framework with these two properties.

Let $\{f_n\}$ be a countable collection of continuous functions from a separable topological space $\{X; \mathcal{U}\}$ into a metric space $\{Y; d_Y\}$. The functions f_n are equibounded at x if the closure in $\{Y; d_Y\}$ of the set $\{f_n(x)\}$ is compact. The functions f_n are equicontinuous at a point $x \in X$ if for every $\varepsilon > 0$, there exists an open set $\mathcal{O} \in \mathcal{U}$ containing x and such that

$$d_Y(f_n(x), f_n(y)) \le \varepsilon \quad \text{for all } y \in \mathcal{O} \text{ and all } n \in \mathbb{N}.$$

Theorem 19.1c. *Let $\{f_n\}$ be a sequence of continuous functions from a separable space $\{X; \mathcal{U}\}$ into a metric space $\{Y; d_Y\}$. Assume that the functions f_n are equibounded and equicontinuous at each $x \in X$. Then there exist a subsequence $\{f_{n'}\} \subset \{f_n\}$ and a continuous function $f : X \to Y$ such that $\{f_{n'}\} \to f$ pointwise in X. Moreover, the convergence is uniform on compact subsets of X.*

V

The $L^p(E)$ Spaces

1 Functions in $L^p(E)$ and their norms

Let $\{X, \mathcal{A}, \mu\}$ be a measure space and let E be a measurable subset of X. A measurable function $f : E \to \mathbb{R}^*$ is said to be in $L^p(E)$ for $p \geq 1$ if $|f|^p$ is integrable on E, i.e., if

$$\|f\|_p \overset{\text{def}}{=} \left(\int_E |f|^p d\mu \right)^{1/p} < \infty. \tag{1.1$_p$}$$

Equivalently, the collection of all such functions is denoted by $L^p(E)$. The quantity $\|f\|_p$ is the *norm* of f in $L^p(E)$. It follows from the definition that $\|f\|_p \geq 0$ and

$$\|f\|_p = 0 \iff f = 0 \quad \text{a.e. in } E. \tag{1.2}$$

Let f and g be in $L^p(E)$ and let $\alpha, \beta \in \mathbb{R}$. Then[1]

$$|\alpha f + \beta g|^p \leq 2^{p-1}(|\alpha|^p |f|^p + |\beta|^p |g|^p), \quad p \geq 1, \quad \text{a.e. in } E.$$

Therefore, a linear combination of any two functions $f, q \in L^p(E)$ remains in $L^p(E)$. Thus $L^p(E)$ is a linear space.

A measurable function $f : E \to \mathbb{R}^*$ is said to be in $L^\infty(E)$ if there exists a positive number M such that $|f(x)| \leq M$ a.e. in E. Set

$$\operatorname*{ess\,sup}_E f = \inf\{k | \mu([f > k]) = 0\},$$
$$\operatorname*{ess\,inf}_E f = \sup\{k | \mu([f < k]) = 0\}. \tag{1.3}$$

---

[1] See (2.3c) in the Problems and Complements.

For $f \in L^\infty(E)$, also set

$$\|f\|_\infty \overset{\text{def}}{=} \underset{E}{\text{ess sup}} |f|. \qquad (1.1)_\infty$$

It follows from the definition that for $\varepsilon > 0$ arbitrarily small,

$$\begin{aligned} \mu(|f| \geq \|f\|_\infty + \varepsilon) &= 0, \\ \mu(|f| \geq \|f\|_\infty - \varepsilon) &> 0. \end{aligned} \qquad (1.4)$$

If $f, g \in L^\infty(E)$ and $\alpha, \beta \in \mathbb{R}$, then $|\alpha f + \beta g| \in L^\infty(E)$. Thus $L^\infty(E)$ is a linear space.

Remark 1.1. The μ-measurability of f is essential in the definition of the spaces $L^p(E)$. For example, let $E \subset [0, 1]$ be the Vitali non-Lebesgue-measurable set constructed in Section 13 of Chapter II. The function

$$f(x) = \begin{cases} 1 & \text{for } x \in E, \\ -1 & \text{for } x \in [0, 1] - E \end{cases}$$

is such that f^2 is Lebesgue integrable in $[0, 1]$, but f it is not in $L^2[0, 1]$.

1.1   The spaces L^p for $0 < p < 1$. A measurable function $f : E \to \mathbb{R}^*$ is in $L^p(E)$ for $0 < p < 1$ if $|f|^p \in L^1(E)$. A norm-like function $f \to \|f\|_p$ might be defined as in $(1.1)_p$.

The collection $L^p(E)$ for $0 < p < 1$ is a linear space. This is a direct consequence of the following lemma.

Lemma 1.1. *Let $0 < p < 1$. Then for nonnegative x, y,*

$$(x + y)^p \leq x^p + y^p. \qquad (1.5)$$

Proof. If x and y are both zero, the inequality is trivial. Assuming that $y > 0$ and setting $t = x/y$, the inequality is equivalent to

$$(1 + t)^p \leq 1 + t^p \quad \text{for } t \geq 0 \quad \text{and} \quad 0 < p < 1. \qquad (1.5)'$$

The function

$$f(t) = (1 + t)^p - (1 + t^p) \quad \text{for } t \geq 0 \quad \text{and} \quad 0 < p < 1$$

vanishes for $t = 0$ and $f' \leq 0$ for $t > 0$. Thus $f \leq 0$ for all $t \geq 0$.   □

1.2   The spaces L^q for $q < 0$. A measurable function $f : E \to \mathbb{R}^*$ is in $L^q(E)$ for $q < 0$ if

$$0 < \int_E |f|^q d\mu = \int_E \frac{1}{|f|^{\frac{p}{1-p}}} d\mu < \infty, \quad \frac{1}{p} + \frac{1}{q} = 1.$$

A norm-like function $f \to \|f\|_q$ might be defined as in $(1.1)_p$. It follows from the definition that if $f \in L^q(E)$ for $q < 0$, then $f \neq 0$ a.e. on E, and $|f| \not\equiv \infty$. If $q < 0$, the set $L^q(E)$ is not a linear space.

2 The Hölder and Minkowski inequalities

Two elements p and q in the extended real numbers $\mathbb{R}^*$ are said to be *conjugate* if $p, q \geq 1$ and

$$\frac{1}{p} + \frac{1}{q} = 1. \tag{2.1}$$

Since $p, q \in \mathbb{R}^*$, if $p = 1$, then $q = \infty$. Likewise, if $q = 1$, then $p = \infty$.

Proposition 2.1. *Let $1 \leq p, q \leq \infty$ be conjugate. Then for all $a, b \in \mathbb{R}$,[2]*

$$|ab| \leq \frac{1}{p}|a|^p + \frac{1}{q}|b|^q. \tag{2.2}$$

Proof. The inequality is obvious if either a or b is zero. Thus we assume that $|a| > 0$ and $|b| > 0$. The inequality is also obvious if either $p = 1$ or $q = 1$. Thus we assume that $1 < p, q < \infty$. The function

$$s \to \left(\frac{s^p}{p} + \frac{1}{q} - s \right), \quad s \geq 0,$$

has an absolute minimum at $s = 1$. Therefore, for all $s > 0$,

$$s \leq \frac{s^p}{p} + \frac{1}{q}, \tag{2.3}$$

and equality holds only if $s = 1$. Choosing

$$s = \frac{|a|}{|b|^{q/p}} \tag{2.3)'}$$

yields

$$\frac{|a|}{|b|^{q/p}} \leq \frac{1}{p} \frac{|a|^p}{|b|^q} + \frac{1}{q}.$$

Multiplying this by $|b|^q$ proves (2.2). $\qquad\square$

Proposition 2.2 (Hölder's inequality[3]). *Let $f \in L^p(E)$ and $g \in L^q(E)$, where $1 \leq p, q \in \mathbb{R}^*$ satisfy (2.1). Then $fg \in L^1(E)$ and*

$$\int_E |fg|d\mu \leq \|f\|_p \|g\|_q. \tag{2.4}$$

Moreover, equality holds only if there exists a constant c such that

$$|f(x)|^p = c|g(x)|^q \quad a.e. \ x \in E.$$

---

[2]When $p = q = 2$, this is the Cauchy–Schwartz inequality. For $p \neq 2$, the inequality is due to Young. An alternative proof of (2.2) can be given using Proposition (14.1c) of the Problems and Complements of Chapter IV. See also [24, pp. 132–133].

[3]O. Hölder, Über einen Mittelwertsatz, *Göttinger Nachr.*, 1889, 38–47.

Proof. We may assume without loss of generality that f and g are nonnegative and that neither is zero a.e. in E. Also, (2.4) is obvious if either $p = 1$ or $q = 1$. If $p, q > 1$, in (2.2) take

$$a = \frac{f}{\|f\|_p}, \qquad b = \frac{g}{\|g\|_q}$$

to obtain

$$\frac{fg}{\|f\|_p \|g\|_q} \leq \frac{1}{p} \frac{f^p}{\|f\|_p^p} + \frac{1}{q} \frac{g^q}{\|g\|_q^q} \quad \text{a.e. in } E.$$

Integrating over E,

$$\frac{\int_E fg d\mu}{\|f\|_p \|g\|_q} \leq \frac{1}{p} + \frac{1}{q} = 1.$$

Equality in (2.3) holds only if $s = 1$. Therefore, the choice of s, indicated in (2.3)$'$, implies that equality in (2.4) holds only if

$$f^p(x) = \frac{\|f\|_p^p}{\|g\|_q^q} g^q(x) \quad \text{for a.e. } x \in E. \qquad \square$$

Proposition 2.3 (the Minkowski inequality). *Let $f, g \in L^p(E)$ for some $1 \leq p \leq \infty$. Then*

$$\|f + g\|_p \leq \|f\|_p + \|g\|_p. \tag{2.5}$$

Moreover, equality holds only if there exists a constant C such that

$$f(x) = Cg(x) \quad \text{for a.e. } x \in E.$$

Proof. The inequality is obvious if $p = 1$ and $p = \infty$. If $1 < p < \infty$,

$$\|f + g\|_p^p = \int_E |f + g|^p d\mu = \int_E |f + g|^{p-1} |f + g| d\mu$$
$$\leq \int_E |f + g|^{p-1} |f| d\mu + \int_E |f + g|^{p-1} |g| d\mu.$$

The integrals on the right-hand side are majorized by Hölder's inequality and give

$$\|f + g\|_p^p \leq \|f + g\|_p^{p-1} (\|f\|_p + \|g\|_p). \qquad \square$$

3   The reverse Hölder and Minkowski inequalities

If $0 < p < 1$, the Hölder inequality (2.4) and the Minkowski inequality (2.5) continue to hold, provided the inequality sign is reversed.

Proposition 3.1 (reverse Hölder inequality). *Let $p \in (0, 1)$ and $q < 0$ satisfy (2.1). Then for every $f \in L^p(E)$ and $g \in L^q(E)$,*

$$\int_E |fg|d\mu \geq \|f\|_p\|g\|_q, \quad \frac{1}{p} + \frac{1}{q} = 1. \tag{3.1}$$

Proof. We may assume that $fg \in L^1(E)$; otherwise, the inequality is trivial. By the Hölder inequality (2.4) applied with the pair of numbers

$$\overline{p} = \frac{1}{p} > 1, \quad \overline{q} = \frac{1}{1-p} > 1, \quad \frac{1}{\overline{p}} + \frac{1}{\overline{q}} = 1,$$

we compute and estimate

$$\|f\|_p = \left(\int_E \frac{|fg|^p}{|g|^p}d\mu\right)^{1/p}$$

$$\leq \left(\int_E |fg|d\mu\right)\left(\int_E \frac{1}{|g|^{\frac{p}{1-p}}}d\mu\right)^{\frac{1-p}{p}}$$

$$= \frac{1}{\|g\|_q}\int_E |fg|d\mu. \qquad \square$$

Proposition 3.2 (reverse Minkowski inequality). *Let $p \in (0, 1)$. Then for all $f, g \in L^p(E)$,*

$$\||f| + |q|\|_p \geq \|f\|_p + \|g\|_p. \tag{3.2}$$

Proof. We may assume that $\||f| + |g|\|_p > 0$. By the reverse Hölder inequality,

$$\||f| + |g|\|_p^p = \int_E (|f| + |g|)^{p-1}(|f| + |g|)d\mu$$

$$\geq \left\{\int_E (|f| + |g|)^{q(p-1)}d\mu\right\}^{1/q}(\|f\|_p + \|g\|_p)$$

$$= \||f| + |g|\|_p^{p/q}(\|f\|_p + \|g\|_p). \qquad \square$$

4   More on the spaces L^p and their norms

4.1   Characterizing the norm $\|f\|_p$ for $1 \leq p < \infty$.

Proposition 4.1. *Let $f \in L^p(E)$ for some $1 \leq p < \infty$. Then*

$$\|f\|_p = \left(\int_E |f|^p d\mu\right)^{1/p} = \sup_{\substack{g \in L^q(E) \\ \|g\|_q = 1}} \int_E fg d\mu, \tag{4.1}$$

where $1 \leq p < \infty$ and $1 < q \leq \infty$ are conjugate.

Proof. We may assume that $f \not\equiv 0$; otherwise, there is nothing to prove. By Hölder's inequality,

$$\sup_{\substack{g \in L^q(E) \\ \|g\|_q = 1}} \int_E fg d\mu \leq \|f\|_p.$$

If $1 < p < \infty$, one verifies that

$$g_* = \frac{|f|^{p-2}f}{\|f\|_p^{p/q}} \in L^q(E) \quad \text{and} \quad \|g_*\|_q = 1.$$

Then

$$\sup_{\substack{g \in L^q(E) \\ \|g\|_q = 1}} \int_E fg d\mu \geq \int_E fg_* d\mu = \|f\|_p.$$

If $p = 1$, the proof is similar for the choice $g_* = \text{sign } f \in L^\infty(E)$. $\square$

4.2 The norm $\|\cdot\|_\infty$ for E of finite measure. Assume that $\mu(E) < \infty$. If $f \in L^q(E)$, then for all $1 \leq p < q$, by the Hölder inequality applied to the pair of functions f and $g \equiv 1$.

$$\|f\|_p \leq \mu(E)^{\frac{q-p}{qp}} \|f\|_q.$$

Therefore, f belongs to every $L^p(E)$ for all $1 \leq p \leq q$. In particular, if $f \in L^\infty(E)$, then $f \in L^p(E)$ for all $p \geq 1$.

Proposition 4.2. *Let $\mu(E) < \infty$ and $f \in L^\infty(E)$. Then*

$$\lim_{p \to \infty} \|f\|_p = \|f\|_\infty. \tag{4.2}$$

Proof. Since E is of finite measure,

$$\limsup_{p \to \infty} \|f\|_p \leq \|f\|_\infty \limsup_{p \to \infty} \mu(E)^{1/p} = \|f\|_\infty.$$

Next, for any $\varepsilon > 0$,

$$\int_E |f|^p d\mu \geq \int_{[|f| > \|f\|_\infty - \varepsilon]} |f|^p d\mu$$
$$\geq (\|f\|_\infty - \varepsilon)^p \mu[|f| > \|f\|_\infty - \varepsilon].$$

From the second part of (1.4), the last term is positive. Therefore, taking the $(\frac{1}{p})$-power and letting $p \to \infty$ gives

$$\liminf_{p \to \infty} \|f\|_p \geq \|f\|_\infty - \varepsilon.$$

$\square$

4.3   The continuous version of the Minkowski inequality.

Proposition 4.3. *Let $\{X, \mathcal{A}, \mu\}$ and $\{Y, \mathcal{B}, \nu\}$ be two complete measure spaces, and let $f \in L^p(X \times Y)$ for some $1 \le p < \infty$. Then*

$$\left(\int_X \left| \int_Y f(x, y) d\nu \right|^p d\mu \right)^{1/p} \le \int_Y \| f(\cdot, y) \|_{p, X} d\nu. \tag{4.3}$$

Proof. Setting $F = \int_Y f(\cdot, y) d\nu$, the left-hand side of (4.3) is $\| F \|_{p, X}$. By (4.1) and Fubini's theorem,

$$\| F \|_{p, X} = \sup_{\substack{g \in L^q(X) \\ \|g\|_{q, X} = 1}} \int_X F g \, d\mu$$

$$= \sup_{\substack{g \in L^q(X) \\ \|g\|_{q, X} = 1}} \int_X \left(\int_Y f(x, y) d\nu \right) g(x) d\mu$$

$$= \sup_{\substack{g \in L^q(X) \\ \|g\|_{q, X} = 1}} \int_Y \left(\int_X f(x, y) g(x) d\mu \right) d\nu$$

$$\le \int_Y \left(\sup_{\substack{g \in L^q(X) \\ \|g\|_{q, X} = 1}} \int_X f(x, y) g(x) d\mu \right) d\nu$$

$$= \int_Y \| f(\cdot, y) \|_{p, X} d\nu. \qquad \square$$

5 $L^p(E)$ for $1 \le p \le \infty$ as normed spaces of equivalence classes

Since $L^p(E)$ is a linear space, it must contain a *zero element* with respect to the operations of addition and multiplication by scalars. Such an element is defined by $f + (-1)f$ for any $f \in L^p(E)$.

A *norm* in $L^p(E)$ is a function $\| \cdot \| : L^p(E) \to \mathbb{R}^+$ satisfying

$$\| f \| = 0 \iff f \text{ is the zero element of } L^p(E), \tag{5.1}$$

$$\| \alpha f \| = |\alpha| \| f \| \quad \text{for all } f \in L^p(E) \quad \text{and} \quad \text{for all } \alpha \in \mathbb{R}, \tag{5.2}$$

$$\| f + g \| \le \| f \| + \| g \| \quad \text{for all } f, g \in L^p(E). \tag{5.3}$$

The norm $\| \cdot \|_p$ defined by $(1.1)_p$ for $p \in [1, \infty)$ and by $(1.1)_\infty$ for $p = \infty$ satisfies (5.2) by the properties of the Lebesgue integral. It also satisfies (5.3) by the Minkowski inequality. However, it does not satisfy (5.1) if the zero element of $L^p(E)$ is meant in the usual sense:

$$f \text{ is the zero element if } f(x) = 0 \text{ for all } x \in E.$$

Indeed, in view of the definition (1.2), the norm $\|\cdot\|_p$ does not distinguish between two elements f and g in $L^p(E)$ that differ on a set of measure zero.

Motivated by this remark, we regard the elements of $L^p(E)$ as *equivalence classes*. If C_f is one such class and f is a representative, then

$$C_f = \left\{ \begin{array}{c} \text{all measurable functions } g : E \to \mathbb{R}^* \text{ such that } |g|^p \\ \text{is integrable on } E \text{ and such that } f = g \text{ a.e. in } E \end{array} \right\}.$$

With such an interpretation, the function $\|\cdot\|_p : L^p(E) \to \mathbb{R}^+$ is a norm in $L^p(E)$, which then becomes a normed linear space.

5.1 $L^p(E)$ for $1 \le p \le \infty$ as a metric topological vector space. The norm $\|\cdot\|_p$ generates a distance in $L^p(E)$ by the formula

$$d(f, g) = \|f - g\|_p \quad (1 \le p \le \infty).$$

One verifies that such a metric is translation invariant, and therefore it generates a translation invariant topology in $L^p(E)$ determined by a base at the origin consisting of the balls

$$[\|f\|_p < \rho] = \{f \in L^p(E) \mid \|f\|_p < \rho\}, \quad \rho > 0.$$

Such a topology is called the *norm topology* of $L^p(E)$. By Minkowski's inequality, for $h, g \in L^p(E)$ and $t \in (0, 1)$,

$$\|tg + (1 - t)h\|_p \le t\|g\|_p + (1 - t)\|h\|_p.$$

Therefore, the balls $[\|f\|_p < \rho]$ are convex, and the norm topology of $L^p(E)$ for $1 \le p \le \infty$ is locally convex.

The unit ball $[\|f\|_p < 1]$ is *uniformly* convex if for every $\varepsilon > 0$, there exists $\delta > 0$ such that for any pair $h, g \in L^p(E)$ such that

$$\|h\|_p = \|g\|_p = 1 \quad \text{and} \quad \|h - g\|_p \ge \varepsilon,$$

it holds that

$$\left\| \frac{h + g}{2} \right\|_p \le 1 - \delta. \tag{5.4}$$

If this occurs, it is said that the norm topology of $L^p(E)$ is uniformly convex or simply that $L^p(E)$ is uniformly convex.

If $p = \infty$, one can construct examples of functions $h, g \in L^\infty(E)$ such that

$$\|h\|_\infty = \|g\|_\infty = 1, \qquad \|h - g\|_\infty = 1, \quad \text{and} \quad \left\| \frac{h + g}{2} \right\|_\infty = 1.$$

In a similar way, if $p = 1$ one can construct a pair of functions $h, g \in L^1(E)$ such that

$$\|h\|_1 = \|g\|_1 = 1, \qquad \|h - g\|_1 \ge \frac{1}{2}, \quad \text{and} \quad \left\| \frac{h + g}{2} \right\|_1 = 1.$$

Thus $L^\infty(E)$ and $L^1(E)$ are not uniformly convex. However, $L^p(E)$ are uniformly convex for all $1 < p < \infty$.[4]

6 A metric topology for $L^p(E)$ when $0 < p < 1$

The norm-like function $f \to \|f\|_p$ defined as in $(1.1)_p$ for $0 < p < 1$ is not a norm. Indeed, (5.3) is violated in view of the reverse Minkowski inequality. By the same token, the function

$$d(f, g) = \|f - g\|_p \quad \text{for } 0 < p < 1$$

is not a metric in $L^p(E)$. A topology in $L^p(E)$ for $0 < p < 1$ could be generated by the balls

$$B_\rho(g) = \{f \in L^p(E) \mid \|f - g\|_p < \rho\}, \quad g \in L^p(E).$$

By the reverse Minkowski inequality, these balls are not convex.

A distance function in $L^p(E)$ for $0 < p < 1$ is introduced by setting

$$d(f, g) = \|f - g\|_p^p = \int_E |f - g|^p d\mu. \tag{6.1}$$

One verifies that $d(\cdot, \cdot)$ satisfies requirements (i)–(iii) of a metric from Section 13 of Chapter I. To verify the triangle inequality (iv), fix $f, g, h \in L^p(E)$ for $0 < p < 1$. Then by (1.5),

$$|f - g|^p \leq |f - h|^p + |h - g|^p \quad \text{a.e. in } E.$$

Integrating over E proves the triangle inequality for the metric in (6.1).

6.1 Open convex subsets of $L^p(E)$ when $0 < p < 1$. Let $\{X, \mathcal{A}, \mu\}$ be $\mathbb{R}^N$ with the Lebesgue measure.

Proposition 6.1 (Day[5]). *Let $L^p(E)$ for $0 < p < 1$ be equipped with the topology generated by the metric in (6.1). Then the only open, convex subsets of $L^p(E)$ are $\emptyset$ and $L^p(E)$ itself.*

Proof. Let $\mathcal{O}$ be a nonempty, open, convex neighborhood of the origin of $L^p(E)$, and let f be an arbitrary element of $L^p(E)$. Since $\mathcal{O}$ is open, it contains some ball B_ρ centered at the origin. Let n be a positive integer such that

$$\frac{1}{n^{1-p}} \int_E |f|^p dx \leq \rho; \quad \text{i.e.,} \quad nf \in B_{n\rho}.$$

---

[4]See Section 15.

[5]M. M. Day, The spaces L^p with $0 < p < 1$, *Bull. Amer. Math. Soc.*, **46** (1940), 816–823.

Partition E into exactly n disjoint, measurable subsets $\{E_1, E_2, \ldots, E_n\}$ such that

$$\int_{E_j} |f|^p dx = \frac{1}{n} \int_E |f|^p dx, \quad j = 1, 2, \ldots, n.$$

Such a partition can be carried out in view of the absolute continuity of the Lebesgue integral. Set

$$h_j = \begin{cases} nf & \text{within } E_j, \\ 0 & \text{otherwise} \end{cases}$$

and compute

$$\int_E |h_j|^p dx = n^p \int_{E_j} |f|^p dx \le \rho.$$

Thus $h_j \in B_\rho \subset \mathcal{O}$ for all $j = 1, 2, \ldots, n$. Since $\mathcal{O}$ is convex,

$$f = \frac{h_1 + h_2 + \cdots + h_n}{n} \in \mathcal{O}.$$

Since $f \in L^p(E)$ is arbitrary, this implies that $\mathcal{O} \equiv L^p(E)$.   □

Corollary 6.2. *The topology generated by the metric* (6.1) *in* $L^p(E)$ *for* $0 < p < 1$ *is not locally convex.*

Remark 6.1. The conclusions of Proposition 6.1 and Corollary 6.2 continue to hold for measure spaces $\{X, \mathcal{A}, \mu\}$ with the following property: Given $E \in \mathcal{A}$, let A denote μ-measurable subsets of E. Then for $f \in L^p(E)$, the functional

$$E \bigcap \mathcal{A} \ni A \longrightarrow \int_E \chi_A |f|^p d\mu, \quad 0 < p < 1, \tag{6.2}$$

takes all the values between 0 and $\|f\|_p^p$.

7   Convergence in $L^p(E)$ and completeness

A sequence $\{f_n\}$ of functions in $L^p(E)$ for some $1 \le p \le \infty$ converges in the sense of $L^p(E)$ to a function $f \in L^p(E)$ if

$$\lim \|f_n - f\|_p = 0. \tag{7.1}$$

This notion of convergence is also called convergence in the *mean* of order p, or in the *norm* $L^p(E)$, or *strong* convergence in $L^p(E)$.

The sequence $\{f_n\}$ is a Cauchy sequence in $L^p(E)$ if for every $\varepsilon > 0$, there exists a positive integer n_ε such that

$$\|f_n - f_m\|_p \le \varepsilon \quad \text{for all } n, m \ge n_\varepsilon. \tag{7.2}$$

If $\{f_n\} \to f$ in $L^p(E)$, then $\{f_n\}$ is a Cauchy sequence. Indeed, from (7.1), it follows that if n, m are sufficiently large, then

$$\| f_n - f_m \|_p \leq \| f_n - f \|_p + \| f_m - f \|_p \leq \varepsilon. \tag{7.3}$$

The next theorem asserts the converse, i.e., that if $\{f_n\}$ is a Cauchy sequence in $L^p(E), 1 \leq p \leq \infty$, it converges, in the sense of $L^p(E)$, to a function $f \in L^p(E)$. In this sense, the spaces $L^p(E)$ for $1 \leq p \leq \infty$ are *complete*.

Theorem 7.1 (Riesz–Fischer[6]). *Let* $\{f_n\}$ *be a Cauchy sequence in* $L^p(E)$ *for some* $1 \leq p \leq \infty$. *There exists* $f \in L^p(E)$ *such that* $\{f_n\} \to f$ *in* $L^p(E)$.

Proof. We assume that $p \in [1, \infty)$, the arguments for $L^\infty(E)$ being similar. For $j \in \mathbb{N}$, let n_j be a positive integer such that

$$\| f_n - f_m \|_p \leq \frac{1}{2^j} \quad \text{for all } n, m \geq n_j. \tag{7.4}$$

Without loss of generality, we may arrange that $n_j < n_{j+1}$ for all $j \in \mathbb{N}$. Formally set

$$f(x) = f_{n_1}(x) + \sum [f_{n_{j+1}}(x) - f_{n_j}(x)] \quad \text{for a.e. } x \in E. \tag{7.5}$$

We claim that (7.5) defines a function $f \in L^p(E)$ and that $\{f_n\} \to f$ in $L^p(E)$. For $m = 1, 2, \ldots$, set

$$g_m(x) = \sum_{j=1}^{m} |f_{n_{j+1}}(x) - f_{n_j}(x)| \quad \text{for a.e. } x \in E.$$

Since $g_m \leq g_{m+1}$, there exists the limit

$$\lim g_m(x) = g(x) \quad \text{for a.e. } x \in E.$$

By Fatou's lemma, Minkowski's inequality, and (7.4),

$$\left(\int_E g^p d\mu \right)^{1/p} \leq \left(\liminf \int_E g_m^p d\mu \right)^{1/p} \leq \sum_{j=1}^{m} \frac{1}{2^j} \leq 1.$$

Thus $g \in L^p(E)$. The a.e. convergence of $\{g_n\}$ implies that the limit

$$\lim_{m \to \infty} \sum_{j=1}^{m} [f_{n_{j+1}}(x) - f_{n_j}(x)]$$

exists for a.e. $x \in E$. Therefore, (7.5) defines a function f measurable in E.

---

[6]F. Riesz, Sur les systémes orthogonaux de fonctions, *C. R. Acad. Sci. Paris*, **144** (1907), 615–619; F. Fischer, Sur la convergence en moyenne, *C. R. Acad. Sci. Paris*, **144** (1907), 1148–1150.

From (7.5) and the definition of g,

$$|f(x)| \le |f_{n_1}(x)| + |g(x)| \quad \text{for a.e. } x \in E.$$

Thus $f \in L^p(E)$. Next, from (7.5)–(7.4) and Minkowski's inequality, it follows that for any positive integer k,

$$\| f_{n_k} - f \|_p \le \sum_{j=k}^{\infty} \| f_{n_{j+1}} - f_{n_j} \|_p \le \frac{1}{2^{k-1}}.$$

Therefore, $\{f_{n_j}\}$ converges to f in $L^p(E)$. In particular, for every $\varepsilon > 0$, there exists a positive integer j_ε such that

$$\| f_{n_j} - f \|_p \le \frac{1}{2}\varepsilon \quad \text{for all } j \ge j_\varepsilon.$$

We finally establish that the entire sequence $\{f_n\}$ converges to f in $L^p(E)$.

Since $\{f_n\}$ is a Cauchy sequence, having fixed $\varepsilon > 0$, there exists a positive integer n_ε such that

$$\| f_n - f_m \|_p \le \frac{1}{2}\varepsilon \quad \text{for all } n, m \ge n_\varepsilon.$$

Therefore, for $n \ge n_\varepsilon$,

$$\| f_n - f \|_p \le \| f_n - f_{n_j} \|_p + \| f_{n_j} - f \|_p \le \varepsilon,$$

provided $j \ge j_\varepsilon$ and $n_j \ge n_\varepsilon$. $\square$

Remark 7.1. The spaces $L^p(E)$ for all $1 \le p \le \infty$ endowed with their norm topology are complete metric spaces. As such, they are of second category; i.e., they are not the countable union of nowhere dense sets.

8   Separating $L^p(E)$ by simple functions

Proposition 8.1. *Let $f \in L^p(E)$ for $1 \le p \le \infty$. For every $\varepsilon > 0$, there exists a simple function $\varphi \in L^p(E)$ such that $\| f - \varphi \|_p \le \varepsilon$.*[7]

Proof. By the decomposition $f = f^+ - f^-$, one may assume that f is nonnegative. Since f is measurable, there exists a sequence $\{\varphi_n\}$ of nonnegative, simple functions such that

$$\varphi_n \le \varphi_{n+1} \quad \text{and} \quad \varphi_n \to f \quad \text{everywhere in } E.$$

If $1 \le p < \infty$, the sequence $(f - \varphi_n)^p$ converges to zero a.e. in E, and it is dominated by the integrable function f^p. Therefore, $\| f - \varphi_n \|_p \to 0$.

---

[7]It is not claimed here that $L^p(E)$ is separable. See Section 18.

If $p = \infty$, the construction of the φ_n implies that

$$\mu\left(\left[x \in E \mid f(x) - \varphi_n(x) > \frac{1}{2^n}\|f\|_\infty\right]\right) = 0.$$

Thus $\|f - \varphi_n\|_\infty \leq 2^{-n}\|f\|_\infty$ for all n. $\qquad\square$

Proposition 8.2. *Let $1 \leq p, q \leq \infty$ be conjugate, and let $g \in L^1(E)$ satisfy*

$$\int_E \varphi g d\mu \leq K\|\varphi\|_p \quad \text{for all simple functions } \varphi$$

for some positive constant K. Then $g \in L^q(E)$ and $\|g\|_q \leq K$.

Proof. If $q = 1$, it suffices to choose $\varphi = \text{sign } g \in L^\infty(E)$. Assuming that $q \in (1, \infty)$, let $\{\varphi_n\}$ denote a sequence of nonnegative, simple functions such that $\varphi_n \leq \varphi_{n+1}$ and $\varphi_n \to |g|^q$. Since

$$0 \leq \varphi_n^{1/q} \leq |g| \in L^1(E),$$

each φ_n is simple and vanishes outside a set of finite measure. Therefore, the functions

$$h_n = \varphi_n^{1/p} \text{ sign } g$$

are simple and in $L^p(E)$. For these choices,

$$\int_E \varphi_n d\mu = \int_E \varphi_n^{1/p}\varphi_n^{1/q} d\mu \leq \int_E \varphi_n^{1/p}|g| d\mu$$

$$= \int_E h_n g d\mu \leq K\|h_n\|_p = K\left(\int_E \varphi_n d\mu\right)^{1/p}.$$

From this and Fatou's lemma,

$$\|g\|_q \leq (\liminf \int_E \varphi_n d\mu)^{1/q} \leq K.$$

Now consider the case $q = \infty$. Having fixed $\varepsilon > 0$, set

$$E_\varepsilon = \{x \in E \text{ such that } |g(x)| \geq K + \varepsilon\}$$

and choose $\varphi = \chi_{E_\varepsilon} \text{ sign } g$. Since $g \in L^1(E)$, the set E_ε is of finite measure and $\varphi \in L^1(E)$. Therefore,

$$(K + \varepsilon)\mu(E_\varepsilon) \leq \left|\int_E \varphi g d\mu\right| \leq K\mu(E_\varepsilon).$$

Thus $\mu(E_\varepsilon) = 0$ for all $\varepsilon > 0$. $\qquad\square$

Corollary 8.3. *Let $1 \leq p, q \leq \infty$ be conjugate, and let $g \in L^1(E)$ satisfy*

$$\int_E f g d\mu \leq K\|f\|_p \quad \text{for all } f \in L^p(E) \cap L^\infty(E)$$

for some positive constant K. Then $g \in L^q(E)$ and $\|g\|_q \leq K$.

9   Weak convergence in $L^p(E)$

Let $1 \le p, q \le \infty$ be conjugate. A sequence of functions $\{f_n\}$ in $L^p(E)$, $1 \le p \le \infty$, converges *weakly* to a function $f \in L^p(E)$ if

$$\lim \int_E f_n g d\mu = \int_E f g d\mu \quad \text{for all } g \in L^q(E).$$

If $\{f_n\}$ converges to f in $L^p(E)$, it also converges weakly to f in $L^p(E)$. Indeed, by the Hölder inequality, for all $g \in L^q(E)$,

$$\left| \int_E (f_n g - f g) d\mu \right| \le \|g\|_q \|f_n - f\|_p.$$

Thus strong convergence implies weak convergence. The converse is false as there exist sequences of functions $\{f_n\}$ in $L^p(E)$ converging weakly to some $f \in L^p(E)$ and not converging to f in the sense of $L^p(E)$.

9.1   A counterexample. The functions $x \to \cos nx$, $n = 1, 2, \ldots$, satisfy

$$\int_0^{2\pi} \cos^2 nx \, dx = \pi \quad \text{for all } n \in \mathbb{N}. \tag{9.1}$$

Therefore, $\{\cos nx\}$ is a sequence of functions in $L^2[0, 2\pi]$ that cannot converge to zero in the sense of $L^2[0, 2\pi]$. However, such a sequence converges to zero weakly in $L^2[0, 2\pi]$. To prove this, first let $g = \chi_{[\alpha, \beta]}$, where $[\alpha, \beta] \subset [0, 2\pi]$. By direct calculation,

$$\int_0^{2\pi} \chi_{[\alpha, \beta]} \cos nx \, dx = \frac{1}{n} \{\sin n\beta - \sin n\alpha\} \to 0 \quad \text{as } n \to \infty.$$

Now let $\{[\alpha_i, \beta_i]\}_{i=1}^m$ be a finite collection of mutually disjoint subintervals of $[0, 2\pi]$ and let φ be a simple function of the form

$$\varphi = \sum_{i=1}^m g_i \chi_{[\alpha_i, \beta_i]}. \tag{9.2}$$

For any such simple function,

$$\lim \int_0^{2\pi} \varphi \cos nx \, dx = 0.$$

Simple functions of the form (9.2) are dense in $L^2[0, 2\pi]$. Thus

$$\lim \int_0^{2\pi} g \cos nx \, dx = 0 \quad \text{for all } g \in L^2[0, 2\pi].$$

10   Weak lower semicontinuity of the norm in $L^p(E)$

Proposition 10.1. *Let $\{f_n\}$ be a sequence of functions in $L^p(E)$ for some $1 \leq p < \infty$ converging weakly to some $f \in L^p(E)$. Then*

$$\liminf \|f_n\|_p \geq \|f\|_p. \tag{10.1}$$

If $p = \infty$, the same conclusion holds if $\{X, \mathcal{A}, \mu\}$ is σ-finite.

Remark 10.1. The previous counterexample shows that the inequality in (10.1) might be strict.

Proof of Proposition 10.1. Assume first that $1 \leq p < \infty$. The function $g = |f|^{p/q} \operatorname{sign} f$ belongs to $L^q(E)$, and by the definition of weak convergence,

$$\lim \int_E f_n g \, d\mu = \int_E f g \, d\mu = \|f\|_p^p.$$

On the other hand, by Hölder's inequality,

$$\left| \int_E f_n g \, d\mu \right| \leq \|f_n\|_p \|g\|_p = \|f_n\|_p \|f\|_p^{p/q}.$$

Therefore,

$$\liminf \|f_n\|_p \|f\|_p^{p/q} \geq \|f\|_p^p.$$

Next, assume that $p = \infty$ and $\mu(E) < \infty$. Fix $\varepsilon > 0$ and set

$$E_\varepsilon = \{x \in E \mid |f(x)| \geq \|f\|_\infty - \varepsilon\} \quad \text{and} \quad g = \chi_{E_\varepsilon} \operatorname{sign} f.$$

Then for such choices,

$$\lim \int_E f_n g \, d\mu = \int_E f g \, d\mu \geq (\|f\|_\infty - \varepsilon) \mu(E_\varepsilon).$$

Also by Hölder's inequality,

$$\left| \int_E f_n g \, dx \right| \leq \|f_n\|_\infty \mu(E_\varepsilon).$$

Since $\mu(E_\varepsilon) > 0$, this implies

$$\liminf \|f_n\|_\infty \geq \|f\|_\infty - \varepsilon \quad \text{for all } \varepsilon > 0.$$

If $p = \infty$ and $\{X, \mathcal{A}, \mu\}$ is σ-finite, let $A_j \subset A_{j+1}$ be a sequence of measurable sets of finite measure whose union is X. Setting $E_j = E \cap A_j$, the previous remarks give

$$\liminf \|f_n\|_{\infty, E} \geq \|f\|_{\infty, E_j} \quad \text{for all } j \in \mathbb{N}. \qquad \square$$

Corollary 10.2. *Let $p \in [1, \infty)$. The function*

$$\|\cdot\|_p : L^p(E) \longrightarrow \mathbb{R}^+$$

is weakly lower semicontinuous. If $p = \infty$, the same conclusion holds if $\{X, \mathcal{A}, \mu\}$ is σ-finite.

11   Weak convergence and norm convergence

Weak convergence does not imply norm convergence, nor does the latter imply weak convergence. The sequence in Section 9.1 provides counterexample to both statements. The next proposition relates these two notions of convergence.

Proposition 11.1 (Radon[8]). *Let $p \in (1, \infty)$, and let $\{f_n\}$ be a sequence of functions in $L^p(E)$ converging weakly to some $f \in L^p(E)$. If also*

$$\lim \|f_n\|_p = \|f\|_p,$$

then $\{f_n\}$ converges to f strongly in $L^p(E)$.

Remark 11.1. The proposition asserts that weak convergence and norm convergence to the *same* function imply strong convergence.

Remark 11.2. The counterexample of Section 9.1 shows that weak convergence and norm convergence do not imply strong convergence. For this to occur, the weak limit is required to coincide with the norm limit.

Remark 11.3. The proposition is false for $p = \infty$. In $(0, 1)$ with the Lebesgue measure, set

$$f_n(x) = \begin{cases} 0 & \text{for } 0 \le x \le \dfrac{1}{n}, \\ 1 & \text{for } \dfrac{1}{n} < x \le 1. \end{cases}$$

Then

$$\{f_n\} \longrightarrow 1 \quad \text{weakly in } L^\infty(0, 1) \quad \text{and} \quad \|f_n\|_\infty \longrightarrow 1.$$

However, $\|f_n - 1\|_\infty = 1$ for all $n \in \mathbb{N}$.

Remark 11.4. The proposition is false for $p = 1$. In $(0, 2\pi)$ with the Lebesgue measure, set

$$f_n(x) = 4 + \sin nx, \quad n \in \mathbb{N}.$$

Then

$$\{f_n\} \longrightarrow 4 \quad \text{weakly in } L^1(0, 2\pi) \quad \text{and} \quad \|f_n\|_1 \longrightarrow \|4\|_1.$$

However, $\|f_n - 4\|_1 = 4$ for all $n \in \mathbb{N}$.[9]

The proof of Proposition 11.1 is based on the following inequalities.

---

[8] J. Radon, Theorie und Anwendungen der absolut additiven Mengenfunktionen, *Sitzungsber. Akad. Wiss. Wien*, **122** (1913), Abt. IIa, 1295–1438; also in F. Riesz, Sur la convergence en moyenne, *Acta Sci. Math. (Szeged)*, **4** (1928), 58–64.

[9] This example was suggested by J. Manfredi.

Lemma 11.2. *Let $p \geq 2$. There exists a positive constant c such that for all $t \in \mathbb{R}$,*

$$|1 + t|^p \geq 1 + pt + c|t|^p. \tag{11.1}$$

Lemma 11.3. *Let $1 < p < 2$. There exists a positive constant c such that for all $t \in \mathbb{R}$,*

$$|1 + t|^p \geq \begin{cases} 1 + pt + c|t|^p & \text{if } |t| \geq 1, \\ 1 + pt + c|t|^2 & \text{if } |t| \leq 1. \end{cases} \tag{11.2}$$

The proof of these lemmas is technical in nature and is given in Section 11.2 of the Problems and Complements.

11.1 Proof of Proposition 11.1 for $p \geq 2$.

In (11.1), put

$$t = \frac{f_n(x) - f(x)}{f(x)} \quad \text{for } f(x) \neq 0. \tag{11.3}$$

Multiplying the inequality so obtained by $|f(x)|^p$ gives

$$|f_n|^p \geq |f|^p + p|f|^{p-2} f(f_n - f) + c|f_n - f|^p.$$

One verifies that such an inequality continues to hold also if $f(x) = 0$. Integrating it over E and taking the limit as $n \to \infty$ yields

$$c \lim \sup \int_E |f_n - f|^p d\mu \leq \lim(\|f_n\|_p^p - \|f\|_p^p)$$

$$- p \lim \int_E |f|^{p-2} f(f_n - f) d\mu = 0. \qquad \square$$

11.2 Proof of Proposition 11.1 for $1 < p < 2$.

For $n \in \mathbb{N}$, introduce the sets

$$E_n = \{x \in E \mid |f_n(x) - f(x)| \geq |f(x)|\}.$$

In (11.2), choose t as in (11.3) and multiply by $|f(x)|^p$ to obtain

$$\begin{aligned} |f_n|^p &\geq |f|^p + p|f|^{p-2} f(f_n - f) + c|f_n - f|^p & \text{in } E_n, \\ |f_n|^p &\geq |f|^p + p|f|^{p-2} f(f_n - f) + c(f_n - f)^2 |f|^{p-2} & \text{in } E - E_n. \end{aligned} \tag{11.4}$$

Integrate the first part over E_n and the second part over $(E - E_n)$, add the resulting inequalities, and let $n \to \infty$ to obtain

$$c \lim \sup \left\{ \int_{E_n} |f_n - f|^p d\mu + \int_{E - E_n} (f_n - f)^2 |f|^{p-2} d\mu \right\}$$

$$\leq \lim\{\|f_n\|_p^p - \|f\|_p^p\} - p \lim \int_E |f|^{p-2} f(f_n - f) d\mu = 0.$$

Thus, in particular,

$$\lim \int_{E_n} |f_n - f|^p d\mu = 0$$

and

$$\lim \int_{E - E_n} (f_n - f)^2 |f|^{p-2} d\mu = 0.$$

From this, the definition of E_n, and Hölder's inequality,

$$\begin{aligned}
\limsup \int_E |f_n - f|^p d\mu \\
\leq \lim \int_{E_n} |f_n - f|^p d\mu + \limsup \int_{E - E_n} |f|^{p-1} |f_n - f| d\mu \\
\leq \lim \int_{E_n} |f_n - f|^p d\mu \\
+ \left(\int_E |f|^p d\mu \right)^{1/2} \lim \left(\int_{E - E_n} |f|^{p-2} |f_n - f|^2 d\mu \right)^{1/2} = 0. \quad \square
\end{aligned}$$

12   Linear functionals in $L^p(E)$

A map $\mathcal{F} : L^p(E) \to \mathbb{R}$ is a linear functional in $L^p(E)$ if for all $f, g \in L^p(E)$ and $\alpha, \beta \in \mathbb{R}$,

$$\mathcal{F}(\alpha f + \beta g) = \alpha \mathcal{F}(f) + \beta \mathcal{F}(g).$$

The functional $\mathcal{F}$ is bounded if there exists a constant K such that

$$|\mathcal{F}(f)| \leq K \|f\|_p \quad \text{for all } f \in L^p(E). \tag{12.1}$$

The norm of $\mathcal{F}$ is the smallest constant K for which (12.1) holds. Therefore,

$$\|\mathcal{F}\| = \sup_{\substack{f \in L^p(E) \\ \|f\|_p \neq 0}} \frac{|\mathcal{F}(f)|}{\|f\|_p} = \sup_{\substack{f \in L^p(E) \\ \|f\|_p = 1}} |\mathcal{F}(f)|. \tag{12.2}$$

Proposition 12.1. *Let $1 < p \leq \infty$ and $1 \leq q < \infty$ be conjugate. Every $g \in L^q(E)$ generates a bounded linear functional in $L^p(E)$ by the formula*

$$\mathcal{F}_g(f) = \int_E fg \, d\mu \quad \text{for all } f \in L^p(E). \tag{12.3}$$

Moreover, $\|\mathcal{F}_g\| = \|g\|_q$. If $p = 1$ and $q = \infty$, the same conclusion holds if $\{X, \mathcal{A}, \mu\}$ is σ-finite.

Remark 12.1. If $p = 1$ and $\{X, \mathcal{A}, \mu\}$ is not σ-finite, the formula (12.3) for a given $g \in L^\infty(E)$ still defines a bounded linear functional in $L^1(E)$. However, the identification $\|\mathcal{F}\| = \|g\|_\infty$ might fail. A counterexample can be constructed using the measure space $\{X, \mathcal{A}, \mu\}$ in Section 3.3 of the Problems and Complements of Chapter II.

Proof of Proposition 12.1. The map $\mathcal{F}_g$ is linear. By Hölder's inequality, it is also bounded. If $1 \leq q < \infty$, Proposition 4.1 identifies the norm $\|\mathcal{F}_g\|$ as the norm $\|g\|_q$.

Now let $q = \infty$ and assume momentarily that E is of finite measure. Having fixed $\varepsilon > 0$, set

$$E_\varepsilon = \{x \in E \mid |g(x)| \geq \|g\|_{\infty,E} - \varepsilon\}, \tag{12.4}$$

and in (12.3) choose $f = \chi_{E_\varepsilon} \operatorname{sign} g \in L^1(E)$. This gives

$$\mathcal{F}_g(f) = \int_{E_\varepsilon} |g| dx \geq \|f\|_{1,E}(\|g\|_{\infty,E} - \varepsilon) \quad \text{for all } \varepsilon > 0.$$

Therefore,

$$\|g\|_{\infty,E} - \varepsilon \leq \|\mathcal{F}\| \leq \|g\|_{\infty,E}.$$

If $\{X, \mathcal{A}, \mu\}$ is σ-finite, let $A_j \subset A_{j+1}$ be a countable collection of measurable sets of finite measure whose union is X. Set $E_j = E \cap A_j$ and define $E_{j,\varepsilon}$ as in (12.4) with E replaced by E_j. Choosing $f = \chi_{E_{j,\varepsilon}} \operatorname{sign} g \in L^1(E)$ in (12.3) gives

$$\|g\|_{\infty,E_j} - \varepsilon \leq \|\mathcal{F}\| \leq \|g\|_{\infty,E}$$

for all $\varepsilon > 0$ and all $j \in \mathbb{N}$. $\qquad\square$

Remark 12.2. Let $p \in (1, \infty)$. The proof of Proposition 12.1 shows that if g is not the zero equivalence class of $L^q(E)$, the norm $\|\mathcal{F}_g\|$ is achieved by computing $\mathcal{F}_g$ at the element

$$g^* = \frac{|g|^{q-1} \operatorname{sign} g}{\|g\|_q^{q/p}} \in L^p(E). \tag{12.5}$$

This formula suggests that $\mathcal{F}_g$ given by (12.3) is in some sense the only functional in $L^p(E)$ generated by the element $g \in L^q(E)$. This is indeed the case and will be proved in Proposition 15.2.

The Riesz representation theorem asserts that if $1 \leq p < \infty$, the functionals in (12.3) are the only bounded linear functionals in $L^p(E)$.

13   The Riesz representation theorem

Theorem 13.1. *Let* $1 < p, q < \infty$ *be conjugate. For every bounded, linear functional* $\mathcal{F}$ *in* $L^p(E)$, *there exists a unique function* $g \in L^q(E)$ *such that* $\mathcal{F}$ *is represented by the formula* (12.3). *Moreover,* $\|\mathcal{F}\| = \|g\|_q$. *If* $p = 1$ *and* $q = \infty$, *the same conclusion holds if* $\{X, \mathcal{A}, \mu\}$ *is* σ-*finite.*

Remark 13.1. The theorem is false if $p = 1$ and if $\{X, \mathcal{A}, \mu\}$ is not σ-finite. A counterexample can be constructed using the measure space in Section 3.3 of the Problems and Complements of Chapter II.

Remark 13.2. The theorem is also false for $p = \infty$. A counterexample is in Section 9.2 of the Problems and Complements of Chapter VI.

13.1   Proof of Theorem 13.1: The case where $\{X, \mathcal{A}, \mu\}$ is finite. Assume first that $\mu(X) < \infty$ and that $E = X$. For every μ-measurable set $A \subset X$, the function χ_A is in $L^p(E)$. The functional $\mathcal{F}$ induces a set function ν defined on the σ-algebra $\mathcal{A}$ by the formula

$$\mathcal{A} \ni A \longrightarrow \nu(A) = \mathcal{F}(\chi_A).$$

Such a set function is finite for all $A \in \mathcal{A}$, vanishes on the empty set, and is countably additive. To establish the last claim, let $\{A_n\}$ be a countable collection of mutually disjoint sets in $\mathcal{A}$. Since $\mu(\bigcup A_n) < \infty$, for every $\varepsilon > 0$, there exists a positive integer n_ε such that

$$\sum_{j > n_\varepsilon} \mu(A_j) < \varepsilon.$$

By the linearity of $\mathcal{F}$,

$$\mathcal{F}\left(\chi_{\bigcup A_n}\right) = \mathcal{F}\left(\chi_{\bigcup_{j=1}^{n_\varepsilon} A_j} + \chi_{\bigcup_{j > n_\varepsilon} A_j}\right)$$

$$= \mathcal{F}\left(\sum_{j=1}^{n_\varepsilon} \chi_{A_j} + \chi_{\bigcup_{j > n_\varepsilon} A_j}\right)$$

$$= \sum_{j=1}^{n_\varepsilon} \mathcal{F}(\chi_{A_j}) + \mathcal{F}\left(\chi_{\bigcup_{j > n_\varepsilon} A_j}\right).$$

From this, since the A_j are disjoint and $p \in [1, \infty)$,

$$\left| \mathcal{F}\left(\chi_{\bigcup A_n}\right) - \sum_{j=1}^{n_\varepsilon} \mathcal{F}(\chi_{A_j}) \right| \leq \|\mathcal{F}\| \left\| \chi_{\bigcup_{j > n_\varepsilon} A_j} \right\|_p$$

$$= \|\mathcal{F}\| \left(\sum_{j > n_\varepsilon} \mu(A_j) \right)^{1/p}$$

$$\leq \|\mathcal{F}\| \varepsilon^{1/p}.$$

Since ε is arbitrary, this implies that

$$\nu\left(\bigcup A_n\right) = \mathcal{F}\left(\chi_{\bigcup A_n}\right) = \sum \mathcal{F}(\chi_{A_n}) = \sum \nu(A_n).$$

Therefore, ν is countably additive and defines a signed measure on $\mathcal{A}$. Since $|\nu(A)| = 0$ whenever $\mu(A) = 0$, the signed measure ν is absolutely continuous with respect to μ. By the Radon–Nikodým theorem, there exists a μ-measurable function $g : X \to \mathbb{R}^*$ such that

$$\mathcal{A} \ni A \longrightarrow \nu(A) = \int_E g\chi_A d\mu.$$

For every simple function

$$\varphi = \sum_{i=1}^{n} \alpha_i \chi_{A_i},$$

and by the linearity of $\mathcal{F}$,

$$\mathcal{F}(\varphi) = \sum_{i=1}^{n} \alpha_i \nu(A_i) = \sum_{i=1}^{n} \alpha_i \int_E g\chi_{A_i} d\mu = \int_E g\varphi d\mu.$$

For all such simple functions,

$$\left| \int_E g\varphi d\mu \right| \le \|\mathcal{F}\| \|\varphi\|_{p,X}.$$

Therefore, $g \in L^q(E)$ by Proposition 8.2, and $\|g\|_q \le \|\mathcal{F}\|$. Since such step functions are dense in $L^p(E)$,

$$\mathcal{F}(f) = \int_E fg d\mu \quad \text{for all } f \in L^p(E).$$

If $g' \in L^q(E)$ identifies the same functional $\mathcal{F}$, then

$$\int_E f(g - g')d\mu = 0 \quad \text{for all } f \in L^p(E).$$

Therefore, $g = g'$ a.e. in E. $\qquad\qquad\qquad\qquad\qquad\qquad\qquad\square$

13.2   Proof of Theorem 13.1: The case where $\{X, \mathcal{A}, \mu\}$ is σ-finite. Let $A_j \subset A_{j+1}$ be a countable collection of sets of finite measure exhausting X, and set $E_j = E \cap A_j$. For each $f \in L^p(E)$ and $j \in \mathbb{N}$, set

$$f_j = \begin{cases} f & \text{on } E_j, \\ 0 & \text{on } E - E_j. \end{cases}$$

For each $j \in \mathbb{N}$, there exists $g_j \in L^q(E_j)$ such that

$$\mathcal{F}(f_j) = \int_{E_j} fg_j d\mu \quad \text{for all } f \in L^p(E).$$

We regard g_j as defined in the whole E by setting them to be equal to zero outside E_j. If $f \in L^p(E)$ vanishes outside E_j,

$$\mathcal{F}(f) = \int_{E_j} fg_j d\mu = \int_{E_{j+1}} fg_{j+1} d\mu.$$

Therefore,

$$\int_{E_j} f(g_j - g_{j+1})d\mu = 0 \quad \text{for all } f \in L^p(E_j)$$

and g_j coincides with g_{j+1} on E_j. The sequence $\{g_j\}$ converges a.e. on E to a measurable function g. The sequence $\{|g_j|\}$ is nondecreasing, and by monotone convergence,

$$\|g\|_q = \lim \|g_j\|_q \le \|\mathcal{F}\|, \quad 1 < q \le \infty.$$

Thus $g \in L^q(E)$. Now given any $f \in L^p(E)$, the sequence $\{f_j g\}$ converges to fg a.e. on E and $|f_j g| \le |fg| \in L^1(E)$. Therefore, by dominated convergence,

$$\int_E fg d\mu = \lim \int_E f_j g d\mu = \lim \mathcal{F}(f_j) = \mathcal{F}(f).$$

The characterization of $\|\mathcal{F}\|$ follows from Proposition 12.1. $\square$

13.3 Proof of Theorem 13.1: The case where $1 < p < \infty$. We assume that $1 < p < \infty$ and place no restrictions on the measure space $\{X, \mathcal{A}, \mu\}$. If $A \subset E$ is of σ-finite measure, there exists a unique $g_A \in L^q(E)$ vanishing on $(E - A)$ such that[10]

$$\mathcal{F}(f|_A) = \int_E f g_A d\mu \quad \text{for all } f \in L^p(E).$$

Moreover, if $B \subset A$ is of σ-finite measure, then $g_B = g_A$ a.e. on B. The set function $A \to \|g_A\|_q$ defined on the subsets of E of σ-finite measure is uniformly bounded since

$$\|g_A\|_q \le \|\mathcal{F}\| \quad \text{for all sets } A \text{ of } \sigma\text{-finite measure.}$$

Denote by M the supremum of $\|g_A\|_q$ as A ranges over such sets, and let $\{A_n\}$ be a sequence of sets of σ-finite measure such that

$$\|g_{A_n}\|_q \le \|g_{A_{n+1}}\|_q \quad \text{and} \quad \lim \|g_{A_n}\|_q = M.$$

The set $A_o = \bigcup A_n$ is of σ-finite measure and $\|g_{A_o}\|_q = M$. Thus the supremum of $\|g_A\|$ is actually achieved at A_o. We regard g_{A_o} as defined in the whole E by setting it to be zero outside A_o. In such a way, $g_{A_o} \in L^q(E)$. Such a function g_{A_o} is the one claimed by the Riesz representation theorem.

If B_o is a set of σ-finite measure containing A_o, then $g_{A_o} = g_{B_o}$ a.e. on A_o. Also, by maximality,

$$\|g_{A_o}\|_q \le \|g_{B_o}\|_q \le \|g_{A_o}\|_q.$$

Therefore, $g_{B_o} = 0$ a.e. on $(B_o - A_o)$ since $1 < q < \infty$.

Given $f \in L^p(E)$, the set $[|f| > 0]$ is of σ-finite measure. Since the set $B_o = [|f| > 0] \bigcup A_o$ also is of σ-finite measure,

$$\mathcal{F}(f) = \int_E f g_{B_o} d\mu = \int_E f g_{A_o} d\mu.   \square$$

---

[10] $f|_A$ is the restriction of f to A defined in the whole E by setting it to be zero outside A.

14   The Hanner and Clarkson inequalities

Proposition 14.1 (Hanner's inequalities[11]). *Let f and g be in $L^p(E)$ for some $1 \le p < \infty$. Then*

$$\|f+g\|_p^p + \|f-g\|_p^p \le (\|f\|_p + \|g\|_p)^p + |\|f\|_p - \|g\|_p|^p \quad \text{for } p \ge 2,$$
(14.1)

$$\|f+g\|_p^p + \|f-g\|_p^p \ge (\|f\|_p + \|g\|_p)^p + |\|f\|_p - \|g\|_p|^p \quad \text{for } p \in [1,2],$$
(14.2)

$$(\|f+g\|_p + \|f-g\|_p)^p + |\|f+g\|_p - \|f-g\|_p|^p$$
$$\ge 2^p(\|f\|_p^p + \|g\|_p^p) \qquad \text{for } p \ge 2,$$
(14.3)

$$(\|f+g\|_p + \|f-g\|_p)^p + |\|f+g\|_p - \|f-g\|_p|^p$$
$$\le 2^p(\|f\|_p^p + \|g\|_p^p) \qquad \text{for } p \in [1,2].$$
(14.4)

Proposition 14.2 (Clarkson's inequalities[12]). *Let $1 < p, q < \infty$ be conjugate, and let $f, g \in L^p(E)$. Then*

$$\left\|\frac{f+g}{2}\right\|_p^p + \left\|\frac{f-g}{2}\right\|_p^p \le \frac{\|f\|_p^p + \|g\|_p^p}{2} \qquad \text{for } p \ge 2, \qquad (14.5)$$

$$\left\|\frac{f+g}{2}\right\|_p^p + \left\|\frac{f-g}{2}\right\|_p^p \ge \frac{\|f\|_p^p + \|g\|_p^p}{2} \qquad \text{for } p \in (1,2], \qquad (14.6)$$

$$\left\|\frac{f+g}{2}\right\|_p^q + \left\|\frac{f-g}{2}\right\|_p^q \ge \left(\frac{\|f\|_p^p + \|g\|_p^p}{2}\right)^{q-1} \qquad \text{for } p \ge 2, \qquad (14.7)$$

$$\left\|\frac{f+g}{2}\right\|_p^q + \left\|\frac{f-g}{2}\right\|_p^q \le \left(\frac{\|f\|_p^p + \|g\|_p^p}{2}\right)^{q-1} \qquad \text{for } p \in (1,2]. \qquad (14.8)$$

If $p = 2$, Hanner's inequalities become the standard parallelogram identity, and for $p = 1$ they coincide with the triangle inequality.

Assuming that $p > 1$ and $p \ne 2$, set

$$\varphi(s; t) = h(s) + k(s)t^p,$$

where for $s \in (0, 1]$ and $t > 0$,

$$h(s) = (1+s)^{p-1} + (1-s)^{p-1},$$
$$k(s) = \{(1+s)^{p-1} - (1-s)^{p-1}\}s^{1-p}.$$

[11]O. Hanner, On the uniform convexity of L^p and ℓ^p, *Ark. Mat.*, **3** (1956), 239–244.

[12]J. A. Clarkson, Uniformly convex spaces, *Trans. Amer. Math. Soc.*, **40** (1936), 396–414.

Lemma 14.3. *Let* $1 < p, q < \infty$ *be conjugate. For every fixed* $t > 0$, *it holds that*

$$\varphi(s; t) \leq |1 + t|^p + |1 - t|^p \quad \text{for } p \in (1, 2],$$
$$\varphi(s; t) \geq |1 + t|^p + |1 - t|^p \quad \text{for } p \geq 2.$$

(14.9)

Moreover for all $t \in [0, 1]$,

$$\left| \frac{1+t}{2} \right|^q + \left| \frac{1-t}{2} \right|^q \leq \left(\frac{1+t^p}{2} \right)^{q-1} \quad \text{for } q \geq 2,$$

$$\left| \frac{1+t}{2} \right|^q + \left| \frac{1-t}{2} \right|^q \geq \left(\frac{1+t^p}{2} \right)^{q-1} \quad \text{for } q \in (1, 2].$$

(14.10)

Proof. First, assume that $t \in (0, 1)$. By direct calculation,

$$\frac{1}{p-1} \frac{d\varphi(s; t)}{ds} = \{(1+s)^{p-2} - (1-s)^{p-2}\} \frac{s^p - t^p}{s^p}.$$

Therefore, if $p \in (1, 2)$, the function $s \to \varphi(s; t)$ increases for $s \in (0, t)$, decreases for $s \in (t, 1]$, and takes its maximum at $s = t$. Analogously, if $p > 2$, the function $s \to \varphi(s; t)$ takes its minimum for $s = t$. Therefore, if $t \in (0, 1)$,

$$\varphi(s; t) \leq \varphi(t; t) = |1 + t|^p + |1 - t|^p \quad \text{for } p \in (1, 2],$$
$$\varphi(s; t) \geq \varphi(t; t) = |1 + t|^p + |1 - t|^p \quad \text{for } p \geq 2.$$

By continuity, these continue to hold also for $t = 1$.

Now assume that $t > 1$. If $p \in (1, 2)$, then $k(s) \leq h(s)$.[13] Therefore,

$$\varphi(s; t) = h(s) + k(s)t^p \leq h(s)t^p + k(s)$$

$$= t^p \left\{ h(s) + k(s)\frac{1}{t^p} \right\} = t^p \varphi\left(s; \frac{1}{t} \right)$$

$$\leq t^p \varphi\left(\frac{1}{t}; \frac{1}{t} \right) = |1 + t|^p + |1 - t|^p.$$

If $p > 2$ and $t > 1$, the argument is similar, starting from the inequality $k(s) \geq h(s)$ for $p > 2$. The inequalities in (14.10) are obvious for $t = 0$ and $t = 1$. To prove the first part of (14.10) for $t \in (0, 1)$, write the second part of (14.9) with q replacing p, and in the resulting inequality take $s = t^p$. Such a choice is admissible since $t \in (0, 1)$. The second part of (14.10) is proved analogously. □

14.1 Proof of Hanner's inequalities. Having fixed f and g in $L^p(E)$, we may assume $\|f\|_p \geq \|g\|_p > 0$. Let $p \in (1, 2)$, and in the first part of (14.9) take

$$t = \frac{|g|}{|f|}, \quad \text{provided } |f| \neq 0.$$

---

[13] The function $s \to \{k(s) - h(s)\}$ vanishes for $s = 1$ and is increasing for $s \in (0, 1)$.

Multiplying the inequality so obtained by $|f|^p$ gives

$$h(s)|f|^p + k(s)|g|^p \leq |f + g|^p + |f - g|^p,$$

and one checks that this inequality continues to hold if $|f| = 0$. Integrating over E this yields

$$h(s)\|f\|_p^p + k(s)\|g\|_p^p \leq \|f + g\|_p^p + \|f - g\|_p^p \qquad (14.11)$$

for all $s \in (0, 1]$. Taking

$$s = \frac{\|g\|_p}{\|f\|_p}$$

proves (14.2). Inequality (14.4) follows from (14.2) by replacing f with $(f + g)$ and g with $(f - g)$. The proofs of (14.1) and (14.3) are analogous, starting from the second part of (14.9). □

14.2 Proof of Clarkson's inequalities. Since (14.11) holds for all $s \in (0, 1]$, taking $s = 1$ proves (14.6). If $p \geq 2$, inequality (14.11) holds with the sign reversed and still holds for all $s \in (0, 1]$. Taking $s = 1$ proves (14.5). To establish (14.7) and (14.8), first observe that

$$\left\| \frac{f \pm g}{2} \right\|_p^q = \left(\int_E \left| \frac{f \pm g}{2} \right|^p d\mu \right)^{\frac{1}{p-1}}$$

$$= \left(\int_E \left| \frac{f \pm g}{2} \right|^{\frac{p}{p-1}(p-1)} d\mu \right)^{\frac{1}{p-1}} = \left\| \left| \frac{f \pm g}{2} \right|^q \right\|_{p-1}.$$

To prove (14.7), since $q \in (1, 2)$, the second part of (14.10) implies the pointwise inequality

$$\left| \frac{f + g}{2} \right|^q + \left| \frac{f - g}{2} \right|^q \geq \left(\frac{|f|^p + |g|^p}{2} \right)^{q-1}. \qquad (14.12)$$

By Minkowski's inequality,

$$\left\| \frac{f + g}{2} \right\|_p^q + \left\| \frac{f - g}{2} \right\|_p^q = \left\| \left| \frac{f + g}{2} \right|^q \right\|_{p-1} + \left\| \left| \frac{f - g}{2} \right|^q \right\|_{p-1}$$

$$\geq \left(\int_E \left(\left| \frac{f + g}{2} \right|^q + \left| \frac{f - g}{2} \right|^q \right)^{p-1} d\mu \right)^{\frac{1}{p-1}}$$

$$\geq \left(\int_E \frac{|f|^p + |g|^p}{2} d\mu \right)^{\frac{1}{p-1}}$$

$$= \left(\frac{\|f\|_p^p + \|g\|_p^p}{2} \right)^{q-1}.$$

Inequality (14.8) is established in the same way by making use of the reverse Minkowski inequality. Since $q > 2$, inequality (14.12) is reversed. Therefore, since $(p - 1) \in (0, 1)$,

$$\left\| \left| \frac{f + g}{2} \right|^q \right\|_p + \left\| \left| \frac{f - g}{2} \right|^q \right\|_p = \left\| \left| \frac{f + g}{2} \right|^q \right\|_{p-1} + \left\| \left| \frac{f - g}{2} \right|^q \right\|_{p-1}$$

$$\leq \left(\int_E \left(\left| \frac{f + g}{2} \right|^q + \left| \frac{f - g}{2} \right|^q \right)^{p-1} d\mu \right)^{\frac{1}{p-1}}$$

$$\leq \left(\int_E \frac{|f|^p + |g|^p}{2} d\mu \right)^{\frac{1}{p-1}}$$

$$= \left(\frac{\|f\|_p^p + \|g\|_p^p}{2} \right)^{\frac{1}{p-1}}. \qquad \square$$

15 Uniform convexity of $L^p(E)$ for $1 < p < \infty$

Proposition 15.1. *The spaces $L^p(E)$ for $1 < p < \infty$ are uniformly convex.*

Proof. It suffices to verify (5.4). Let $f, g \in L^p(E)$, satisfying

$$\|f\|_p = \|g\|_p = 1 \quad \text{and} \quad \|f - g\|_p \geq \varepsilon > 0.$$

By Clarkson's inequalities,

$$\left\| \frac{f + g}{2} \right\|_p^p \leq 1 - \frac{\varepsilon^p}{2^p} \quad \text{if } p \geq 2,$$

$$\left\| \frac{f + g}{2} \right\|_p^q \leq 1 - \frac{\varepsilon^q}{2^q} \quad \text{if } p \in (1, 2]. \qquad \square$$

A remarkable fact is that the Riesz representation theorem characterizing all the bounded linear functionals in $L^p(E)$ depends only on the uniform convexity of $L^p(E)$ and, in particular, is independent of the Radon–Nikodým theorem. We begin by using the uniform convexity to make the assertion of Remark 12.2 precise.

Proposition 15.2. *Let $1 < p, q < \infty$ be conjugate. For a nonzero $g \in L^q(E)$, let g^* be defined by (12.5). If $\mathcal{F}_1$ and $\mathcal{F}_2$ are two bounded linear functionals in $L^p(E)$ satisfying*

$$\mathcal{F}_i(g^*) = \|\mathcal{F}_i\| = 1, \quad i = 1, 2,$$

for some fixed $g \in L^q(E)$, then $\mathcal{F}_1 = \mathcal{F}_2$.

Proof. If $\mathcal{F}_1 \neq \mathcal{F}_2$, there exists $f \in L^p(E)$ such that $\mathcal{F}_1(f) \neq \mathcal{F}_2(f)$. Set

$$\varphi = \frac{2f}{\mathcal{F}_1(f) - \mathcal{F}_2(f)} - \frac{\mathcal{F}_1(f) + \mathcal{F}_2(f)}{\mathcal{F}_1(f) - \mathcal{F}_2(f)} g^* \in L^p(E).$$

One verifies that $\mathcal{F}_1(\varphi) = 1$ and $\mathcal{F}_2(\varphi) = -1$. Now let $t \in (0, 1)$ and compute

$$1 + t = \mathcal{F}_1(g^* + t\varphi) \leq \|\mathcal{F}_1\| \|g^* + t\varphi\|_p = \|g^* + t\varphi\|_p,$$
$$1 + t = \mathcal{F}_2(g^* - t\varphi) \leq \|\mathcal{F}_2\| \|g^* - t\varphi\|_p = \|g^* - t\varphi\|_p.$$

Assume first that $p \geq 2$. Then from these inequalities and Clarkson's inequality (14.7),

$$(1 + t)^q \leq \left(\frac{\|g^* + t\varphi\|_p^p + \|g^* - t\varphi\|_p^p}{2} \right)^{q-1}$$
$$\leq \left\| \frac{(g^* + t\varphi) + (g^* - t\varphi)}{2} \right\|_p^q + \left\| \frac{(g^* + t\varphi) - (g^* - t\varphi)}{2} \right\|_p^q$$
$$= \|g^*\|_p^q + t^q \|\varphi\|_p^q$$
$$= 1 + t^q \|\varphi\|_p^q$$

since $\|g^*\|_p = 1$. Similarly, if $1 < p \leq 2$, using Clarkson's inequality (14.6),

$$(1 + t)^p \leq \frac{\|g^* + t\varphi\|_p^p + \|g^* - t\varphi\|_p^p}{2}$$
$$\leq \left\| \frac{(g^* + t\varphi) + (g^* - t\varphi)}{2} \right\|_p^p + \left\| \frac{(g^* + t\varphi) - (g^* - t\varphi)}{2} \right\|_p^p$$
$$= \|g^*\|_p^p + t^p \|\varphi\|_p^p$$
$$= 1 + t^p \|\varphi\|_p^p.$$

Consider the last of these. Expanding the left-hand side with respect to t about $t = 0$ gives

$$pt + O(t^2) \leq t^p \|\varphi\|_p^p.$$

Dividing by t and letting $t \to 0$ gives a contradiction unless $\|\varphi\|_p = 0$. A similar contradiction occurs if $p \geq 2$. $\square$

16 The Riesz representation theorem by uniform convexity

Theorem 16.1. *Let $1 < p, q < \infty$ be conjugate. To every bounded, linear functional $\mathcal{F}$ in $L^p(E)$, there corresponds a unique function $g \in L^q(E)$ such that $\mathcal{F}$ is represented by the formula (12.3). Moreover, $\|\mathcal{F}\| = \|g\|_q$. If $p = 1$ and $q = \infty$, the same conclusion holds if $\{X, \mathcal{A}, \mu\}$ is σ-finite.*

16.1 Proof of Theorem 13.1: The case where $1 < p < \infty$. Without loss of generality, we may assume that $\|\mathcal{F}\| = 1$. By the definition (12.2) of $\|\mathcal{F}\|$, there exists a sequence $\{f_n\}$ of functions in $L^p(E)$ such that

$$\|f_n\|_p = 1, \quad |\mathcal{F}(f_n)| \geq \frac{1}{2}, \quad \text{and} \quad \lim |\mathcal{F}(f_n)| = 1.$$

By possibly replacing f_n with $-f_n$, we may assume without loss of generality that $\mathcal{F}(f_n) > 0$ for all $n \in \mathbb{N}$.

We claim that $\{f_n\}$ is a Cauchy sequence in $L^p(E)$. Proceeding by contradiction, if not, there exists some $\varepsilon > 0$ such that $\|f_m - f_n\|_p \geq \varepsilon$ for infinitely many indices m and n. The uniform convexity of $L^p(E)$ then implies that there exists $\delta = \delta(\varepsilon) \in (0, 1)$ such that

$$\left\| \frac{f_m + f_n}{2} \right\|_p \leq 1 - \delta,$$

for infinitely many indices m and n. Letting $m, n \to \infty$ along such indices,

$$2 = \lim\{\mathcal{F}(f_m) + \mathcal{F}(f_n)\} = \lim \mathcal{F}(f_m + f_n)$$
$$\leq \lim \|f_m + f_n\|_p \leq 2(1 - \delta).$$

The contradiction proves that $\{f_n\}$ is a Cauchy sequence in $L^p(E)$, and we let f denote its limit. By construction, $\|f\|_p = 1$. Set

$$g = |f|^{p/q} \operatorname{sign} f \quad \text{and} \quad g^* = |g|^{q-1} \operatorname{sign} g = f.$$

By construction, $g \in L^q(E)$ and $g^* \in L^p(E)$ and

$$\|f\|_p = \|g\|_q^{q/p} = \|g^*\|_p = 1.$$

Let $\mathcal{F}_g$ be the bounded linear functional in $L^p(E)$ generated by such a $g \in L^q(E)$ by formula (12.3). By construction, the two functionals $\mathcal{F}$ and $\mathcal{F}_g$ satisfy

$$\mathcal{F}(g^*) = \mathcal{F}_g(g^*) = \|\mathcal{F}\| = \|\mathcal{F}_g\| = 1.$$

Therefore, $\mathcal{F} = \mathcal{F}_g$ by Proposition 15.2. $\qquad\square$

16.2   The case where $p = 1$ and E is of finite measure. Without loss of generality, we may assume that $\|\mathcal{F}\| = 1$. If $\mu(E) < \infty$, then $L^p(E) \subset L^1(E)$ for all $p \geq 1$. In particular, for all $f \in L^p(E)$,

$$|\mathcal{F}(f)| \leq \|f\|_1 \leq \mu(E)^{1/q} \|f\|_p.$$

Therefore, for each fixed $p \in (1, \infty)$, the map $\mathcal{F}$ may be identified with a bounded linear functional in $L^p(E)$. By Theorem 16.1, for any such p, there exists a unique function $g_p \in L^q(E)$ such that

$$\mathcal{F}(f) = \mathcal{F}_{g_p}(f) = \int_E f g_p d\mu \quad \text{for all } f \in L^p(E). \tag{16.1}$$

Moreover,

$$\|g_p\|_q = \|\mathcal{F}_{g_p}\| = \sup_{\substack{f \in L^p(E) \\ \|f\|_p = 1}} |\mathcal{F}_{g_p}(f)|$$

$$= \sup_{\substack{f \in L^p(E) \\ \|f\|_p = 1}} |\mathcal{F}(f)| \leq \sup_{\substack{f \in L^p(E) \\ \|f\|_p = 1}} \|\mathcal{F}\| \|f\|_1$$

$$\leq \sup_{\substack{f \in L^p(E) \\ \|f\|_p = 1}} \|f\|_p \mu(E)^{1/q} = \mu(E)^{1/q}.$$

Now let $1 < p_1 < p_2 < \infty$, and let

$$g_{p_i} \in L^{q_i}(E), \quad i = 1, 2, \qquad \frac{1}{p_i} + \frac{1}{q_i} = 1, \quad 1 < p_i, q_i < \infty,$$

be the two functions that identify $\mathcal{F}_{g_{p_1}}$ and $\mathcal{F}_{g_{p_2}}$. If φ is a simple function, compute

$$\mathcal{F}(\varphi) = \mathcal{F}_{g_{p_i}}(\varphi) = \int_E \varphi g_{p_i} d\mu, \quad i = 1, 2,$$

from (16.1). From these, by difference $(g_{p_1} - g_{p_2}) \in L^1(E)$, and

$$\int_E (g_{p_1} - g_{p_2})\varphi d\mu = 0 \quad \text{for all simple functions } \varphi.$$

Since the simple functions are dense in $L^1(E)$, this implies that $g_{p_1} = g_{p_2}$. Therefore, there exists a function $g \in L^q(E)$ for all $q \in [1, \infty)$ such that

$$\mathcal{F}(f) = \int_E f g d\mu \quad \text{for all } f \in L^p(E).$$

For such a function g,

$$\left| \int_E f g d\mu \right| = |\mathcal{F}(f)| \leq \|f\|_1 \quad \text{for all } f \in L^1(E) \bigcap L^\infty(E).$$

By Proposition 8.2, this implies that $g \in L^\infty(E)$ and $\|g\|_\infty \leq 1$. Therefore, by density, (12.3) gives a representation of $\mathcal{F}$ for all $f \in L^1(E)$. Also,

$$1 = \|\mathcal{F}\| = \sup_{\substack{f \in L^1(E) \\ \|f\|_1 = 1}} \int_E f g d\mu \leq \|g\|_\infty \leq 1. \qquad \square$$

16.3   The case where $p = 1$ and $\{X, \mathcal{A}, \mu\}$ is σ-finite. Let $A_n \subset A_{n+1}$ be a countable collection of sets of finite measure whose union is X. Set

$$E_n = E \bigcap (A_{n+1} - A_n),$$

and let $\mathcal{F}_n$ be the restriction of $\mathcal{F}$ to $L^1(E_n)$.

A function $\varphi \in L^1(E_n)$ may be regarded as an element of $L^1(E)$ by possibly defining it to be zero in $(E - E_n)$. In this sense, $L^1(E_n) \subset L^1(E)$ and $\mathcal{F} = \mathcal{F}_n$ within $L^1(E_n)$.

Since E_n has finite measure, there exists $g_n \in L^\infty(E_n)$ such that

$$\mathcal{F}_n(\varphi) = \int_{E_n} g_n \varphi d\mu \quad \text{for all } \varphi \in L^1(E_n)$$

and $\|g_n\|_{\infty, E_n} = 1$. By extending g_n to be zero in $(E - E_n)$, we regard it as an element of $L^\infty(E)$ and set

$$g = \sum g_n \chi_{E_n}.$$

Then for every $f \in L^1(E)$,

$$\mathcal{F}(f) = \mathcal{F}\left(\sum f \chi_{E_n}\right) = \sum \mathcal{F}(f \chi_{E_n})$$

$$= \sum \int_{E_n} g_n f \chi_{E_n} d\mu = \int_E g f d\mu. \qquad \square$$

17   Bounded linear functional in $L^p(E)$ for $0 < p < 1$

Let $\{X, \mathcal{A}, \mu\}$ be $\mathbb{R}^N$ with the Lebesgue measure, and let $E \subset \mathbb{R}^N$ be Lebesgue measurable. The topology generated in $L^p(E)$ for $0 < p < 1$ by the metric (6.1) is not locally convex. A consequence is that there are no linear, bounded maps $T : L^p(E) \to \mathbb{R}$ except the identically zero map.

Proposition 17.1 (Day[14]). *Let E be a Lebesgue measurable subset of $\mathbb{R}^N$, and let $L^p(E)$ for $0 < p < 1$ be equipped with the topology generated by the metric (6.1). Then the only bounded, linear functional on $L^p(E)$ is the identically zero functional.*

Proof. Let T be a bounded linear functional in $L^p(E)$ for some $0 < p < 1$. Since T is continuous and linear, the preimage of any interval must be open and convex in $L^p(E)$. However, by Proposition 6.1, $L^p(E)$ for $0 < p < 1$ with the indicated topology does not have any open convex sets except the empty set and $L^p(E)$ itself. Let $(-\alpha, \alpha)$ for some $\alpha > 0$ be an interval of the origin of $\mathbb{R}$. Then

$$T^{-1}(-\alpha, \alpha) = L^p(E) \quad \text{for all } \alpha > 0.$$

Thus $T \equiv 0$. $\qquad \square$

Remark 17.1. The conclusion of Proposition 17.1 continues to hold for measure spaces $\{X, \mathcal{A}, \mu\}$ satisfying (6.2).

17.1   An alternate proof of Proposition 17.1. The alternate proof below is independent of the lack of open, convex neighborhoods of the origin in the metric topology of $L^p(E)$ and uses only (6.2).[15]

Let T be a continuous, linear functional in $L^p(E)$, and let $f \in L^p(E)$ be such that $T(f) \neq 0$. There exists $A \in E \cap \mathcal{A}$ such that

$$\int_E \chi_A |f|^p d\mu = \frac{1}{2} \|f\|_p^p.$$

Set $f = f_o$ and

$$f_{o,1} = f_o \chi_A, \qquad f_{o,2} = f_o \chi_{E-A}.$$

---

[14]M. M. Day, The spaces L^p with $0 < p < 1$, *Bull. Amer. Math. Soc.*, **46** (1940), 816–823.
[15]This proof was provided by A. E. Nussbaum.

Therefore, $f_o = f_{o,1} + f_{o,2}$, and

$$\|f_o\|_p^p = \|f_{o,1}\|_p^p + \|f_{o,2}\|_p^p, \qquad \|f_{o,1}\|_p^p = \|f_{o,2}\|_p^p = \frac{1}{2}\|f_o\|_p^p.$$

Since T is linear, $T(f_o) = T(f_{o,1}) + T(f_{o,2})$, and

$$|T(f_o)| \le |T(f_{o,1})| + |T(f_{o,2})|.$$

Therefore, either

$$|T(f_{o,1})| \ge \frac{1}{2}|T(f_o)| \quad \text{or} \quad |T(f_{o,2})| \ge \frac{1}{2}|T(f_o)|.$$

Assume that the first holds true and set $f_1 = 2f_{o,1}$. For this choice,

$$|T(f_1)| \ge |T(f_o)|$$

and

$$\|f_1\|_p^p = 2^p\|f_{o,1}\|_p^p = 2^{p-1}\|f_o\|_p^p.$$

Now repeat this construction with f_o replaced by f_1 and generate a function f_2 such that

$$|T(f_2)| \ge |T(f_o)|$$

and

$$\|f_2\|_p^p = 2^{p-1}\|f_1\|_p^p = 2^{2(p-1)}\|f_o\|_p^p.$$

By iteration, generate a sequence of functions $\{f_n\}$ in $L^p(E)$ such that

$$|T(f_n)| \ge |T(f_o)| \quad \text{and} \quad \|f_n\|_p^p = 2^{n(p-1)}\|f_o\|_p^p.$$

Since $p \in (0, 1)$, the sequence $\{f_n\}$ converges to zero in the metric topology of $L^p(E)$. However, $T(f_n)$ does not converge to zero. $\qquad\qquad\square$

18   If $E \subset \mathbb{R}^N$ and $p \in [1, \infty)$, then $L^p(E)$ is separable

Let dx denote the Lebesgue measure in $\mathbb{R}^N$, and let $E \subset \mathbb{R}^N$ be Lebesgue measurable. The $\mathbb{R}^N$-structure of E affords further defining properties to $L^p(E)$. For example, by Proposition 8.1, the collection of simple functions is dense in $L^p(E)$ for E a measurable subset of any measure space $\{X, \mathcal{A}, \mu\}$. When $E \subset \mathbb{R}^N$ is Lebesgue measurable and $1 \le p < \infty$, the space $L^p(E)$ contains a dense *countable* collection of simple functions.

Theorem 18.1. *Let $E \subset \mathbb{R}^N$ be measurable. Then $L^p(E)$ for $1 \le p < \infty$ is separable; i.e., it contains a dense countable set.*

Proof. Let $\{Q_1, Q_2, \ldots\}$ denote the collection of closed diadic cubes in $\mathbb{R}^N$. For a fixed positive integer n, let S_n denote the family of simple functions defined on E and taking constant, rational values on the first n cubes; i.e.,

$$S_n = \left\{ \begin{array}{c} \varphi : E \to \mathbb{R} \text{ of the form } \sum_{i=1}^{n} f_i \chi_{Q_i \cap E}, \\ \text{where the numbers } f_i \text{ are rational} \end{array} \right\}.$$

Each S_n is countable and the union $S = \bigcup S_n$ is a countable family of simple functions. □

Lemma 18.2. *S is dense in $L^p(E)$ for $1 \le p < \infty$; i.e., for every $f \in L^p(E)$ and every $\varepsilon > 0$, there exists $\varphi \in S$ such that $\|f - \varphi\|_p \le \varepsilon$.*

Proof. Assume first that E is bounded and that $f \in L^p(E) \bigcap L^\infty(E)$. By Lusin's theorem f is quasi-continuous. Therefore, having fixed $\varepsilon > 0$, there exists a closed set $E_\varepsilon \subset E$ such that

$$\mu(E - E_\varepsilon) \le \frac{\varepsilon^p}{4^p \|f\|_\infty^p},$$

and f is uniformly continuous in E_ε. In particular, there exists $\delta > 0$ such that

$$|f(x) - f(y)| \le \frac{\varepsilon}{4\mu(E)^{1/p}}$$

for all $x, y \in E_\varepsilon$ such that $|x - y| < \delta$. Since E_ε is bounded, having determined $\delta > 0$, there exist a finite number of closed diadic cubes with pairwise-disjoint interior and with diameter less than δ whose union covers E_ε, say, for example,

$$Q_1, Q_2, \ldots, Q_{n_\delta}, \quad \overset{\circ}{Q}_i \bigcap \overset{\circ}{Q}_j \ne \emptyset \quad \text{for } i \ne j.$$

Select $x_i \in Q_i \bigcap E_\varepsilon$ and then choose a rational number f_i such that

$$|f_i - f(x_i)| \le \frac{\varepsilon}{4\mu(E)^{1/p}}.$$

Set

$$\varphi(x) = \sum_{i=1}^{n_\delta} f_i \chi_{Q_i \cap E}$$

and compute

$$\int_E |f - \varphi|^p dx = \int_{E_\varepsilon} |f - \varphi|^p dx + \int_{E-E_\varepsilon} |f - \varphi|^p dx$$

$$= \sum_{i=1}^{n_\delta} \int_{Q_i \cap E_\varepsilon} |(f - f(x_i)) + (f(x_i) - f_i)|^p dx$$

$$+ \int_{E-E_\varepsilon} |f - \varphi|^p dx$$

$$\le \frac{1}{2^p} \varepsilon^p + 2^p \|f\|_\infty^p \mu(E - E_\varepsilon) \le \varepsilon^p.$$

Now assume that E is unbounded. Since $f \in L^p(E)$,

$$\int_E |f|^p dx = \sum \int_{E \cap \{n < |x| \le n+1\}} |f|^p dx < \infty.$$

Therefore, having fixed $\varepsilon > 0$, there exists an index n_ε so large that

$$\int_{E \cap \{|x| \ge n_\varepsilon\}} |f|^p dx \le \frac{1}{4^p} \varepsilon^p.$$

Also, since $f \in L^p(E)$,

$$n^p \mu([|f| \ge n]) \le \int_E |f|^p dx = \|f\|_p^p.$$

Therefore, for every $\delta > 0$, there exists a positive integer n_δ such that

$$\mu([|f| \ge n]) \le \delta \quad \text{for all } n \ge n_\delta.$$

By the Vitali theorem on the absolute continuity of the integral, having fixed $\varepsilon > 0$, there exists $\delta > 0$ such that

$$\int_{[|f| \ge n]} |f|^p dx \le \frac{1}{4^p} \varepsilon^p \quad \text{for all } n \ge n_\delta.$$

Setting

$$E_n = E \cap \left\{ [|x| \ge n] \bigcup [|f| \ge n] \right\},$$

it follows that

$$\|f\|_{p, E_n} \le \frac{1}{2} \varepsilon \quad \text{for all } n \ge \max\{n_\varepsilon; n_\delta\}.$$

The set $(E - E_n)$ is bounded and the restriction of f to such a set is bounded. Therefore, there exists a simple function $\varphi \in S$ vanishing outside $(E - E_n)$ such that $\|f - \varphi\|_{p, E - E_n} \le \frac{1}{2} \varepsilon$. □

Remark 18.1. Let E be a Lebesgue-measurable subset of $\mathbb{R}^N$. Then the spaces $L^p(E)$ for all $1 \le p < \infty$ satisfy the second axiom of countability.[16]

---

[16]See Proposition 13.2 of Chapter I.

18.1 $L^\infty(E)$ is not separable. Let E be a Lebesgue measurable subset of $\mathbb{R}^N$.

Proposition 18.3. *$L^\infty(E)$ is not separable; i.e., it does not contain a dense sequence $\{f_n\}$.*

Proof. For $y \in E$ and $s > 0$, let $B_s(y)$ denote the ball of radius s centered at y. If $s \neq r$,

$$\|\chi_{E \cap B_s(y)} - \chi_{E \cap B_r(y)}\|_{\infty, E} = 1.$$

The collection

$$\{\chi_{E \cap B_s(y)} | s > 0, \, y \in E\} \subset L^\infty(E)$$

is uncountable, and it cannot be separated by a countable sequence $\{f_n\}$ of functions in $L^\infty(E)$. $\square$

Corollary 18.4. *$L^\infty(E)$ satisfies the first but not the second axiom of countability.*

19   Selecting weakly convergent subsequences

We continue to assume that E is a Lebesgue-measurable subset of $\mathbb{R}^N$ and dx is the Lebesgue measure in $\mathbb{R}^N$. A sequence $\{f_n\}$ of functions in $L^p(E)$ is *bounded in $L^p(E)$* if there exists a constant K such that

$$\|f_n\|_p \leq K \quad \text{for all } n \in \mathbb{N}.$$

Proposition 19.1. *Let $1 < p \leq \infty$. Out of every sequence $\{f_n\}$ bounded in $L^p(E)$, one may extract a subsequence $\{f_{n'}\} \subset \{f_n\}$ weakly convergent in $L^p(E)$.*

Remark 19.1. This is false for $p = 1$. A counterexample is provided by the sequence in Section 10.10 of the Problems and Complements of Chapter III.

Proof of Proposition 19.1. Let $q \in [1, \infty)$ be the conjugate of $p \in (1, \infty]$. The corresponding $L^q(E)$ is separable, and we let $\{g_n\}$ be a countable collection of simple functions dense in $L^q(E)$. For any such simple function g_j, set

$$\mathcal{F}_n(g_j) = \int_E g_j f_n dx.$$

By the assumption and Hölder's inequality,

$$|\mathcal{F}_n(g_j)| \leq K \|g_j\|_q.$$

The sequence of numbers $\{\mathcal{F}_n(g_1)\}$ is bounded, and we extract a convergent subsequence $\{\mathcal{F}_{n,1}(g_1)\}$. Analogously, the sequence $\{\mathcal{F}_{n,1}(g_2)\}$ is bounded, and we extract a convergent subsequence $\{\mathcal{F}_{n,2}(g_2)\}$. Proceeding in this fashion, for each $m \in \mathbb{N}$, we may select a sequence $\{\mathcal{F}_{n,m}\}$ such that

$$\lim_{n \to \infty} \mathcal{F}_{n,m}(g_j) \text{ exists} \quad \text{for all } j = 1, 2, \ldots, m.$$

The diagonal sequence $\mathcal{F}_{n'} = \mathcal{F}_{n,n}$ is such that

$$\lim_{n' \to \infty} \mathcal{F}_{n'}(g_j) \text{ exists } \quad \text{for all simple functions } g_j \in \{g_n\}.$$

Having fixed $g \in L^q(E)$ and $\varepsilon > 0$, let $g_j \in \{g_n\}$ be such that $\|g - g_j\|_q < \varepsilon$. Since $\{F_{n'}(g_j)\}$ is a Cauchy sequence, there exists an index n_ε such that

$$|\mathcal{F}_{n'}(g_j) - \mathcal{F}_{m'}(g_j)| \leq \varepsilon \quad \text{for all } n', m' \geq n_\varepsilon.$$

From this,

$$\begin{aligned}
|\mathcal{F}_{n'}(g) - \mathcal{F}_{m'}(g)| &\leq |\mathcal{F}_{n'}(g - g_j)| + |\mathcal{F}_{m'}(g - g_j)| \\
&\quad + |\mathcal{F}_{n'}(g_j) - \mathcal{F}_{m'}(g_j)| \\
&\leq 2K\|g - g_j\|_q + \varepsilon \leq (2K + 1)\varepsilon.
\end{aligned}$$

Thus for all $g \in L^q(E)$, the sequence $\{\mathcal{F}_{n'}(g)\}$ is a Cauchy sequence and hence convergent. Setting

$$\mathcal{F}(g) = \lim \mathcal{F}_{n'}(g) \quad \text{for all } g \in L^q(E)$$

defines a bounded, linear functional in $L^q(E)$. Since $q \in [1, \infty)$, by the Riesz representation theorem, there exists $f \in L^p(E)$ such that

$$\mathcal{F}(g) = \int_E gf\, dx \quad \text{for all } g \in L^q(E).$$

Therefore,

$$\lim \int_E g f_{n'}\, dx = \int_E gf\, dx \quad \text{for all } g \in L^q(E). \qquad \square$$

20 Continuity of the translation in $L^p(E)$ for $1 \leq p < \infty$

Denote by dx the Lebesgue measure in $\mathbb{R}^N$, and let $E \subset \mathbb{R}^N$ be a Lebesgue-measurable subset of $\mathbb{R}^N$. For a given $f \in L^p(E)$ and $h \in \mathbb{R}^N$, let $T_h f$ denote the translation of f; i.e.,

$$T_h f(x) = \begin{cases} f(x + h) & \text{if } x + h \in E, \\ 0 & \text{if } x + h \in \mathbb{R}^N - E. \end{cases}$$

The following proposition asserts that the translation operation is continuous in the norm topology of $L^p(E)$ for all $p \in [1, \infty)$.

Proposition 20.1. *Let E be a Lebesgue-measurable set in $\mathbb{R}^N$, and let $f \in L^p(E)$ for some $1 \leq p < \infty$. For every $\varepsilon > 0$, there exists $\delta = \delta(\varepsilon)$ such that*

$$\sup_{|h| \leq \delta} \|T_h f - f\|_p \leq \varepsilon.$$

Proof. Assume first that E is bounded, i.e., contained in a ball B_R centered at the origin and radius R, for some $R > 0$ sufficiently large. Indeed, without loss of generality, we may assume that $E = B_R$ by defining f to be zero outside E. For a subset $\mathcal{E} \subset E$ and a vector $\eta \in \mathbb{R}^N$, set

$$\mathcal{E} - \eta = \{x \in \mathbb{R}^N | x + \eta \in \mathcal{E}\}.$$

Having fixed $\varepsilon > 0$, let δ_p be the number claimed by Vitali's theorem on the absolute continuity of the integral and for which

$$\int_{\mathcal{E}} |f|^p dx \leq \frac{1}{2^{p+1}} \varepsilon^p,$$

for every measurable subset $\mathcal{E} \subset E$ such that $\mu(\mathcal{E}) \leq \delta_p$. Since the Lebesgue measure is translation invariant, for any vector $\eta \in \mathbb{R}^N$ and for any such set $\mathcal{E}$,

$$\mu\left\{(\mathcal{E} - \eta) \bigcap E\right\} \leq \delta_p.$$

Therefore, for any such set $\mathcal{E}$,

$$\int_{\mathcal{E}} |T_\eta f - f|^p dx \leq 2^{p-1} \left\{\int_{\mathcal{E}} |f|^p dx + \int_{(\mathcal{E}-\eta)\bigcap E} |f|^p dx\right\} \leq \frac{1}{2}\varepsilon^p.$$

Since f is measurable, by Lusin's theorem it is quasi-continuous. Therefore, having fixed the positive number

$$\sigma = \frac{\delta_p}{2 + \mu(E)},$$

there exists a closed set $E_\sigma \subset E$ such that $\mu(E - E_\sigma) \leq \sigma$ and f is uniformly continuous in E_σ. In particular, there exists $\delta_\sigma > 0$ such that

$$|T_h f(x) - f(x)|^p \leq \frac{1}{2\mu(E)} \varepsilon^p \quad \text{for all vectors } |h| < \delta_\sigma,$$

provided both x and $(x + h)$ belong to E_σ. For any such vector h,

$$\int_{E_\sigma \bigcap (E_\sigma - h)} |T_h f - f|^p dx \leq \frac{1}{2}\varepsilon^p.$$

For any vector η of length $|\eta| < \sigma$, we compute[17]

$$\mu(E - (E_\sigma - \eta)) = \mu((E + \eta) - E_\sigma)$$
$$\leq \mu(E - E_\sigma) + \mu((E + \eta) - E)$$
$$\leq \sigma + \sigma\mu(E).$$

---

[17] Since $E = B_R$, we may estimate $\mu\{(E + \eta) - E\} \leq \sigma\mu(B_R)$.

From this,

$$\mu\left(E - \left[E_\sigma \bigcap (E_\sigma - \eta)\right]\right) \le \mu(E - E_\sigma) + \mu(E - (E_\sigma - \eta))$$

$$\le \sigma(2 + \mu(E)) = \delta_p.$$

If h is a vector in $\mathbb{R}^N$ of length $|h| \le \delta = \min\{\sigma; \delta_\sigma\}$, we estimate

$$\int_E |T_h f - f|^p dx \le \int_{E_\sigma \bigcap (E_\sigma - h)} |T_h f - f|^p dx$$

$$+ \int_{E - \{E_\sigma \bigcap (E_\sigma - h)\}} |T_h f - f|^p dx$$

$$= \int_{E_\sigma \bigcap (E_\sigma - h)} |T_h f - f|^p dx + \frac{1}{2}\varepsilon^p \le \varepsilon^p.$$

If E is unbounded, having fixed $\varepsilon > 0$, there exists R so large that for all $|h| < 1$,

$$\int_{E \bigcap \{|x| > 2R\}} |T_h f - f|^p dx \le 2^p \int_{E \bigcap \{|x| > R\}} |f|^p dx \le \frac{1}{2^p}\varepsilon^p.$$

For such a fixed R, there exists $\delta = \delta(\varepsilon)$ such that

$$\sup_{|h| < \delta} \|T_h f - f\|_{p, E \bigcap \{|x| < 2R\}} \le \frac{1}{2}\varepsilon.$$

Therefore,

$$\sup_{|h| < \delta} \|T_h f - f\|_{p, E} \le \sup_{|h| < \delta} \|T_h f - f\|_{p, E \bigcap \{|x| > 2R\}}$$

$$+ 2\|f\|_{p, E \bigcap \{|x| > R\}} \le \varepsilon. \qquad \square$$

Remark 20.1. The proposition is false for $p = \infty$. Counterexamples can be constructed as in Section 18.1.

21 Approximating functions in $L^p(E)$ with functions in $C^\infty(E)$

Let E be an open set in $\mathbb{R}^N$, and let f be a real-valued function f defined in E. The support of f is the closure in $\mathbb{R}^N$ of the set $[|f| > 0]$, and we write

$$\mathrm{supp}\{f\} = \overline{[|f| > 0]}.$$

A function $f : E \to \mathbb{R}$ is of *compact support* in E if $\mathrm{supp}\{f\}$ is compact and contained in E. Set

$$C^\infty(E) = \left\{ \begin{array}{c} \text{the collection of all infinitely differentiable} \\ \text{functions } f : E \to \mathbb{R} \end{array} \right\},$$

$$C_o^\infty(E) = \left\{ \begin{array}{c} \text{the collection of all infinitely differentiable} \\ \text{functions } f : E \to \mathbb{R} \text{ of compact support in } E \end{array} \right\}.$$

A measurable function $f : \mathbb{R}^N \to \mathbb{R}$ is in $L^p_{\text{loc}}(\mathbb{R}^N)$ if $f \in L^p(E)$ for every bounded measurable set $E \subset \mathbb{R}^N$. We regard a function $f \in L^p(E)$ as defined in the whole $\mathbb{R}^N$ by extending it to be zero in $(\mathbb{R}^N - E)$. In this sense, a function $f \in L^p(E)$ is in $L^p_{\text{loc}}(\mathbb{R}^N)$. The converse is, in general, false.

Given a Lebesgue-measurable set $E \subset \mathbb{R}^N$ and $f \in L^p(E)$, we may assume by a similar extension procedure that E is open.

A function $f \in L^p(E)$ for $1 \leq p < \infty$ can be approximated in its norm topology by functions in $C^\infty(E)$ by means of the Friedrichs[18] *mollifying* kernels. Set

$$J(x) = \begin{cases} k \exp\left\{ \dfrac{-1}{1 - |x|^2} \right\} & \text{if } |x| < 1, \\ 0 & \text{if } |x| \geq 1, \end{cases}$$

where k is a positive constant chosen so that $\|J(x)\|_{1,\mathbb{R}^N} = 1$. For $\varepsilon > 0$, we let

$$J_\varepsilon(x) = \frac{1}{\varepsilon^N} J\left(\frac{x}{\varepsilon}\right).$$

The kernels J and J_ε are in $C^\infty_o(\mathbb{R}^N)$. Moreover, for all $\varepsilon > 0$,

$$\int_{\mathbb{R}^N} J_\varepsilon(x)dx = 1 \quad \text{and} \quad J_\varepsilon(x) = 0 \quad \text{for } |x| \geq \varepsilon.$$

The convolution of the mollifying kernels J_ε with a function $f \in L^1_{\text{loc}}(\mathbb{R}^N)$ is defined by

$$x \to (J_\varepsilon * f)(x) = \int_{\mathbb{R}^N} J_\varepsilon(x - y)f(y)dy. \tag{21.1}$$

Since $J_\varepsilon \in C^\infty_o(\mathbb{R}^N)$, the convolution $(J_\varepsilon * f)$ is in $\in C^\infty(\mathbb{R}^N)$. In this sense, $(J_\varepsilon * f)$ is a regularization or mollification of f.

For each fixed $x \in \mathbb{R}^N$, the domain of integration in (21.1) is the ball centered at x and radius ε. If $f \in L^p(E)$ and $x \in \partial E$, then (21.1) is well defined since f is extended to be zero outside E. If the support of f is contained in E and has positive distance from ∂E, then[19]

$$(J_\varepsilon * f) \in C^\infty_o(E), \quad \text{provided } \varepsilon < \text{dist}\{\text{supp}\{f\}; \partial E\}.$$

Proposition 21.1. *Let* $f \in L^p(E)$ *for some* $1 \leq p < \infty$. *Then* $(J_\varepsilon * f) \in L^p(E)$ *and*

$$\|(J_\varepsilon * f)\|_p \leq \|f\|_p. \tag{21.2}$$

The mollifications $(J_\varepsilon * f)$ *approximate* f *in the following sense:*

$$\lim_{\varepsilon \to 0} \|(J_\varepsilon * f) - f\|_p = 0. \tag{21.3}$$

[18]K. O. Friedrichs, The identity of weak and strong extensions of differential operators, *Trans. Amer. Math. Soc.*, **55** (1944), 132–151.

[19]The support of f is the closure of the set $[|f| > 0]$.

If $f \in C(E)$, for every compact subset $\mathcal{K} \subset E$,

$$\lim_{\varepsilon \to 0} (J_\varepsilon * f)(x) = f(x) \quad \text{uniformly in } \mathcal{K}. \tag{21.4}$$

Proof. By Hölder's inequality,

$$|(J_\varepsilon * f)(x)| = \left| \int_{\mathbb{R}^N} J_\varepsilon(x - y) f(y) dy \right|$$

$$\leq \left(\int_{\mathbb{R}^N} J_\varepsilon(x - y) dy \right)^{1/q} \left(\int_{\mathbb{R}^N} J_\varepsilon(x - y)|f|^p(y) dy \right)^{1/p}$$

$$= \left(\int_{\mathbb{R}^N} J_\varepsilon(x - y)|f|^p(y) dy \right)^{1/p}.$$

Taking the p-power of both sides and integrating over $\mathbb{R}^N$ with respect to the x-variables gives

$$\int_{\mathbb{R}^N} |J_\varepsilon * f|^p dx \leq \int_{\mathbb{R}^N} |f|^p dy.$$

This proves (21.2). To establish (21.3), write

$$|(J_\varepsilon * f)(x) - f(x)|$$

$$= \left| \int_{\mathbb{R}^N} J_\varepsilon(x - y)\{f(y) - f(x)\} dy \right| \leq \int_{|\eta| < \varepsilon} J_\varepsilon(\eta)|f(x + \eta) - f(x)| d\eta$$

$$\leq \left(\int_{\mathbb{R}^N} J_\varepsilon(\eta) d\eta \right)^{1/q} \left(\int_{\mathbb{R}^N} J_\varepsilon(\eta)|f(x + \eta) - f(x)|^p d\eta \right)^{1/p}.$$

Taking the p-power and integrating $\mathbb{R}^N$ with respect to the x-variables gives

$$\|(J_\varepsilon * f) - f\|_{p, \mathbb{R}^N} \leq \sup_{|\eta| < \varepsilon} \|T_\eta f - f\|_{p, \mathbb{R}^N}.$$

Now (21.3) follows from Proposition 20.1. To prove (21.4), fix a compact subset $\mathcal{K} \subset E$ and for all $x \in \mathcal{K}$, write

$$|(J_\varepsilon * f)(x) - f(x)| = \left| \int_{\mathbb{R}^N} J_\varepsilon(x - y)\{f(y) - f(x)\} dy \right|$$

$$\leq \sup_{|x-y| < \varepsilon} |f(x) - f(y)|. \qquad \square$$

Proposition 21.2. *Let E be an open subset of $\mathbb{R}^N$. Then $C_o^\infty(E)$ is dense in $L^p(E)$ for $p \in [1, \infty)$.*

Proof. Having fixed $\varepsilon > 0$, let K_ε be a compact subset of E such that

$$\int_{E-K_\varepsilon} |f|^p dy \leq \frac{1}{2^p} \varepsilon^p.$$

Let $2\delta_\varepsilon = \text{dist}\{K_\varepsilon; \partial E\}$ and set

$$f_\varepsilon = \begin{cases} f(x) & \text{for } x \in K_\varepsilon, \\ 0 & \text{for } x \in E - K_\varepsilon. \end{cases}$$

If $\delta < \delta_\varepsilon$, the functions $f_\delta = f_\varepsilon * J_\delta$ have compact support in E. By Proposition 21.1, there exists δ'_ε such that

$$\|f_\varepsilon - f_\delta\|_p \leq \frac{1}{2}\varepsilon \quad \text{for all } \delta < \delta'_\varepsilon.$$

Therefore, for all such δ,

$$\|f - f_\delta\|_p \leq \|f_\varepsilon - f_\delta\|_p + \|f\|_{p, E-K_\varepsilon} < \varepsilon. \qquad \square$$

22 Characterizing precompact sets in $L^p(E)$

Let E denote an open set in $\mathbb{R}^N$, and let dx denote the Lebesgue measure in $\mathbb{R}^N$. By Proposition 17.6 of Chapter I, compact subsets of a complete metric space are characterized in terms of total boundedness. This permits one to characterize precompact subsets of $L^p(E)$ for $1 \leq p < \infty$. For $\delta > 0$, set

$$E_\delta = \left\{ x \in E \,\big|\, \text{dist}\{x; \partial E\} > \delta \text{ and } |x| < \frac{1}{\delta} \right\}$$

and assume that δ is so small that E_δ is nonempty.

Theorem 22.1. *A bounded subset K of $L^p(E)$ for $1 \leq p < \infty$ is precompact in $L^p(E)$ if and only if for every $\varepsilon > 0$, there exists $\delta > 0$ such that for all vectors $h \in \mathbb{R}^N$ of length $|h| < \delta$ and for all $u \in K$,*

$$\|T_h u - u\|_p < \varepsilon \quad \text{and} \quad \|u\|_{p, E-E_\delta} < \varepsilon. \tag{22.1}$$

Proof of the sufficient condition. Fix $\varepsilon > 0$ and choose δ so that (22.1) is satisfied. The proof consists of constructing an ε-net for K.

Let J_η be the η-mollifying kernel. If $\eta \leq \delta$, for a.e. $x \in E_\delta$,

$$|(J_\eta * u)(x) - u(x)| = \left| \int_{|y| < \eta} J_\eta(y)[u(x+y) - u(x)]dy \right|$$

$$\leq \int_{|y| < \eta} J_\eta(y)|(T_{-y}u - u)(x)|dy.$$

From this and the condition of the theorem,

$$\|J_\eta * u - u\|_{p, E_\delta} \leq \sup_{|h| < \eta} \|T_h u - u\|_{p, E_\delta} \leq \varepsilon.$$

The family

$$\{J_\eta * u \,|\, u \in K\} \quad \text{for a fixed } \eta \in (0, \delta)$$

is equibounded and equicontinuous in $\overline{E}_\delta$. Indeed,

$$|(J_\eta * u)(x)| \leq \|J_\eta\|_q \|u\|_p \quad \text{for all } x \in \overline{E}_\delta.$$

Moreover, for every vector $\xi \in \mathbb{R}^N$,

$$|J_\eta * u(x + \xi) - J_\eta * u(x)| \leq \|J_\eta\|_q \|T_\xi u - u\|_p.$$

Therefore, having fixed an arbitrary $\sigma > 0$, there exists $\tau > 0$ such that for all vectors $|\xi| < \tau$,

$$|J_\eta * u(x + \xi) - J_\eta * u(x)| \leq \|J_\eta\|_q \sigma \quad \text{for all } u \in K.$$

Thus by the Ascoli–Arzelá theorem, the set $\{J_\eta * u \mid u \in \mathcal{K}\}$ is precompact in $C(\overline{E}_\delta)$ and admits a finite ε-net $\{\varphi_1, \varphi_2 \ldots, \varphi_m\}$ of functions in $C(\overline{E}_\delta)$.[20] Precisely, for every $u \in K$, there exists some φ_j such that

$$|\varphi_j(x) - (J_\eta * u)(x)| < \varepsilon \quad \text{for all } x \in \overline{E}_\delta. \tag{22.2}$$

Define

$$\overline{\varphi}_j = \begin{cases} \varphi_j(x) & \text{for } x \in \overline{E}_\delta, \\ 0 & \text{otherwise.} \end{cases}$$

We claim that the collection $\{\overline{\varphi}_1, \overline{\varphi}_2, \ldots, \overline{\varphi}_m\}$ is a finite ε-net for K in $L^p(E)$. Indeed, if $u \in \mathcal{K}$ and φ_j is chosen to satisfy (22.2),

$$\|u - \overline{\varphi}_j\|_p \leq \|u\|_{p, E-E_\delta} + \|J_\eta * u - u\|_{p, E_\delta} + \|J_\eta * u - \varphi_j\|_{p, E_\delta}.$$

The proof is concluded by suitably redefining ε. □

Proof of the necessary condition. The existence of ε-nets for all $\varepsilon > 0$ implies that for all $\varepsilon \in (0, 1)$, there exists $\delta > 0$ such that

$$\|u\|_{p, E-E_\delta} \leq \varepsilon \quad \text{for all } u \in K.$$

To prove this, let $\{\varphi_1, \varphi_2, \ldots, \varphi_n\}$ be a finite $\frac{1}{2}\varepsilon$-net. Then for every $u \in K$, there exists some φ_j such that for all $\delta > 0$,

$$\|u\|_{p, E-E_\delta} \leq \|\varphi_j\|_{p, E-E_\delta} + \frac{1}{2}\varepsilon.$$

Now we may choose δ so that

$$\|\varphi_j\|_{p, E-E_\delta} \leq \frac{1}{2}\varepsilon \quad \text{for all } j = 1, 2, \ldots, n.$$

---

[20]See Section 19.1 of Chapter IV.

Fix $\varepsilon > 0$ and $\delta > 0$ so small that

$$\|u\|_p \leq \|u\|_{p,E_\delta} + \frac{1}{2}\varepsilon \quad \text{for all } u \in K.$$

Next, for all $u \in K$,

$$\|T_h u - u\|_p \leq 2\varepsilon + \|T_h u - u\|_{p,E_\delta}$$

and

$$\|T_h u - u\|_{p,E_\delta} \leq \|T_h u - T_h \varphi_j\|_{p,E_\delta} + \|T_h \varphi_j - \varphi_j\|_{p,E_\delta} + \|u - \varphi_j\|_{p,E_\delta}$$
$$\leq 2\varepsilon + \|T_h \varphi_j - \varphi_j\|_{p,E_\delta}. \qquad \square$$

PROBLEMS AND COMPLEMENTS

1   FUNCTIONS IN $L^p(E)$ AND THEIR NORMS

1.1   THE SPACES ℓ_p FOR $1 \leq p \leq \infty$. Let $\mathbf{a} = \{a_1, a_2, \ldots, a_n, \ldots\}$ denote a sequence of real numbers, and set

$$\begin{aligned}
\|\mathbf{a}\|_p &= \left(\sum |a_n|^p\right)^{1/p} &&\text{if } 1 \leq p < \infty, \\
\|\mathbf{a}\|_\infty &= \sup |a_n| &&\text{if } p = \infty.
\end{aligned} \tag{1.1c}$$

Denote by ℓ_p the set of all sequences $\mathbf{a}$ such that $\|\mathbf{a}\|_p < \infty$. One verifies that ℓ_p is a linear space for all $1 \leq p \leq \infty$ and that (1.1c) is a norm. Moreover, ℓ_∞ satisfies analogues of (1.3) and (1.4).

1.2   THE SPACES ℓ_p FOR $0 < p < 1$. Introduce these spaces and trace their main properties. In particular, ℓ_p for $0 < p < 1$ is a linear space.

2   THE INEQUALITIES OF HÖLDER AND MINKOWSKI

Corollary 2.1c. *Let $1 < p, q < \infty$ be conjugate. Then for all $a, b \in \mathbb{R}$ and all $\varepsilon > 0$,*

$$|ab| \leq \frac{\varepsilon^p}{p}|a|^p + \frac{1}{\varepsilon^q q}|b|^q. \tag{2.1c}$$

Proof. Apply (2.2) with a replaced by εa and b replaced by $\varepsilon^{-1}b$. $\square$

Corollary 2.2c. *Let $\{a_1, a_2, \ldots, a_n\}$ be an n-tuple of positive numbers satisfying*

$$\alpha_i \in (0, 1) \quad and \quad \sum_{i=1}^{n} \frac{1}{\alpha_i} = 1.$$

Then for any n-tuple of real numbers $\{\xi_1, x_2, \ldots, \xi_n\}$,

$$\prod_{i=1}^{n} |\xi_i| \leq \sum_{i=1}^{n} \alpha_i |\xi_i|^{\alpha_i}.$$

Proof. The inequality holds for $n = 1, 2$ by (2.2). Proceed by induction.   □

2.1   State and prove a variant of Corollaries 2.1c and 2.2c when p is permitted to be 1 or some of the α_i are permitted to be 1.

2.2   VARIANTS OF THE HÖLDER AND MINKOWSKI INEQUALITIES.

Corollary 2.3c. *Let $f_i \in L^{p_i}(E)$ for $i = 1, 2, \ldots, n$, where*

$$p_i > 1 \quad and \quad \sum_{i=1}^{n} \frac{1}{p_i} = 1.$$

Then

$$\int_E \prod_{i=1}^{n} |f_i| d\mu \leq \prod_{i=1}^{n} \| f_i \|_{p_i}.$$

State a variant of Corollary 2.3c when $p_i = 1$ for some $i \in \{1, 2, \ldots, n\}$.

Corollary 2.4c. *Let $1 \leq p, q \leq \infty$ be conjugate. Then for any two elements $\mathbf{a} \in \ell_p$ and $\mathbf{b} \in \ell_q$,*

$$\sum_{i=1}^{n} |a_i b_i| \leq \| \mathbf{a} \|_p \| \mathbf{b} \|_q.$$

Moreover, equality holds if and only if there exists a positive constant c such that $|a_i|^p = c|b_i|^q$ for all $i \in \mathbb{N}$.

Corollary 2.5c. *Let $\mathbf{a}$ and $\mathbf{b}$ be any two elements of ℓ_p. Then for $1 \leq p \leq \infty$,[21]*

$$\| \mathbf{a} + \mathbf{b} \|_p \leq \| \mathbf{a} \|_p + \| \mathbf{b} \|_p.$$

2.3   SOME AUXILIARY INEQUALITIES.

Lemma 2.6c. *Let x and y be any two positive numbers. Then for $1 \leq p < \infty$,*

$$|x - y|^p \leq |x^p - y^p|,$$
$$(x + y)^p \geq x^p + y^p,$$ (2.2c)
$$(x + y)^p \leq 2^{p-1}(x^p + y^p).$$ (2.3c)

---

[21]H. Minkowski, *Geometrie der Zahlen*, Teubner, Leipzig, 1896 and 1910; reprinted by Chelsea, New York, 1953.

Proof. We may assume that $x \geq y$ and $p > 1$. Then

$$
\begin{aligned}
x^p - y^p &= \int_0^1 \frac{d}{ds}\{sx + (1 - s)y\}^p ds \\
&= p(x - y) \int_0^1 \{sx + (1 - s)y\}^{p-1} ds \\
&= p(x - y) \int_0^1 \{s(x - y) + y\}^{p-1} ds \\
&\geq p(x - y) \int_0^1 s^{p-1}(x - y)^{p-1} ds = (x - y)^p.
\end{aligned}
$$

This proves the first part of (2.2c). The second part follows from the first since

$$
(x + y)^p - y^p \geq (x + y - y)^p.
$$

To prove (2.3c), we may assume that both x and y are nonzero and that $p > 1$. Also, by replacing x with $\frac{x}{y}$, we may assume that $y = 1$. Consider the function

$$
f(x) = \frac{(1 + x)^p}{1 + x^p} \quad \text{defined for } x > 0.
$$

Using the second part of (2.2c), one verifies that $f(x) \geq 1$. Also,

$$
f(0) = 1, \quad \lim_{x \to \infty} f(x) = 1, \quad \text{and} \quad f'(x) = \frac{p(1 + x)^{p-1}}{(1 + x^p)^2}(1 - x^{p-1}).
$$

Therefore, f has its only critical point at $x = 1$. It follows that

$$
1 \leq f(x) \leq f(1) = 2^{p-1} \quad \text{for all } x > 0. \qquad \square
$$

3   THE REVERSE HÖLDER AND MINKOWSKI INEQUALITIES

3.1   Prove the reverse Hölder and Minkowski inequalities for ℓ_p when $0 < p < 1$.

4   MORE ON THE SPACES L^p AND THEIR NORMS

4.1   A VARIANT OF PROPOSITION 4.1.

Corollary 4.1c. *Let $\{f_n\}$ be a sequence of functions in $L^p(E)$ for some $p \geq 1$. Then*

$$\left(\int_E \left| \sum f_n \right|^p d\mu \right)^{\frac{1}{p}} \le \sum \| f_n \|_p, \tag{4.1c}$$

$$\left(\sum \left| \int_E f_n d\mu \right|^p \right)^{\frac{1}{p}} \le \int_E \left(\sum |f_n|^p \right)^{\frac{1}{p}} d\mu. \tag{4.2c}$$

4.2   The spaces ℓ_p satisfy analogues of Propositions 4.1 and 4.2.

Proposition 4.2c. *Let E be of finite measure and let $f \in L^p(E)$ for all $1 \le p < \infty$. Assume that there exists a constant K such that*

$$\| f \|_p \le K \quad \text{for all } 1 \le p < \infty.$$

Then $f \in L^\infty(E)$ and $\| f \|_\infty \le K$.

6   A METRIC TOPOLOGY FOR $L^p(E)$ WHEN $0 < p < 1$

6.1   In ℓ_p for $0 < p < 1$, introduce the distance

$$d(\mathbf{a}, \mathbf{b}) = \sum |a_i - b_i|^p.$$

This is a metric in ℓ_p that generates a translation invariant topology in ℓ_p.

7   CONVERGENCE IN $L^p(E)$ AND COMPLETENESS

7.1   A sequence $\{f_n\}$ of functions in $L^\infty(E)$ converges in $L^\infty(E)$ to a function $f \in L^\infty(E)$ if and only if there exists a set $\mathcal{E} \subset E$ of measure zero such that $\{f_n\} \to f$ uniformly in $E - \mathcal{E}$.

7.2   A sequence $\{f_n\}$ of functions in $L^p(E)$ converges to some f in $L^p(E)$ if and only if every subsequence $\{f_{n'}\} \subset \{f_n\}$ contains in turn a subsequence $\{f_{n''}\}$ converging to f in $L^p(E)$.

7.3 ℓ_p is complete for all $1 \le p \le \infty$.

7.4   POINTWISE CONVERGENCE AND NORM CONVERGENCE.

Proposition 7.1c. *Let $\{f_n\}$ be a sequence of functions in $L^p(E)$ for $p \in [1, \infty)$ converging a.e. in E to a function $f \in L^p(E)$. Then $\{f_n\}$ converges to f in $L^p(E)$ if and only if $\lim \| f_n \|_p = \| f \|_p$.*

Proof. The implication $\Longleftarrow$ follows by dominated convergence from the inequality

$$|f_n - f|^p \le 2^{p-1}(|f|^p + |f_n|^p).$$

The implication $\Longrightarrow$ follows from the inequality[22]

$$\left| |f_n|^p - |f|^p \right| \le C_p(|f_n - f|^p + |f|^{p-1}|f_n - f|)$$

for a constant C_p depending only upon p.   □

---

[22] Proved by a variant of the arguments for the first part of (2.2c).

7.5 Let $L^p(E)$ for $0 < p < 1$ be endowed with the metric topology generated by the metric in (6.1). With respect to such a topology, $L^p(E)$ is a complete metric space. In particular, $L^p(E)$ is of the second category.

9 WEAK CONVERGENCE IN $L^p(E)$

9.1 REMARKS ON WEAK CONVERGENCE.

Proposition 9.1c. *Let E be a Lebesgue-measurable subset of $\mathbb{R}^N$ of finite measure, and let $\{f_n\}$ be a sequence of functions in $L^p(E)$ such that*

> $\{f_n\}$ *converges weakly in $L^p(E)$ to some $f \in L^p(E)$,*
>
> $\{f_n\}$ *converges a.e. in E to some g.*

Then $f = g$ a.e. in E.

Proof. Having fixed $\varepsilon > 0$ from the a.e. convergence and the Egorov theorem, there exists a closed set $E_\varepsilon \subset E$ such that $\mu(E - E_\varepsilon) \le \varepsilon$ and $\{f_n\}$ converges to g uniformly in E_ε.

Since f is measurable, it is also quasi-continuous by Lusin's theorem. Therefore, by taking a smaller ε if necessary, the closed set E_ε can be chosen to ensure that also f is bounded in E_ε. Then

$$\lim \int_{E_\varepsilon} f_n(g - f)d\mu = \int_{E_\varepsilon} g(g - f)d\mu.$$

On the other hand, by the weak convergence,

$$\lim \int_{E_\varepsilon} f_n(g - f)d\mu = \lim \int_E f_n(g - f)\chi_{E_\varepsilon}d\mu$$

$$= \int_{E_\varepsilon} f(g - f)d\mu.$$

Thus

$$\int_{E_\varepsilon} (g - f)^2 d\mu = 0 \quad \text{for all } \varepsilon > 0. \qquad \square$$

Corollary 9.2c. *Let E be a Lebesgue-measurable subset of $\mathbb{R}^N$ of finite measure, and let $\{f_n\}$ be a sequence of functions in $L^p(E)$ such that*

> $\{f_n\}$ *converges weakly in $L^p(E)$ to some $f \in L^p(E)$,*
>
> $\{f_n\}$ *converges in measure to some g.*

Then $f = g$ a.e. in E.

9.2   COMPARING THE VARIOUS NOTIONS OF CONVERGENCE. Let $\{X, \mathcal{A}, \mu\}$ be $\mathbb{R}^N$ with the Lebesgue measure, and let $E \subset \mathbb{R}^N$ be measurable and of finite measure. Denote by {a.e.} the set of all sequences $\{f_n\}$ of functions from E into $\mathbb{R}^*$ convergent a.e. in E. Analogously, denote by {meas}, $\{L^p(E)\}$, and $\{w\text{-}L^p(E)\}$ the set of all sequences $\{f_n\}$ of measurable functions defined in E and convergent, respectively, in measure, strongly, and weakly in $L^p(E)$.

By the remarks in Section 9 and the counterexample of Section 9.1,

$$\{L^p(E)\} \subset \{w\text{-}L^p(E)\} \quad \text{and} \quad \{w\text{-}L^p(E)\} \not\subset \{L^p(E)\}. \tag{9.1c}$$

By Proposition 4.2 of Chapter III and the counterexample following it,

$$\{a.e.\} \subset \{meas\} \quad \text{and} \quad \{meas\} \not\subset \{a.e.\}. \tag{9.2c}$$

Weak convergence in $L^p(E)$ does not imply convergence in measure, nor does convergence in measure imply weak convergence in $L^p(E)$; i.e.,

$$\{w\text{-}L^p(E)\} \not\subset \{meas\} \quad \text{and} \quad \{meas\} \not\subset \{w\text{-}L^p(E)\}. \tag{9.3c}$$

The sequence $\{\cos nx\}$ for $x \in [0, 2\pi]$ converges weakly to zero but not in measure. Indeed,

$$\mu \left\{ x \in [0, 2\pi] \mid |\cos nx| \geq \frac{1}{2} \right\} = \frac{4}{3}\pi.$$

This proves the first part of (9.3c). For the second part, consider the sequence

$$f_n(x) = \begin{cases} n & \text{for } x \in \left[0, \dfrac{1}{n}\right], \\ 0 & \text{for } x \in \left(\dfrac{1}{n}, 1\right]. \end{cases} \tag{9.4c}$$

Such a sequence converges to zero in measure and not weakly in $L^p[0, 1]$ since

$$\lim \int_0^1 f_n \, dx = 1.$$

Almost everywhere convergence does not imply weak convergence in $L^p(E)$. Strong convergence in $L^p(E)$ does not imply a.e. convergence; i.e.,

$$\{a.e.\} \not\subset \{w\text{-}L^p(E)\} \quad \text{and} \quad \{L^p(E)\} \not\subset \{a.e.\}. \tag{9.5c}$$

Indeed, the sequence $\{f_n\}$ in (9.4c) converges a.e. to zero and does not converge to zero weakly in $L^p(E)$. To prove the second part of (9.5c), consider the sequence $\{\varphi_{nm}\}$ introduced in (4.1) of Chapter III. Such a sequence converges to zero in $L^p[0, 1]$ for all $1 \leq p < \infty$ and does not converge to zero a.e. in $[0, 1]$.

The relationships between these notions of convergence can be organized in Figure 9.1c in form of inclusion of sets where each square represents convergent sequences.[23]

---

[23]This discussion on the various notions of convergence and the figure have been taken from the 1974 lectures on real analysis by C. Pucci at the University of Florence, Italy.

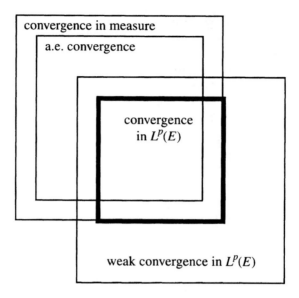

Figure 9.1c.

9.3 Consistent with Figure 9.1c are sequences $\{f_n\}$ of measurable functions in E convergent to zero a.e. in E and weakly in $L^2(E)$ but not convergent in $L^2(E)$. For example, in $[0, 1]$,

$$f_n(x) = \begin{cases} \sqrt{n} & \text{for } x \in \left[0, \dfrac{1}{n}\right], \\ 0 & \text{for } x \in \left(\dfrac{1}{n}, 1\right]. \end{cases} \tag{9.6c}$$

9.4 Exhibit a sequence $\{f_n\}$ of measurable functions in $[0, 1]$ convergent to zero in measure and weakly in $L^2[0, 1]$ but not convergent a.e. in $[0, 1]$. (*Hint*: See (4.1) in Chapter III.)

9.5 **WEAK CONVERGENCE IN ℓ_p.** Let $1 \le p, q \le \infty$ be conjugate. Let $\{\mathbf{a}_n\}$ be a sequence of elements is ℓ_p, and let $\mathbf{b}$ an element of ℓ_q; i.e.,

$$\begin{aligned} \mathbf{a}_n &= \{a_{1n}, a_{2n}, \dots, a_{mn}, \dots\} \in \ell_p, \qquad \frac{1}{p} + \frac{1}{q} = 1. \\ \mathbf{b} &= \{b_1, b_2, \dots, b_m, \dots\} \in \ell_q, \end{aligned}$$

The sequence $\{\mathbf{a}_n\}$ converges weakly in ℓ_p to some $\mathbf{a} \in \ell_p$ if

$$\lim \sum a_{in} b_i = \sum a_i b_i \quad \text{for all } \mathbf{b} \in \ell_q.$$

Strong convergence implies weak convergence; the converse is false. Construct counterexamples. Discuss and compare the various notions of convergence for ℓ_p.

11   WEAK CONVERGENCE AND NORM CONVERGENCE

11.1   When $p = 2$, the proof of Proposition 11.1 is particularly simple and elegant. Indeed,

$$\| f_n - f \|_2^2 = \| f_n \|_2^2 + \| f \|_2^2 - 2 \int_E f_n f d\mu.$$

11.2   PROOF OF LEMMAS 11.2 and 11.3. For $t \neq 0$ and $x = t^{-1}$, consider the function

$$f(t) = \frac{|1 + t|^p - 1 - pt}{|t|^p} = |1 + x|^p - |x|^p - p|x|^{p-2}x = \varphi(x).$$

It suffices to prove that $\varphi(x) \geq c$ for all $x \in \mathbb{R}$ for some $c > 0$. If $x \in (-1, 0]$, we estimate directly

$$\varphi(x) \geq (1 - |x|)^p + (p - 1)|x|^{p-1} \geq \frac{1}{2^p} \min\{1; (p - 1)\}.$$

If $x \in (0, 1]$,

$$
\begin{aligned}
(1 + x)^p - x^p &= \int_0^1 \frac{d}{ds}(s + x)^p ds \\
&= p \int_0^1 (s + x)^{p-1} ds \\
&= -p \int_0^1 (s + x)^{p-1} \frac{d}{ds}(1 - s) ds \\
&= px^{p-1} + p(p - 1) \int_0^1 (1 - s)(s + x)^{p-2} ds \\
&\geq px^{p-1} + \min\left\{\frac{p}{2^p}; \frac{p(p - 1)}{4}\right\}.
\end{aligned}
$$

Therefore, for $|x| \leq 1$,

$$\varphi(x) \geq c_p(p - 1) \quad \text{for some positive constant } c_p. \tag{11.1c}$$

The case where $p \geq 2$ and $|x| > 1$. By direct calculation,

$$
\begin{aligned}
|1 + x|^p - |x|^p &= \int_0^1 \frac{d}{ds}|s + x|^p ds \\
&= p \int_0^1 |s + x|^{p-1} \text{sign}(s + x) ds.
\end{aligned}
$$

From this, for $x \geq 1$ and $p \geq 2$, making use of the second part of (2.2c),

$$|1 + x|^p - |x|^p \geq p \int_0^1 (x^{p-1} + s^{p-1}) ds \geq p|x|^{p-2}x + 1.$$

For $x \leq -1$ and $p \geq 2$, making use of the first part of (2.2c),

$$
\begin{aligned}
|1 + x|^p - |x|^p &= -p \int_0^1 (|x| - s)^{p-1} ds \\
&\geq -p \int_0^1 (|x|^{p-1} - s^{p-1}) ds \\
&= p|x|^{p-2} x + 1.
\end{aligned}
$$

The case where $1 < p < 2$ *and* $|x| > 1$. Assume first that $t \in (0, 1)$. Then by repeated integration by parts,

$$
\begin{aligned}
(1 + t)^p - 1 &= -\int_0^t \frac{d}{ds}\{1 + (t - s)\}^p ds \\
&= p \int_0^t \{1 + (t - s)\}^{p-1} ds \\
&= pt + p(p - 1) \int_0^t s\{1 + (t - s)\}^{p-2} ds \\
&\geq pt + \frac{p(p - 1)}{4} t^2.
\end{aligned}
$$

Therefore,

$$
\frac{(1 + t)^p - 1 - pt}{t^2} \geq \frac{p(p - 1)}{4} \quad \text{for all } t \in (0, 1).
$$

A similar calculation holds for $t \in (-1, 0)$ with the same bound below.   □

13   THE RIESZ REPRESENTATION THEOREM

13.1   Denote by **a** an element of ℓ_p and by **b** an element of ℓ_q, where $1 \leq p, q \leq \infty$ are conjugate; i.e.,

$$
\begin{aligned}
\mathbf{a} &= \{a_1, a_2, \ldots, a_n, \ldots\} \in \ell_p, & \frac{1}{p} + \frac{1}{q} &= 1. & (13.1c)
\end{aligned}
$$
$$
\mathbf{b} = \{b_1, b_2, \ldots, b_n, \ldots\} \in \ell_q,
$$

Every element $\mathbf{b} \in \ell_q$ induces a bounded linear functional on ℓ_p by the formula

$$
T(\mathbf{a}) = \sum a_i b_i \quad \text{for all } \mathbf{a} \in \ell_p. \tag{13.2c}
$$

Theorem 13.1c. *Let* $1 \leq p < \infty$. *For every bounded, linear functional* $\mathcal{F}$ *in* ℓ_p, *there exists a unique element* $\mathbf{b} \in \ell_q$ *as in* (13.1c) *such that* $\mathcal{F}(\mathbf{a})$ *can be represented as in* (13.2c) *for all* $\mathbf{a} \in \ell_p$.

Proof. A suitable adaptation of the proof of Theorem 13.1.   □

18   IF $E \subset \mathbb{R}^N$ AND $p \in [1, \infty)$, THEN $L^p(E)$ IS SEPARABLE

18.1   The spaces ℓ_p for $1 \leq p < \infty$ are separable, whereas ℓ_∞ is not separable.

18.2 $BV[a, b]$ is not separable.[24]

18.3   Let E be a measurable subset in $\mathbb{R}^N$, and let $L^p(E)$ for $0 < p < 1$ be endowed with the metric topology generated by the metric in (6.1). With respect to such a topology, $L^p(E)$ is separable. In particular, it satisfies the second axiom of countability.

21   APPROXIMATING FUNCTIONS IN $L^p(E)$ WITH FUNCTIONS IN $C^\infty(E)$

Approximating a function $f \in L^1_{\text{loc}}(\mathbb{R}^N)$ with smooth functions can be realized by forming the convolution of f with kernels other than the Friedrichs mollifying kernels J_ε. We mention two such kernels here. Their advantage with respect to the Friedrichs kernels is that they satisfy specific partial differential equations, and therefore they are more suitable in applications related to such equations. Their disadvantage is that they are not compactly supported. Therefore, even if f is of compact support in $\mathbb{R}^N$, its approximations will not be.

21.1   CALORIC EXTENSIONS OF FUNCTIONS IN $L^p(\mathbb{R}^N)$. For $x \in \mathbb{R}^N$ and $t > 0$, set

$$\Gamma(x - y; t) = \frac{1}{(4\pi t)^{N/2}} e^{-\frac{|x-y|^2}{4t}}. \tag{21.1c}$$

21.1(A). Formally set

$$\Delta_x = \text{div } \nabla_x = \sum_{j=1}^{N} \frac{\partial^2}{\partial x_j^2}, \tag{21.2c}$$

and define Δ_y similarly. Verify by direct calculation that for all $x, y \in \mathbb{R}^N$ and $t > 0$, it holds that

$$\Gamma_t - \Delta_x \Gamma = 0 \quad \text{and} \quad \Gamma_t - \Delta_y \Gamma = 0. \tag{21.3c}$$

This partial differential equation is called the *heat equation*. The variables x are referred to as the *space variables*, and t is referred to as the *time*.

A function $(x, t) \to u(x, t)$ that satisfies the heat equation in a space–time open set $E \subset \mathbb{R}^N \times \mathbb{R}$, is said to be *caloric* in E. For example, $(x, t) \to \Gamma(x - y; t)$ is caloric in $\mathbb{R}^N \times \mathbb{R}^+$ for all $y \in \mathbb{R}^N$.

21.1(B). Verify that

$$\int_{\mathbb{R}^N} \Gamma(x - y; t)dy = \int_{\mathbb{R}^N} \Gamma(x - y; t)dx = 1 \tag{21.4c}$$

----

[24] See Section 1.8 of the Problems and Complements of Chapter IV.

for all $x, y \in \mathbb{R}^N$ and all $t > 0$. *Hint*: Introduce the change of variables

$$\frac{x - y}{2\sqrt{t}} = \eta \quad \text{for a fixed } y \in \mathbb{R}^N \quad \text{and} \quad t > 0 \tag{21.5c}$$

and use Section 14.2 of the Problems and Complements of Chapter III.

21.1(c). Let E be an open set in $\mathbb{R}^N$ and interpret functions in $L^p(E)$ as functions in $L^p(\mathbb{R}^N)$ by extending them to be zero in $\mathbb{R}^N - E$. For a function $f \in L^p(E)$ and $t > 0$, set

$$f_t(x) = (\Gamma * f)(x) = \int_{\mathbb{R}^N} \Gamma(x - y; t) f(y) dy. \tag{21.6c}$$

Such a function $(x, t) \to f_t(x)$ is caloric in $\mathbb{R}^N \times \mathbb{R}^+$ and is called the *caloric extension* of f in the upper half-space $\mathbb{R}^N \times \mathbb{R}^+$. Such an extension is a mollification of f since $x \to f_t(x) \in C^\infty(\mathbb{R}^N)$.

Proposition 21.1c. *Let $f \in L^p(E)$ for some $1 \le p < \infty$. Then $\Gamma * f \in L^p(E)$ and*

$$\|\Gamma * f\|_p \le \|f\|_p. \tag{21.7c}$$

The mollifications $\Gamma * f$ approximate f in the following sense:

$$\lim_{t \to 0} \|f_t - f\|_p = 0. \tag{21.8c}$$

Moreover, if $f \in C(E)$ and f is bounded in $\mathbb{R}^N$, then for every compact subset $\mathcal{K} \subset E$,

$$\lim_{t \to 0} f_t(x) = f(x) \quad \text{uniformly in } \mathcal{K}. \tag{21.9c}$$

Proof. Statements (21.7c) and (21.8c) are proved as in Proposition 21.1. To prove (21.9c), fix a compact set $\mathcal{K} \subset E$ and a positive number ε_o. There exists a compact set $\mathcal{K}_{\varepsilon_o}$ such that

$$\mathcal{K} \subset \mathcal{K}_{\varepsilon_o} \subset E \quad \text{and} \quad \text{dist}\{\mathcal{K}; \mathcal{K}_{\varepsilon_o}\} \ge \varepsilon_o.$$

For all $x \in \mathcal{K}$ and all $\varepsilon \in (0, \varepsilon_o)$, write

$$|f_t(x) - f(x)| = \left| \int_{\mathbb{R}^N} \Gamma(x - y; t)(f(y) - f(x)) dy \right|$$

$$\le \left| \int_{|x-y| \le \varepsilon} \Gamma(x - y; t)(f(y) - f(x)) dy \right|$$

$$+ \left| \int_{|x-y| > \varepsilon} \Gamma(x - y; t)(f(y) - f(x)) dy \right|$$

$$\le \sup_{|x-y| \le \varepsilon; x \in \mathcal{K}} |f(y) - f(x)| \int_{\mathbb{R}^N} \Gamma(x - y; t) dy$$

$$+ 2\|f\|_\infty \int_{|x-y| > \varepsilon} \Gamma(x - y; t) dy$$

$$\le \omega(\varepsilon) + 2\|f\|_\infty \int_{|x-y| > \varepsilon} \Gamma(x - y; t) dy,$$

where $\omega_o(\cdot)$ is the uniform modulus of continuity of f in $\mathcal{K}_{\varepsilon_o}$. The last integral is transformed by the change of variables (21.5c) and gives

$$\int_{|x-y|>\varepsilon} \Gamma(x-y;t)dy = \frac{1}{\pi^{N/2}} \int_{|\eta|>\frac{\varepsilon}{2\sqrt{t}}} e^{-|\eta|^2} d\eta.$$

From this, for $\varepsilon \in (0, \varepsilon_o)$ fixed,

$$\lim_{t\to 0} \int_{|x-y|>\varepsilon} \Gamma(x-y;t)dy = 0.$$

Now letting $t \to 0$ in the previous inequality gives

$$\lim_{t\to 0} |f_t(x) - f(x)| = \omega_o(\varepsilon) \quad \text{for all } \varepsilon \in (0, \varepsilon_o). \qquad \square$$

Remark 21.1c. The assumption that f be bounded in $\mathbb{R}^N$ can be removed. Indeed, a similar approximation would hold if f grows as $|x| \to \infty$ not faster than $e^{\gamma|x|^2}$, for a positive constant γ.[25]

21.2 HARMONIC EXTENSIONS OF FUNCTIONS IN $L^p(\mathbb{R}^N)$. For $x, y \in \mathbb{R}^N$ and $t > 0$, set

$$H(x-y;t) = \frac{1}{\omega_{N+1}} \frac{2t}{[|x-y|^2 + t^2]^{\frac{N+1}{2}}}. \qquad (21.10c)$$

Formally set

$$\Delta_{(x,t)} = \Delta_x + \frac{\partial^2}{\partial t^2},$$

and define $\Delta_{(y,t)}$ similarly. Verify by direct calculation that for all $x, y \in \mathbb{R}^N$ and all $t > 0$,

$$\Delta_{(x,t)}H = \Delta_{(y,t)}H = 0. \qquad (21.11c)$$

This is the *Laplace equation* in the variables (x, t). A function that satisfies the Laplace equation in an open set $E \subset \mathbb{R}^{N+1}$ is called *harmonic* in E. As an example, $(x, t) \to H(x-y;t)$ is harmonic in $\mathbb{R}^N \times \mathbb{R}^+$ for all $y \in \mathbb{R}^N$.

21.2(a). Verify that for all $x, y \in \mathbb{R}^N$ and all $t > 0$,

$$\int_{\mathbb{R}^N} H(x-y;t)dy = \int_{\mathbb{R}^N} H(x-y;t)dx = 1. \qquad (21.12c)$$

Hint: The change of variables $(x - y) = t\eta$ transforms these integrals in

$$\frac{2}{\omega_{N+1}} \int_{\mathbb{R}^N} \frac{d\eta}{(1+|\eta|^2)^{\frac{N+1}{2}}} = 2\frac{\omega_N}{\omega_{N+1}} \int_0^\infty \frac{\rho^{N-1}d\rho}{(1+\rho^2)^{\frac{N+1}{2}}} = 1.$$

---

[25] See [11, Chapter V].

Use also Section 14.1 of the Complements of Chapter III.

21.2(b). Let E be an open set in $\mathbb{R}^N$, and regard functions in $L^p(E)$ as functions in $L^p(\mathbb{R}^N)$ by extending them to be zero in $\mathbb{R}^N - E$. For a function $f \in L^p(E)$ and $t > 0$, set

$$f_t(x) = (H * f)(x) = \int_{\mathbb{R}^N} H(x - y; t) f(y) dy. \tag{21.13c}$$

Such a function $(x, t) \to f_t(x)$ is harmonic in $\mathbb{R}^N \times \mathbb{R}^+$ and is called the *harmonic extension* of f in the upper half-space $\mathbb{R}^N \times \mathbb{R}^+$. Such extension is a mollification of f since $x \to f_t(x) \in C^\infty(\mathbb{R}^N)$.

The integral in (21.13c) is also called the *Poisson integral* of f.[26]

Proposition 21.2c. *Let* $f \in L^p(E)$ *for some* $1 \le p < \infty$. *Then* $H * f \in L^p(E)$ *and*

$$\|H * f\|_p \le \|f\|_p. \tag{21.14c}$$

The mollifications $H * f$ *approximate* f *in the following sense:*

$$\lim_{t \to 0} \|f_t - f\|_p = 0. \tag{21.15c}$$

Moreover, if $f \in C(E)$ *and* f *is bounded in* $\mathbb{R}^N$, *then for every compact subset* $\mathcal{K} \subset E$,

$$\lim_{t \to 0} f_t(x) = f(x) \quad \textit{uniformly in } \mathcal{K}. \tag{21.16c}$$

22   CHARACTERIZING PRECOMPACT SETS IN $L^p(E)$

22.1   A closed, bounded subset C of ℓ_p for $1 \le p < \infty$ is compact if and only if for every $\varepsilon > 0$, there exists an index n_ε such that

$$\sum_{n=n_\varepsilon}^{\infty} |x_n|^p \le \varepsilon \quad \text{for all } x = \{x_n\} \in C.$$

22.2   The closed unit ball of $L^p(E)$ or ℓ_p or $C(\overline{E})$ is not compact since is not sequentially compact.

---

[26]The construction of the kernel $H(x - y; t)$ is in [11, Chapter II, Section 8].

VI

Banach Spaces

1   Normed spaces

Let X be a vector space and let Θ be its zero element. A norm on X is a function $\|\cdot\| : X \to \mathbb{R}^+$ satisfying the following:

(i) $\|x\| = 0$ if and only if $x = \Theta$.

(ii) $\|x + y\| \leq \|x\| + \|y\|$ for all $x, y \in X$.

(iii) $\|\lambda x\| = |\lambda| \|x\|$ for all $\lambda \in \mathbb{R}$ and $x \in X$.

Every norm on X defines a translation-invariant metric by the formula

$$d(x, y) = \|x - y\|.$$

This, in turn, generates a translation-invariant topology in X. We denote by $\{X; \|\cdot\|\}$ the corresponding metric space.

By Proposition 14.1 of Chapter I, the sum $+ : X \times X \to X$ is continuous with respect to such a topology. Also the product $\bullet : \mathbb{R} \times X \to X$ is continuous with respect to the metric topology of X. This follows from requirement (iii) for $\|\cdot\|$ to be a norm. Therefore, the norm $\|\cdot\|$ induces a topology on X by which $\{X; \|\cdot\|\}$ is a *topological* vector space. By requirements (ii) and (iii) of a norm, the balls in $\{X; \|\cdot\|\}$ are convex. Therefore, such a topology is locally convex.

Remark 1.1. Requirement (iii) distinguishes between metrics and norms. While every norm is a metric, there exist metrics that do not satisfy (iii). For example, the metric d_o in (13.1) of Chapter I does not satisfy (iii) even if d does.

The pair $\{X; \|\cdot\|\}$, is called a *normed* space, and the topology generated by $\|\cdot\|$ is the norm topology of X. If $\{X; \|\cdot\|\}$ is complete, it is called a *Banach space*.

The spaces $\mathbb{R}^N$ for all $N \in \mathbb{N}$ endowed with their Euclidean norm are Banach spaces.

The spaces $L^p(E)$ for $1 \le p \le \infty$ are Banach spaces.

The spaces ℓ_p for $1 \le p \le \infty$ are Banach spaces.

Let E be a bounded open set in $\mathbb{R}^N$. Then $C(\overline{E})$ is a Banach space by the norm

$$C(\overline{E}) \ni f \longrightarrow \|f\| = \sup_{\overline{E}} |f|. \tag{1.1}$$

This is also called the sup-norm and generates the topology of uniform convergence.[1]

Let $[a, b]$ be an interval of $\mathbb{R}$. Then the space $BV[a, b]$ of the functions of bounded variations in $[a, b]$ is a Banach space by the norm

$$BV[a, b] \ni f \longrightarrow \|f\| = |f(a)| + \mathcal{V}_f[a, b], \tag{1.2}$$

where $\mathcal{V}_f[a, b]$ is the variation of f in $[a, b]$.[2]

If X_o is a subspace of X, we denote by $\{X_o; \|\cdot\|\}$ the normed space X_o with the norm inherited from $\{X; \|\cdot\|\}$. If X_o is a closed subspace of X, then $\{X_o; \|\cdot\|\}$ also is a Banach space.

The same vector space X can be endowed with different norms. The notion of equivalence of two norms $\|\cdot\|_1$ and $\|\cdot\|_2$ on the same vector space X can be inferred from the notion of equivalence of the corresponding metrics.[3]

All norms in a finite-dimensional, Hausdorff, topological vector space are equivalent.[4] Every finite-dimensional subspace of a normed space is closed.

1.1   Seminorms and quotients. Let X be a vector space. A nonnegative function $p : X \to \mathbb{R}$ is a *seminorm* if it satisfies requirements (ii) and (iii) of a norm. A seminorm p is a norm on X if and only if it satisfies requirement (i) as well. As an example, the function $p : C[0, 1] \to \mathbb{R}$ defined by $p(f) = |f(\frac{1}{2})|$ is a seminorm in $C[0, 1]$.

The *kernel* of p is defined by

$$\ker\{p\} = \{x \in X \,|\, p(x) = 0\}.$$

Since p is nonnegative, the triangle inequality implies that $\ker\{p\}$ is a subspace of X. If p is a norm $\ker\{p\} = \{\Theta\}$ and if $p \equiv 0$, then $\ker\{p\} = X$. As an example, let $\{X; \|\cdot\|\}$ be a normed space and let X_o be a subspace of X. The distance from an element $x \in X$ to X_o is defined by

$$d(x, X_o) = \inf_{y \in X_o} \|x - y\|.$$

---

[1] See Section 15 of Chapter I.

[2] See Sections 1.8–1.9 of the Problems and Complements of Chapter IV.

[3] See Section 13.2 of Chapter I and related problems. See also Sections 1.1 and 1.4 of the Problems and Complements.

[4] See Section 12 of Chapter I.

This is a seminorm on X whose kernel is X_o.

A seminorm p on X and its kernel $\ker\{p\}$ induce an equivalence relation in X by stipulating that two elements x, $y \in X$ are equivalent if and only if $p(x - y) = 0$. Equivalently, x is equivalent to y if and only if $p(x) = p(y)$. The quotient space $X/\ker\{p\}$ consists of the equivalence classes of elements $x' = x + \ker\{p\}$.

The operation of linear combination of any two elements x' and y' of $X/\ker\{p\}$ can be introduced by operating with representatives out of these equivalence classes and by verifying that such an operation is independent of the choice of these representatives. This turns $X/\ker\{p\}$ into a vector space whose zero element is the equivalence class $\ker\{p\}$. Moreover, the function $p : X \to \mathbb{R}$ may be redefined as a map p' from $X/\ker\{p\}$ into $\mathbb{R}$ by setting

$$p'(x') = p'(x + \ker\{p\}) = p(x).$$

One verifies that $p' : X/\ker\{p\} \to \mathbb{R}$ is now a norm on $X/\ker\{p\}$, by which $\{X/\ker\{p\}; \, p'\}$ is a normed space.

2 Finite- and infinite-dimensional normed spaces

A vector space X is of *finite dimension* if it has a finite Hamel basis and is of *infinite dimension* if any Hamel basis is infinite.[5] The unit sphere of a normed space $\{X; \|\cdot\|\}$ centered at the origin of X is defined as

$$\mathcal{S}_1 = \{x \in X \,|\, \|x\| = 1\}.$$

Let $\mathcal{S}_1$ be the unit sphere centered at the origin of $\mathbb{R}^N$, and let π be a hyperplane through the origin of $\mathbb{R}^N$. Then there exists at least one element $x_o \in \mathcal{S}_1$ whose distance from π is 1.

The analogue in an infinite-dimensional normed space $\{X; \|\cdot\|\}$ is stated as follows. One fixes a closed, proper subspace $\{X_o, \|\cdot\|\}$ and seeks an element x_o such that $\|x_o\| = 1$ and

$$d(x_o, X_o) = \inf_{x \in X_o} \|x_o - x\| = 1. \tag{2.1}$$

Unlike the finite-dimensional case, such an element, in general, does not exist, as shown by the following counterexample.

2.1 A counterexample. Let X be the subspace of $C[0, 1]$ endowed with the sup-norm of those functions vanishing at 0. Also, let $X_o \subset X$ be the subspace of X of those functions with vanishing integral average over $[0, 1]$. One verifies that X_o endowed with the sup-norm is a closed, proper subspace of X.

Proposition 2.1. *There exists no function $f \in X$ such that $\|f\| = 1$ and*

$$\|f - g\| \geq 1 \quad \text{for all functions } g \in X_o. \tag{2.2}$$

---

[5]See Section 9.6 of the Problems and Complements of Chapter I.

Remark 2.1. Establishing (2.2) is equivalent to (2.1). Indeed, (2.1) implies (2.2), whereas (2.2) implies that $d(f, X_o) \geq 1$. However, $g \equiv 0$ is in X_o and $\|f\| = 1$. Therefore, $d(f, X_o) = 1$.

Proof of Proposition 2.1. Assume that such a function f exists. For every fixed $h \in X - X_o$, set

$$g = f - ch, \quad \text{where } c = \frac{\int_0^1 f(t)dt}{\int_0^1 h(t)dt}.$$

Then $g \in X_o$ and

$$1 \leq \|f - (f - ch)\| = |c|\|h\|;$$

i.e.,

$$\left| \int_0^1 h(t)dt \right| \leq \left| \int_0^1 f(t)dt \right| \sup_{t \in [0,1]} |h(t)|.$$

Choosing $h = t^{1/n} \in X - X_o$ gives

$$\frac{n}{n+1} \leq \left| \int_0^1 f(t)dt \right| \quad \text{for all } n \in \mathbb{N}.$$

Letting $n \to \infty$, this implies that

$$1 \leq \int_0^1 |f(t)|dt.$$

This, however, is impossible since

$$f \text{ is continuous,} \quad \sup_{t \in [0,1]} |f(t)| = 1, \quad \text{and} \quad f(0) = 0. \qquad \square$$

2.2   The Riesz lemma. While, in general, an element $x_o \in \mathcal{S}_1$ at distance 1 from a given subspace does not exist, there exist elements of norm 1 and at a distance arbitrarily close to 1 from the given subspace.

Lemma 2.2 (Riesz[6]). *Let* $\{X; \|\cdot\|\}$ *be a normed space, and let* $\{X_o; \|\cdot\|\}$ *be a closed, proper subspace of* $\{X; \|\cdot\|\}$. *For every* $\varepsilon \in (0, 1)$, *there exists* $x_\varepsilon \in X$ *such that* $\|x_\varepsilon\| = 1$ *and*

$$\|x_\varepsilon - x\| \geq 1 - \varepsilon \quad \text{for all } x \in X_o. \tag{2.3}$$

---

[6]F. Riesz, Über lineare Funktionalgleichungen, *Acta Math.*, **41** (1917), 71–98.

Proof. Fix a point $x_o \in X - X_o$ and let d be the distance from x_o to X_o; i.e.,

$$d = \inf_{x \in X_o} \|x - x_o\|.$$

Since X_o is a proper subspace of X and $\{X_o; \|\cdot\|\}$ is closed, $d > 0$. There exists an element $x_1 \in X_o$ such that

$$d \leq \|x_o - x_1\| \leq d + \frac{\varepsilon d}{1 - \varepsilon}.$$

The element claimed by the lemma is

$$x_\varepsilon = \frac{x_o - x_1}{\|x_o - x_1\|}, \quad \|x_\varepsilon\| = 1.$$

To prove (2.3), first observe that for every $x \in X_o$,

$$\widetilde{x} = x_1 + \|x_o - x_1\|x \in X_o.$$

Then for every $x \in X_o$,

$$\|x_\varepsilon - x\| = \left\| \frac{x_o - x_1}{\|x_o - x_1\|} - x \right\| = \frac{1}{\|x_o - x_1\|} \|x_o - \widetilde{x}\|$$

$$\geq \frac{d}{\|x_o - x_1\|} \geq 1 - \varepsilon. \qquad \square$$

2.3   Finite-dimensional spaces. A locally compact, Hausdorff, topological vector space is of finite dimension (Proposition 12.3 of Chapter I). In normed spaces $\{X; \|\cdot\|\}$, the Riesz lemma permits one to give an independent proof.

Proposition 2.3. *Let $\{X; \|\cdot\|\}$ be a normed space. If the unit sphere $\mathcal{S}_1 = \{\|x\| = 1\}$ is compact, then X is finite dimensional.*

Proof. If not, choose $x_1 \in \mathcal{S}_1$ and consider the subspace X_1 spanned by x_1. It is a proper subspace of X, and by Corollary 12.2 of Chapter I, it is closed. Therefore, by the Riesz lemma, there exists $x_2 \in \mathcal{S}_1$ such that $\|x_1 - x_2\| \geq \frac{1}{2}$.

The space $X_2 = \text{span}\{x_1, x_2\}$ is a proper, closed subspace of $\{X; \|\cdot\|\}$. Therefore, there exists $x_3 \in \mathcal{S}_1$ such that $\|x_2 - x_3\| \geq \frac{1}{2}$.

Proceeding in this fashion, we generate a sequence $\{x_n\}$ of elements in $\mathcal{S}_1$ such that $\|x_n - x_m\| \geq \frac{1}{2}$ for $n \neq m$. Such a sequence contains no convergent subsequences, contradicting the compactness of $\mathcal{S}_1$. $\qquad \square$

Corollary 2.4. *A normed space $\{X; \|\cdot\|\}$ is of finite dimension if and only if $\mathcal{S}_1$ is compact.*

3   Linear maps and functionals

Let $\{X; \|\cdot\|_X\}$ and $\{Y; \|\cdot\|_Y\}$ be normed spaces. A linear map $T : X \to Y$ is bounded if there exists a positive number M such that

$$\|T(x)\|_Y \leq M\|x\|_X \quad \text{for all } x \in X.$$

The norm of T is defined as the smallest of such M; i.e.,

$$\|T\| = \sup_{\substack{x \in X \\ x \neq \Theta}} \frac{\|T(x)\|_Y}{\|x\|_X} = \sup_{\substack{x \in X \\ \|x\|_X = 1}} \|T(x)\|_Y. \tag{3.1}$$

Proposition 3.1. *Let T be a linear map from a normed space $\{X; \|\cdot\|_X\}$ into a normed space $\{Y; \|\cdot\|_Y\}$.*

   (i) *If T is bounded, it is uniformly continuous.*

   (ii) *If T is continuous at some $x_o \in X$, it is bounded.*

Proof. If T is bounded, for any pair of elements $x, y \in X$,

$$\|T(x) - T(y)\|_Y \leq \|T\|\|x - y\|_X.$$

This proves (i). To prove (ii), we may assume that $x_o = \Theta$. There exists a ball B_ε of radius ε and centered at the origin of X such that

$$\|T(y)\|_Y \leq 1 \quad \text{for all } y \in B_\varepsilon.$$

If x is an element of X not in the ball B_ε, put

$$x_\varepsilon = \frac{\varepsilon}{2}\frac{x}{\|x\|_X} \in B_\varepsilon.$$

By the linearity of T,

$$\frac{\varepsilon}{2\|x\|_X}\|T(x)\|_Y = \|T(x_\varepsilon)\|_Y \leq 1.$$

Therefore,

$$\|T(x)\|_Y \leq \frac{2}{\varepsilon}\|x\|_X$$

for all $x \in X$. $\qquad\qquad\qquad\square$

   Let $\mathcal{B}(X; Y)$ denote the collection of all the bounded linear maps from $\{X; \|\cdot\|_X\}$ into $\{Y; \|\cdot\|_Y\}$. Such a collection has the structure of a vector space for the operations of sum and product by scalars; i.e.,

$$(T_1 + T_2)(x) = T_1(x) + T_2(x), \qquad \alpha T(x) = T(\alpha x)$$

for all $x \in X$ and all $\alpha \in \mathbb{R}$. It also has the structure of a normed space by the norm in (3.1). Therefore, a topology and the various notions of convergence in $\mathcal{B}(X; Y)$ are defined in terms of such a norm.

One also verifies that the operations

$$+ : \mathcal{B}(X; Y) \times \mathcal{B}(X; Y) \longrightarrow \mathcal{B}(X; Y),$$
$$\bullet : \mathbb{R} \times \mathcal{B}(X; Y) \longrightarrow \mathcal{B}(X; Y)$$

are continuous with respect to the corresponding product topology. Therefore, $\mathcal{B}(X; Y)$ is a topological vector space by the topology generated by the norm in (3.1).

The next proposition gives a sufficient condition for the normed space $\mathcal{B}(X; Y)$ to be a Banach space.

Proposition 3.2. *Let $\{Y; \| \cdot \|_Y\}$ be a Banach space. Then $\mathcal{B}(X; Y)$ endowed with the norm (3.1) is also a Banach space.*

Proof. Let $\{T_n\}$ be a Cauchy sequence in $\mathcal{B}(X; Y)$; i.e., for any fixed $\varepsilon > 0$, there exists an index n_ε such that

$$\|T_n - T_m\| \le \varepsilon \quad \text{for all } n, m \ge n_\varepsilon.$$

For each fixed $x \in X$, the sequence $\{T_n(x)\}$ is a Cauchy sequence in Y. Indeed,

$$\|T_n(x) - T_m(x)\|_Y \le \|T_n - T_m\| \|x\|_X.$$

Since $\{Y; \| \cdot \|_Y\}$ is complete, $\{T_n(x)\}$ has a limit, which is denoted by $T(x)$. By construction, for every $x \in X$ and all $n \ge n_\varepsilon$,

$$\|T_n(x) - T(x)\|_Y = \lim_m \|T_n(x) - T_m(x)\|_Y \le \varepsilon \|x\|_X.$$

The map $T : X \to Y$ so constructed is linear. It is also bounded since

$$\|T\| = \sup_{\substack{x \in X \\ \|x\|_X = 1}} \lim \|T_n(x)\|_Y \le \|T_{n_\varepsilon}\| + \varepsilon.$$

This implies that $T \in \mathcal{B}(X; Y)$ and that $\{T_n\}$ converges to T in the norm of $\mathcal{B}(X; Y)$. Indeed,

$$\|T_n - T\| = \sup_{\substack{x \in X \\ \|x\|_X = 1}} \|T_n(x) - T(x)\|_Y \le \varepsilon$$

provided $n \ge n_\varepsilon$. $\square$

If the target space $\{Y; \| \cdot \|_Y\}$ is the set of the real numbers endowed with the Euclidean norm, then T is said to be a *functional*. The space $\mathcal{B}(X; \mathbb{R})$ of all the bounded linear functionals in X is denoted by X^* and is called the dual space to X.

Corollary 3.3. *The dual X^* of a normed space $\{X; \| \cdot \|_X\}$ is a Banach space.*

4   Examples of maps and functionals

Let E be a bounded open set in $\mathbb{R}^N$, and let dx denote the Lebesgue measure in $\mathbb{R}^N$. Given a function

$$E \times E \ni (x, y) \longrightarrow K(x, y) \in L^1(E \times E),$$

formally set

$$T(f)(x) = \int_E K(x, y) f(y) dy. \tag{4.1}$$

If $K(\cdot, \cdot) \in L^q(E \times E)$ for some $1 < q < \infty$, then (4.1) defines a bounded linear map

$$T : L^p(E) \longrightarrow L^q(E), \qquad \frac{1}{p} + \frac{1}{q} = 1$$

with norm

$$\|T\| = \|K(\cdot, \cdot)\|_{q, E \times E}.$$

Now assume that $K(\cdot, \cdot)$ satisfies either

$$\|T\|_\infty \stackrel{\text{def}}{=} \sup_{x \in E} \int_E |K(x, y)| dy < \gamma \quad \text{or}$$

$$\|T\|_1 \stackrel{\text{def}}{=} \sup_{y \in E} \int_E |K(x, y)| dx < \gamma \tag{4.2}$$

for some positive constant γ. An example of such a function is

$$K(x, y) = \frac{1}{|x - y|^{N-1}}. \tag{4.3}$$

Indeed,[7]

$$\sup_{y \in E} \int_E |K(x, y)| dx \leq N \kappa_N^{\frac{N-1}{N}} \mu(E)^{\frac{1}{N}}. \tag{4.4}$$

If $K(\cdot, \cdot)$ satisfies the first part of (4.2), then (4.1) defines a bounded, linear map

$$T : L^\infty(E) \longrightarrow L^\infty(E) \quad \text{with norm } \|T\|_\infty.$$

If $K(\cdot, \cdot)$ satisfies the second part of (4.2), then (4.1) defines a bounded, linear map

$$T : L^1(E) \longrightarrow L^1(E) \quad \text{with norm } \|T\|_1.$$

---

[7] See Proposition 23.2 of Chapter VIII. Here κ_N is the volume of the unit ball in $\mathbb{R}^N$.

If $K(\cdot, \cdot)$ is as in (4.3), the corresponding map is the Riesz potential[8]

$$T(f)(x) = \int_E \frac{f(y)}{|x - y|^{N-1}} dy. \tag{4.5}$$

Such a potential can be regarded as a bounded linear map from $L^1(E)$ into $L^1(E)$ or from $L^\infty(E)$ into $L^\infty(E)$.

Denote by $C^1([0, 1])$ the space of all continuously differentiable functions in $[0, 1]$ with the topology inherited from the sup-norm on $C[0, 1]$. The differentiation map[9]

$$T(f) = f' : C^1([0, 1]) \longrightarrow C([0, 1]).$$

is linear and not bounded. Let $C[0, 1]$ be equipped with the sup-norm. The map

$$T(f)(t) = \int_0^t f(s)ds, \quad t \in [0, 1],$$

from $C[0, 1]$ into itself is linear and continuous but not onto.

4.1   Functionals. The dual of $\mathbb{R}^N$ equipped with its Euclidean norm is isometrically isomorphic to $\mathbb{R}^N$, i.e., $\mathbb{R}^{N^*} = \mathbb{R}^N$, up to an isomorphism.

By the Riesz representation theorem, the dual of $L^p(E)$ for $1 \le p < \infty$ is isomorphic $L^q(E)$, where p and q are conjugate; i.e.,

$$L^p(E)^* = L^q(E), \quad 1 \le p < \infty, \quad \frac{1}{p} + \frac{1}{q} = 1, \tag{4.6}$$

up to an isomorphism. Similarly,

$$\ell_p^* = \ell_q, \quad 1 \le p < \infty, \quad \frac{1}{p} + \frac{1}{q} = 1. \tag{4.7}$$

4.2   Linear functionals on $C(\overline{E})$. Let $E \subset \mathbb{R}^N$ be bounded and open. For a fixed $x_o \in E$, set

$$T(f) = f(x_o) \quad \text{for all } f \in C(\overline{E}). \tag{4.8}$$

This is called the *evaluation* map at x_o, and it identifies a bounded linear functional $T \in C(\overline{E})^*$, with norm $\|T\| = 1$. Now let μ be a σ-finite Borel measure in $\mathbb{R}^N$, and set

$$T(f) = \int_E f d\mu \quad \text{for all } f \in C(\overline{E}). \tag{4.9}$$

This identifies a functional $T \in C(\overline{E})^*$ with norm $\|T\| = \mu(E)$.

---

[8] See Section 23 Chapter VIII.

[9] Differentiation is an *interior* operation. Here it is meant that $(0, 1) \ni t \to f'(t)$ has a continuous extension in $[0, 1]$.

Remark 4.1. It will be shown that, in a sense to be made precise, these are essentially *all* the bounded linear functionals on $C(\overline{E})$.[10] Also, the evaluation map (4.8) can be represented as in (4.9) if μ is the Dirac mass δ_{x_o} concentrated at x_o.

5   Kernels of maps and functionals

Let T be a map from $\{X; \|\cdot\|_X\}$ into $\{Y; \|\cdot\|_Y\}$. The kernel of T is defined as

$$\ker\{T\} = \{x \in X | T(x) = \Theta_Y\}.$$

If T is linear, then $\ker\{T\}$ is a linear subspace of X. If $T \in \mathcal{B}(X; Y)$ and $\{X; \|\cdot\|_X\}$ is a Banach space, then $\ker\{T\}$ is a closed subspace of X.

Now let $T : \{X; \|\cdot\|\} \to \mathbb{R}$ be a linear functional. It is not assumed that T is bounded, nor is it assumed that $\{X; \|\cdot\|\}$ is a Banach space. The mere linearity of T permits one to derive information on the structure of X in terms of the kernel of T.

Proposition 5.1. *Let X_o be a closed subspace of X such that the quotient space X/X_o is one dimensional. Then there exists a linear functional $T : X \to \mathbb{R}$ such that $X_o = \ker\{T\}$.*

Conversely, let $T : X \to \mathbb{R}$ be a not-identically-zero, linear functional. Then the quotient space $X/\ker\{T\}$ is one dimensional.

Proof. To prove the first statement, choose $x \in X - X_o$ such that $\mathrm{dist}\{x; X_o\} > 0$, and write

$$X - X_o = \bigcup\{\lambda x | \lambda \in \mathbb{R}\}.$$

Every element of $y \in X$ can be written as

$$y = x_o + \lambda x \quad \text{for some } x_o \in X_o \quad \text{and} \quad \text{some } \lambda \in \mathbb{R}.$$

Then set

$$T(y) = \lambda.$$

To establish the converse statement, fix $x \in X - \ker\{T\}$. Such a choice is possible since $T \not\equiv 0$. To show that $X = \mathrm{span}\{x; X_o\}$, pick any element $y \in X$ and compute

$$T\left(y - \frac{T(y)}{T(x)}x\right) = T(y) - \frac{T(y)}{T(x)}T(x) = 0.$$

Therefore,

$$y - \frac{T(y)}{T(x)}x \in \ker\{T\}.$$

Thus y is a linear combination of x and an element of $\ker\{T\}$. $\square$

---

[10]See Sections 2–6 of Chapter VII, as well as Theorem 6.1c in Chapter VII's Problems and Complements.

Corollary 5.2. *Let* $\{X; \|\cdot\|\}$ *be a normed space. For any nonzero linear functional* $T : X \to \mathbb{R}$, *the normed space* $\{X; \|\cdot\|\}$ *can be written as the direct sum of* $\ker\{T\}$ *and the one-dimensional space spanned by an element in* $X - \ker\{T\}$.

Corollary 5.3. *Let* T *be a not-identically-zero, linear functional on a normed space* $\{X; \|\cdot\|\}$. *Then* T *is continuous if and only if* $\ker\{T\}$ *is not dense in* X. *As a consequence,* T *is continuous if and only if* $\ker\{T\}$ *is nowhere dense in* X.

Corollary 5.4. *Let* T *be a not-identically-zero, bounded, linear functional in a normed space* $\{X; \|\cdot\|\}$. *Then for every* $x \in X - \ker\{T\}$,

$$\|T\| = \frac{|T(x)|}{\mathrm{dist}\{x; \ker\{T\}\}}.$$

6 Equibounded families of linear maps

If $\{X; \|\cdot\|\}$ is a Banach space, it is of the second category. In such a case, the uniform boundedness principle can be applied to families of bounded linear maps in $\mathcal{B}(X; Y)$ in the following form.

Proposition 6.1. *Let* $\mathcal{T}$ *be a family of bounded linear maps from a Banach space* $\{X; \|\cdot\|_X\}$ *into a normed space* $\{Y; \|\cdot\|_Y\}$. *Assume that the elements of* $\mathcal{T}$ *are pointwise equibounded in* X; *i.e., for every* $x \in X$, *there exists a positive number* $F(x)$ *such that*

$$\|T(x)\|_Y \leq F(x) \quad \text{for all } T \in \mathcal{T}. \tag{6.1}$$

Then the elements of $\mathcal{T}$ *are uniformly equibounded in* $\mathcal{B}(X; Y)$; *i.e., there exists a positive number* M *such that*

$$\|T(x)\|_Y \leq M\|x\|_X \quad \text{for all } T \in \mathcal{T} \quad \text{and} \quad \text{all } x \in X. \tag{6.2}$$

Proof. The functions $\|T(x)\|_Y : X \to \mathbb{R}$ satisfy the assumptions of the Banach–Steinhaus theorem. Therefore, there exists a positive F and a ball $B_\varepsilon(x_o) \subset X$ of radius ε and centered at some $x_o \in X$ such that

$$\|T(y)\|_Y \leq F \quad \text{for all } T \in \mathcal{T} \quad \text{and} \quad \text{all } y \in B_\varepsilon(x_o).$$

Given any $x \in X$, the element

$$y = x_o + \frac{\varepsilon}{2\|x\|_X} x$$

belongs to $B_\varepsilon(x_o)$. Therefore,

$$\|T(x)\|_Y \leq 2\frac{F + \|T(x_o)\|_Y}{\varepsilon}\|x\|_X \leq \frac{4F}{\varepsilon}\|x\|_X$$

for all $T \in \mathcal{T}$. $\qquad\square$

Corollary 6.2. *Let* $\{X; \|\cdot\|\}$ *be a Banach space. Then a pointwise-bounded family of elements in* X^* *is equibounded in* X.

6.1   Another proof of Proposition 6.1. The proposition can be proved without appealing to category arguments. It suffices to establish that the elements of $\mathcal{T}$ are equiuniformly bounded in some open ball $B_\varepsilon(x_o) \subset X$. The proof proceeds by contradiction, assuming that such a ball does not exist.[11] Fix any such ball $B_1(x_o)$. There exist $x_1 \in B_1(x_o)$ and $T_1 \in \mathcal{T}$ such that $\|T_1(x_1)\|_Y > 1$. By continuity, there exists a ball $B_{\varepsilon_1}(x_1) \subset X$ such that $\|T_1(x)\|_Y > 1$ for all $x \in B_{\varepsilon_1}(x_1)$. By taking ε_1 sufficiently small, we may ensure that

$$\overline{B}_{\varepsilon_1}(x_1) \subset B_\varepsilon(x_o) \quad \text{and} \quad \varepsilon_1 < 1.$$

There exist $x_2 \in B_{\varepsilon_1}(x_1)$ and $T_2 \in \mathcal{T}$ such that $\|T_2(x_2)\|_Y > 2$. By continuity, there exists a ball $B_{\varepsilon_2}(x_2) \subset X$ such that $\|T_2(x)\|_Y > 2$ for all $x \in B_{\varepsilon_2}(x_2)$. By taking ε_2 sufficiently small, we may ensure that

$$\overline{B}_{\varepsilon_2}(x_2) \subset B_{\varepsilon_1}(x_1) \quad \text{and} \quad \varepsilon_2 < \frac{1}{2}.$$

Proceeding in this fashion, we construct a sequence $\{x_n\}$ of elements of X, a countable family of balls $B_{\varepsilon_n}(x_n) \subset X$, and a sequence $\{T_n\}$ of elements of $\mathcal{T}$ such that

$$\overline{B}_{\varepsilon_{n+1}}(x_{n+1}) \subset B_{\varepsilon_n}(x_n), \quad \varepsilon_n < \frac{1}{n},$$

and

$$\|T_n(x)\|_Y > n \quad \text{for all } x \in B_{\varepsilon_n}(x_n).$$

The sequence $\{x_n\}$ is a Cauchy sequence, and its limit x must belong to the closure of all $B_{\varepsilon_n}(x_n)$. Therefore, $\|T_n(x)\|_Y > n$ for all $n \in \mathbb{N}$. This contradicts the assumption that the maps in $\mathcal{T}$ are pointwise equibounded. $\square$

7   Contraction mappings

A map T from a normed space $\{X; \|\cdot\|\}$ into itself is a *contraction* if there exists $t \in (0, 1)$ such that

$$\|T(x) - T(y)\| \le t\|x - y\| \quad \text{for all } x, y \in X.$$

Theorem 7.1 (Banach[12]). *Let T be a contraction from a Banach space $\{X; \|\cdot\|\}$ into itself. Then T has a unique fixed point; i.e, there exists a unique $x_o \in X$ such that $T(x_o) = x_o$.*

---

[11]W. F. Osgood, Nonuniform convergence and the integration of series term by term, *Amer. J. Math.*, **19** (1897), 155–190.

[12]See [2].

Proof. Starting from an arbitrary $x_1 \in X$, define the sequence

$$x_{n+1} = T(x_n).$$

Then

$$\|T(x_{n+1}) - T(x_n)\| \le t^n \|x_2 - x_1\|.$$

Equivalently,

$$\|T(x_n) - x_n\| \le t^{n-1} \|x_2 - x_1\|.$$

Therefore, the sequences $\{x_n\}$ and $\{T(x_n)\}$ are both Cauchy sequences. If x_o is the limit of $\{x_n\}$, then by continuity,

$$\|T(x_o) - x_o\| = \lim \|T(x_n) - x_n\| = 0.$$

Thus $T(x_o) = x_o$. If $\overline{x}$ were another fixed point,

$$\|T(\overline{x}) - T(x_o)\| \le t\|\overline{x} - x_o\| = t\|T(\overline{x}) - T(x_o)\|.$$

Thus $\overline{x} = x_o$.   □

Remark 7.1. The theorem continues to hold in a complete metric space with the proper variants.

7.1 Applications to some Fredholm integral equations. Let E be an open set in $\mathbb{R}^N$, and formally consider the integral equation[13]

$$f(x) = \int_E K(x, y) f(y) dy + h(x). \tag{7.1}$$

Assume that the *kernel* $K(x, y)$ satisfies the second part of (4.2). Given a function $h \in L^1(E)$, one seeks a function $f \in L^1(E)$ satisfying (7.1) for a.e. $x \in E$.

Proposition 7.2. *Assume that the constant γ in (4.2) is less than 1. Then the integral equation (7.1) has a unique solution.*

Proof. The solution is the unique fixed point of the map

$$T(f) = \int_E K(x, y) f(y) dy + h,$$

provided $T : L^1(E) \to L^1(E)$ is a contraction. For $f, g \in L^1(E)$,

$$\|T(f - g)\|_1 \le \gamma \|f - g\|_1.$$ □

Remark 7.2. In the case of the Riesz kernel (4.3), the assumption is satisfied if $\mu(E)$ is sufficiently small.

Remark 7.3. If the kernel $K(x, y)$ satisfies the first part of (4.2), one gives a function $h \in L^\infty(E)$ and seeks a function $f \in L^\infty(E)$ satisfying (7.1). The integral equation (7.1) could be set in $L^2(E)$, provided $K \in L^2(E \times E)$.

---

[13]I. Fredholm, Sur une classe d'équations fonctionnelles, *Acta Math.*, **27** (1903), 365–390. See also [40], [53], and [11, Chapter IV].

8   The open mapping theorem

A map T from a topological space $\{X; \mathcal{U}\}$ into a topological space $\{Y; \mathcal{V}\}$ is called *open* if it maps open sets of $\mathcal{U}$ into open sets of $\mathcal{V}$.

If T is one-to-one and open, T^{-1} is continuous.

An open map $T : X \to Y$ that is continuous, one-to-one, and onto is a homeomorphism between $\{X; \mathcal{U}\}$ and $\{Y; \mathcal{V}\}$.

The next theorem, called the *open mapping theorem*, states that continuous linear maps between Banach spaces are open mappings.

Theorem 8.1 (open mapping theorem). *A bounded linear map T from a Banach space $\{X; \| \cdot \|_X\}$ onto a Banach space $\{Y; \| \cdot \|_Y\}$ is an open mapping. If T is also one-to-one, it is a homeomorphism between $\{X; \| \cdot \|_X\}$ and $\{Y; \| \cdot \|_Y\}$.*

Remark 8.1. The requirement that T be linear cannot be removed. Let $T(x) = e^x \cos x : \mathbb{R} \to \mathbb{R}$, where $\mathbb{R}$ is endowed with the Euclidean norm. Then T is continuous but not open since the image of $(-\infty, 0)$ is not open.

Let B_ρ denote the open ball in X of radius ρ and centered at the origin of X. Also, denote by $\mathcal{B}_\varepsilon$ the open ball in Y of radius ε and centered at the origin of Y.

Lemma 8.2. *$T(B_1)$ contains a ball $\mathcal{B}_\varepsilon$ for some $\varepsilon > 0$.*

Proof. Since T is linear and onto,

$$X = \bigcup n B_{1/2} \quad \text{implies that} \quad Y = \bigcup n T(B_{1/2}).$$

Since $\{Y; \| \cdot \|_Y\}$ is of the second category, $T(B_{1/2})$ is not nowhere dense, and its closure contains an open ball $\mathcal{B}_{2\varepsilon}(\overline{y})$ in Y centered at some $\overline{y}$. From the inclusions

$$\overline{T(B_{1/2})} - \overline{y} \subset \overline{T(B_{1/2})} - \overline{T(B_{1/2})}$$
$$\subset 2\overline{T(B_{1/2})}$$
$$\subset \overline{T(B_1)},$$

it follows that the ball $\mathcal{B}_{2\varepsilon} \subset \overline{T(B_1)}$. The linearity of T also implies that

$$\mathcal{B}_{2\varepsilon/2^n} \subset \overline{T(B_{1/2^n})} \quad \text{for all } n \in \mathbb{N}. \tag{8.1}$$

We next show that $\mathcal{B}_\varepsilon \subset T(B_1)$. Fix $y \in \mathcal{B}_\varepsilon$. Since $y \in \overline{T(B_{1/2})}$, there exists $x_1 \in B_{1/2}$ such that

$$\|y - T(x_1)\|_Y < \frac{1}{2}\varepsilon.$$

In particular, the element $y - T(x_1)$ belongs to $\mathcal{B}_{\varepsilon/2}$. Therefore, by (8.1), for $n = 2$, there exists $x_2 \in B_{1/4}$ such that

$$\|y - T(x_1) - T(x_2)\|_Y < \frac{1}{4}\varepsilon.$$

Proceeding in this fashion, we find a sequence $\{x_n\}$ of elements of X such that $x_n \in B_{1/2^n}$ and

$$\left\| y - \sum_{j=1}^{n} T(x_n) \right\|_Y \leq \frac{1}{2^n} \varepsilon.$$

Since $\|x_n\|_X < 2^{-n}$, the series $\sum x_n$ is absolutely convergent and identifies an element x in the ball B_1 by the formula

$$x = \sum x_n \quad \text{and} \quad \|x\|_X \leq \sum \|x_n\|_X < 1.$$

Since T is linear and continuous,

$$T(x) = T\left(\sum x_n\right) = \sum T(x_n) = y.$$

Thus $y \in T(B_1)$. Since y is an arbitrary element of $\mathcal{B}_\varepsilon$, this implies that $\mathcal{B}_\varepsilon \subset T(B_1)$. $\qquad\square$

Proof of Theorem 8.1. Let $\mathcal{O} \subset X$ be open. To establish that $T(\mathcal{O})$ is open in Y, for every fixed $\xi \in \mathcal{O}$ and $\eta = T(\xi)$, we exhibit a ball $\mathcal{B}_\varepsilon(\eta)$ contained in $T(\mathcal{O})$. Since $\mathcal{O}$ is open, there exists an open ball $B_\sigma(\xi) \subset \mathcal{O}$. Then by Lemma 8.2, $T(B_\sigma(\xi))$ contains a ball $\mathcal{B}_\varepsilon(\eta)$. $\qquad\square$

8.1 Some applications. The open mapping theorem may be applied to finding conditions for two different norms on the same vector space to generate the same topology.

Proposition 8.3. *Let $\| \cdot \|_1$ and $\| \cdot \|_2$ be two norms on the same vector space X, by which $\{X; \| \cdot \|_1\}$ and $\{X; \| \cdot \|_2\}$ are both Banach spaces. Assume that there exists a positive constant C_1 such that*

$$\|x\|_2 \leq C_1 \|x\|_1 \quad \textit{for all } x \in X. \tag{8.2}_1$$

Then there exists a positive constant C_2 such that

$$\|x\|_1 \leq C_2 \|x\|_2 \quad \textit{for all } x \in X. \tag{8.2}_2$$

Proof. The identity map $T(x) = x$ from $\{X; \| \cdot \|_1\}$ onto $\{X; \| \cdot \|_2\}$ is linear and one-to-one. By $(8.2)_1$, it is also continuous. Therefore, it is a homeomorphism. In particular, the inverse T^{-1} is linear and continuous and hence bounded. $\qquad\square$

8.2 The closed graph theorem. Let T be a linear map from a Banach space $\{X; \| \cdot \|_X\}$ into a Banach space $\{Y; \| \cdot \|_Y\}$. The graph $\mathcal{G}_T$ of T is a subset of $X \times Y$ defined by

$$\mathcal{G}_T = \bigcup \{(x, T(x)) | x \in X\}.$$

The closure of $\mathcal{G}_T$ is meant in the sense of the product topology on $X \times Y$. In particular, the graph $\mathcal{G}_T$ is closed if and only if whenever $\{x_n\}$ is a Cauchy sequence in $\{X; \|\cdot\|_X\}$ and $\{T(x_n)\}$ is a Cauchy sequence in $\{Y; \|\cdot\|_Y\}$, then

$$\lim T(x_n) = T(\lim x_n). \tag{8.3}$$

This would hold if T were continuous. The next theorem, called the *closed graph theorem*, states the converse; i.e., (8.3) implies that T is continuous.

Theorem 8.4 (closed graph theorem). *Let T be a linear map from a Banach space $\{X; \|\cdot\|_X\}$ into a Banach space $\{Y; \|\cdot\|_Y\}$. Then if $\mathcal{G}_T$ is closed, T is continuous.*

Proof. On X, introduce a new norm $\|\cdot\|$ by

$$\|x\| = \|x\|_X + \|T(x)\|_Y \quad \text{for all } x \in X.$$

One verifies that this is a norm on X, and if $\mathcal{G}_T$ is closed, $\{X; \|\cdot\|\}$ is complete. Therefore, if $\mathcal{G}_T$ is closed, $\{X; \|\cdot\|\}$ is a Banach space. Since $\|x\|_X \leq \|x\|$, by Proposition 8.3, there exists a positive constant C such that

$$\|x\|_X + \|T(x)\|_Y \leq C\|x\|_X \quad \text{for all } x \in X.$$

This implies that $\|T(x)\|_Y \leq C\|x\|_X$. Thus T is bounded and hence continuous. $\qquad\square$

Remark 8.2. The assumption that both $\{X; \|\cdot\|_X\}$ and $\{Y; \|\cdot\|_Y\}$ be Banach spaces is essential for the closed graph theorem to hold. Indeed, without such a completeness requirement, there is no relationship between the continuity of a linear map $T : X \to Y$ and the closedness of its graph $\mathcal{G}_T$, as illustrated by the following two counterexamples.

Let $C[0, 1]$ be endowed with the sup-norm. Also, let $C^1[0, 1] \subset C[0, 1]$ be equipped with the topology inherited from the norm topology of $C[0, 1]$. The map $T = \frac{d}{dt}$ from $C^1[0, 1]$ into $C[0, 1]$ is linear and has a closed graph but is discontinuous.

Regard $C[0, 1]$ as a topological subspace of $L^2[0, 1]$. The identity map from $C[0, 1]$ into $L^2[0, 1]$ is continuous but its graph is not closed.

9   The Hahn–Banach theorem

Let X be a linear vector space over the reals. A sublinear, homogeneous, real-valued map from X into $\mathbb{R}$ is a function $p : X \to \mathbb{R}$ satisfying

$$\begin{aligned} p(x + y) &\leq p(x) + p(y) \quad \text{for all } x, y \in X \quad \text{and} \\ p(\lambda x) &= \lambda p(x) \quad \text{for all } x \in X \quad \text{and} \quad \text{all } \lambda > 0. \end{aligned} \tag{9.1}$$

The *dominated extension theorem* states that a linear map T_o defined on a linear vector subspace X_o of X and dominated by a sublinear map p can be extended into a linear map $T : X \to \mathbb{R}$ in such a way that the extended map is dominated by p in the whole X and agrees with T_o on X_o.

While the main applications of this extension procedure are in Banach spaces, it is worth noting that such an extension is algebraic in nature and topology independent. In particular, while X is required to have a vector structure, it is not required to be a topological vector space. Likewise, no topological assumptions, such as continuity, are placed on the sublinear map p or on the linear functional T_o defined on X_o.

Theorem 9.1 (Hahn–Banach[14]). *Let X be a real vector space, and let $p : X \to \mathbb{R}$ be sublinear and homogeneous. Then every linear functional $T_o : X_o \to \mathbb{R}$ defined on a subspace X_o of X and satisfying*

$$T_o(x) \le p(x) \quad \text{for all } x \in X_o \tag{9.2}$$

admits an extension $T : X \to \mathbb{R}$ such that

$$T(x) \le p(x) \quad \text{for all } x \in X$$

and

$$T(x) = T_o(x) \quad \text{for all } x \in X_o.$$

Proof. If $X_o \neq X$, choose $\eta \in X - X_o$ and let X_η be the linear span of X_o and η; i.e.,

$$X_\eta = \{x + \lambda\eta | x \in X_o \text{ and } \lambda \in \mathbb{R}\}. \tag{9.3}$$

First, we extend T_o to a linear functional T_η defined in X_η, coinciding with T_o on X_o, and dominated by p on X_η. If T_η is such an extension, then for all $\lambda \in \mathbb{R}$ and all $x \in X_o$, by linearity,

$$T(x + \lambda\eta) = T_o(x) + \lambda T_\eta(\eta).$$

Therefore, to construct T_η, it suffices to specify its value at η. Since T_o is dominated by the sublinear map p on X_o, for all $x, y \in X_o$,

$$T_o(x) + T_o(y) = T_o(x + y) \le p(x + y) \le p(x - \eta) + p(y + \eta).$$

Therefore,

$$T_o(x) - p(x - \eta) \le p(y + \eta) - T_o(y)$$

and

$$\sup_{x \in X_o} \{T_o(x) - p(x - \eta)\} \le \inf_{y \in X_o} \{p(y + \eta) - T_o(y)\}.$$

---

[14]H. Hahn, Über lineare Gleichungen in linearen Räumen, *J. Math.*, **157** (1927), 214–229; S. Banach, Sur les lignes rectifiables et les surfaces dont l'aire est finie, *Fund. Math.*, **7** (1925), 225–237.

Define $T_\eta(\eta) = \alpha$, where α is any number satisfying

$$\sup_{x \in X_o} \{T_o(x) - p(x - \eta)\} \le \alpha \le \inf_{y \in X_o} \{p(y + \eta) - T_o(y)\}. \tag{9.4}$$

By construction, $T_\eta : X_\eta \to \mathbb{R}$ is linear and it coincides with T_o on X_o.

It remains to show that such an extended functional is dominated by p in X_η; i.e.,

$$T(x + \lambda\eta) = T_o(x) + \lambda\alpha \le p(x + \lambda\eta)$$

for all $x \in X_o$ and all $\lambda \in \mathbb{R}$. If $\lambda > 0$, by using the upper inequality in (9.4),

$$
\begin{aligned}
\lambda\alpha + T_o(x) &= \lambda \left\{ \alpha + T_o \left(\frac{x}{\lambda} \right) \right\} \\
&\le \lambda \left\{ p \left(\frac{x}{\lambda} + \eta \right) - T_o \left(\frac{x}{\lambda} \right) + T_o \left(\frac{x}{\lambda} \right) \right\} \\
&\le p(x + \lambda\eta).
\end{aligned}
$$

If $\lambda < 0$, the same conclusion holds by using the lower inequality in (9.4).

If $X_\eta \ne X$, the construction can be repeated by extending T_η to a larger subspace of X. The extension of T_o to the whole X can now be concluded by transfinite induction.

Introduce the set $\mathcal{E}$ of all the dominated extensions of T_o, i.e., the set of pairs, $\{T_\eta; X_\eta\}$, where T_η is a linear functional defined on X_η and satisfying

$$T_\eta(x) \le p(x) \quad \text{for all } x \in X_\eta$$

and

$$T_\eta(x) = T_o(x) \quad \text{for all } x \in X_o.$$

On the set $\mathcal{E}$, introduce an ordering relation by stipulating that $\{T_\xi; X_\xi\} \le \{T_\eta; X_\eta\}$ if and only if $X_\xi \subset X_\eta$ and $T_\eta = T_\xi$ on X_ξ. One verifies that such a relation is a partial ordering on $\mathcal{E}$.

Every linearly ordered subset $\mathcal{E}' \subset \mathcal{E}$ has an upper bound. Indeed, denoting by $\{T_\sigma; X_\sigma\}$ the elements of $\mathcal{E}'$, setting

$$X' = \bigcup X_\sigma \quad \text{and} \quad T' = T_\sigma \quad \text{on } X_\sigma,$$

provides an upper bound for $\mathcal{E}'$. It follows by Zorn's lemma that $\mathcal{E}$ has a maximal element $\{T; \widetilde{X}\}$. For such a maximal element, $\widetilde{X} = X$. Indeed, otherwise $X - \widetilde{X}$ would be not empty and the extension process could be repeated, contradicting the maximality of $\{T; \widetilde{X}\}$. □

10   Some consequences of the Hahn–Banach theorem

The main applications of the Hahn–Banach theorem occur in normed spaces $\{X; \|\cdot\|\}$ and for a suitable choice of the dominating sublinear function $p : X \to \mathbb{R}$.

For example, such a function p could be a norm or a seminorm in X.

If X_o is a subspace of $\{X; \|\cdot\|\}$, we let $\{X_o; \|\cdot\|\}$ denote the corresponding normed space and let X_o^* denote the collection of all bounded, linear functionals defined in X_o.

Typically, given $T_o \in X_o^*$ dominated by some sublinear function p, one seeks to extend it to an element $T \in X^*$.

Proposition 10.1. *Let $\{X; \|\cdot\|\}$ be a normed space, and let X_o be a subspace of X. To every $T_o \in X_o^*$, there corresponds $T \in X^*$ such that*

$$\|T\| = \|T_o\| \quad and \quad T = T_o \quad on\ X_o.$$

Proof. Apply the Hahn–Banach theorem with $p(x) = \|T_o\|\|x\|$. This gives an extension T defined in X and satisfying

$$|T(x)| \leq \|T_o\|\|x\| \quad \text{for all } x \in X.$$

Therefore, $\|T\| \leq \|T_o\|$. On the other hand, $\|T\| \geq \|T_o\|$ since T is an extension of T_o. $\qquad\square$

Proposition 10.2. *Let $\{X; \|\cdot\|\}$ be a normed space. For every $x_o \in X$ and $x_o \neq \Theta$, there exists $T \in X^*$ such that $\|T\| = 1$ and $T(x_o) = \|x_o\|$.*

Proof. Having fixed $x_o \in X$, let X_o be the span of x_o; i.e.,

$$X_o = \{\lambda x_o | \lambda \in \mathbb{R}\}.$$

Then on X_o consider the functional

$$T_o(\lambda x_o) = \lambda\|x_o\|,$$

and as a function p, take the norm $\|\cdot\|$. By the Hahn–Banach theorem, T_o can be extended into a linear functional T defined in the whole X and satisfying

$$T(x) \leq \|x\| \qquad \text{for all } x \in X \quad \text{and}$$
$$T(\lambda x_o) = \lambda\|x_o\| \quad \text{for all } \lambda \in \mathbb{R}.$$

The first of these implies that $T(\pm x) \leq \|x\|$. Therefore, $\|T\| \leq 1$. The second with $\lambda = \pm 1$ implies $\|T\| = 1$. $\qquad\square$

Corollary 10.3. *Let $\{X; \|\cdot\|\}$ be a normed space. Then X^* separates the points of X; i.e., for any pair x, y of distinct points of X, there exists $T \in X^*$ such that $T(x) \neq T(y)$.*

Proof. Apply Proposition 10.2 to the element $x - y$. $\qquad\square$

Corollary 10.4. *Let $\{X; \|\cdot\|\}$ be a normed space. Then for every $x \in X$,*

$$\|x\| = \sup_{\substack{T \in X^* \\ T \neq 0}} \frac{|T(x)|}{\|T\|} = \sup_{\substack{T \in X^* \\ \|T\|=1}} |T(x)|.$$

Proposition 10.5. *Let* $\{X; \|\cdot\|\}$ *be a normed space, and let* X_o *be a linear subspace of* X. *Assume that there exists an element* $\eta \in X - X_o$ *that has positive distance from* X_o; *i.e.,*

$$\inf_{x \in X_o} \|x - \eta\| \geq \delta > 0.$$

There exists $T \in X^*$ *such that* $\|T\| \leq 1$, $T(\eta) = \delta$, *and* $T(x) = 0$ *for all* $x \in X_o$.

Proof. Let X_η be the span of X_o and η as in (9.3). On X_η, define a linear functional by the formula

$$T_\eta(\lambda\eta + x) = \lambda\delta \quad \text{for all } x \in X_o \quad \text{and} \quad \text{all } \lambda \in \mathbb{R}.$$

From the definition of δ, for all $\lambda \neq 0$ and all $x \in X_o$,

$$\lambda\delta \leq |\lambda| \left\| \eta + \frac{x}{\lambda} \right\| = \|\lambda\eta + x\|.$$

Therefore, denoting by $y = \lambda\eta + x$ the generic element of X_η,

$$T_\eta(y) \leq \|y\| \quad \text{for all } y \in X_\eta.$$

By the Hahn–Banach theorem, T_η has an extension T defined in the whole X and satisfying $T(x) \leq \|x\|$ for all $x \in X$. Therefore, $\|T\| \leq 1$. Moreover, $T(x) = 0$ for all $x \in X_o$ and $T(\eta) = \delta$. $\square$

Remark 10.1. The assumptions of Proposition 10.5 are verified if X_o is a proper, closed, linear subspace of X.

Corollary 10.6. *Let* $\{X; \|\cdot\|\}$ *be a normed space, and let* X_o *be a linear subspace of* X. *If* X_o *is not dense in* X, *there exists a nonzero functional* $T \in X^*$ *such that* $T(x) = 0$ *for all* $x \in X_o$.

Proof. If $\overline{X}_o \neq X$, there exists $\eta \in X - X_o$ such that $\text{dist}\{\eta; X_o\} > 0$. $\square$

Proposition 10.7. *Let* $\{X; \|\cdot\|\}$ *be a normed space. Then if* X^* *is separable,* $\{X; \|\cdot\|\}$ *also is separable.*

Proof. Let $\{T_n\}$ be a sequence dense in X^*. For each T_n, choose $x_n \in X$ such that

$$\|x_n\| = 1 \quad \text{and} \quad |T_n(x_n)| \geq \frac{1}{2}\|T_n\|.$$

Now let X_o be the set of all finite linear combinations of elements of $\{x_n\}$ with rational coefficients. The set X_o is countable, and we claim that it is dense in X.

Proceeding by contradiction, if X_o is not dense in X, the closure $\overline{X}_o$ is a linear, closed, proper subspace of X. By Corollary 10.6, there exists a nonzero functional $T \in X^*$ vanishing on X_o. Since $\{T_n\}$ is dense in X^*, there exists a subsequence $\{T_{n_j}\}$ convergent to T. For such a sequence,

$$\|T - T_{n_j}\| \geq |(T - T_{n_j})(x_{n_j})| \geq \frac{1}{2}\|T_{n_j}\|.$$

Thus $\{T_{n_j}\} \to 0$, contradicting that T is nonzero. $\square$

Remark 10.2. The converse of Proposition 10.7 is, in general, false; i.e., X separable does not imply that X^* is separable. As an example, consider $L^1(E)$, where E is an open set in $\mathbb{R}^N$ equipped with the Lebesgue measure. By the Riesz representation theorem, $L^1(E)^* = L^\infty(E)$. However, $L^1(E)$ is separable and $L^\infty(E)$ is not.

10.1 Tangent planes. Let $\{X; \|\cdot\|\}$ be a normed space. For a fixed $T \in X^*$ and $\alpha \in \mathbb{R}$, consider the set

$$[T = \alpha] = \{x \in X | T(x) = \alpha\}.$$

Introduce analogously the sets $[T > \alpha]$ and $[T < \alpha]$. In analogy with linear functionals in Euclidean spaces, $[T = \alpha]$ is called a *hyperplane* in X, which divides X into two disjoint *half-spaces* $[T > \alpha]$ and $[T < \alpha]$.

Now let C be a set in X, and let $x_o \in C$. A hyperplane $[T = \alpha]$ for some $\alpha \in \mathbb{R}$ is a *tangent* hyperplane to C at x_o if $x_o \in [T = \alpha]$ and

$$T(x) \le T(x_o) \quad \text{for all } x \in C.$$

With this terminology, Proposition 10.2 can be rephrased in the following geometric form.

Proposition 10.2′. *The unit ball of X has a tangent plane at any one of its boundary points.*

11 Separating convex subsets of X

Applications of the Hahn–Banach theorem hinge upon specifying a suitable, dominating, homogeneous, sublinear function $p : X \to \mathbb{R}$. The Minkowski functional is one such a function. Let $\{X; \|\cdot\|\}$ be a normed space, and let C be a convex, open neighborhood of the origin of X. Then for every $x \in X$, there exists some positive number t such that $x \in tC$. Define

$$\mu_C(x) = \inf\{t > 0 | x \in tC\}. \tag{11.1}$$

If C is the unit ball in $\{X; \|\cdot\|\}$, then $\mu_{B_1}(x) = \|x\|$ for all $x \in X$. If C is unbounded, then $\mu_C(x)$ vanishes for $x = 0$ and for infinitely many nonzero elements of X. It follows from the definition that $\lambda \mu_C(x) = \mu_C(\lambda x)$ for all $\lambda > 0$.

Proposition 11.1. *The map $x \to \mu_C(x)$ is sublinear in X.*

Proof. Fix $x, y \in X$, and let t and s be positive numbers such that $\mu_C(x) < t$ and $\mu_C(y) < s$. For such choices, $t^{-1}x \in C$ and $s^{-1}y \in C$. Then since C is convex,

$$\frac{1}{s+t}(x+y) = \frac{t}{s+t}t^{-1}x + \frac{s}{s+t}s^{-1}y \in C.$$

Therefore, $\mu(x+y) \le s+t$. □

By Corollary 10.3, the collection X^* of the bounded linear functionals in X separates points. The next proposition asserts that X^* contains enough elements to separate disjoint, convex sets, provided at least one is open.

Proposition 11.2. *Let C_1 and C_2 be two disjoint, convex subsets of X, and assume that C_1 is open. There exist $T \in X^*$ and $\alpha \in \mathbb{R}$ such that*

$$T(x) < \alpha \le T(y) \quad \text{for all } x \in C_1 \quad \text{and} \quad \text{all } y \in C_2. \tag{11.2}$$

Proof. Fix some $x_1 \in C_1$ and $x_2 \in C_2$, and set

$$C = C_1 - C_2 + x_o, \quad \text{where } x_o = x_2 - x_1.$$

Then C is an open, convex set containing the origin. Since C_1 and C_2 are disjoint, $x_o \notin C$. On the one-dimensional span of x_o, define a bounded, linear functional by $T_o(\lambda x_o) = \lambda$. Such a functional is pointwise bounded above on the span of x_o by the Minkowski functional μ_C relative to the set C. Indeed, for $\lambda \ge 0$,

$$T_o(\lambda x_o) = \lambda \le \lambda \mu_C(x_o) = \mu_C(\lambda x_o).$$

If $\lambda < 0$, then $T_o(\lambda x_o) = \lambda \le \mu_C(\lambda x_o)$. Therefore, by the Hahn–Banach theorem, there exists $T \in X^*$ that coincides with T_o on the span of x_o and such that

$$T(x) \le \mu_C(x) \quad \text{for all } x \in X.$$

For any $x \in C_1$ and $y \in C_2$, the element $(x - y + x_o)$ is in C. Therefore,

$$\mu_C(x - y + x_o) < 1 \quad \text{since } C \text{ is open.}$$

Using $T(x_o) = 1$, we compute

$$T(x - y + x_o) = T(x) - T(y) + 1 \le \mu(x - y + x_o) < 1.$$

Thus $T(x) < T(y)$. The existence of α satisfying (11.2) follows since $T(C_1)$ and $T(C_2)$ are convex subsets of $\mathbb{R}$ and $T(C_1)$ is open by the open mapping theorem. □

The requirement that C_1 be open can be modified, as indicated in the following separation proposition.

Proposition 11.3. *Let C_1 and C_2 be two disjoint, convex subsets of X. Assume that C_1 is compact and C_2 is closed. There exist $T \in X^*$ and $\alpha \in \mathbb{R}$ such that*

$$T(x) < \alpha < T(y) \quad \text{for all } x \in C_1 \quad \text{and} \quad \text{all } y \in C_2. \tag{11.3}$$

Proof. There exists an open ball B_ε centered at the origin of X such that $C_1 + B_\varepsilon$ is open and convex and does not intersect C_2. Then work with C_1 replaced by $C_1 + B_\varepsilon$. □

Corollary 11.4. *Let C be a closed, convex subset of a normed space $\{X; \|\cdot\|\}$. Then C is the intersection of all the closed half-spaces $[T \geq \alpha]$ that contain C.*

Remark 11.1. The topological structure of C_1 and C_2 is essential for the separation statement of either Proposition 11.2 or 11.3 to hold, as shown by the following counterexample.

Let C_o and C_1 be the two convex subsets of $L^2[0, 1]$ defined by

$$C_o = \{f \in C[0, 1] \text{ vanishing only at } t = 0\},$$
$$C_1 = \{f \in C[0, 1] \text{ vanishing only at } t = 1\}.$$

Both are dense in $L^2[0, 1]$, but they are not separated by any bounded linear functional.

12   Weak topologies

Let $\{X; \|\cdot\|\}$ be a normed space, and let X^* be its dual. The topology generated on X by its norm $\|\cdot\|$ is called the *strong* topology of X.

Since any $T \in X^*$ is a continuous linear map from X into $\mathbb{R}$, the inverse image of any open set in $\mathbb{R}$ is open in the strong topology of X. Set

$$\mathcal{B} = \left\{ \begin{array}{c} \text{the collection of the finite intersections of the inverse images} \\ T^{-1}(O), \text{ where } T \in X^* \text{ and } O \text{ are open subsets of } \mathbb{R} \end{array} \right\}.$$

The collection $\mathcal{B}$ is a base for a topology $\mathcal{W}$ on X called the *weak topology* of $\{X; \|\cdot\|\}$.[15] The weak topology $\mathcal{W}$ on X is the weakest topology for which all the functionals $T \in X^*$ are continuous.[16] One verifies that the operations of sum and multiplication by scalars,

$$+ : X \times X \to X, \qquad \bullet : \mathbb{R} \times X \to X,$$

are continuous with respect to such a topology.[17] Thus $\{X; \mathcal{W}\}$ is a topological vector space.

The topology of $\mathcal{W}$ is translation invariant and is determined by a local base at the origin of X. A neighborhood of the origin in the topology $\mathcal{W}$ contains an element of $\mathcal{B}$ of the form

$$\mathcal{O} = \bigcap_{j=1}^{n} T_j^{-1}(-\alpha_j, \alpha_j) \tag{12.1}$$

---

[15]The collection $\mathcal{B}$ satisfies requirements (i) and (ii) of Section 4 of Chapter I to be a base for a topology. The corresponding topology is constructed by the procedure of Proposition 4.1 of Chapter I.

[16]The procedure is similar to the construction of a product topology in the Cartesian product of topological spaces as the weakest topology for which all the projections are continuous (see Section 4.1 of Chapter I).

[17]See Proposition 12.1c of the Problems and Complements.

for some finite n and $\alpha_j > 0$. Equivalently,

$$\mathcal{O} = \{x \in X \,||T_j(x)| < \alpha_j \text{ for } j = 1, \ldots, n\}. \tag{12.1}'$$

The collection of open sets of the form (12.1) forms a local base $\mathcal{B}_o$ at the origin for the weak topology $\mathcal{W}$. These open sets are convex since T_j are linear. Thus $\mathcal{W}$ is a locally convex topology.

A sequence $\{x_n\}$ of elements of X converges weakly to 0 if and only if every weak neighborhood of the origin contains all but finitely many elements of $\{x_n\}$.

From (12.1)$'$, it follows that $\{x_n\}$ converges weakly to 0 if and only if $\{T(x_n)\} \to 0$ for all $T \in X^*$. More generally, $\{x_n\}$ converges weakly to some $x_o \in X$ if and only if $\{T(x_n)\} \to T(x_o)$ for all $T \in X^*$.

A sequence $\{x_n\}$ converges *strongly* to some $x_o \in X$ if $\{x_n\} \to x_o$ in the original, strong topology of $\{X; \|\cdot\|\}$.

Strong convergence implies weak convergence. The converse is false as there exist spaces $\{X; \|\cdot\|\}$ and sequences $\{x_n\}$ of elements of X, weakly convergent and not strongly convergent to some $x_o \in X$.[18]

Thus, in general, the weak topology $\mathcal{W}$ contains, roughly speaking, strictly fewer open sets than those of the strong topology.

The two topologies are also markedly different in terms of local boundedness.

A set $E \subset X$ is weakly bounded if and only if for every $\mathcal{O} \in \mathcal{B}_o$, there exists some positive number t depending upon $\mathcal{O}$ such that $E \subset t\mathcal{O}$.

In view of the structure (12.1)–(12.1)$'$ of the open sets of $\mathcal{B}_o$, a set $E \subset X$ is bounded in the weak topology of X if and only if for every $T \in X^*$, there exists a positive number γ_T such that

$$|T(x)| \leq \gamma_T \quad \text{for all } x \in E.$$

Proposition 12.1. *Let X be infinite dimensional. Then every weak neighborhood of the origin contains an infinite-dimensional subspace X_o.*

Proof. Having fixed a weak neighborhood of the origin, we may assume is of the form (12.1)–(12.1)$'$ and set

$$X_o = \bigcap_{j=1}^{n} \ker\{T_j\}.$$

This is a subspace of X and $X_o \subset \mathcal{O}$. To prove that X_o is infinite dimensional, consider the map $F : X \to \mathbb{R}^n$ defined by

$$X \ni x \to F(x) = (T_1(x), T_2(x), \ldots, T_n(x)) \in \mathbb{R}^n.$$

Such a map is linear and continuous and its kernel is X_o. It is also a one-to-one map between the quotient space X/X_o and $\mathbb{R}^n$. Thus $\dim\{X\} \leq n + \dim\{X_o\}$. $\quad\square$

---

[18] See Section 9.1 of Chapter V.

Corollary 12.2. *Let X be infinite dimensional. Then every weak neighborhood of the origin is unbounded.*

In particular, a ball B_ρ open in the strong topology of $\{X; \|\cdot\|\}$ cannot contain any weak neighborhood of the origin.

12.1 Weakly and strongly closed convex sets. Sets that are closed in the weak topology are also closed in the strong topology. The converse, while false in general, holds for convex sets.

Proposition 12.3 (Mazur[19]). *Let E be a convex subset of a normed space $\{X; \|\cdot\|\}$. Then the weak closure of E coincides with its strong closure.*

Proof. Denote by $\overline{E}_w$ and $\overline{E}_s$, respectively, the closure of E in the weak and strong topologies. Since $\mathcal{W}$ is weaker than the strong topology, $\overline{E}_s \subset \overline{E}_w$. For the converse inclusion, it suffices to show that $X - \overline{E}_s$ is a weakly open set. This, in turn, would follow if every point $x_o \in X - \overline{E}_s$ admits a weakly open neighborhood not intersecting $\overline{E}_s$.

Having fixed $x_o \in X - \overline{E}_s$, consider the two disjoint, closed, convex sets x_o and $\overline{E}_s$. Since x_o is compact, by Proposition 11.3, there exists $T \in X^*$ such that

$$T(x_o) < \alpha < T(x) \quad \text{for all } x \in \overline{E}_s.$$

Therefore, the half-space $[T < \alpha]$ is a weakly open neighborhood of x_o that does not intersect $\overline{E}_s$. $\qquad \square$

Corollary 12.4. *Let $\{X; \|\cdot\|\}$ be a normed space. Then any weakly closed subspace $X_o \subset X$ is also strongly closed.*

Corollary 12.5. *Let $\{X; \|\cdot\|\}$ be a normed space, and let $\{x_n\}$ be a sequence of elements of X converging weakly to some $x \in X$. Then there exists a sequence $\{y_m\}$ of elements of X such that each y_m is the convex combination of finitely many x_n; i.e.,*

$$y_m = \sum_{j=1}^{n_m} \alpha_j x_{n_j}, \quad \text{where } \alpha_j > 0 \quad \text{and} \quad \sum_{j=1}^{n_m} \alpha_j = 1,$$

and $\{y_m\} \to x$ strongly.[20]

Proof. Let $c(\{x_n\})$ be the convex hull of $\{x_n\}$, and denote by $\overline{c(\{x_n\})}_w$ its weak closure. By assumption, the weak limit x belongs to $\overline{c(\{x_n\})}_w$. The conclusion follows since weak and strong closures coincide. $\qquad \square$

---

[19]S. Mazur, Über konvexe Mengen in linearen normierten Räumen, *Stud. Math.*, **4** (1933), 70–84. This reference contains Corollaries 12.4 and 12.5 as well.

[20]In the context of $L^p(E)$ spaces, this corollary has been established in S. Banach and S. Saks, Sur la convergence forte dans les espaces L^p. *Stud. Math.*, **2** (1930), 51–57. When $p = 2$, the coefficients α_j can be given an elegant form (see Section 12.1 of the Problems and Complements).

13 Reflexive Banach spaces

Let $\{X; \|\cdot\|\}$ be a normed space. By Corollary 3.3, its dual X^* endowed with the norm (3.1) is itself a Banach space.

The collection of all bounded linear functionals $f : X^* \to \mathbb{R}$ is denoted by X^{**} and is called the *double dual* of the *second dual* of X. It is itself a Banach space by the norm

$$\|f\| = \sup_{\substack{T \in X^* \\ T \neq 0}} \frac{|f(T)|}{\|T\|} = \sup_{\substack{T \in X^* \\ \|T\|=1}} |f(T)|. \tag{13.1}$$

Every element $x \in X$ identifies an element $f_x \in X^{**}$ by the formula

$$X^* \ni T \longrightarrow f_x(T) = T(x). \tag{13.2}$$

Let $\widetilde{X}$ denote the collection of all such functionals; i.e.,

$$\widetilde{X} = \left\{ \begin{array}{c} \text{the collection of all functionals } f_x \in X^{**} \\ \text{of the form (13.2) as } x \text{ ranges over } X \end{array} \right\}.$$

From Corollary 10.4 and (13.1), it follows that $\|f_x\| = \|x\|$. Therefore, the injection map

$$X \ni x \longrightarrow f_x \in \widetilde{X} \subset X^{**} \tag{13.3}$$

is an isometric isomorphism between X and $\widetilde{X}$.

In general, not all of the bounded linear functionals $f : X^* \to \mathbb{R}$ are derived from the injection map (13.3); in other words, the inclusion $\widetilde{X} \subset X^{**}$ is, in general, strict.[21]

A Banach space $\{X; \|\cdot\|\}$ is *reflexive* if $\widetilde{X} = X^{**}$, i.e., if all the bounded linear functionals $f \in X^{**}$ are derived from the injection map (13.3).

In such a case, $X = X^{**}$ up to the isometric isomorphism in (13.3).

By the Riesz representation theorem, the spaces $L^p(E)$ are reflexive for all $1 < p < \infty$.

The spaces $L^1(E)$ and $L^\infty(E)$ are not reflexive since the dual of $L^\infty(E)$ is strictly larger than $L^1(E)$.[22]

Also, the spaces ℓ_p are reflexive for all $1 < p < \infty$. The spaces ℓ_1 and ℓ_∞ are not reflexive.

Proposition 13.1. *Let X_o be a closed, linear, proper subspace of a reflexive Banach space $\{X; \|\cdot\|\}$. Then X_o is reflexive.*

Remark 13.1. The assumption that X_o is closed is essential. Indeed, if $E \subset \mathbb{R}^N$ is bounded and Lebesgue measurable, $L^\infty(E)$ is a nonreflexive linear subspace of $L^p(E)$ for all $1 \leq p < \infty$.

---

[21] The double dual of $L^\infty(E)$ is strictly larger than $L^\infty(E)$. See Section 9.2 of the Problems and Complements.

[22] Section 9.2 of the Problems and Complements.

Proof of Proposition 13.1. By Proposition 10.1, every $x_o^* \in X_o^*$ can be regarded as the restriction to X_o of some $x^* \in X^*$. Now fix $f_o \in X_o^{**}$, and for all $x^* \in X^*$, set

$$f'(x^*) = f_o(x^*|_{x_o}).$$

One verifies that this is a bounded, linear functional in X^*. Since X is reflexive, there exists some $x_o \in X$ such that $f' = f_{x_o}$ by the injection map (13.3).

To establish the proposition, it suffices to show that $x_o \in X_o$. If not, there exists $T \in X^*$ such that $T(x) = 0$ for all $x \in X_o$ and $T(x_o) \neq 0$. Therefore, such a T, when restricted to X_o is the zero element of X_o^*. Then

$$0 \neq T(x_o) = f_{x_o}(T) = f_o(T|_{X_o}) = 0.$$

The contradiction proves the assertion.   □

The following statements are a consequence of the definitions modulo isometric isomorphisms.

Proposition 13.2. *A reflexive normed space is weakly complete.*

If $\{X; \| \cdot \|_X\}$ and $\{Y; \| \cdot \|_Y\}$ are isometrically isomorphic Banach spaces, X is reflexive if and only if Y is reflexive.

A Banach space $\{X; \| \cdot \|\}$ is reflexive if and only if X^* is reflexive.

14   Weak compactness

Let $\{X; \| \cdot \|\}$ be a normed space, and let X^* be its dual. A subset $E \subset X$ is weakly closed if $X - E$ is weakly open. From the construction of the weak topology on X and the notion of weak convergence, it follows that E is closed if and only if every weakly convergent sequence $\{x_n\}$ of elements of E converges weakly to an element $x \in E$.

A set $E \subset X$ is weakly bounded if and only if for every $T \in X^*$, there exists a constant γ_T such that

$$|T(x)| \leq \gamma_T \quad \text{for all } x \in E. \tag{14.1}$$

Proposition 14.1. *A set $E \subset X$ is weakly bounded if and only if it is strongly bounded.*

Proof. If E is strongly bounded, there exists a constant R such that $\|x\| \leq R$ for all $x \in E$. Then for all $T \in X^*$,

$$|T(x)| \leq \|T\|\|x\| \leq \|T\|R.$$

Thus E is weakly bounded. Now assume that E is weakly bounded so that (14.1) holds. Let f_x be the injection map (13.3). Then for each fixed $T \in X^*$, (14.1) takes the form

$$|f_x(T)| \leq \gamma_T \quad \text{for all } x \in E.$$

The family $\{f_x\}$ for $x \in E$ is a collection of bounded linear maps from the Banach space X^* into $\mathbb{R}$, which are pointwise uniformly bounded in X^*. Therefore, by Proposition 6.1, they are equiuniformly bounded; i.e., there is a positive constant C such that

$$|T(x)| \le C \quad \text{for all } x \in E \quad \text{and} \quad \text{all } T \in X^*.$$

Thus $\|x\| \le C$ for all $x \in E$. $\square$

Corollary 14.2. *Let E be a weakly compact subset of a normed space $\{X; \|\cdot\|\}$. Then E is strongly bounded.*

Proof. If $\mathcal{O}$ is a weakly open neighborhood of the origin, the collection $\{n\mathcal{O}\}$ is a weakly open covering for E since $X = \bigcup n\mathcal{O}$. Therefore, $E \subset t\mathcal{O}$ for some $t > 0$. Thus E is weakly bounded and hence strongly bounded. $\square$

14.1 Weak sequential compactness.

Corollary 14.3. *Let $\{X; \|\cdot\|\}$ be a normed space, and let $\{x_n\}$ be a sequence of elements of X weakly convergent to some $x \in X$. There exists a positive constant C such that $\|x_n\| \le C$ for all n. Moreover,*

$$\|x\| \le \liminf \|x_n\|; \tag{14.2}$$

i.e., in a normed linear space, the norm $\|\cdot\| : X \to \mathbb{R}$ is a weakly lower semicontinuous function.[23]

Proof. The uniform upper bound of $\|x_n\|$ follows from Proposition 14.1. For all $T \in X^*$, by the definition of a weak limit,

$$|T(x)| \le \liminf |T(x_n)| \le \|T\| \liminf \|x_n\|.$$

The conclusion now follows from Corollary 10.4. $\square$

Proposition 14.4. *Let $\{X; \|\cdot\|\}$ be a reflexive Banach space. Then every bounded sequence $\{x_n\}$ of elements of X contains a weakly convergent subsequence $\{x_{n'}\}$.*[24]

Proof. Let X_o be the closed linear span of $\{x_n\}$. Such a subspace is separable since the finite linear combinations of elements of $\{x_n\}$ with rational coefficients is a countable dense subset.

Since X_o is reflexive, its double-dual X_o^{**}, being isometrically isomorphic to X_o, is also separable. Then by Proposition 10.7, also X_o^* is separable.

Let $\{T_n\}$ be a countable, dense subset of X_o^*. The sequence $\{T_1(x_n)\}$ is bounded in $\mathbb{R}$, and we may extract a convergent subsequence $\{T_1(x_{n_1})\}$. The sequence $\{T_2(x_{n_1})\}$ is bounded in $\mathbb{R}$, and we may extract a convergent subsequence $\{T_2(x_{n_2})\}$.

---

[23]Compare with Proposition 10.1 of Chapter V.
[24]Compare with Proposition 19.1 of Chapter V.

Proceeding in this fashion, at the kth step, we extract a subsequence $\{x_{n_k}\}$ such that

$$\{T_j(x_{n_k})\} \text{ is convergent} \quad \text{for all } j = 1, 2, \ldots k.$$

The diagonal sequence $\{x_{n'}\} = \{x_{n_n}\}$ is such that

$$\{T_j(x_{n'})\} \text{ is convergent} \quad \text{for all } j \in \mathbb{N}.$$

Since $\{T_n\}$ is dense in X_o^*, the sequences $\{T(x_{n'})\}$ are convergent for all $T \in X_o^*$. Every $T \in X_o^*$ can be regarded as the restriction to X_o of some element of X^*. Therefore, $\{T(x_{n'})\}$ is convergent for all $T \in X^*$. In particular, for all $T \in X^*$ there exists $\alpha_T \in \mathbb{R}$ such that

$$\lim T(x_{n'}) = \alpha_T.$$

By the identification map (13.3), each $x_{n'}$ identifies a functional $f_{x_{n'}} \in X^{**}$. Therefore, the previous limit can be rewritten as

$$\lim f_{x_{n'}}(T) = \alpha_T \quad \text{for all } T \in X^*.$$

This process identifies an element $h \in X^{**}$ by the formula

$$h(T) = \lim f_{x_{n'}}(T) \quad \text{for all } T \in X^*.$$

Since X is reflexive, there exists $x \in X$ such that $h = f_x$. Now we claim that $\{x_{n'}\} \to x$ weakly in X. Indeed, for any fixed $T \in X^*$,

$$\lim T(x_{n'}) = \lim f_{x_{n'}}(T) = f_x(T) = T(x). \qquad \square$$

Corollary 14.5. *Let* $\{X; \|\cdot\|\}$ *be a reflexive Banach space. A subset* $C \subset X$ *is weakly sequentially compact if and only if it is both bounded and weakly closed.*

As an example, consider the space $L^p(E)$, where E is a Lebesgue-measurable subset of $\mathbb{R}^N$ and $1 < p < \infty$. By Proposition 14.4 and Corollary 14.3, the unit ball $\|f\|_p \leq 1$ is weakly sequentially compact. However, the unit sphere $\{\|f\|_p = 1\}$ is not weakly sequentially compact. For example, the unit sphere of $L^2[0, 2\pi]$ is bounded but not weakly closed and therefore is not sequentially compact (see Section 9.1 of Chapter V).

The unit ball of $L^p(E)$ for $1 < p < \infty$ is not sequentially compact in the strong topology of $L^p(E)$ since it does not satify the necessary and sufficient conditions for compactness given in Section 22 of Chapter V. Also, compare with Proposition 2.3.

15 The weak* topology

The dual X^* of a normed space $\{X; \|\cdot\|\}$ is a Banach space and as such can be endowed with the corresponding weak topology, i.e., the weakest topology for which all the elements of X^{**} are continuous. The elements of X^{**} are continuous with respect to the norm topology of X^*. Then one weakens the topology of X^* so as to keep the continuity of all elements of X^*.

The weak* topology on X^* is the weakest topology that renders continuous all the functionals $f_x \in X^{**}$ of the form (13.3), i.e., those that are in a natural one-to-one correspondence with the elements of X. If $\{X; \|\cdot\|\}$ is a reflexive Banach space, then $X = X^{**}$ up to an isometric isomorphism, and the weak topology of X^* coincides with its weak* topology.

The collection $\mathcal{W}^*$ of weak* open sets in X^* is constructed starting from the base

$$\mathcal{B}^* = \left\{ \begin{array}{c} \text{the collection of the finite intersections of the inverse images} \\ f_x^{-1}(O), \text{ where } x \in X \text{ and } O \text{ are open subsets of } \mathbb{R} \end{array} \right\}.$$

One verifies that the operations of sum and multiplication by scalars,

$$+ : X^* \times X^* \longrightarrow X^*, \qquad \bullet : \mathbb{R} \times X^* \longrightarrow X^*,$$

are continuous with respect to the topology of $\mathcal{W}^*$. Thus $\{X^*; \mathcal{W}^*\}$ is a topological vector space. The topology of $\mathcal{W}^*$ is translation invariant, and it is determined by a local base at the origin of X^*. A weak* open neighborhood of the origin contains an element of $\mathcal{B}^*$ of the form

$$\mathcal{O}^* = \bigcap_{j=1}^{n} f_{x_j}^{-1}(-\alpha_j, \alpha_j) \tag{15.1}$$

for some finite n and $\alpha_j > 0$. Equivalently,

$$\mathcal{O}^* = \{T \in X^* \mid |T(x_j)| < \alpha_j \text{ for } j = 1, \ldots, n\}. \tag{15.1}'$$

Since these open sets are convex, $\mathcal{W}^*$ is a locally convex topology.

A sequence $\{T_n\}$ of elements of X^* converges weakly* to 0 if and only if every weak neighborhood of the origin contains all but finitely many elements of $\{T_n\}$. From (15.1)', it follows that $\{T_n\} \to 0$ if and only if $\{T_n(x)\} \to 0$ for all $x \in X$. More generally, $\{T_n\} \to T_o$ if and only if $\{T_n(x)\} \to T_o(x)$ for all $x \in X$.

Weak convergence implies weak* convergence. The converse is false. Thus, in general, the weak* topology $\mathcal{W}^*$ contains, roughly speaking, fewer open sets than those of the weak topology generated by X^{**}.

16   The Alaoglu theorem

Theorem 16.1 (Alaoglu[25]). *Let $\{X; \|\cdot\|\}$ be a normed space and let X^* denote its dual. The closed unit ball in X^*, i.e.,*

$$B^* = \{T \in X^* \mid \|T\| \leq 1\},$$

is weak compact.*

---

[25]L. Alaoglu, Weak topologies of normed linear spaces. *Ann. Math.*, **41** (1940), 252–267.

Proof. If $T \in B^*$, then $T(x) \in [-\|x\|, \|x\|]$ for all $x \in X$. Consider the Cartesian product

$$P = \prod_{x \in X} [-\|x\|, \|x\|].$$

A point in P is a function $f : X \to \mathbb{R}$ such that $f(x) \in [-\|x\|, \|x\|]$, and P is the collection of all such functions. The set B^* is a subset of P and as such inherits the product topology of P. On the other hand, as a subset of X^* it also inherits the weak* topology of X^*.

Lemma 16.2. *These two topologies coincide on B^*.*

Proof. Every weak* open neighborhood of a point $T_o \in X^*$ contains an open set of the form

$$\mathcal{O} = \left\{ \begin{array}{l} T \in X^* \mid |T(x_j) - T_o(x_j)| < \delta \text{ for some } \delta > 0 \\ \text{for finitely many } x_j, \, j = 1, 2, \ldots, n \end{array} \right\}.$$

Likewise, every neighborhood of a point $T_o \in P$ open in the product topology of P contains an open set of the form

$$\mathcal{V} = \left\{ \begin{array}{l} f \in P \mid |f(x_j) - T_o(x_j)| < \delta \text{ for some } \delta > 0 \\ \text{for finitely many } x_j, \, j = 1, 2, \ldots, n \end{array} \right\}.$$

These open sets form a base for the corresponding topologies. Since $B^* = P \cap X^*$,

$$\mathcal{O} \cap B^* = \mathcal{V} \cap B^*.$$

These intersections form a base for the corresponding relative topologies inherited by B^*. Therefore, the weak* topology and the weak topology coincide on B^*. $\square$

Lemma 16.3. *B^* is closed in its relative product topology.*

Proof. Let f_o be in the closure of B^* in the relative product topology. Fix $x, y \in X$ and $\alpha, \beta \in \mathbb{R}$ and consider the three points

$$x_1 = x, \qquad x_2 = y, \qquad x_3 = \alpha x + \beta y.$$

For $\varepsilon > 0$, the sets

$$V_\varepsilon = \{ f \in P \mid |f(x_j) - f_o(x_j)| < \varepsilon \text{ for } j = 1, 2, 3 \}$$

are open neighborhoods of f_o. Since they intersect B^*, there exists $T \in B^*$ such that

$$|f_o(x) - T(x)| < \varepsilon, \qquad |f_o(y) - T(y)| < \varepsilon,$$

and since T is linear,

$$|f_o(\alpha x + \beta y) - \alpha T(x) - \beta T(y)| < \varepsilon.$$

From this,

$$|f_o(\alpha x + \beta y) - \alpha f(x) - \beta f(y)| < (1 + |\alpha| + |\beta|)\varepsilon$$

for all $\varepsilon > 0$. Thus f_o is linear, and it belongs to B^*. $\square$

Proof of Theorem 16.1, concluded. Each $[-\|x\|, \|x\|]$ as a bounded, closed interval in $\mathbb{R}$ equipped with its Euclidean topology, is compact. Therefore, by Tychonov's theorem, P is compact in its product topology. Then B^*, as a closed subset of a compact space, is compact in its relative product topology and hence in its relative weak* topology. $\square$

Corollary 16.4. *Let* $\{X; \|\cdot\|\}$ *be a reflexive Banach space. Then its unit ball is weakly compact.*

Corollary 16.5. *Let* $\{X; \|\cdot\|\}$ *be a reflexive Banach space. Then a subset of the unit ball of X is weakly compact if and only if it is weakly closed.*

Remark 16.1. The weak* compactness of the unit ball B^* of X^* does not imply that B^* is compact in the norm topology of X^*. As an example, let $X = L^2(E)$. Then

$$L^2(E) = L^2(E)^* = L^2(E)^{**}$$

up to isometric isomorphisms. The unit ball of $L^2(E)^*$ is weak* compact but not compact in the strong norm topology.

Remark 16.2. The weak* compactness of the unit ball of $\{\|x\| \leq 1\}$ of a normed space does not imply that the unit sphere $\{\|x\| = 1\}$ is weak* compact. For example, the unit sphere of $L^2[0, 2\pi]$ is bounded but not weak* closed, and therefore it is not weak* compact.

17   Hilbert spaces

Let X be a vector space over $\mathbb{R}$. A *scalar* or *inner* product on X over $\mathbb{R}$ is a function $\langle \cdot, \cdot \rangle : X \times X \to \mathbb{R}$ satisfying the following:[26]

(i) $\langle x, y \rangle = \langle y, x \rangle$ for all $x, y \in X$.

(ii) $\langle x_1 + x_2, y \rangle = \langle x_1, y \rangle + \langle x_2, y \rangle$ for all $x_1, x_2, y \in X$.

(iii) $\langle \lambda x, y \rangle = \lambda \langle x, y \rangle$ for all $x, y \in X$ and all $\lambda \in \mathbb{R}$.

(iv) $\langle x, x \rangle \geq 0$ for all $x \in X$.

---

[26]If X is a vector space over the field $\mathbf{C}$ of the complex numbers, the requirements for $\langle \cdot, \cdot \rangle$ to be an inner product are analogous, except that (i) now becomes $\langle x, y \rangle = \overline{\langle y, x \rangle}$. We will limit ourselves to vector spaces X over $\mathbb{R}$.

(v) $\langle x, x \rangle = 0$ if and only if $x = \Theta$.

A vector space X equipped with a scalar product $\langle \cdot, \cdot \rangle$ is called a *pre-Hilbert* space. Set

$$\langle x, x \rangle = \|x\|^2. \tag{17.1}$$

Using properties (i)–(v) of a scalar product, one verifies that the function $\| \cdot \| : X \to \mathbb{R}$ defines a norm in X. Therefore, a pre-Hilbert space is a normed space $\{X; \| \cdot \|\}$ by the norm in (17.1).

17.1 The Schwarz inequality. Let X be a pre-Hilbert space for a scalar product $\langle \cdot, \cdot \rangle$. Then for all $x, y \in X$,

$$\langle x, y \rangle \le \|x\| \|y\|, \tag{17.2}$$

and equality holds if and only if $x = \lambda y$ for some $\lambda \in \mathbb{R}$. Indeed, for all $x, y \in X$ and $\lambda \in \mathbb{R}$,

$$0 \le \|x - \lambda y\|^2 = \langle x - \lambda y, x - \lambda y \rangle,$$
$$= \|x\|^2 - 2\lambda \langle x, y \rangle + \lambda^2 \|y\|^2.$$

From this,

$$2\lambda \langle x, y \rangle \le \|x\|^2 + \lambda^2 \|y\|^2.$$

Inequality (17.2) is trivial if $y = \Theta$. Otherwise, choose $\lambda = \|x\|/\|y\|$.

17.2 The parallelogram identity. Let X be a pre-Hilbert space for a scalar product $\langle \cdot, \cdot \rangle$. Then for all $x, y \in X$,[27]

$$\|x + y\|^2 + \|x - y\|^2 = 2(\|x\|^2 + \|y\|^2). \tag{17.3}$$

From the properties of a scalar product,

$$\|x + y\|^2 = \|x\|^2 + 2\langle x, y \rangle + \|y\|^2,$$
$$\|x - y\|^2 = \|x\|^2 - 2\langle x, y \rangle + \|y\|^2.$$

Adding these identities yields (17.3).[28]

A Hilbert space is a pre-Hilbert space that is complete with respect to the topology generated by the norm (17.1). Equivalently, a Hilbert space is a Banach space whose norm is generated by an inner product $\langle \cdot, \cdot \rangle$.

---

[27] If X is $\mathbb{R}^2$ with the Euclidean norm, any two elements $x, y \in \mathbb{R}^2$ can be regarded up to a translation as the sides of a parallelogram whose diagonals are $(x + y)$ and $(x - y)$. In such a case, (17.3) reduces to the law of parallelograms in elementary plane geometry, which, in turn, lends its name to (17.3) in general pre-Hilbert spaces.

[28] See Section 17.1 of the Problems and Complements.

Let $\{X, \mathcal{A}, \mu\}$ be a measure space and let $E \in \mathcal{A}$. Then $L^2(E)$ is a Hilbert space for the scalar product

$$\langle f, g \rangle = \int_E fg d\mu \quad \text{for all } f, g \in L^2(E).$$

The space of sequences of real numbers in ℓ_2 is a Hilbert space for the inner product[29]

$$\langle \mathbf{a}, \mathbf{b} \rangle = \sum a_i b_i \quad \text{for all } \mathbf{a}, \mathbf{b} \in \ell_2.$$

We will denote by H a Hilbert space for the inner product $\langle \cdot, \cdot \rangle$.

18 Orthogonal sets, representations, and functionals

Two elements $x, y \in H$ are said to be *orthogonal* if $\langle x, y \rangle = 0$, and in such a case we write $x \perp y$. In $L^2(0, 2\pi)$, the two elements $t \to \sin t, \cos t$ are orthogonal.

An element $x \in H$ is said to be orthogonal to a set $H_o \subset H$ if $x \perp y$ for all $y \in H_o$, and in such a case we write $x \perp H_o$.

Proposition 18.1 (Riesz[30]). *Let H_o be a closed, convex, proper subset of H. Then for every $x \in H - H_o$, there exists a unique $x_o \in H$ such that*

$$\inf_{y \in H_o} \|x - y\| = \|x_o - x\|. \tag{18.1}$$

Proof. Let $\{y_n\}$ be a sequence in H_o such that

$$\delta \overset{\text{def}}{=} \inf_{y \in H_o} \|x - y\| = \lim \|y_n - x\|. \tag{18.2}$$

Since H_o is convex for any two elements $y_n, y_m \in \{y_n\}$,

$$\frac{y_n + y_m}{2} \in H_o.$$

Therefore, applying the parallelogram identity

$$\begin{aligned}
\|y_n - y_m\|^2 &= \|(y_n - x) + (x - y_m)\|^2 \\
&= 2\|y_n - x\|^2 + 2\|y_m - x\|^2 - \|y_n + y_m - 2x\|^2 \\
&= 2\|y_n - x\|^2 + 2\|y_m - x\|^2 - 4\left\|\frac{y_n + y_m}{2} - x\right\|^2 \\
&\leq 2\|y_n - x\|^2 + 2\|y_m - x\|^2 - 4\delta^2.
\end{aligned}$$

---

[29] See Section 1.1 of the Problems and Complements of Chapter V.

[30] F. Riesz, Zur Theorie des Hilbertschen Raumes, *Acta Sci. Math.* (*Szeged*), **7** (1934), 34–38.

Therefore, $\{y_n\}$ is a Cauchy sequence, and since H_o is closed, it converges to some $x_o \in H_o$, which satisfies (18.2).

If x_o and x_o' both satisfy (18.1), then by the parallelogram identity,

$$\|x_o' - x_o\|^2 = 2\|x_o' - x\|^2 + 2\|x_o - x\|^2$$
$$- 4\left\|\frac{x_o + x_o'}{2} - x\right\|^2 \leq 0. \qquad \square$$

Let H_o be a subset of H. The orthogonal complement $H_o^{\perp}$ of H_o is defined as the collection of all $x \in H$ such that $x \perp H_o$.

Proposition 18.2. *Let H_o be a closed, proper subspace of H. Then $H = H_o \oplus H_o^{\perp}$; i.e., every $x \in H$ can be represented in a unique way as*

$$x = x_o + \eta \quad \text{for some } x_o \in H_o \quad \text{and} \quad \eta \in H_o^{\perp}. \qquad (18.3)$$

Proof. If $x \in H_o$, it suffices to take $x_o = x$ and $y = \Theta$. If $x \in H - H_o$, let x_o be the unique element claimed by Proposition 18.1, and let δ be defined as in (18.2). Now set

$$x = x_o + \eta, \quad \text{where } \eta = x - x_o.$$

To prove that $\eta \perp H_o$, fix any $y \in H_o$ and consider the function

$$\mathbb{R} \ni t \longmapsto \|\eta + ty\|^2 = \|\eta\|^2 + 2t\langle y, \eta \rangle + t^2\|y\|^2.$$

Such a function takes its minimum for $t = 0$. Indeed, $h(0) = \delta^2$ and

$$\inf_{t \in \mathbb{R}} \|\eta - ty\|^2 = \inf_{t \in \mathbb{R}} \|x - (x_o + ty)\|^2$$
$$\geq \inf_{y \in H_o} \|x - y\|^2 = \delta^2.$$

Therefore, $h'(0) = 0$, i.e., $\langle \eta, y \rangle = 0$ for all $y \in H_o$.

If the representation (18.3) were not unique, there would exist $x_o' \neq x_o$ and $\eta' \neq \eta$ such that

$$x = x_o' + \eta', \qquad x_o' \in H_o, \quad \text{and} \quad \eta' \in H_o^{\perp}. \qquad (18.3)'$$

Then by difference,

$$x_o - x_o' \in H_o, \qquad \eta - \eta' \in H_o^{\perp}, \quad \text{and} \quad x_o - x_o' = \eta' - \eta.$$

Thus $x_o - x_o'$ and $\eta - \eta'$ are perpendicular to themselves, and therefore both must be equal to Θ. $\qquad \square$

18.1   Bounded linear functionals on H. Every $y \in H$ identifies a bounded linear functional $T_y \in H^*$ by the formula

$$T_y(x) = \langle y, x \rangle \quad \text{for all } x \in H. \tag{18.4}$$

Moreover, $\|T_y\| = \|y\|$. The next proposition asserts that these are the only bounded, linear functionals on H.

Proposition 18.3. *For every $T \in H^*$, there exists a unique $y \in H$ such that T can be represented as in (18.4).*

Proof. The conclusion is trivial if $T \equiv 0$. If $T \not\equiv 0$, its kernel H_o is a closed, proper subspace of H. Select a nontrivial element $\eta \in H - H_o$ and observe that for all $x \in H$,

$$T(\eta)x - T(x)\eta \in H_o;$$

i.e.,

$$T(x)\eta = T(\eta)x + \eta_o \quad \text{for some } \eta_o \in H_o.$$

From this, by taking the inner product of both sides by η,

$$T(x) = \langle y, x \rangle, \quad \text{where } y = T(\eta)\frac{\eta}{\|\eta\|^2}.$$

Since $x \in H$ is arbitrary, this implies that also $\|T\| = \|y\|$.

If y_1 and y_2 were to identify the same functional T,[31]

$$\langle y_1 - y_2, x \rangle = 0 \quad \text{for all } x \in H.$$

Thus $\|y_1 - y_2\| = 0$. $\square$

19   Orthonormal systems

A set S of elements of H is said to be orthogonal if any two distinct elements x and y of S are orthogonal. The set S is *orthonormal* if it is orthogonal and all its elements have norm 1. In such a case, S is called and orthonormal system. In $\mathbb{R}^N$ with its Euclidean norm, an orthonormal system is given by any n-tuple of mutually orthogonal unit vectors.

In $L^2(0, 2\pi)$, an orthonormal system is given by

$$\frac{1}{\sqrt{2\pi}}, \quad \frac{1}{\sqrt{\pi}}\cos t, \quad \frac{1}{\sqrt{\pi}}\cos 2t, \quad \frac{1}{\sqrt{\pi}}\cos 3t, \quad \ldots, \\ \frac{1}{\sqrt{\pi}}\sin t, \quad \frac{1}{\sqrt{\pi}}\sin 2t, \quad \frac{1}{\sqrt{\pi}}\sin 3t, \quad \ldots. \tag{19.1}$$

---

[31] Roughly speaking, T is identified by the unique "direction" orthogonal to the null space of T. Compare with Proposition 5.1.

In ℓ_2, an orthonormal system is given by

$$\begin{aligned}
\mathbf{e}_1 &= \{1, 0, 0, \ldots, 0_m, 0, \ldots\}, \\
\mathbf{e}_2 &= \{0, 1, 0, \ldots, 0_m, 0, \ldots\}
\end{aligned}$$
$$\cdots$$
$$\mathbf{e}_m = \{0, 0, 0, \ldots, 1_m, 0, \ldots\},$$
$$\cdots = \cdots.$$

(19.2)

Lemma 19.1. *Let S be an orthonormal system in H. Any two elements x and y in S are at mutual distance $\sqrt{2}$; i.e.,*

$$\|x - y\| = \sqrt{2} \quad \text{for all } x, y \in S, \quad x \neq y. \tag{19.3}$$

Proof. For any $x, y \in S$ and $x \neq y$, compute

$$\|x - y\|^2 = \langle x - y, x - y \rangle = \|x\|^2 - 2\langle x, y \rangle + \|y\|^2.$$

The conclusion follows since $\|x\| = \|y\| = 1$. $\qquad\qquad\qquad\qquad\qquad\square$

19.1 The Bessel inequality.

Proposition 19.2. *Let H be a Hilbert space, and let S be an orthonormal system in H. Then for any n-tuple $\{\mathbf{u}_1, \mathbf{u}_2, \ldots, \mathbf{u}_n\}$ of elements of S,*[32]

$$\sum_{i=1}^{n} \langle \mathbf{u}_i, x \rangle^2 \leq \|x\|^2 \quad \text{for all } x \in H. \tag{19.4}$$

Moreover, for any fixed $x \in H$, the inner product $\langle \mathbf{u}, x \rangle$ vanishes except for at most countably many $\mathbf{u} \in S$, and

$$\sum_{\mathbf{u} \in S} \langle \mathbf{u}, x \rangle^2 \leq \|x\|^2 \quad \text{for all } x \in H, \tag{19.5}$$

Proof. For any such n-tuple and any $x \in H$,

$$\begin{aligned}
0 &\leq \left\| x - \sum_{i=1}^{n} \langle \mathbf{u}_i, x \rangle \mathbf{u}_i \right\|^2 \\
&= \left\langle x - \sum_{i=1}^{n} \langle \mathbf{u}_i, x \rangle \mathbf{u}_i, \, x - \sum_{i=1}^{n} \langle \mathbf{u}_i, x \rangle \mathbf{u}_i \right\rangle \\
&= \|x\|^2 - \sum_{i=1}^{n} \langle \mathbf{u}_i, x \rangle^2.
\end{aligned}$$

---

[32] S need not be countable.

This establishes (19.4). To prove (19.5), fix $x \in H$ and observe that for any $m \in \mathbb{N}$, the set

$$\left\{ \mathbf{u} \in S \text{ such that } \frac{1}{2^{m+1}} \|x\|^2 \leq \langle \mathbf{u}, x \rangle^2 < \frac{1}{2^m} \|x\|^2 \right\}$$

contains at most finitely many elements. Therefore, the collection of those $\mathbf{u} \in S$ such that $\langle \mathbf{u}, x \rangle \neq 0$ is countable, and (19.5) holds.   □

19.2 Separable Hilbert spaces.

Proposition 19.3. *Let H be a separable Hilbert space. Then any orthonormal system S is H is countable.*

Proof. Let H_o be a countable subset of H dense in H, and let S be an orthonormal system in H. For every $\mathbf{u} \in S$, there exists $x(\mathbf{u}) \in H_o$ such that

$$\|\mathbf{u} - x(\mathbf{u})\| < \frac{\sqrt{2}}{3}. \tag{19.6}$$

If $\mathbf{u}_1$ and $\mathbf{u}_2$ are distinct elements of S, then any two elements $x(\mathbf{u}_1)$ and $x(\mathbf{u}_2)$ in H_o for which (19.6) holds are distinct. Indeed,

$$\begin{aligned}
\sqrt{2} &= \|\mathbf{u}_1 - \mathbf{u}_2\| \\
&\leq \|\mathbf{u}_1 - x(\mathbf{u}_1)\| + \|\mathbf{u}_2 - x(\mathbf{u}_2)\| + \|x(\mathbf{u}_1) - x(\mathbf{u}_2)\| \\
&\leq 2\frac{\sqrt{2}}{3} + \|x(\mathbf{u}_1) - x(\mathbf{u}_2)\|.
\end{aligned}$$

Thus S can be put in a one-to-one correspondence with a subset of H_o.   □

20 Complete orthonormal systems

An orthonormal system S in H is said to be *complete* if

$$\langle x, \mathbf{u} \rangle = 0 \quad \text{for all } \mathbf{u} \in S \quad \text{implies that} \quad x = \Theta. \tag{20.1}$$

The orthonormal system in (19.2) is complete in ℓ_2.[33]

Proposition 20.1. *Let S be a complete orthonormal system in H. Then for every $x \in H$,*

$$x = \sum_{\mathbf{u} \in S} \langle x, \mathbf{u} \rangle \mathbf{u} \quad (\text{representation of } x). \tag{20.2}$$

Moreover,

$$\|x\|^2 = \sum_{\mathbf{u} \in S} |\langle x, \mathbf{u} \rangle|^2 \quad (\text{Parseval's identity}). \tag{20.3}$$

---

[33] It can be shown that the orthonormal system in (19.1) is complete in $L^2(0, 2\pi)$.

Proof. As **u** ranges over S, only countably many of the numbers $\langle x, \mathbf{u} \rangle$ are not zero, and we order them in some fashion $\{\langle x, \mathbf{u}_n \rangle\}$. By the Bessel inequality, the series

$$\sum \langle x, \mathbf{u}_n \rangle^2$$

converges. Therefore, for any two positive integers $n < m$,

$$\left\| \sum_{i=n}^{m} \langle x, \mathbf{u}_i \rangle \mathbf{u}_i \right\|^2 = \sum_{i=n}^{m} \langle x, \mathbf{u}_i \rangle^2 \longrightarrow 0 \quad \text{as } m, n \to \infty.$$

This implies that the sequence

$$\left\{ \sum_{i=1}^{n} \langle x, \mathbf{u}_i \rangle \mathbf{u}_i \right\}$$

is a Cauchy sequence in H and has a limit

$$y = \sum \langle x, \mathbf{u}_i \rangle \mathbf{u}_i = \sum_{\mathbf{u} \in S} \langle x, \mathbf{u} \rangle \mathbf{u}.$$

For any $\mathbf{u} \in S$,

$$\langle x - y, \mathbf{u} \rangle = \lim \left\langle x - \sum_{i=1}^{m} \langle x, \mathbf{u}_i \rangle \mathbf{u}_i, \mathbf{u} \right\rangle = 0.$$

Thus $\langle x - y, \mathbf{u} \rangle = 0$ for all $\mathbf{u} \in S$, and since S is complete, $x = y$.

From (20.2), by taking the inner product with respect to x,

$$\|x\|^2 = \lim \left\langle x, \sum_{i=1}^{n} \langle x, \mathbf{u}_i \rangle \mathbf{u}_i \right\rangle = \lim \sum_{i=1}^{n} \langle x, \mathbf{u}_i \rangle^2. \qquad \square$$

20.1   Equivalent notions of complete systems. Let S be an orthonormal system in H. If (20.3) holds for all $x \in H$, then S is complete. Indeed, if not, there would be an element $x \in H$ such that $\langle x, \mathbf{u} \rangle = 0$ for all $\mathbf{u} \in S$ and $x \neq \Theta$. However, if (20.3) holds, $x = \Theta$.

The proof of Proposition 20.1 shows that the notion (20.1) of a complete system implies (20.2), and this, in turn, implies (20.3). We have just observed that (20.3) implies the notion (20.1) of a complete system. Thus (20.1), (20.2), and (20.3) are equivalent, and each could be taken as a definition of complete system.

20.2   Maximal and complete orthonormal systems. An orthonormal system S in H is *maximal* if it is not properly contained in any other orthonormal system of H. From the definitions, it follows that an orthonormal system S in H is complete if and only if it is maximal.

The family Σ of all orthonormal systems in H is partially ordered by set inclusion. Moreover, every linearly ordered subset $\Sigma' \subset \Sigma$ has an upper bound given by the union of all orthonormal systems in Σ'. Therefore, by Zorn's lemma, H has a maximal orthonormal system.

Zorn's lemma provides an abstract notion of existence of a maximal orthonormal system in H. Of greater interest is the actual construction of a complete system.

20.3   The Gram–Schmidt orthonormalization process. Let $\{x_n\}$ be a countable collection of linearly independent elements of H, and set[34]

$$\mathbf{u}_1 = \frac{x_1}{\|x_1\|} \quad \text{and} \quad \mathbf{u}_{n+1} = \frac{x_{n+1} - \displaystyle\sum_{i=1}^{n} \langle x_{n+1}, \mathbf{u}_i \rangle \mathbf{u}_i}{\left\| x_{n+1} - \displaystyle\sum_{i=1}^{n} \langle x_{n+1}, \mathbf{u}_i \rangle \mathbf{u}_i \right\|}$$

for $n = 1, 2, \ldots$. These are well defined since $\{x_n\}$ are linearly independent. One verifies that $\{\mathbf{u}_n\}$ forms an orthonormal system and $\operatorname{span}\{\mathbf{u}_n\} = \operatorname{span}\{x_n\}$.

If H is separable, this procedure can be used to generate a maximal orthonormal system S in H, independent of Zorn's lemma.

20.4   On the dimension of a separable Hilbert space. If H is separable, any complete orthonormal system S is either finite or infinite countable. Assume first S is infinite countable and index its elements as $\{\mathbf{u}_n\}$. By Parseval's identity, any element $x \in H$ generates an element of ℓ_2 by the formula

$$\{a_n\} = \{\langle x, \mathbf{u}_n \rangle\}.$$

Vice versa, any element $\mathbf{a} \in \ell_2$ generates a unique element $x \in H$ by the formula

$$x = \sum a_i \mathbf{u}_i.$$

Let x and y be two elements in H, and let $\mathbf{a}$ and $\mathbf{b}$ be their corresponding elements in ℓ_2. By the same reasoning,

$$\langle x, y \rangle_H = \langle \mathbf{a}, \mathbf{b} \rangle_{\ell_2}.$$

This implies that the isomorphism between H and ℓ_2 is an isometry.

Thus any separable Hilbert space with an infinite-countable orthonormal system S is isometrically isomorphic to ℓ_2. Equivalently, any complete, infinite-countable, orthonormal system S of an Hilbert space H can be put in a one-to-one correspondence with the system (9.2), which forms a complete orthonormal system of ℓ_2.

We say that the dimension of a separable Hilbert with an infinite-countable orthonormal system is $\aleph_o$, i.e., the cardinality of $\{\mathbf{e}_n\}$. If S is finite, say, for example, $\{\mathbf{u}_1, \mathbf{u}_2, \ldots, \mathbf{u}_N\}$, then by the same procedure, H is isometrically isomorphic to $\mathbb{R}^N$, and its dimension is N.

PROBLEMS AND COMPLEMENTS

1   NORMED SPACES

1.1   Let E be a bounded, open subset of $\mathbb{R}^N$. The space $C(\overline{E})$ endowed with the norm of $L^p(E)$ is not a Banach space.

---

[34]E. Schmidt, Entwicklung willkürlicher Funktionen nach Systemen vorgeschriebener *Math. Ann.*, **63** (1907), 433–476.

1.2   Every normed space is homeomorphic to its open unit ball.

1.3   A normed space $\{X; \|\cdot\|\}$ is complete if and only the intersection of a countable family of nested, closed balls is nonempty.

1.4   The L^2-norm and the sup-norm on $C[0, 1]$ are not equivalent. In particular, $C[0, 1]$ is not complete in $L^p[0, 1]$ for all $1 \le p < \infty$.
   The norms of $L^p(E)$ and $L^q(E)$ for $1 \le q < p < \infty$ are not equivalent.

1.5   The next proposition provides a criterion for a normed space to be a Banach space.

Proposition 1.1c. *A normed space $\{X; \|\cdot\|\}$ is complete if and only if every absolutely convergent series converges to an element of X.*[35]

Proof ($\Longrightarrow$). Let $\sum x_n$ be an absolutely convergent series in $\{X; \|\cdot\|\}$; i.e., $\sum \|x_n\| \le M$ for some positive number M. Then $\{\sum_{j=1}^{n} x_j\}$ is a Cauchy sequence sequence and hence convergent to some $x \in X$. $\square$

Proof ($\Longleftarrow$). Let $\{x_n\}$ be a Cauchy sequence in $\{X; \|\cdot\|\}$. For each positive integer j, there exists an index n_j such that

$$\|x_n - x_m\| \le \frac{1}{2^j} \quad \text{for all } n, m \ge n_j.$$

Starting from the subsequence $\{x_{n_j}\}$, set $y_j = (x_{n_{j+1}} - x_{n_j})$. The series $\sum y_j$ is absolutely convergent, and we let x denote its limit. Thus the subsequence $\{x_{n_j}\}$ converges to x. Since $\{x_n\}$ is a Cauchy sequence, the whole sequence converges to x. $\square$

2   FINITE- AND INFINITE-DIMENSIONAL NORMED SPACES

2.1   An infinite-dimensional Banach space $\{X; \|\cdot\|\}$ cannot have a countable Hamel basis. (*Hint:* Baire's category theorem.)

2.2   Let ℓ_o be the collection of all sequences of real numbers $\{c_n\}$ with only finitely many nonzero elements. There is no norm on ℓ_o by which ℓ_o would be a Banach space.[36]

2.3 $L^p(E)$ and ℓ_p are of infinite dimension for all $1 \le p \le \infty$, and their dimension is larger than $\aleph_o$.

---

[35] In the context of $L^p(E)$, the criterion has been used in the proof of Theorem 7.1 of Chapter V.
[36] See also Section 9.7 of the Problems and Complements of Chapter I.

2.4   Let E be a bounded open set in $\mathbb{R}^N$, and denote by $C^1(\overline{E})$ the set of all continuously differentiable functions $f : \overline{E} \to \mathbb{R}$ with finite norm

$$\|f\| = \sup_{x \in \overline{E}} |f(x)| + \sum_{i=1}^{N} \sup_{x \in \overline{E}} |f_{x_i}(x)|. \tag{2.1c}$$

With this norm, $C^1(\overline{E})$ is a closed linear subspace of $C(\overline{E})$. Moreover, by the Ascoli–Arzelà theorem, the unit ball of $C^1(\overline{E})$ is compact in $C(\overline{E})$. Thus $C^1(\overline{E})$ as a subspace of $C(\overline{E})$ is finite dimensional.

3   LINEAR MAPS AND FUNCTIONALS

3.1   A linear map T from $\{X; \|\cdot\|_X\}$ into $\{Y; \|\cdot\|_Y\}$ is continuous if and only if it maps sequences $\{x_n\}$ converging to Θ_X into bounded sequences of $\{Y; \|\cdot\|_Y\}$.

3.2   Two normed spaces $\{X; \|\cdot\|_1\}$ and $\{X; \|\cdot\|_2\}$ are homeomorphic if and only if there exist positive constants $0 < c_o \le 1 \le c_1$ such that

$$c_o\|x\|_1 \le \|x\|_2 \le c_1\|x\|_1 \quad \text{for all } x \in X.$$

3.3   Any linear functional on a finite-dimensional normed space is continuous.

3.4   A linear map $T : C(\overline{E}) \to \mathbb{R}$ is a positive functional if $T(f) \ge 0$ whenever $f \ge 0$. A positive linear functional on $C(\overline{E})$ is bounded. Thus positivity implies continuity. Moreover, any two of the conditions

  (i) $\|T\| = 1$,

  (ii) $T(1) = 1$,

  (iii) $T \ge 0$

imply the remaining one.

3.5   Let $\{X; \|\cdot\|\}$ be an infinite-dimensional Banach space. There exists a discontinuous, linear map $T : X \to X$.

Having fixed a Hamel basis $\{x_\alpha\}$ for X, after a possible renormalization, we may assume that $\|x_\alpha\| = 1$ for all α.

Every element $x \in X$ can be represented in a unique way as the *finite* linear combination of elements of $\{x_\alpha\}$; i.e., for every $x \in X$, there exists a unique m-tuple of real numbers $\{c_1, c_2, \ldots, c_m\}$ for some $m \in \mathbb{N}$ such that

$$x = \sum_{j=1}^{m} c_j x_{\alpha_j}. \tag{3.1c}$$

Since X is of infinite dimension, the index α ranges over some set A such that $\text{card}(A) \ge \text{card}(\mathbb{N})$. Out of $\{x_\alpha\}$, select a countable collection $\{x_n\}_{n \in \mathbb{N}} \subset \{x_\alpha\}$. Then set

$$T(x_n) = nx_n \quad \text{and} \quad T(x_\alpha) = \Theta \quad \text{if } \alpha \notin \mathbb{N}.$$

For $x \in X$, having determined its representation (3.1c), also set

$$T(x) = \sum_{j=1}^{m} c_j T(x_{\alpha_j}).$$

In view of the uniqueness of the representation (3.1c), this defines a linear map from X into X. Such a map, however, is discontinuous since

$$\|T(x_n)\| = n\|x_n\| \longrightarrow \infty \quad \text{as } n \to \infty.$$

3.6   Let $\{X; \|\cdot\|\}$ be an infinite-dimensional Banach space. There exists a discontinuous, linear functional $T : X \to \mathbb{R}$.

3.7   The conclusion of Section 3.6 of the Problems and Complements is, in general, false for metric spaces. For example, if X is a set, the discrete metric generates the discrete topology on X. With respect to such a topology, there exist no discontinuous maps $T : X \to \mathbb{R}$. However, there exist metric, not-normed spaces that admit discontinuous linear functionals.

As an example, consider $L^p(E)$ for $0 < p < 1$ endowed with the metric (6.1) of Chapter V. If E is a Lebesgue measurable subset of $\mathbb{R}^N$ and μ is the Lebesgue measure, then every nontrivial, linear functional on $L^p(E)$ is discontinuous.[37]

6   EQUIBOUNDED FAMILIES OF LINEAR MAPS

Let $\{X; \|\cdot\|_X\}$ and $\{Y; \|\cdot\|_Y\}$ be Banach spaces.

6.1   Let $T(x, y) : X \times Y \to \mathbb{R}$ be a functional linear and continuous in each of the two variables. Then T is linear and bounded with respect to both variables.

6.2   Let $\{T_n\}$ be a sequence in $\mathcal{B}(X; Y)$ such that the limit of $\{T_n(x)\}$ exists for all $x \in X$. Then $T(x) = \lim T_n(x)$ defines a bounded, linear map from $\{X; \|\cdot\|_X\}$ into $\{Y; \|\cdot\|_Y\}$.

6.3   Let $\{T_n\}$ be a sequence in $\mathcal{B}(X; Y)$ such that $\|T_n\| \leq C$ for some positive constant C and all $n \in \mathbb{N}$. Let X_o be the set of $x \in X$ for which $\{T_n(x)\}$ converges. Then X_o is a closed subspace of $\{X; \|\cdot\|_X\}$.

6.4   Let T_1 and T_2 be elements of $\mathcal{B}(X; X)$ and define

$$T_1 T_2(x) = T_1(T_2(x)) \quad \text{for all } x \in X.$$

Then $T_1 T_2 \in \mathcal{B}(X; X)$ and $\|T_1 T_2\| \leq \|T_1\|\|T_2\|$.

Let $\{X; \|\cdot\|\}$ be a Banach spaces, and let $T \in \mathcal{B}(X; X)$ satisfy $\|T\| < 1$. Then $(I + T)^{-1}$ exists as an element of $\mathcal{B}(X; X)$, and

$$(I + T)^{-1} = I + \sum(-1)^n T^n.$$

---

[37]These remarks on existence and nonexistence of unbounded linear functionals in metric spaces were suggested by Allen Devinatz.

Hint: Since $\|T\| < 1$, the series $\sum \|T^n\| \le \sum \|T\|^n$ converges. Since $\mathcal{B}(X; X)$ is a Banach space, $\sum (-1T)^n$ converges to an element $\mathcal{B}(X; X)$.

6.5 Let E be a bounded open set in $\mathbb{R}^N$. Having fixed a function $h \in L^\infty(E)$, consider the problem of finding $f \in L^\infty(E)$ such that

$$h(x) = f(x) + \int_E \frac{1}{|x - y|^{N-1}} f(y) dy. \tag{6.1c}$$

Setting

$$T(f)(x) = \int_E \frac{1}{|x - y|^{N-1}} f(y) dy,$$

one verifies that T is a bounded linear map from $L^\infty(E)$ into itself. Then (6.1c) can be rewritten concisely as

$$h = (I + T)f; \quad \text{i.e., formally} \quad f = (I + T)^{-1}h.$$

Give conditions on E so that such a formal solution is actually justified and exhibits the solution f explicitly.

6.6 Let E be a Lebesgue-measurable subset of $\mathbb{R}^N$ of finite measure, and let $1 \le p, q \le \infty$ be conjugate. Then if $q > p$, the space $L^q(E)$ is of the first category in $L^p(E)$. (*Hint*: $L^q(E)$ is the union of $[\|g\|_q \le n]$.)

8   THE OPEN MAPPING THEOREM

$\{X; \|\cdot\|_X\}$ and $\{Y; \|\cdot\|_Y\}$ are Banach spaces.

8.1 $T \in \mathcal{B}(X; Y)$ is a homeomorphism if and only if there exist positive constants $c_1 \le c_2$ such that

$$c_1 \|x\|_X \le \|T(x)\|_Y \le c_2 \|x\|_X \quad \text{for all } x \in X.$$

8.2 A map $T \in \mathcal{B}(X; Y)$ has closed graph if and only if its domain is closed.

8.3 The sup-norm on $C[0, 1]$ generates a strictly stronger topology than the L^2-norm.

8.4 Let $T \in \mathcal{B}(X; Y)$. If $T(X)$ is of second category in Y, then T is onto.

9   THE HAHN–BANACH THEOREM

Let X be a vector space over the field $\mathbf{C}$ of the complex numbers. The norm $\|\cdot\|$ is defined as in requirements (i)–(iii) of Section 1, except that $\lambda \in \mathbf{C}$. In such a case, $|\lambda|$ is the modulus of λ as element of $\mathbf{C}$.

Denote by $X_\mathbb{R}$ the vector space X when multiplication is restricted to scalars in $\mathbb{R}$. A linear functional $T : X \to \mathbf{C}$ is separated into its real and imaginary part by

$$T(x) = T_\mathbb{R}(x) + iT_i(x). \tag{9.1c}$$

where the maps $T_{\mathbb{R}}$ and T_i are functionals from $X_{\mathbb{R}}$ into $\mathbb{R}$. Since $T : X \to \mathbf{C}$ is linear, $T(ix) = iT(x)$ for all $x \in X$. From this, compute

$$T(ix) = T_{\mathbb{R}}(ix) + iT_i(ix),$$
$$iT(x) = iT_{\mathbb{R}}(x) - T_i(x).$$

This implies that

$$T_i(x) = -T_{\mathbb{R}}(ix) \quad \text{for all } x \in X. \tag{9.2c}$$

Thus $T : X \to \mathbf{C}$ is identified by its real part $T_{\mathbb{R}}$ regarded as a linear functional from $X_{\mathbb{R}}$ into $\mathbb{R}$. Vice versa, any such real-valued functional $T_{\mathbb{R}} : X_{\mathbb{R}} \to \mathbb{R}$ identifies a linear functional $T : X \to \mathbf{C}$ by formulas (9.1c)–(9.2c).

9.1   THE COMPLEX HAHN–BANACH THEOREM.

Theorem 9.1c. *Let X be a complex vector space, and let $p : X \to \mathbb{R}$ be a seminorm on X. Then every linear functional $T_o : X_o \to \mathbf{C}$ defined on a subspace X_o of X and satisfying* [38]

$$|T_o(x)| \le p(x) \quad \text{for all } x \in X_o$$

admits an extension $T : X \to \mathbf{C}$ such that

$$|T(x)| \le p(x) \quad \text{for all } x \in X$$

and

$$T(x) = T_o(x) \quad \text{for all } x \in X_o.$$

Proof. Denote by $X_{o,\mathbb{R}}$ the real subspace of X_o and by $T_{o,\mathbb{R}} : X_{o,\mathbb{R}} \to \mathbb{R}$ the real part of T. By the representation (9.1c)–(9.2c), it suffices to extend $T_{o,\mathbb{R}}$ into a linear map $T_{\mathbb{R}} : X_{\mathbb{R}} \to \mathbb{R}$. This follows from the Hahn–Banach theorem since

$$T_{o,\mathbb{R}}(x) \le |T(x)| \le p(x) \quad \text{for all } x \in X. \qquad \square$$

9.2   LINEAR FUNCTIONALS IN $L^\infty(E)$.   The Riesz representation theorem for the bounded linear functionals in $L^p(E)$ fails for $p = \infty$. A counterexample can be constructed as follows.

Let T_o be the linear, bounded functional on $C[-1, 1]$ defined by

$$T_o(f) = f(0) \quad \text{for all } f \in C[-1, 1].$$

---

[38] H. F. Bohnenblust and A. Sobczyk, Extensions of functionals on complex linear spaces, *Bull. Amer. Math. Soc.*, **44** (1938), 91–93; G. Soukhomlinoff, Über Fortsetzung von linearen Funktionalen in linearen komplexen Räumen und linearen Quaternionräumen, *Recueil (Sb.) Math. Moscou N. S.*, **3** (1938), 353–358.

The boundedness of T_o is meant in the sense of $L^\infty[-1, 1]$ with norm

$$\|T_o\| = \sup_{\substack{\varphi \in C[-1,1] \\ \|\varphi\|_\infty = 1}} |T_o(\varphi)|.$$

Then by the Hahn–Banach theorem, T_o can be extended to a bounded linear functional T in $L^\infty[-1, 1]$ coinciding with T_o on $C[-1, 1]$ and such that $\|T\| = \|T_o\|$. For such an extension, there exists no function $g \in L^1[-1, 1]$ such that

$$T(f) = \int_{-1}^{1} fg\,dx \quad \text{for all } f \in L^\infty[-1, 1].$$

12 WEAK TOPOLOGIES

Proposition 12.1c. *Let $\{X; \|\cdot\|\}$ be a normed space, and let $\mathcal{W}$ denote its weak topology. Also, denote by $\mathcal{O}$ a weak neighborhood of the origin of X. Then the following hold:*

(i) *For every $V \in \mathcal{W}$ and $x_o \in V$, there exists $\mathcal{O}$ such that $x_o + \mathcal{O} \subset V$.*

(ii) *For every $V \in \mathcal{W}$ and $x_o \in V$, there exists $\mathcal{O}$ such that $\mathcal{O} + \mathcal{O} + x_o \subset V$.*

(iii) *For every $\mathcal{O}$, there exists $\mathcal{O}'$ such that $\lambda\mathcal{O}' \subset \mathcal{O}$ for all $|\lambda| \leq 1$.*

Proof of (i). An open weak neighborhood V of $x_o \in X$ contains an open set of the form

$$V_o = \bigcap_{j=1}^{m} T_j^{-1}(T_j(x_o) - \alpha_j, T_j(x_o) + \alpha_j)$$

for some finite m and $\alpha_j > 0$. Equivalently,

$$V_o = \{x \in X \mid |T_j(x - x_o)| < \alpha_j \text{ for } j = 1, \dots, n\}.$$

The neighborhood of the origin,

$$\mathcal{O} = \{x \in X \mid |T_j(x)| < \alpha_j \text{ for } j = 1, \dots, n\},$$

is such that $x_o + \mathcal{O} \subset V_o$. $\qquad\square$

The remaining statements are proved similarly.

12.1 Let $\{X, \mathcal{A}, \mu\}$ be a measure space, and let $E \in \mathcal{A}$.

Proposition 12.2c. *Let $\{f_n\}$ be a sequence of functions in $L^2(E)$ weakly convergent to some $f \in L^2(E)$. There exists a subsequence $\{f_{n_j}\}$ such that setting*

$$\varphi_j = \frac{f_{n_1} + f_{n_2} + \cdots + f_{n_m}}{m},$$

the sequence $\{\varphi_m\}$ converges to f strongly in $L^2(E)$.

Proof. By possibly replacing f_n with $f_n - f$, we may assume that $f = 0$. Fix $n_1 = 1$. Since $\{f_n\} \to 0$ weakly in $L^2(E)$, there exists an index n_2 such that

$$\left| \int_E f_{n_1} f_{n_2} d\mu \right| \leq \frac{1}{2}.$$

Then there is an index n_3 such that

$$\left| \int_E f_{n_1} f_{n_3} d\mu \right| \leq \frac{1}{3} \quad \text{and} \quad \left| \int_E f_{n_2} f_{n_3} d\mu \right| \leq \frac{1}{3}.$$

Proceeding in this fashion, we extract out of $\{f_n\}$ a subsequence $\{f_{n_j}\}$ such that

$$\left| \int_E f_{n_\ell} f_{n_k} d\mu \right| \leq \frac{1}{k} \quad \text{for all } \ell = 1, 2, \ldots, k - 1.$$

Denoting by M the upper bound of $\|f_n\|_2$, compute

$$\int_E \varphi_m^2 d\mu = \frac{1}{m^2} \int_E (f_{n_1} + f_{n_2} + \cdots + f_{n_m})^2 d\mu$$

$$\leq \frac{1}{m^2} \left(mM^2 + 2 + 4\frac{1}{2} + \cdots + 2m\frac{1}{m} \right)$$

$$\leq \frac{M^2 + 2}{m} \longrightarrow 0 \quad \text{as } m \to \infty. \qquad \square$$

12.2   In a finite-dimensional normed linear space, the notions of weak and strong convergence are the same.

12.3   Construct counterexamples for the following statements:

- A weakly closed subset of a normed linear is also strongly closed. The converse is false.

- A strongly sequentially compact subset of a normed linear space is also weakly sequentially compact. The converse is false.

12.4   A normed linear space is weakly complete if and only if it is complete in its strong topology. A weakly dense set in $\{X; \|\cdot\|\}$ is also strongly dense.

12.5   **INFINITE-DIMENSIONAL NORMED SPACES.** Let $\{X; \|\cdot\|\}$ be an infinite-dimensional Banach space. There exists a countable collection $\{X_n\}$ of infinite-dimensional, closed subspaces of X such that $X_{n+1} \subset X_n$ with strict inclusion. For example, one might take a nonzero functional $T_1 \in X^*$ and set $X_1 = \ker\{T_1\}$. Such a subspace is infinite dimensional. (*Hint:* Proposition 5.1). Then select a nonzero functional $T_2 \in X_1^*$, set $X_2 = \ker\{T_2\}$, and proceed by induction.

For each n, select an element $x_n \in X_{n+1} - X_n$ so that $\|x_n\| = 2^{-n}$. This generates a sequence $\{x_n\}$ of linearly independent elements of X whose span X_o is isomorphic to ℓ_∞. Indeed, the representation

$$\ell_\infty \ni \{c_n\} \longrightarrow \sum c_n x_n \in X$$

is an isomorphism between X_o and ℓ_∞. Since the dimension of ℓ_∞ is not less than the cardinality of $\mathbb{R}$, the dimension of an infinite-dimensional Banach space is at least the cardinality of $\mathbb{R}$.[39] Compare with Section 2.1 of the Problems and Complements.

12.6 A Banach space $\{X; \|\cdot\|\}$ is finite dimensional if and only if every linear subspace is closed.

12.7 The weak topology of an infinite-dimensional normed space X is not normable; i.e., there exists no norm $\|\cdot\|$ on X that generates the weak topology.

13 REFLEXIVE BANACH SPACES

13.1 Let C be a weakly closed subset of a reflexive Banach space. Then the functional $h(x) = \|x\|$ takes its minimum on C; i.e., there exists $x_o \in C$ such that $\inf_{x \in C} \|x\| = \|x_o\|$.

14 WEAK COMPACTNESS

14.1 LINEAR FUNCTIONALS ON SUBSPACES OF $C(\overline{E})$. Let E be a bounded, open set in $\mathbb{R}^N$, and let $C(\overline{E})$ denote the space of the continuous functions in $\overline{E}$ equipped with the sup-norm. Let X_o be a subspace of $C(\overline{E})$, and regard $C(\overline{E})$ as a subspace of $L^2(E)$. If X_o is closed in the topology of $L^2(E)$, it is also closed in the topology of $C(\overline{E})$. Moreover, there exist positive constants $C_o \leq C_1$ such that

$$C_o \|f\|_\infty \leq \|f\|_2 \leq C_1 \|f\|_\infty.$$

Let $T_{o,y} \in X_o^*$ be the evaluation map at y; i.e.,

$$T_{o,y}(f) = f(y) \quad \text{for all } f \in X_o.$$

By the Hahn–Banach theorem, there exists a functional $T_y \in L^2(E)^*$ such that $T_y = T_{o,y}$ on X_o. By the Riesz representation of the bounded linear functionals in $L^2(E)$, there exists a function $K(\cdot, y) \in L^2(E)$ such that

$$f(y) = \int_E K(x, y) f(x) d\mu \quad \text{for all } f \in X_o. \tag{14.1c}$$

Proposition 14.1c. *The unit ball of X_o is compact.*

Proof. The closed unit ball $\overline{B}_{o,1}$ of X_o is also weakly closed. Since $L^2(E)$ is reflexive, X_o also is reflexive. Therefore, $\overline{B}_{o,1}$ is bounded and weakly closed and hence sequentially compact. In particular, every sequence $\{f_n\}$ in $\overline{B}_{o,1}$ contains, in turn, a subsequence $\{f_{n'}\}$ weakly convergent to some $f \in \overline{B}_{o,1}$. By (14.1c), such a sequence converges pointwise to f, and

$$\|f_n\|_\infty \leq (\text{const}) \|f_n\|_2 \leq (\text{const})'$$

---

[39] See Sections 9.6 and, in particular, 9.8 of the Problems and Complements of Chapter I.

for a constant independent of n. Therefore, by the Lebesgue dominated convergence theorem, $\{f_{n'}\} \to f$ strongly in $L^2(E)$.

Thus every sequence $\{f_n\}$ in $\overline{B}_{o,1}$ contains, in turn, a strongly convergent subsequence. This implies that $\overline{B}_{o,1}$ is compact in the strong topology inherited from $L^2(E)$. $\qquad\square$

Proposition 14.2c. *Every subspace $X_o \subset C(\overline{E})$ closed in $L^2(E)$ is finite dimensional.*

17 HILBERT SPACES

17.1   The parallelogram identity is equivalent to the existence of an inner product on a vector space X in the following sense.

A scalar product $\langle \cdot, \cdot \rangle$ on a vector space X generates a norm $\| \cdot \|$ on X that satisfies the parallelogram identity. Vice versa, let $\{X; \| \cdot \|\}$ be a normed space whose norm $\| \cdot \|$ satisfies the parallelogram identity. Then setting

$$\langle x, y \rangle = \frac{1}{4}(\|x + y\|^2 - \|x - y\|^2)$$

defines a scalar product in X.

17.2   If $p \neq 2$, then $L^p(E)$ is not a Hilbert space.

17.3   Let $\{x_n\}$ and $\{y_n\}$ be Cauchy sequences in a Hilbert space H. Then $\{\langle x_n, y_n \rangle\}$ is a Cauchy sequence in $\mathbb{R}$.

18 ORTHOGONAL SETS, REPRESENTATIONS, AND FUNCTIONALS

18.1   Let $\{x_1, x_2, \ldots, x_n\}$ be an n-tuple of orthogonal elements in a Hilbert space H. Then[40]

$$\left\| \sum_{i=1}^{n} x_i \right\|^2 = \sum_{i=1}^{n} \|x_i\|^2.$$

18.2   Let E be a subset of H. Then $E^\perp$ is a linear subspace of H and $(E^\perp)^\perp$ is the smallest closed, linear subspace of H containing E.

18.3   Every closed convex set of H has a unique element of least norm.

18.4   Let E be a bounded open set in $\mathbb{R}^N$, and let $f \in C(\overline{E})$. Denote by $\mathcal{P}_n$ the collections of polynomials of degree at most n in the coordinate variables. There exists a unique $P_o \in \mathcal{P}_n$ such that

$$\int_E |f - P|^2 dx \geq \int_E |f - P_o|^2 dx \quad \text{for all } P \in \mathcal{P}_n.$$

---

[40]Pythagoras's theorem.

19   ORTHONORMAL SYSTEMS

19.1   Let H be a Hilbert space, and let S be an orthonormal system in H. Then for any pair x, y of elements in H,

$$\sum_{\mathbf{u} \in S} |\langle \mathbf{u}, x \rangle| |\langle \mathbf{u}, y \rangle| \leq \|x\| \|y\|.$$

19.2   Let S be an orthonormal system in H, and denote by H_o the closure of the linear span of S. The projection of an element $x \in H$ into H_o is defined as

$$x_o = \sum_{\mathbf{u} \in S} \langle x, \mathbf{u} \rangle \mathbf{u}.$$

Such a formula uniquely defines x_o. Moreover, $x_o \in H_o$ and $(x - x_o) \perp H_o$ if and only if $x \in H_o$.

VII

Spaces of Continuous Functions, Distributions, and Weak Derivatives

1 Spaces of continuous functions

Let E be an open set in $\mathbb{R}^N$, and let f be a real-valued function f defined in E. The support of f is the closure in $\mathbb{R}^N$ of the set $[|f| > 0]$, and we write

$$\operatorname{supp}\{f\} = \overline{[|f| > 0]}.$$

A function $f : E \to \mathbb{R}$ is of *compact support* in E if $\operatorname{supp}\{f\}$ is compact and contained in E. Set

$$C_o(E) = \left\{ \begin{array}{c} \text{the collection of all continuous functions} \\ f : E \to \mathbb{R} \text{ compactly supported in } E \end{array} \right\}. \qquad (1.1)$$

Also set

$$C_o^\infty(E) = \left\{ \begin{array}{c} \text{the collection of all infinitely differentiable} \\ \text{functions } f : E \to \mathbb{R} \text{ of compact support in } E \end{array} \right\}. \qquad (1.2)$$

These are linear vector spaces that become normed spaces by the norm

$$\|f\| = \sup_{x \in E} |f(x)|. \qquad (1.3)$$

However, neither is a Banach space by the norm in (1.3). Indeed, it will be shown that a topology by which $C_o^\infty(E)$ is complete is not metrizable (see Section 12.2). Continuity of functionals on $C_o(E)$ is meant with respect to the topology generated by the norm in (1.3). Precisely, a linear map $T : C_o(E) \to \mathbb{R}$ is continuous if there exists a positive constant γ_T such that

$$|T(f)| \le \gamma_T \|f\| \quad \text{for all } f \in C_o(E). \qquad (1.4)$$

The norm $\|T\|$ of T is the smallest constant γ_T for which (1.4) holds.

1.1 Partition of unity. Let E be a subset of $\mathbb{R}^N$, and let $\mathcal{U}$ be an open covering of E. A countable collection $\{\varphi_n\}$ of functions $\varphi \in C_o^\infty(\mathbb{R}^N)$ is a locally finite *partition of the unity* for E subordinate to the open covering $\mathcal{U}$ if the following hold:

(i) For any compact set K, all but finitely many of the functions φ_n are identically zero on K.

(ii) All of the functions φ_j satisfy $0 \leq \varphi_j(x) \leq 1$ for all $x \in E$.

(iii) For any given $\varphi_j \in \{\varphi_n\}$, there exists an open set $\mathcal{O} \in \mathcal{U}$ such that the support of φ_j is contained in $\mathcal{O}$.

(iv) $\sum \varphi_n(x) = 1$ for all $x \in E$.

Proposition 1.1. *Let E be a subset of $\mathbb{R}^N$. For every open covering $\mathcal{U}$ of E, there exists a partition of unity for E subordinate to $\mathcal{U}$.*

Proof. Consider the collection of balls $B_{r_i}(x_j)$ centered at points $x_j \in E$ of rational coordinates and rational radii r_i and contained in some $\mathcal{O} \in \mathcal{U}$. The union of the balls $B_{\frac{1}{2}r_i}(x_j)$ covers E. For each such ball, construct a function

$$\psi_{ij} \in C_o^\infty(B_{r_i}(x_j)), \quad 0 \leq \psi_{ij} \leq 1, \quad \psi_{ij} \equiv 1 \quad \text{on } B_{\frac{1}{2}r_i}(x_j).$$

The ψ_{ij} can be constructed, for example, by mollifying the characteristic functions of $B_{\frac{2}{3}r_i}(x_j)$. The countable family $\{\psi_{ij}\}$ is ordered in some fashion, say, for example,

$$\{\psi_1, \psi_2, \ldots, \psi_n, \ldots\}.$$

Then the elements of the collection $\{\varphi_n\}$ are constructed by

$$\varphi_1 = \psi_1,$$
$$\varphi_2 = (1 - \psi_1)\psi_2,$$
$$\cdots$$
$$\varphi_{j+1} = (1 - \psi_1)(1 - \psi_2) \cdots (1 - \psi_j)\psi_{j+1},$$
$$\cdots$$

Such a collection satisfies requirements (i) and (iii) of a partition of unity. To verify requirements (ii) and (iv), observe that the functions $\{\varphi_n\}$ satisfy the identities

$$\sum_{j=1}^m \varphi_j = 1 - (1 - \psi_1)(1 - \psi_2) \cdots (1 - \psi_m)$$

for all $m \in \mathbb{N}$. This holds for $m = 1$ and is verified by induction for all $m \in \mathbb{N}$ by making use of the definition of the φ_n.

To verify (ii) and (iv), observe that for each $x \in E$, there exists some ψ_{ij} such that $\psi_{ij}(x) = 1$. $\square$

2   Bounded linear functionals on $C_o(\mathbb{R}^N)$

We will give a precise description of the dual $C_o(\mathbb{R}^N)^*$ of $C_o(\mathbb{R}^N)$, i.e., the collections of all bounded linear functionals on $C_o(\mathbb{R}^N)$. One such functional is the *evaluation map* at some fixed $x_o \in \mathbb{R}^N$, i.e.,

$$T_{x_o}(f) = f(x_o) \quad \text{for all } f \in C_o(\mathbb{R}^N). \tag{2.1}$$

One verifies that $T_{x_o} \in C_o(\mathbb{R}^N)^*$ and $\|T_{x_o}\| = 1$. Another example is constructed by fixing a finite Radon measure μ in $\mathbb{R}^N$ and by setting

$$T_\mu(f) = \int_{\mathbb{R}^N} f d\mu \quad \text{for all } f \in C_o(\mathbb{R}^N). \tag{2.2}$$

One verifies that $T_\mu \in C_o(\mathbb{R}^N)^*$ and $\|T_\mu\| = \mu(\mathbb{R}^N)$. Such a functional is *positive* in the sense that $T_\mu(f) \geq 0$ whenever $f \geq 0$. The evaluation map T_{x_o} in (2.1) is of the form (2.2) if the Radon measure μ is the Dirac mass δ_{x_o} concentrated at x_o.

More generally, having fixed two finite Radon measures μ_1 and μ_2 in $\mathbb{R}^N$ and real numbers α, β, the signed measure

$$\mu = \alpha\mu_1 + \beta\mu_2$$

identifies a bounded linear functional $T_\mu \in C_o(\mathbb{R}^N)^*$ by the formula

$$T_\mu(f) = \alpha \int_E f d\mu_1 + \beta \int_E f d\mu_2 \quad \text{for all } f \in C_o(\mathbb{R}^N). \tag{2.3}$$

One verifies that $\|T_\mu\| = |\mu|(\mathbb{R}^N)$, where $|\mu|$ is the total variation of the signed measure μ.

2.1   Remarks on functionals of the type (2.2) **and** (2.3)**.** For the integral in (2.2) to be well defined, f has to be μ-measurable; i.e., the sets $[f > c]$ must be μ-measurable for all $c \in \mathbb{R}$. These sets are open since f is continuous. Therefore, for (2.2) to identify a bounded, linear functional on $C_o(\mathbb{R}^N)$, it suffices that the measure μ be defined only on the smallest σ-algebra containing the open sets and that it be finite.

A Radon measure μ is a Borel measure; i.e., it is defined on a σ-algebra $\mathcal{A}$ containing the Borel sets. Thus to identify a bounded, linear functional T_μ on $C_o(\mathbb{R}^N)$ by formula (2.2), it suffices to consider the restriction of μ to the Borel σ-algebra $\mathcal{B}$. Alternatively, (2.2) identifies a bounded, positive, linear functional on $C_o(\mathbb{R}^N)$ whenever μ is a finite Radon measure whose domain of definition is exactly the σ-algebra $\mathcal{B}$ of the Borel sets.

Since μ is finite it is also regular; i.e., the measure of every Borel set E can be approximated by the μ-measure of closed sets included in E or by the μ-measure of open sets containing E.[1]

---

[1] See Section 15 of Chapter II.

2.2   Characterizing $C_o(\mathbb{R}^N)^*$. The Riesz representation theorem asserts that every $T \in C_o(\mathbb{R}^N)^*$ is of the form (2.3); i.e., it is identified by a linear combination of two finite Radon measures in $\mathbb{R}^N$ defined on the Borel sets.

Theorem 2.1. *Let $T \in C_o(\mathbb{R}^N)^*$. There exist two finite Radon measures μ_1 and μ_2 in $\mathbb{R}^N$ such that the functional T can be represented as*[2]

$$T(f) = \int_{\mathbb{R}^N} f \, d\mu_1 - \int_{\mathbb{R}^N} f \, d\mu_2 \quad \text{for all } f \in C_o(\mathbb{R}^N). \qquad (2.4)$$

Moreover, $\|T\| = |\mu|(\mathbb{R}^N)$, where $|\mu|$ is the total variation of the signed measure $\mu = \mu_1 - \mu_2$. Finally, the restrictions of μ_1 and μ_2 to the Borel σ-algebra $\mathcal{B}$ are unique.

The proof involves several independent facts regarding positive functionals on $C_o(\mathbb{R}^N)$ and their characterization. These are collected in Sections 3–6. As a result of these, the proof of Theorem 2.1 is given in Section 6.2.

3   Positive linear functionals on $C_o(\mathbb{R}^N)$

A linear map $T : C_o(\mathbb{R}^N) \to \mathbb{R}$ is *positive* if $T(f) \geq 0$ whenever $f \geq 0$. Since T is linear,

$$f, g \in C_o(\mathbb{R}^N) \quad \text{and} \quad f \geq g \quad \text{imply that} \quad T(f) \geq T(g). \qquad (3.1)$$

Positive, linear functionals $T : C_o(\mathbb{R}^N) \to \mathbb{R}$, while in general not bounded, are *locally bounded* in the following sense.

Proposition 3.1. *Let T be a positive, linear functional on $C_o(\mathbb{R}^N)$. For every compact set $K \subset \mathbb{R}^N$, there is a positive constant γ_K such that*

$$|T(f)| \leq \gamma_K \|f\| \quad \text{for all } f \in C_o(\mathbb{R}^N), \quad \text{supp}\{f\} \subset K. \qquad (3.2)$$

Proof. Having fixed a compact set $K \subset \mathbb{R}^N$, choose a function

$$\varphi \in C_o(\mathbb{R}^N), \quad 0 \leq \varphi \leq 1, \quad \text{and} \quad \varphi \equiv 1 \quad \text{on } K.$$

Given $f \in C_o(\mathbb{R}^N)$ with $\text{supp}\{f\} \subset K$, the two functions $\|f\|\varphi \pm f$ are both nonnegative and are in $C_o(\mathbb{R}^N)$. Therefore, by the linearity of T,

$$\pm T(f) \leq \|f\|T(\varphi). \qquad \square$$

---

[2]The first form of this theorem is in F. Riesz, *Sur les opérations fonctionnelles linéaires, C. R. Acad. Sci. Paris*, **149** (1909), 974–977. Through various extensions, it is known to hold for $C_o(X)$, where X is a locally compact Hausdorff topological space; see E. Hewitt and K. A. Ross, *Abstract Harmonic Analysis*, Vol. I, Springer-Verlag, Berlin, 1963, Section 11. We have chosen to present it for $X = \mathbb{R}^N$ as this setting contains all the main ideas. See also Section 6 of the Problems and Complements.

As an example, consider a positive functional of the type

$$T(f) = \int_{\mathbb{R}^N} f \, d\mu \quad \text{for all } f \in C_o(\mathbb{R}^N), \tag{3.3}$$

where μ is a Radon measure in $\mathbb{R}^N$. If μ is not finite, then T is only locally bounded in the sense of (3.2).

For (3.3) to identify a positive, linear functional on $C_o(\mathbb{R}^N)$, it suffices that μ be defined only on the Borel σ-algebra $\mathcal{B}$. For example, if μ is the Lebesgue measure in $\mathbb{R}^N$, it is only its restriction to $\mathcal{B}$ that identifies a positive, linear functional on $C_o(\mathbb{R}^N)$ by the formula (3.3). Also, for (3.3) to identify a positive, linear functional on $C_o(\mathbb{R}^N)$, the measure μ must be locally finite. Thus (3.3) identifies a positive, linear functional on $C_o(\mathbb{R}^N)$ if and only if μ is a Radon measure in $\mathbb{R}^N$ defined on $\mathcal{B}$.

For an open set $\mathcal{O} \subset \mathbb{R}^N$, set

$$\Gamma_{\mathcal{O}} = \left\{ \begin{array}{c} \text{the collection of all functions } f \in C_o(\mathbb{R}^N) \\ \text{such that } 0 \leq f \leq 1 \text{ and supp}\{f\} \subset \mathcal{O} \end{array} \right\}. \tag{3.4}$$

For a compact set $K \subset \mathbb{R}^N$, set

$$\Gamma_K = \left\{ \begin{array}{c} \text{the collection of all functions} \\ f \in C_o(\mathbb{R}^N) \text{ such that } f \geq \chi_K \end{array} \right\}. \tag{3.5}$$

Proposition 3.2. *Let μ be a Radon measure in $\mathbb{R}^N$ defined on $\mathcal{B}$, and let T be the positive, linear functional on $C_o(\mathbb{R}^N)$ defined by (3.3). Then*

$$\mu(\mathcal{O}) = \sup_{f \in \Gamma_{\mathcal{O}}} T(f) \quad \text{for all open sets } \mathcal{O} \subset \mathbb{R}^N, \tag{3.6}$$

$$\mu(K) = \inf_{f \in \Gamma_K} T(f) \quad \text{for all compact sets } K \subset \mathbb{R}^N. \tag{3.7}$$

Proof. Having fixed an open set $\mathcal{O}$ in $\mathbb{R}^N$, let $\{K_n\}$ be a countable collection of expanding, compact sets invading $\mathcal{O}$ and at positive mutual distance; i.e.,[3]

$$K_n \subset K_{n+1}, \quad \text{dist}\{K_{n+1}; K_n\} = \delta_n > 0, \quad \bigcup K_n = \mathcal{O}.$$

Also, let $\{f_n\}$ be a sequence of functions in $C_o(\mathbb{R}^N)$ such that

$$f_n = 1 \quad \text{on } K_n, \quad \text{supp}\{f_n\} \subset K_{n+1}, \quad \text{and} \quad 0 \leq f_n \leq 1.$$

Such functions are in $\Gamma_{\mathcal{O}}$. Therefore, by the dominated convergence theorem and (3.3),

$$\sup_{f \in \Gamma_{\mathcal{O}}} T(f) \leq \mu(\mathcal{O}) = \lim \int_{K_n} d\mu$$

$$\leq \lim \int_{K_{n+1}} f_n \, d\mu \leq \sup_{f \in \Gamma_{\mathcal{O}}} T(f).$$

---

[3]For example, one might take $K_n = \{x \in \mathcal{O} \mid \text{dist}\{x, \partial\mathcal{O} \geq \frac{1}{n}\}\} \bigcap \{|x| \leq n\}$ for positive integers n so large that $K_n \neq \emptyset$. Such a construction is possible since the distance function from the boundary of a bounded, open set is Lipschitz continuous (see Lemma 15.3 of Chapter III).

This establishes (3.6). To prove (3.7), fix a compact set K in $\mathbb{R}^N$ and construct a countable collection $\{K_n\}$ of compact sets at positive mutual distance and shrinking to K; i.e.,

$$K_{n+1} \subset K_n, \qquad \text{dist}\{K_{n+1}; K_n\} = \delta_n > 0, \qquad \bigcap K_n = K.$$

Also, let $\{f_n\}$ be a sequence of functions in $C_o(\mathbb{R}^N)$ such that

$$f_n = 1 \text{ on } K_{n+1}, \qquad \text{supp}\{f_n\} \subset K_n, \quad \text{and} \quad 0 \leq f_n \leq 1.$$

Such functions are in Γ_K. By the dominated convergence theorem,

$$\mu(K) = \inf \mu(K_n) = \lim \int_{K_n} d\mu.$$

Therefore, having fixed $\varepsilon > 0$, there exists an index n_ε such that

$$\mu(K) + \varepsilon \geq \mu(K_n) \quad \text{for all } n \geq n_\varepsilon.$$

From this and (3.3), for all $n \geq n_\varepsilon$,

$$\inf_{f \in \Gamma_K} T(f) + \varepsilon \geq \mu(K) + \varepsilon \geq \mu(K_n)$$

$$\geq \int_{K_n} f_n d\mu \geq \inf_{f \in \Gamma_K} T(f).$$

This implies (3.7) since $\varepsilon > 0$ is arbitrary. □

The next theorem characterizes all of the positive linear functionals in $C_o(\mathbb{R}^N)$ as those of the form (3.3) for some Radon measure μ defined on the Borel σ-algebra. It also gives an operative way of computing μ on open sets $\mathcal{O} \subset \mathbb{R}^N$ and on compact sets $K \subset \mathbb{R}^N$ by formulas (3.6) and (3.7).

Theorem 3.3. *Let T be a positive, linear functional on $C_o(\mathbb{R}^N)$. There exists a unique Radon measure μ defined on the Borel σ-algebra $\mathcal{B}$ such that T is represented as in (3.3). Moreover, such a measure μ satisfies (3.6) and (3.7).*

Proof (uniqueness). Let μ_1 and μ_2 be two Radon measures that identify the same positive, linear functional T on $C_o(\mathbb{R}^N)$. Since they both satisfy (3.6) and (3.7), they coincide on the open sets and on the compact subsets of $\mathbb{R}^N$. Thus they coincide on the Borel sets.[4] □

---

[4]See Theorem 11.1 of Chapter II.

4   Proof of Theorem 3.3: Constructing the measure μ

Having fixed a positive, linear functional $T : C_o(\mathbb{R}^N) \to \mathbb{R}$, define a nonnegative set function λ on the open sets of $\mathbb{R}^N$ by setting $\lambda(\emptyset) = 0$ and

$$\lambda(\mathcal{O}) = \sup_{f \in \Gamma_{\mathcal{O}}} T(f) \quad \text{for all, nonempty, open sets } \mathcal{O} \subset \mathbb{R}^N. \tag{4.1}$$

The collection $\mathcal{Q}$ of all nonvoid, open sets in $\mathbb{R}^N$ complemented with the empty set $\emptyset$ forms a sequential covering for $\mathbb{R}^N$. The set function λ defined on the sequential covering $\mathcal{Q}$ generates an outer measure μ_e by the formula

$$\mu_e(E) = \inf \left\{ \sum \lambda(\mathcal{O}_n) | \mathcal{O}_n \in \mathcal{Q} \text{ and } E \subset \bigcup \mathcal{O}_n \right\} \tag{4.2}$$

for all sets $E \subset \mathbb{R}^N$.

Lemma 4.1. *The set function $\lambda : \mathcal{Q} \to \mathbb{R}^*$ is monotone, is countably subadditive, and coincides with μ_e on the open sets $\mathcal{O}$.*

Proof. The monotonicity is a direct consequence of the definition (4.1). Let $\{\mathcal{O}_n\}$ be a countable collection of open sets in $\mathbb{R}^N$, and set

$$\mathcal{O} = \bigcup \mathcal{O}_n.$$

Having fixed $f \in \Gamma_{\mathcal{O}}$, the collection $\{\mathcal{O}_n\}$ is an open covering for $\text{supp}\{f\}$.

Let $\{\varphi_n\}$ be a partition of the unity for $\text{supp}\{f\}$ subordinate to the covering $\{\mathcal{O}_n\}$. By construction,

$$f \varphi_n \in \Gamma_{\mathcal{O}_n} \quad \text{and} \quad f = \sum f \varphi_n.$$

Since $\text{supp}\{f\}$ is a compact subset of $\mathcal{O}$, this sum involves only finitely many nonidentically zero terms. By the linearity of T and the definition of $\lambda(\mathcal{O}_n)$,

$$T(f) = \sum T(f \varphi_n) \le \sum \lambda(\mathcal{O}_n).$$

Since $f \in \Gamma_{\mathcal{O}}$ is arbitrary, by the definition of $\lambda(\mathcal{O})$,

$$\lambda \left(\bigcup \mathcal{O}_n \right) = \sup_{f \in \Gamma_{\mathcal{O}}} T(f) \le \sum \lambda(\mathcal{O}_n).$$

The definition implies that $\mu_e(\mathcal{O}) \le \lambda(\mathcal{O})$. On the other hand, since λ is countably subadditive,

$$\mu_e(\mathcal{O}) \ge \inf \left\{ \lambda \left(\bigcup \mathcal{O}_n \right) | \mathcal{O}_n \text{ open and } \mathcal{O} \subset \bigcup \mathcal{O}_n \right\} \ge \lambda(\mathcal{O})$$

since λ is monotone. $\qquad \square$

The outer measure μ_e generates in turn a measure μ in $\mathbb{R}^N$ defined on the σ-algebra $\mathcal{A}$ of all sets E satisfying

$$\mu_e(A) \geq \mu_e\left(A \bigcap E\right) + \mu_e(A - E) \tag{4.3}$$

for all sets $A \subset \mathbb{R}^N$.

Proposition 4.2. *The open sets $\mathcal{O}$ are μ-measurable and the σ-algebra $\mathcal{A}$ contains the Borel σ-algebra $\mathcal{B}$. Moreover, μ satisfies (3.6) and (3.7).*

Proof. An open set $\mathcal{O}$ is μ-measurable if it satisfies (4.3) for all sets $A \subset \mathbb{R}^N$ of finite outer measure. Assume first that A is itself open. Then $A \bigcap \mathcal{O}$ is open, and

$$\mu_e\left(A \bigcap \mathcal{O}\right) = \lambda\left(A \bigcap \mathcal{O}\right).$$

From the definition (4.1), for any $\varepsilon > 0$, there exists $f \in \Gamma_{A \cap \mathcal{O}}$ such that

$$T(f) \geq \lambda\left(A \bigcap \mathcal{O}\right) - \varepsilon.$$

The set $A - \mathrm{supp}\{f\}$ is open, and

$$\mu_e(A - \mathrm{supp}\{f\}) = \lambda(A - \mathrm{supp}\{f\}).$$

There exists $g \in \Gamma_{A - \mathrm{supp}\{f\}}$ such that

$$T(g) \geq \lambda(A - \mathrm{supp}\{f\}) - \varepsilon.$$

Then $f + g \in \Gamma_A$, and by the linearity of T,

$$\begin{aligned}
\mu_e(A) = \lambda(A) &\geq T(f + g) = T(f) + T(g) \\
&\geq \lambda\left(A \bigcap \mathcal{O}\right) + \lambda(A - \mathrm{supp}\{f\}) - 2\varepsilon \\
&\geq \mu_e\left(A \bigcap \mathcal{O}\right) + \mu_e(A - \mathcal{O}) - 2\varepsilon.
\end{aligned}$$

If A is any subset of $\mathbb{R}^N$ of finite outer measure, having fixed $\varepsilon > 0$, there exists an open set A_ε such that

$$A \subset A_\varepsilon \quad \text{and} \quad \mu_e(A) \geq \mu_e(A_\varepsilon) - \varepsilon.$$

From this,

$$\begin{aligned}
\mu_e(A) &\geq \mu_e(A_\varepsilon) - \varepsilon \\
&\geq \mu_e\left(A_\varepsilon \bigcap \mathcal{O}\right) + \mu_e(A_\varepsilon - \mathcal{O}) - \varepsilon \\
&\geq \mu_e\left(A \bigcap \mathcal{O}\right) + \mu_e(A - \mathcal{O}) - \varepsilon.
\end{aligned}$$

Since $\mathcal{A}$ contains the open sets, it contains the smallest σ-algebra containing the open sets. Thus $\mathcal{B} \subset \mathcal{A}$.

From the construction of μ from μ_e, the measure μ coincides with μ_e on $\mathcal{A}$. Therefore,

$$\mu(\mathcal{O}) = \mu_e(\mathcal{O}) = \lambda(\mathcal{O}) \quad \text{for all open sets } \mathcal{O}.$$

Thus, in particular, μ satisfies (3.6).

Now let $K \subset \mathbb{R}^N$ be compact. Since K is a Borel set, $\mu_e(K) = \mu(K)$ and

$$\mu(K) = \inf \left\{ \sum \lambda(\mathcal{O}_n) | \mathcal{O}_n \text{ open and } K \subset \bigcup \mathcal{O}_n \right\}$$
$$\leq \inf \{\lambda(\mathcal{O}) | \mathcal{O} \text{ open and } K \subset \mathcal{O}\}.$$

Fix $f \in \Gamma_K$ and consider the open set $[f > 1 - \varepsilon]$. Every $g \in \Gamma_{[f > 1 - \varepsilon]}$ satisfies

$$g \leq \frac{1}{1 - \varepsilon} f.$$

Therefore,

$$\mu(K) \leq \lambda([f > 1 - \varepsilon]) = \sup_{g \in \Gamma_{[f > 1 - \varepsilon]}} T(g) \leq \frac{1}{1 - \varepsilon} T(f).$$

Since $f \in \Gamma_K$ is arbitrary,

$$\mu(K) \leq \frac{1}{1 - \varepsilon} \inf_{f \in \Gamma_K} T(f) \quad \text{for all } \varepsilon \in (0, 1).$$

Since $\mu(K) = \mu_e(K)$, it follows from (4.2) that for any $\varepsilon > 0$, there exists an open set $\mathcal{O}n$ such that

$$K \subset \mathcal{O} \quad \text{and} \quad \mu(K) \geq \lambda(\mathcal{O}) - \varepsilon.$$

There exists $f_K \in \Gamma_{\mathcal{O}}$ such that $f \geq 1$ on K. From this and (4.1),

$$\mu(K) \geq \sup_{f \in \Gamma_{\mathcal{O}}} T(f) - \varepsilon \geq T(f_K) - \varepsilon \geq \inf_{f \in \Gamma_K} T(f) - \varepsilon$$

for all $\varepsilon \in (0, 1)$. $\qquad \square$

The process by which the measure μ is constructed from the outer measure μ_e generates a σ-algebra $\mathcal{A}$ that might be strictly larger than the Borel σ-algebra. We restrict μ to $\mathcal{B}$.

5   Proof of Theorem 3.3: Representing T as in (3.3)

Having fixed $f \in C_o(\mathbb{R}^N)$, we may assume after a possible renormalization that $\|f\| = 1$. For a fixed $n \in \mathbb{N}$ and $j = 0, 1, \ldots, n$ set

$$K_o = \text{supp}\{f\} \quad \text{and} \quad K_j = \left[f \geq \frac{j}{n} \right] \quad \text{for } j = 1, 2, \ldots, n.$$

For $j = 1, \ldots, n$, also set

$$f_j = \begin{cases} \dfrac{1}{n} & \text{for } x \in K_j. \\[2mm] f(x) - \dfrac{j-1}{n} & \text{for } x \in K_{j-1} - K_j, \\[2mm] 0 & \text{for } x \in \mathbb{R}^N - K_j. \end{cases}$$

One verifies that

$$f_j \in C_o(\mathbb{R}^N) \quad \text{and} \quad \frac{1}{n}\chi_{K_j} \le f_j \le \frac{1}{n}\chi_{K_{j-1}}.$$

From this and by the properties of the integral in $d\mu$,

$$\frac{1}{n}\mu(K_j) \le \int_{\mathbb{R}^N} f_j d\mu \le \frac{1}{n}\mu(K_{j-1}),$$

and from the properties (3.6) and (3.7) of the constructed measure μ,

$$\frac{1}{n}\mu(K_j) \le T(f) \le \frac{1}{n}\mu(K_{j-1}).$$

By construction,

$$f = \sum_{j=1}^{n} f_j.$$

Therefore,

$$\frac{1}{n}\sum_{j=1}^{n}\mu(K_j) \le \int_{\mathbb{R}^N} f d\mu \le \frac{1}{n}\sum_{j=1}^{n}\mu(K_{j-1})$$

and also

$$\frac{1}{n}\sum_{j=1}^{n}\mu(K_j) \le T(f) \le \frac{1}{n}\sum_{j=1}^{n}\mu(K_{j-1}).$$

Therefore, by subtraction,

$$\left| T(f) - \int_{\mathbb{R}^N} f d\mu \right| \le \frac{1}{n}(\mu(K_o) - \mu(K_n)).$$

Since T is locally bounded, (3.7) implies that $\mu(K_o) < \infty$ so that the right-hand side tends to zero as $n \to \infty$. $\qquad\square$

6 Characterizing bounded linear functionals on $C_o(\mathbb{R}^N)$

6.1   Locally bounded linear functionals on $C_o(\mathbb{R}^N)$. A linear map $T : C_o(\mathbb{R}^N)$ $\to \mathbb{R}$ is *locally bounded* if for every compact set $K \subset \mathbb{R}^N$, there exists a positive constant γ_K such that

$$|T(f)| \leq \gamma_K \|f\| \quad \text{for all } f \in C_o(\mathbb{R}^N), \quad \text{supp}\{f\} \subset K. \tag{6.1}$$

By Proposition 3.1, positive, linear functionals on $C_o(\mathbb{R}^N)$ are locally bounded. Here it is not assumed that T is positive.

The next proposition asserts that locally bounded, linear functionals on $C_o(\mathbb{R}^N)$ can be expressed as the difference of two positive, linear functionals on $C_o(\mathbb{R}^N)$.

Proposition 6.1. *Let* $T : C_o(\mathbb{R}^N) \to \mathbb{R}$ *be linear and locally bounded. There exist two positive, linear functionals* T^+ *and* T^- *on* $C_o(\mathbb{R}^N)$ *such that*

$$T = T^+ - T^-. \tag{6.2}$$

Proof. For $f \in C_o(\mathbb{R}^N)$ and $f \geq 0$, set

$$T^+(f) = \sup\{T(h)|h \in C_o(\mathbb{R}^N), \ 0 \leq h \leq f\}. \tag{6.3}$$

One verifies that for all $\alpha > 0$,

$$T^+(\alpha f) = \alpha T^+(f).$$

To show that T^+ is linear, fix two nonnegative functions f_1 and f_2 in $C_o(\mathbb{R}^N)$ and select two functions $h_1, h_2 \in C_o(\mathbb{R}^N)$ such that

$$0 \leq h_1 \leq f_1 \quad \text{and} \quad 0 \leq h_2 \leq f_2. \tag{6.4}$$

Then

$$h_1 + h_2 \in C_o(\mathbb{R}^N) \quad \text{and} \quad 0 \leq h_1 + h_2 \leq f_1 + f_2.$$

From this,

$$T^+(f_1 + f_2) \geq T^+(f_1) + T^+(f_2).$$

To prove the reverse inequality, select $h \in C_o(\mathbb{R}^N)$ such that

$$0 \leq h \leq f_1 + f_2 \tag{6.5}$$

and set

$$h_1 = \min\{h; f_1\} \quad \text{and} \quad h_2 = h - h_1.$$

One verifies that h_1 and h_2 satisfy (6.4). Therefore,

$$T(h) = T(h_1) + T(h_2) \leq T^+(f_1) + T^+(f_2).$$

Since the function h satisfying (6.5) is arbitrary, this implies that

$$T^+(f_1 + f_2) \leq T^+(f_1) + T^+(f_2).$$

Thus T^+ is well defined and linear on nonnegative functions $f \in C_o(\mathbb{R}^N)$. For $f \in C_o(\mathbb{R}^N)$ with no sign restriction, set

$$T^+(f) = T^+(f^+) - T^+(f^-).$$

By construction, T^+ is linear and positive on the whole $C_o(\mathbb{R}^N)$. Finally, set

$$T^- = T^+ - T.$$

Then T^- is linear and positive and realizes the decomposition in (6.2).   □

Theorem 6.2. *Let $T : C_o(\mathbb{R}^N) \to \mathbb{R}$ be linear and locally bounded. There exist two Radon measures μ_1 and μ_2 in $\mathbb{R}^N$ defined on the Borel σ-algebra $\mathcal{B}$ such that*

$$T(f) = \int_{\mathbb{R}^N} f d\mu_1 - \int_{\mathbb{R}^N} f d\mu_2 \quad \text{for all } f \in C_o(\mathbb{R}^N). \tag{6.6}$$

Remark 6.1. The two Radon measures need not be finite. However, they are finite on bounded sets and (6.5) is well defined since $f \in C_o(\mathbb{R}^N)$.

6.2 Bounded linear functionals on $C_o(\mathbb{R}^N)$. If $T : C_o(\mathbb{R}^N) \to \mathbb{R}$ is linear and bounded, it is, in particular, locally bounded, and we decompose it as in (6.2). By the construction in (6.3), the linear, positive functional T^+ is bounded and $\|T^+\| \leq \|T\|$. Thus

$$T^+(f) \leq \|T\|\|f\| \quad \text{for all } f \in C_o(\mathbb{R}^N).$$

Let μ_1 be the Radon measure corresponding to T^+. Let B_n be the ball of radius n about the origin of $\mathbb{R}^N$, and let f_n be a function in $C_o(\mathbb{R}^N)$ satisfying

$$f_n = 1 \quad \text{on } B_n, \quad \text{supp}\{f_n\} \subset B_{n+1}, \quad \text{and} \quad 0 \leq f_n \leq 1.$$

For such a function,

$$\mu_1(B_n) \leq \int_{\mathbb{R}^N} f_n d\mu_1 = T^+(f_n) \leq \|T\| \quad \text{for all } n \in \mathbb{N}.$$

From this and Fatou's lemma,

$$\mu_1(\mathbb{R}^N) \leq \|T\|.$$

Thus the Radon measure corresponding to T^+ must be finite. From the decomposition (6.2), it follows that T^- also is bounded, and a similar argument shows that the Radon measure μ_2 corresponding to T^- is also bounded.

7 A topology for $C_o^\infty(E)$ for an open set $E \subset \mathbb{R}^N$

An N-dimensional multiindex α of size $|\alpha|$ is an N-tuple of nonnegative integers whose sum is $|\alpha|$; i.e.,

$$\alpha = (\alpha_1, \alpha_2, \ldots, \alpha_N), \quad \alpha_i \in \mathbb{N} \bigcup \{0\}, \quad |\alpha| = \sum_{j=1}^{N} \alpha_j.$$

If all the components of α are zero, α is the null multiindex.

Let E be an open set in $\mathbb{R}^N$. For a function $f \in C^{|\alpha|}(E)$, the derivatives $D^\alpha f$ are defined by

$$D^\alpha f = \frac{\partial^{\alpha_1 + \alpha_2 + \cdots + \alpha_N} f}{\partial x_1^{\alpha_1} \partial x_2^{\alpha_2} \cdots \partial x_N^{\alpha_N}}.$$

If some of the components of α are zero, say, for example, if $\alpha_j = 0$, then $\partial^{\alpha_j} f / \partial x_j^{\alpha_j} = f$. If α is the null multiindex, we set $D^\alpha f = f$.

On $C_o^\infty(E)$, introduce the norms

$$p_j(f) = \max_{x \in E} \{|D^\alpha f(x)|; \ |\alpha| \leq j\}, \quad j = 0, 1, 2, \ldots . \tag{7.1}$$

It follows from the definition that

$$p_j(f) \leq p_{j+1}(f) \quad \text{for all } f \in C_o^\infty(E), \quad j = 0, 1, 2, \ldots .$$

Introduce the neighborhoods of the origin of $C_o^\infty(E)$,

$$\mathcal{O}_j = \left\{ f \in C_o^\infty(E) | p_j(f) < \frac{1}{j+1} \right\}, \quad j = 0, 1, 2, \ldots ,$$

and the neighborhoods of a given $\varphi \in C_o^\infty(E)$,

$$B_{\varphi, j} = \varphi + \mathcal{O}_j, \quad j = 0, 1, 2, \ldots .$$

By construction,

$$\mathcal{O}_{j+1} \subset \mathcal{O}_j \quad \text{and} \quad B_{\varphi, j+1} \subset B_{\varphi, j}, \quad j = 0, 1, 2, \ldots .$$

Lemma 7.1. *For any $f \in \mathcal{O}_j$, there exists an index ℓ_j such that*

$$f + \mathcal{O}_\ell \subset \mathcal{O}_j \quad \text{for all } \ell \geq \ell_j.$$

Proof. Having fixed $f \in \mathcal{O}_j$, there exists $\varepsilon > 0$ such that

$$p_j(f) \leq \frac{1}{j+1} - \varepsilon.$$

Let ℓ_j be a positive integer satisfying

$$\ell_j \geq \max \left\{ j + 1; \frac{1}{\varepsilon} \right\}.$$

Then for every $g \in \mathcal{O}_\ell$, for $\ell \geq \ell_j$,

$$p_j(f + g) \leq p_j(f) + p_j(g) \leq \frac{1}{j+1} - \varepsilon + \frac{1}{\ell+1} < \frac{1}{j+1}.$$

Thus $f + g \in \mathcal{O}_j$. $\qquad\square$

Proposition 7.2. *The collection* $\mathcal{B} = \{B_{\varphi,j}\}$ *as* φ *ranges over* $C_o^\infty(E)$ *and* $j = 0, 1, 2, \ldots$ *forms a base for a topology* $\mathcal{U}$ *of* $C_o^\infty(E)$.

Proof. We verify that such a collection verifies requirements (i) and (ii) of Section 4 of Chapter I to be a base for a topology. By construction, every φ belongs to at least one of the sets in $\mathcal{B}$. It suffices to verify that for any two sets $B_{\varphi,i}$ and $B_{\psi,j}$ out of $\mathcal{B}$ with nonempty intersection, there exist an element $B_{\eta,\ell} \in \mathcal{B}$ such that

$$B_{\eta,\ell} \subset B_{\varphi,i} \bigcap B_{\psi,j}.$$

Fix $\eta \in B_{\varphi,i} \bigcap B_{\psi,j}$. Then

$$\eta - \varphi \in \mathcal{O}_i \quad \text{and} \quad \eta - \psi \in \mathcal{O}_j.$$

There exist positive integers ℓ_i and ℓ_j such that

$$\eta - \varphi + \mathcal{O}_\ell \subset \mathcal{O}_i \quad \text{and} \quad \eta - \psi + \mathcal{O}_\ell \subset \mathcal{O}_j$$

for all $\ell \geq \max\{\ell_i; \ell_j\}$. For all such ℓ,

$$\eta + \mathcal{O}_\ell \subset \varphi + \mathcal{O}_i \quad \text{and} \quad \eta + \mathcal{O}_\ell \subset \psi + \mathcal{O}_j. \qquad\square$$

The topology $\mathcal{U}$ satisfies the first axiom of countability and is translation invariant.

Lemma 7.3. *For all* $\delta \in (0, 1]$, *the sets* $\delta\mathcal{O}_j$ *are open, i.e., are elements of* $\mathcal{U}$.

Proof. Proceeding as in the proof of Lemma 7.1, for every $f \in \delta\mathcal{O}_j$, there exists an index $\ell(\delta, j)$ such that

$$f + \mathcal{O}_\ell \subset \delta\mathcal{O}_j \quad \text{for all } \ell > \ell(\delta, j).$$

Therefore, by the construction procedure of the topology $\mathcal{U}$ from the base $\mathcal{B}$, the set $\delta\mathcal{O}_j$ is open. $\qquad\square$

Corollary 7.4. *The sets* $\delta\mathcal{O}_j$ *for all* $\delta \neq 0$ *are open, convex, symmetric neighborhoods of the origin of* $C_o^\infty(E)$.

The space $C_o^\infty(E)$ endowed with the topology $\mathcal{U}$ is denoted by $D(E)$.

Proposition 7.5. $D(E)$ *is a topological vector space.*

Proof. Let φ_1 and φ_2 be in $D(E)$, and let U be an open set containing $\varphi_1 + \varphi_2$. There exists an open neighborhood of the origin $\mathcal{O}_j$ such that

$$\varphi_1 + \varphi_2 + \mathcal{O}_j \subset U \in \mathcal{U}.$$

Since $\mathcal{O}_j$ is convex,

$$\left(\varphi_1 + \frac{1}{2}\mathcal{O}_j\right) + \left(\varphi_2 + \frac{1}{2}\mathcal{O}_j\right) \subset \varphi_1 + \varphi_2 + \mathcal{O}_j \subset U.$$

This implies that the sum

$$+ : \mathcal{D}(E) \times \mathcal{D}(E) \longrightarrow D(E)$$

is continuous with respect to the indicated topology. Fix $\varphi_o \in D(E)$, a real number λ_o, and an open neighborhood of the origin $\mathcal{O}_j$. To show that the product

$$\bullet : \mathbb{R} \times D(E) \longrightarrow D(E)$$

is continuous, one needs to show that for all $\varepsilon > 0$ there exists a positive number $\delta = \delta(\mathcal{O}_j, \varphi_o, \lambda_o, \varepsilon)$ such that

$$\lambda\varphi \in \lambda_o\varphi_o + \varepsilon\mathcal{O}_j \quad \text{for all } |\lambda - \lambda_o| < \delta$$

and for all $\varphi \in \varphi_o + \delta\mathcal{O}_j$.

The element φ_o belongs to $\sigma\mathcal{O}_j$ for some $\sigma > 0$. Having fixed $\varepsilon > 0$, we choose the number δ from

$$\lambda\varphi - \lambda_o\varphi_o = \lambda(\varphi - \varphi_o) + (\lambda - \lambda_o)\varphi_o$$
$$\subset \lambda\delta\mathcal{O}_j + \delta\sigma\mathcal{O}_j \subset \varepsilon\mathcal{O}_j$$

for a suitable choice of δ. □

8   A metric topology for $C_o^\infty(E)$

As an alternative construction of a topology for $C_o^\infty(E)$, introduce the metric[5]

$$d(f, g) = \sum \frac{1}{2^j} \frac{p_j(f - g)}{1 + p_j(f - g)}, \tag{8.1}$$

where the p_j are defined in (7.1). Since each of the p_j is translation invariant, $d(\cdot, \cdot)$ is also translation invariant and generates a translation-invariant topology in $C_o^\infty(E)$. The sum and the multiplications by scalars are continuous with respect to such a topology.[6] Thus $C_o^\infty(E)$ equipped with the topology generated by $d(\cdot, \cdot)$ is a metric, topological, vector space.

---

[5]Compare with (15.2) of Chapter I.

[6]The continuity of the sum follows from Proposition 14.1 of Chapter I. The continuity of the product is a direct consequence of the definition of p_j and $d(\cdot, \cdot)$.

8.1   Equivalence of these topologies. A base for the metric topology of $C_o^\infty(E)$ is the collection of open balls

$$B_\rho(\varphi) = \{f \in C_o^\infty(E) | d(f, \varphi) < \rho\}$$

as φ ranges over $C_o^\infty(E)$ and ρ ranges over the rational numbers in $(0, 1)$.

Lemma 8.1. *For every element $\varphi + \mathcal{O}_j \in \mathcal{B}$, there exists a ball $B_\rho(\varphi)$ for some rational number $\rho \in (0, 1)$ contained in $\varphi + \mathcal{O}_j$.*

Proof. The topologies generated by the base $\mathcal{B}$ and the one generated by the metric $d(\cdot, \cdot)$ are both translation invariant. Therefore, it suffices to assume that $\varphi = 0$. The ball B_ρ centered at the origin of $C_o^\infty(E)$ and radius

$$\rho = \frac{1}{2^{j+1}(j + 1)}$$

is contained in $\mathcal{O}_j$. Indeed, for every $f \in B_\rho$,

$$\sum \frac{1}{2^i} \frac{p_i(f)}{1 + p_i(f)} < \frac{1}{2^{j+1}(j + 1)}.$$

From this,

$$\frac{p_j(f)}{1 + p_j(f)} < \frac{1}{2(j + 1)}; \quad \text{i.e.,} \quad p_j(f) < \frac{1}{j + 1}.$$

Thus $f \in \mathcal{O}_j$.   □

Lemma 8.2. *Every ball $B_\rho(\varphi)$ contains an element $\varphi + \mathcal{O}_j$ of the base $\mathcal{B}$.*

Proof. Assume that $\varphi = 0$ and let ℓ be a nonnegative integer such that $\rho > 4(\ell + 1)^{-1}$. Then for every $f \in \mathcal{O}_\ell$,

$$d(f, 0) = \sum \frac{1}{2^j} \frac{p_j(f)}{1 + p_j(f)}$$

$$\leq \sum_{j=1}^{\ell} \frac{1}{2^j} \frac{p_j(f)}{1 + p_j(f)} + \frac{1}{2^\ell}$$

$$\leq \frac{2}{\ell + 1} + \frac{1}{2^\ell} < \rho.$$

provided ℓ is sufficiently large. Thus $\mathcal{O}_\ell \subset B_\rho$.   □

The mutual inclusion asserted by these lemmas implies that the two topologies coincide.

8.2 $D(E)$ is not complete. While it is convenient to describe the topology of $D(E)$ by a metric, Cauchy sequences in $D(E)$ need not converge to an element of $C_o^\infty(E)$. For example, let $E = \mathbb{R}$. Having fixed some $f \in C_o^\infty(0, 1)$, consider the sequence

$$f_n(x) = \sum_{j=1}^{n} \frac{1}{j} f(x - j).$$

One verifies that each $f_n \in C_o^\infty(\mathbb{R})$ and that $\{f_n\}$ is a Cauchy sequence in $D(\mathbb{R})$. However, $\{f_n\}$ does not converge to a function in $C_o^\infty(\mathbb{R})$.

As for an example in bounded domains, let $E = B_1$ be the open unit ball centered at the origin of $\mathbb{R}^N$. The functions

$$f_n(x) = \begin{cases} \exp\left\{ \dfrac{n^2}{|nx|^2 - (n-1)^2} \right\} & \text{for } |x| < \dfrac{n-1}{n}, \\ 0 & \text{for } |x| \geq \dfrac{n-1}{n} \end{cases}$$

are in $C_o^\infty(B_1)$ and form a Cauchy sequence in $D(B_1)$. However, their limit is not in $C_o^\infty(B_1)$.

An indirect proof of the noncompleteness of $D(E)$ is given in Section 10.1 by a category-type argument.

9   A topology for $C_o^\infty(K)$ for a compact set $K \subset E$

Let K be a compact subset of E, and denote by $C_o^\infty(K)$ the vector space of all functions $f \in C_o^\infty(E)$ whose support is contained in K. On $C_o^\infty(K)$, introduce the norms

$$p_{K;j}(f) = \max_{x \in K}\{|D^\alpha f(x)|; |\alpha| \leq j\}, \quad j = 0, 1, 2, \ldots. \tag{9.1}$$

It follows from the definition that

$$p_{K;j}(f) \leq p_{K;j+1}(f) \quad \text{for all } f \in C_o^\infty(K), \quad j = 0, 1, 2, \ldots.$$

Introduce the neighborhoods of the origin of $C_o^\infty(K)$,

$$\mathcal{O}_{K;j} = \left\{ f \in C_o^\infty(K) | p_{K;j}(f) < \frac{1}{j+1} \right\}, \quad j = 0, 1, 2, \ldots,$$

and the neighborhoods of a given $\varphi \in C_o^\infty(K)$,

$$B_{K;\varphi,j} = \varphi + \mathcal{O}_{K;j}, \quad j = 0, 1, 2, \ldots.$$

By construction,

$$\mathcal{O}_{K;j+1} \subset \mathcal{O}_{K;j} \quad \text{and} \quad B_{\varphi;j+1} \subset B_{\varphi;j}, \quad j = 0, 1, 2, \ldots.$$

The next assertions are proved as in the previous section.

Proposition 9.1.

(i) *For any $f \in \mathcal{O}_{K:j}$, there exists an index ℓ_j such that*

$$f + \mathcal{O}_{k:\ell} \subset \mathcal{O}_{k:j} \quad \text{for all } \ell \geq \ell_j.$$

(ii) *The collection $\mathcal{B}_K = \{B_{K:\varphi,j}\}$ as φ ranges over $C_o^\infty(K)$ and j ranges over $\{0, 1, 2, \ldots\}$ forms a base for a topology $\mathcal{U}_K$ of $C_o^\infty(K)$.*

(iii) *The topology $\mathcal{U}_K$ satisfies the first axiom of countability and is translation invariant.*

(iv) *For all $\delta \neq 0$, the sets $\delta \mathcal{O}_{K:j}$ are open.*

(v) *Let $\mathcal{D}(K)$ denote the space $C_o^\infty(K)$ equipped with the topology $\mathcal{U}_K$. Then $\mathcal{D}(K)$ is a topological vector space.*

9.1 A metric topology for $C_o^\infty(K)$. A topology in $C_o^\infty(K)$ can be constructed by the the metric

$$d_K(f, g) = \sum \frac{1}{2^j} \frac{p_{K:j}(f - g)}{1 + p_{K:j}(f - g)}. \tag{9.2}$$

Since each of the $p_{K:j}$ is translation invariant, d_K is also translation invariant and generates a translation-invariant topology in $C_o^\infty(K)$. Moreover, the sum and the multiplication by scalars are continuous with respect to such a topology. Therefore, $C_o^\infty(K)$ equipped with the topology generated by d_K is a metric, topological vector space.

The equivalence of $\mathcal{U}_K$ with the metric topology generated by d_K can be established as in Lemmas 8.1 and 8.2.

9.2 $\mathcal{D}(K)$ is complete. The vector space $C_o^\infty(K)$ equipped with the topology $\mathcal{U}_K$ generated by the base $\mathcal{B}_K$ of Proposition 9.1 or by the metric d_K is denoted by $\mathcal{D}(K)$.

The notion of convergence in $\mathcal{D}(K)$ can be given in terms of the metric in (9.2); i.e., a sequence $\{f_n\}$ of functions in $\mathcal{D}(K)$ converges to some $f \in \mathcal{D}(K)$ if and only if, for every N-dimensional multiindex α,

$$D^\alpha f_n \longrightarrow D^\alpha f \quad \text{uniformly in } K.$$

Since all the f_n are in $C_o^\infty(E)$, the indicated convergence is uniform in every compact subset $K' \subset E$ containing K. Moreover, $D^\alpha f_n \to 0$ in $E - K$.

With respect to such a notion of convergence $\mathcal{D}(K)$ is complete.

Let $\{f_n\}$ be a Cauchy sequence in $\mathcal{D}(K)$. Then having fixed a compact set K' contained in E and containing K, each of the sequences $\{D^\alpha f_n\}$ converges in the uniform topology of $C(K')$ to some function f_α continuous in K' and vanishing outside K. By working with difference quotients, one identifies $D^\alpha f = f_\alpha$.

10 Relating the topology of $D(E)$ to the topology of $\mathcal{D}(K)$

Let $\mathcal{U}$ be the collection of open sets making up the topology of $D(E)$. For a fixed compact subset $K \subset E$, let $\mathcal{U}_K$ be the collection of open sets making up the topology of $\mathcal{D}(K)$.

Proposition 10.1. *$\mathcal{U}_K$ is the restriction of $\mathcal{U}$ to $\mathcal{D}(K)$; i.e.,*

$$\mathcal{U}_K = \mathcal{U} \bigcap \mathcal{D}(K). \tag{10.1}$$

Proof. The proof is based on the construction procedure of a topology from a given base and the definition of the base neighborhoods of the origin $\mathcal{O}_{K;j}$ of $\mathcal{D}(K)$. We first establish the inclusion

$$\mathcal{U}_K \subset \mathcal{U} \bigcap \mathcal{D}(K). \tag{10.2}$$

A set $\mathcal{O}_K$ is an element of $\mathcal{U}_K$ if and only if, for every $\varphi \in \mathcal{O}_K$, there exists $\mathcal{O}_{K;j_\varphi}$ for some $j_\varphi \in \{0, 1, 2, \dots\}$ such that

$$\varphi + \mathcal{O}_{K;j_\varphi} \subset \mathcal{O}_K.$$

On the other hand, $\mathcal{O}_{K;j_\varphi} \subset \mathcal{O}_{j_\varphi}$, and

$$\mathcal{O}_{K;j_\varphi} = \mathcal{O}_{j_\varphi} \bigcap \mathcal{D}(K).$$

Therefore,

$$\varphi + \mathcal{O}_{K;j_\varphi} = (\varphi + \mathcal{O}_{j_\varphi}) \bigcap \mathcal{D}(K).$$

The set $\mathcal{O}_K$ open in $\mathcal{D}(K)$ is the union of all open sets $(\varphi + \mathcal{O}_{K;j_\varphi})$ as φ ranges over $\mathcal{O}_K$. The corresponding union

$$\mathcal{O} = \bigcup_{\varphi \in \mathcal{O}_K} (\varphi + \mathcal{O}_{j_\varphi})$$

is an open set of $\mathcal{U}$. Thus given an open set $\mathcal{O}_K$ in $\mathcal{D}(K)$, there exists an open set $\mathcal{O}$ in $D(E)$ such that

$$\mathcal{O}_K = \mathcal{O} \bigcap \mathcal{D}(K).$$

This implies the inclusion (10.2). For the converse inclusion, let $\mathcal{O} \in \mathcal{U}$ and pick $\varphi \in \mathcal{O} \bigcap \mathcal{D}(K)$. There exists $\mathcal{O}_{j_\varphi}$ such that $(\varphi + \mathcal{O}_{j_\varphi}) \subset \mathcal{O}$. By the construction of $\mathcal{O}_{K;j}$,

$$\mathcal{O}_{K;j_\varphi} = \mathcal{O}_{j_\varphi} \bigcap \mathcal{D}(K).$$

Therefore,

$$(\varphi + \mathcal{O}_{K:j_\varphi}) \subset \mathcal{O} \bigcap \mathcal{D}(K)$$

and

$$\bigcup_{\varphi \in \mathcal{O} \bigcap \mathcal{D}(K)} (\varphi + \mathcal{O}_{K:j_\varphi}) = \mathcal{O} \bigcap \mathcal{D}(K).$$

By the construction of the topology $\mathcal{U}_K$ from the base $\mathcal{B}_K$, this, in turn, implies that $\mathcal{O} \bigcap \mathcal{D}(K)$ is an open set in $\mathcal{D}(K)$. $\qquad\square$

10.1   Noncompleteness of $D(E)$. Each of the spaces $\mathcal{D}(K)$ is a closed subspace of $D(E)$. Moreover, $\mathcal{D}(K)$ does not contain any open set of $D(E)$. This follows from the definition of the open sets $\mathcal{O}_j$ in $D(E)$ and the open sets $\mathcal{O}_{K:j}$ in $\mathcal{D}(K)$. Therefore, $\mathcal{D}(K)$ is nowhere dense in $D(E)$.

Let $\{K_n\}$ be a countable collection of nested, compact subsets of E, exhausting E; i.e.,

$$K_n \subset K_{n+1}, \quad \bigcup K_n = E. \tag{10.3}$$

By the definition of $C_o^\infty(E)$ and $C_o^\infty(K)$,

$$D(E) = \bigcup \mathcal{D}(K_n). \tag{10.4}$$

If $D(E)$ were complete, it would be the countable union of nowhere dense sets, against the Baire category theorem.

11   The Schwartz topology of $\mathcal{D}(E)$

Roughly speaking, the topology of $D(E)$ allows for too many sequences $\{f_n\}$ in $C_o^\infty(E)$ to be Cauchy. Equivalently, it does not contain sufficiently many open sets to restrict the Cauchy sequences to the ones with limit in $C_o^\infty(E)$. This accounts for the lack of completeness of $D(E)$.

A topology $\mathcal{W}$ by which $C_o^\infty(E)$ would be complete would have to be stronger than $\mathcal{U}$; i.e., $\mathcal{U} \subset \mathcal{W}$. Since the topology of eack $\mathcal{D}(K)$ is completely determined by $\mathcal{U}$, by formula (10.1), such a stronger topology $\mathcal{W}$, would have to preserve such a property. Specifically, when restricted to $\mathcal{D}(K)$, it should not generate new open sets in $\mathcal{D}(K)$. Such a topology $\mathcal{W}$ will be generated by the following base $\mathcal{V}$.

First, let $\mathcal{V}_o$ denote the collection of open base neighborhoods of the origin. A set V containing the origin is in $\mathcal{V}_o$ if and only if the following hold:

(i) V is convex and $\delta V \subset V$ for all $\delta \in (0, 1]$.

(ii) $V \bigcap \mathcal{D}(K) \in \mathcal{U}_K$ for all compact subsets $K \subset E$.

The collection $\mathcal{V}_o$ is not empty since $\mathcal{O}_j \in \mathcal{V}_o$ for all $j = 0, 1, 2, \ldots$. The base neighborhoods of a given $\varphi \in C_o^\infty(E)$ are of the form

$$\varphi + V \quad \text{for some } V \in \mathcal{V}_o.$$

Proposition 11.1 (Schwartz[7]). *The collection* $\mathcal{V} = \{\varphi + V\}$ *as* φ *ranges over* $C_o^\infty(E)$ *and* V *ranges over* $\mathcal{V}_o$ *forms a base for a topology* $\mathcal{W}$ *on* $C_o^\infty(E)$.

Proof. Let $(\varphi_1 + V_1)$ and $(\varphi_2 + V_2)$ be any two elements of $\mathcal{V}$ with nonempty intersection. Choose

$$\eta \in (\varphi_1 + V_1) \bigcap (\varphi_2 + V_2),$$

and let K be a compact subset of E such that

$$\text{supp}\{\varphi_1\} \bigcup \text{supp}\{\varphi_2\} \bigcup \text{supp}\{\eta\} \subset K.$$

Then $\eta - \varphi_i \in \mathcal{D}(K)$ for $i = 1, 2$ and

$$\eta - \varphi_1 \in V_1 \bigcap \mathcal{D}(K), \qquad \eta - \varphi_2 \in V_2 \bigcap \mathcal{D}(K).$$

Since $V_i \bigcap \mathcal{D}(K)$ are open in $\mathcal{D}(K)$, there exists $\varepsilon > 0$ such that

$$\eta - \varphi_i \in (1 - \varepsilon) V_i \bigcap \mathcal{D}(K) \subset (1 - \varepsilon) V_i.$$

Since the sets V_i are convex,

$$\eta - \varphi_i + \varepsilon V_i \subset (1 - \varepsilon) V_i + \varepsilon V_i.$$

From this,

$$\eta + \varepsilon V_i \subset \varphi_i + V_i, \quad i = 1, 2,$$

and

$$\eta + \varepsilon \left(V_1 \bigcap V_2 \right) \subset (\varphi_1 + V_1) \bigcap (\varphi_2 + V_2).$$

Since $\mathcal{V}_o$ is closed under intersection, $V_1 \bigcap V_2 \in \mathcal{V}_o$. $\square$

The continuity of the sum and multiplication by scalars with respect to such a topology is proved as in Section 7. Thus $C_o^\infty(E)$ equipped with the topology $\mathcal{W}$ is a topological vector space and is denoted by $\mathcal{D}(E)$.

Proposition 11.2. *The restriction of* $\mathcal{W}$ *to* $\mathcal{D}(K)$ *is* $\mathcal{U}_K$.

Proof. The inclusion $\mathcal{U}_K \subset \mathcal{W} \bigcap \mathcal{D}(K)$ is proved as in Proposition 10.1. For the converse inclusion, let $W \in \mathcal{W}$ and select any $\varphi \in W \bigcap \mathcal{D}(K)$. By the construction of $\mathcal{W}$ from the base $\mathcal{V}$, there exists $V \in \mathcal{V}_o$ such that $(\varphi + V) \subset W$. From this,

$$(\varphi + V) \bigcap \mathcal{D}(K) = \varphi + V \bigcap \mathcal{D}(K).$$

---

[7] See [49].

Since $V \cap \mathcal{D}(K)$ is open in $\mathcal{D}(K)$, by the construction procedure of $\mathcal{U}_K$ from the base $\mathcal{B}_K$, the set

$$(\varphi + V) \cap \mathcal{D}(K) \in \mathcal{U}_K.$$

Therefore, $W \cap \mathcal{D}(K) \in \mathcal{U}_K$. $\square$

Remark 11.1. A given set $\mathcal{O}_K$ open in $\mathcal{D}(K)$ might result from the intersection of $\mathcal{D}(K)$ and several open sets in $\mathcal{W}$. Thus, in particular, the enlargement of the topology $\mathcal{U}$ to $\mathcal{W}$ does not affect its restrictions to any of the $\mathcal{D}(K)$.

The topology $\mathcal{W}$ contains more open sets than $\mathcal{U}$; i.e., the inclusion $\mathcal{U} \subset \mathcal{W}$ is strict. This is established in Section 12.2.

12 $\mathcal{D}(E)$ is complete

A set $B \subset \mathcal{D}(E)$ is bounded if and only if, for every open set $W \in \mathcal{W}$ containing the origin, there exists a positive number δ such that

$$B \subset \delta W. \tag{12.1}$$

From the construction of $\mathcal{W}$ from the base $\mathcal{V}$, it suffices to verify (12.1) for the open base neighborhoods $V \in \mathcal{V}_o$.

Proposition 12.1. *Let B be a bounded subset of $\mathcal{D}(E)$. There exists a compact subset $K \subset E$ such that $B \subset \mathcal{D}(K)$.*

Proof. Proceeding by contradiction, assume that such a K does not exist. Let $\{K_n\}$ be a countable collection of compact, nested subsets of E exhausting E as in (10.3). For each $n \in \mathbb{N}$, there exist

$$f_n \in B \quad \text{and} \quad x_n \in K_{n+1} - K_n,$$

such that $|f_n(x_n)| > 0$. Therefore, if the conclusion of the proposition is violated, there exists a sequence $\{f_n\}$ of functions in B and a sequence $\{x_n\}$ of points in E without limit points in E such that $|f_n(x_n)| > 0$.

Let V be the collection of functions $f \in \mathcal{D}(E)$ satisfying

$$|f(x_n)| < \frac{1}{n}|f_n(x_n)| \quad \text{for all } n \in \mathbb{N}. \tag{12.2}$$

Such a set is convex and contains the origin and $\delta V \subset V$ for all $\delta \in (0, 1]$.

Let K be a compact subset of E. Since only finitely many points $\{x_n\}$ are in K, there exists $\delta_K > 0$ such that

$$V \cap \mathcal{D}(K) = \delta_K \mathcal{O}_{K;0}.$$

Therefore, $V \cap \mathcal{D}(K) \in \mathcal{U}_K$. Since K is arbitrary, $V \in \mathcal{V}_o$. On the other hand, for such a V, (12.2) implies that B is not contained in δV for any $\delta > 0$. $\square$

12.1 Cauchy sequences in $\mathcal{D}(E)$. Let $\{X; \mathcal{U}\}$ be a vector, topological space, and let $\mathcal{B}_o$ be a base at the origin for $\mathcal{U}$. The notion of a Cauchy sequence in $\{X; \mathcal{U}\}$ can be given with no reference to a metric.

A sequence $\{x_n\}$ in $\{X; \mathcal{U}\}$ is a Cauchy sequence if for any element $B \in \mathcal{B}_o$, there exists a positive integer n_B such that

$$x_n - x_m \in B \quad \text{for all } n, m \ge n_B.$$

If $\mathcal{U}$ is generated by a translation-invariant metric $d(\cdot, \cdot)$, then this notion coincides with the usual one given in terms of metrics.

A sequence $\{f_n\}$ of functions in $C_o^\infty(E)$ is a Cauchy sequence in $\mathcal{D}(E)$ if for every base neighborhood of the origin $V \in \mathcal{V}_o$, there exists a positive integer n_V such that

$$f_n - f_m \in V \quad \text{for all } n, m \ge n_V.$$

It follows from the definition that a Cauchy sequence in $\mathcal{D}(E)$ is a bounded set in $\mathcal{D}(E)$. In this sense, the enlargement of the topology of $C_o^\infty(E)$ from $\mathcal{U}$ to $\mathcal{W}$ places stringent restrictions on a sequence $\{f_n\}$ to be Cauchy. Indeed, if $\{f_n\}$ is a Cauchy sequence, it is bounded, and therefore, by Proposition 12.1, $\{f_n\} \subset \mathcal{D}(K)$ for some compact subset $K \subset E$. Thus if $\{f_n\}$ is a Cauchy sequence in $\mathcal{D}(E)$, the notion of convergence coincides with that of the metric topology of $\mathcal{D}(K)$ since the restriction of $\mathcal{W}$ to $\mathcal{D}(K)$ is precisely the metric topology $\mathcal{U}_K$.

Corollary 12.2. *$\mathcal{D}(E)$ is complete.*

Proof. If $\{f_n\}$ is a Cauchy sequence in $\mathcal{D}(E)$, it is a Cauchy sequence in $\mathcal{D}(K)$ for some compact $K \subset E$. Since $\mathcal{D}(K)$ is complete, $\{f_n\}$ converges in the metric topology of $\mathcal{D}(K)$ to some $f \in \mathcal{D}(K)$.   □

Corollary 12.3. *A sequence $\{f_n\}$ of functions in $\mathcal{D}(E)$ is a Cauchy sequence if and only if there exists a compact subset $K \subset E$ and a function $f \in \mathcal{D}(K)$ such that*

$$D^\alpha f_n \longrightarrow D^\alpha f \quad \text{uniformly in } K$$

for all multiindices α.

12.2 The topology of $\mathcal{D}(E)$ is not metrizable. There exists no metric $d(\cdot, \cdot)$ on $C_o^\infty(E)$ that generates the same topology $\mathcal{W}$. If such a metric were to exist, then $\mathcal{D}(E)$ would be a complete metric space and hence of the second category.

Let $\{K_n\}$ be a countable collection of compact, nested subsets of E exhausting E as in (10.3). Then by (10.4), $\mathcal{D}(E)$ would be the countable union of nowhere-dense, closed subsets of $\mathcal{D}(E)$, thereby violating Baire's category theorem.

Corollary 12.4. *The inclusion $\mathcal{U} \subset \mathcal{W}$ is strict.*

Proof. The topology $\mathcal{U}$ is metric. If $\mathcal{W} = \mathcal{U}$, the topology $\mathcal{W}$ would be metric.   □

13   Continuous maps and functionals

A set $B \subset \mathcal{D}(K)$ is bounded if and only if for every neighborhood of the origin $\mathcal{O}_K$, there exists a positive number δ such that

$$B \subset \delta \mathcal{O}_K.$$

From the definition of the base neighborhoods $\mathcal{O}_{K;j}$ of the origin and the construction of the topology $\mathcal{U}_K$ from the base $\mathcal{B}_K$, it follows that it suffices to verify such an inclusion for the open sets $\mathcal{O}_{K;j}$. This, in turn, implies that $B \subset \mathcal{D}(K)$ is bounded if and only if there exist positive numbers λ_j such that

$$p_{K;j}(f) < \lambda_j \quad \text{for all } f \in B. \tag{13.1}$$

13.1   Distributions on E. Let $T : \mathcal{D}(E) \to \mathbb{R}$ be a linear functional. If T is continuous, it must be bounded; i.e., it maps bounded neighborhoods of the origin of $\mathcal{D}(E)$ into bounded intervals about the origin of $\mathbb{R}$.[8]

Since a bounded neighborhood of the origin of $\mathcal{D}(E)$ is a bounded neighborhood of the origin of $\mathcal{D}(K)$ for some compact subset $K \subset E$, a linear functional $T : \mathcal{D}(E) \to \mathbb{R}$ is continuous if and only if its restriction to every $\mathcal{D}(K)$ is a continuous functional $T_K : \mathcal{D}(K) \to \mathbb{R}$. Since $\mathcal{D}(K)$ is a metric space, the continuity of T can be characterized in terms of sequences.

Proposition 13.1. *A linear functional $T : \mathcal{D}(E) \to \mathbb{R}$ is continuous if and only if for every sequence $\{f_n\}$ converging to zero in the sense of $\mathcal{D}(E)$, the sequence $\{T(f_n)\} \to 0$ in $\mathbb{R}$.*

Corollary 13.2. *A linear functional $T : \mathcal{D}(E) \to \mathbb{R}$ is continuous if and only if for every compact subset $K \subset E$, and for every sequence $\{f_n\} \subset \mathcal{D}(K)$ converging to zero in $\mathcal{D}(K)$, the sequence $\{T(f_n)\} \to 0$ in $\mathbb{R}$.*

A distribution on E is a continuous linear functional $T : \mathcal{D}(E) \to \mathbb{R}$. Its action on $\varphi \in \mathcal{D}(E)$ is denoted by either of the symbols

$$T[\varphi] = \langle T, \varphi \rangle. \tag{13.2}$$

The latter is also referred to as the distribution *pairing* of T and φ.

The linear space of all the continuous, linear functionals T on $\mathcal{D}(E)$ is denoted by $\mathcal{D}'(E)$. The topology on $\mathcal{D}'(E)$ is the weak* topology induced by $\mathcal{D}(E)$. Each element $\varphi \in \mathcal{D}(E)$ can be identified with a linear functional acting on $\mathcal{D}'(E)$ by the formula (13.2). The weak* topology of $\mathcal{D}'(E)$ is the weakest topology on $\mathcal{D}'(E)$ by which all elements of $\mathcal{D}(E)$ define a continuous linear functional on $\mathcal{D}'(E)$ by formula (13.2).

With respect to such a topology, sets of the type

$$\mathcal{O} = \left\{ \begin{array}{c} \text{collection of } T \in \mathcal{D}'(E) \text{ such that } \langle T, \varphi \rangle \in (\alpha, \beta) \\ \text{for some } \alpha, \beta \in \mathbb{R} \text{ and some } \varphi \in \mathcal{D}(E) \end{array} \right\} \tag{13.3}$$

---

[8] See Proposition 10.3 of Chapter I.

are open. A base for the weak* topology of $\mathcal{D}'(E)$ is

$$\mathcal{B} = \{\text{finite intersections of sets of the type (13.3)}\}.$$

This induces a notion of convergence in $\mathcal{D}'(E)$. A sequence $\{T_n\}$ of elements in $\mathcal{D}'(E)$ converges to some $T \in \mathcal{D}'(E)$ if and only if

$$\lim \langle T_n - T, \varphi \rangle = 0 \quad \text{for all } \varphi \in \mathcal{D}(E). \tag{13.4}$$

13.2   Continuous linear maps $T : \mathcal{D}(E) \to \mathcal{D}(E)$.

Proposition 13.3. *A linear map $T : \mathcal{D}(E) \to \mathcal{D}(E)$ is continuous if and only if it is bounded.*

Remark 13.1. The conclusion of the proposition holds true for maps T between metric spaces. This is the content of Propositions 10.3 and 14.2 of Chapter I.

The point of the proof is to observe that boundedness of T implies that T is restricted to some $\mathcal{D}(K)$, which is a metric space.

Proof of Proposition 13.3. The continuity of T implies that T is bounded. To prove the converse, if $B \subset \mathcal{D}(E)$ is bounded, it is a bounded subset of $\mathcal{D}(K)$ for some compact subset $K \subset E$. Since $T(B)$ is bounded in $\mathcal{D}(E)$, it is a bounded subset of $\mathcal{D}(K')$ for some compact subset $K' \subset E$. Since both $\mathcal{D}(K)$ and $\mathcal{D}(K')$ are metric spaces, the restriction of T to $\mathcal{D}(K)$ is continuous. $\qquad\square$

Proposition 13.4. *The differentiation maps $D^\alpha : \mathcal{D}(E) \to \mathcal{D}(E)$ are continuous for every multiindex α.*

Proof. It suffices to prove that D^α is bounded. Let $B \subset \mathcal{D}(E)$ be bounded, and let λ_j be the positive numbers claimed by the characterization (13.1) of the bounded subset of $\mathcal{D}(E)$. Then

$$p_{K;j}(D^\alpha f) < \lambda_{j+|\alpha|}.$$

Thus $D^\alpha(B) \subset B'$ for some bounded set $B' \subset \mathcal{D}(E)$. $\qquad\square$

14   Distributional derivatives

Every function $f \in L^1_{\text{loc}}(E)$ can be identified with an element of $\mathcal{D}'(E)$ by the formula

$$\int_E f\varphi dx = \langle f, \varphi \rangle, \quad \varphi \in \mathcal{D}(E). \tag{14.1}$$

Thus, in particular, $L^1_{\text{loc}}(E) \subset \mathcal{D}'(E)$ up to an identification map. More generally, a Radon measure μ in $\mathbb{R}^N$ can be identified with an element of $\mathcal{D}'(E)$ by the formula

$$\int_E \varphi d\mu = \langle \mu, \varphi \rangle, \quad \varphi \in \mathcal{D}(E). \tag{14.1}'$$

Thus Radon measures are distributions. As an example, fix $x_o \in E$ and consider the operation δ_{x_o} that associates with each $\varphi \in \mathcal{D}(E)$ its value at x_o; i.e.,

$$\langle \delta_{x_o}, \varphi \rangle = \varphi(x_o) \quad \text{for all } \varphi \in \mathcal{D}(E). \tag{14.2}$$

One verifies that δ_{x_o} is linear and continuous in $\mathcal{D}(E)$ and therefore is an element of $\mathcal{D}'(E)$. The distribution δ_{x_o} is called the Dirac mass concentrated at x_o and coincides up to identification with with a Radon measure in $\mathbb{R}^N$—precisely, the Dirac delta-measure introduced in Section 3.2 of Chapter II.

14.1 Derivatives of distributions. Let α be a multiindex, and let $f \in C^{|\alpha|}(E)$. Then by integration by parts,

$$\int_E D^\alpha f \varphi dx = (-1)^{|\alpha|} \int_E f D^\alpha \varphi dx \quad \text{for all } \varphi \in \mathcal{D}(E).$$

This motivates the following definition of derivative of a distribution.

The derivative $D^\alpha T$ of a distribution T is a distribution acting on $\varphi \in \mathcal{D}(E)$ as

$$\langle D^\alpha T, \varphi \rangle = (-1)^{|\alpha|} \langle T, D^\alpha \varphi \rangle \quad \text{for all } \varphi \in \mathcal{D}(E). \tag{14.3}$$

Such a definition coincides with the classical one when $T \in C^{|\alpha|}(E)$. One also verifies the formula

$$D^\alpha(D^\beta T) = D^\beta(D^\alpha T),$$

valid for any pair of multiindices α, β.

14.2 Some examples. Let α be a multiindex of size $|\alpha|$. For $f \in L^1_{\text{loc}}(E)$, the derivative $D^\alpha f$ is that distribution acting on $\varphi \in \mathcal{D}(E)$ as

$$\langle D^\alpha f, \varphi \rangle = (-1)^{|\alpha|} \int_E f D^\alpha \varphi dx.$$

The distributional derivative of the function $f(x) = |x|$ for $x \in \mathbb{R}$ is the Heaviside graph $H(\cdot)$; i.e.,

$$f_x(x) = H(x) = \begin{cases} 1 & \text{if } x > 0, \\ [0, 1] & \text{if } x = 0, \\ -1 & \text{if } x < 0. \end{cases}$$

Now taking the distributional derivative of H gives

$$H'[\varphi] = -\int_{\mathbb{R}} H(s)\varphi'(s)ds$$

$$= \int_{-\infty}^0 \varphi'ds - \int_0^\infty \varphi'ds = 2\varphi(0).$$

Therefore,

$$|x|'' = H'(x) = 2\delta_o \quad \text{in the sense of } \mathcal{D}'(\mathbb{R}).$$

14.3   Miscellaneous remarks. Two distributions in $\mathcal{D}'(E)$ are equal if and only if

$$\langle T_1, \varphi \rangle = \langle T_2, \varphi \rangle \quad \text{for all } \varphi \in \mathcal{D}(E).$$

If a distribution $T \in \mathcal{D}'(E)$ coincides with a function in $C^k(E)$ for some positive integer k, then the distributional derivatives $D^\alpha T$ for all multiindices $|\alpha| \leq k$ coincide with the classical derivatives of T.

The product of two distributions is, in general, not defined. However, if $\psi \in \mathcal{D}(E)$ and $T \in \mathcal{D}'(E)$, the product ψT is defined as that distribution acting on $\varphi \in \mathcal{D}(E)$ as

$$\langle \psi T, \varphi \rangle = \langle T, \psi \varphi \rangle \quad \text{for all } \varphi \in \mathcal{D}(E).$$

Let J_ε be the mollifying kernel of Section 21 of Chapter V, and let $T \in \mathcal{D}'(\mathbb{R}^N)$. The convolution

$$T_\varepsilon(x) = T * J_\varepsilon(\cdot - x),$$

is defined as that distribution acting on $\varphi \in \mathcal{D}(\mathbb{R}^N)$ as

$$\langle T * J_\varepsilon(\cdot - x), \varphi \rangle = \langle T, \varphi * J_\varepsilon(x - \cdot) \rangle. \tag{14.4}$$

To justify such a formula, assume first that $T \in L^1_{\text{loc}}(\mathbb{R}^N)$. Then for every $\varphi \in \mathcal{D}(\mathbb{R}^N)$,

$$\begin{aligned}
\langle T, \varphi * J_\varepsilon(x - \cdot) \rangle &= \int_{\mathbb{R}^N} T(x) \int_{\mathbb{R}^N} \varphi(y) J_\varepsilon(x - y) dy dx \\
&= \int_{\mathbb{R}^N} \left(\int_{\mathbb{R}^N} T(x) j_\varepsilon(x - y) dx \right) \varphi(y) dy \\
&= \langle T * J_\varepsilon(\cdot - x), \varphi \rangle.
\end{aligned}$$

Remark 14.1. The convolution of a distribution $T \in \mathcal{D}'(\mathbb{R}^N)$ with a kernel $K \in L^1_{\text{loc}}(\mathbb{R}^N)$ is defined by the formula

$$\langle T * K(\cdot - x), \varphi \rangle = \langle T, \varphi * K(x - \cdot) \rangle \quad \text{for all } \varphi \in \mathcal{D}(\mathbb{R}^N), \tag{14.4$'$}$$

provided that proper assumptions are made on K to ensure that $K * \varphi \in \mathcal{D}(\mathbb{R}^N)$. For example, $T * D^\alpha J_\varepsilon(\cdot - x)$ is well defined by formula (14.4)$'$ with $K = D^\alpha J_\varepsilon$. One verifies that $x \to T * J_\varepsilon(\cdot - x)$ is a function in $C^\infty(\mathbb{R}^N)$ by the formula

$$D^\alpha (T * J_\varepsilon(\cdot - x)) = T * D^\alpha J_\varepsilon(\cdot - x).$$

Moreover, $T_\varepsilon \to T$ in of $\mathcal{D}'(\mathbb{R}^N)$. Thus a distribution $T \in \mathcal{D}'(\mathbb{R}^N)$ can be approximated in the weak* topology by functions in $C^\infty(\mathbb{R}^N)$. The approximation is meant in the sense of (13.4).

15 Fundamental Solutions

For a multiindex α, we let a_α denote an N-tuple of real numbers labelled with the entries of α; i.e.,

$$a_\alpha = (a_{\alpha_1}, a_{\alpha_2}, \ldots, a_{\alpha_N}), \quad a_{\alpha_j} \in \mathbb{R}, \quad j = 1, 2, \ldots, N.$$

A linear differential operator $\mathcal{L}$ of order m with constant coefficients and its adjoint $\mathcal{L}^*$ are formally defined by

$$\mathcal{L} = \sum_{|\alpha| \leq m} a_\alpha D^\alpha, \qquad \mathcal{L}^* = \sum_{|\alpha| \leq m} (-1)^{|\alpha|} a_\alpha D^\alpha.$$

Let $f \in \mathcal{D}'(E)$ be fixed and formally consider the partial differential equation

$$\mathcal{L}(u) = f \quad \text{in } E. \tag{15.1}$$

A distribution $u \in \mathcal{D}'(E)$ is a solution of (15.1) if

$$u[\mathcal{L}^*(\varphi)] = f[\varphi] \quad \text{for all } \varphi \in \mathcal{D}(E). \tag{15.2}$$

If $f = \delta_x$, a solution to (15.1) is called a *fundamental solution* of the operator $\mathcal{L}$ with pole at x. When regarding x as varying over E, a fundamental solution of $\mathcal{L}$ is a family of distributions parameterized with $x \in E$ satisfying

$$\mathcal{L}_y u = \delta_x \quad \text{in the sense } u[\mathcal{L}_y^*(\varphi)] = \varphi(x) \quad \text{for all } \varphi \in \mathcal{D}(E).$$

Here $\mathcal{L}_y$ and $\mathcal{L}_y^*$ are the differential operators $\mathcal{L}$ and $\mathcal{L}^*$, where the derivatives are taken with respect to the variables y.

A linear differential operator of any fixed order m with constant coefficients admits a fundamental solution.[9] Here we compute the fundamental solution of two particular differential operators.

15.1   The fundamental solution of the wave operator. Let $E = \mathbb{R}^2$, fix $(\xi, \eta) \in \mathbb{R}^2$, and set

$$u(x, y) = \begin{cases} 1 & \text{if } x > \xi \text{ and } y > \eta, \\ 0 & \text{otherwise.} \end{cases} \tag{15.3}$$

Compute in $\mathcal{D}'(\mathbb{R}^2)$

$$\frac{\partial^2 u}{\partial x \partial y}[\varphi] = \iint_{\mathbb{R}^2} u \varphi_{xy} dx dy$$

$$= \int_\eta^\infty \int_\xi^\infty \varphi_{xy} dx dy = \varphi(\xi, \eta).$$

---

[9]L. Ehrenpreis, Solutions of some problems of division, *Amer. J. Math.*, **76** (1954), 883–903; B. Malgrange, Existence at approximation des solutions des équations aux dérivée partielles et des équations de convolution, *Ann. Inst. Fourier*, **6** (1955/1956), 271–355.

Therefore,

$$\frac{\partial^2 u}{\partial x \partial y} = \delta_{(\xi,\eta)} \quad \text{in } \mathcal{D}'(E).$$

For fixed $(\xi, \eta) \in \mathbb{R}^2$, now consider the function

$$u(x, y) = \begin{cases} \dfrac{1}{2} & \text{if } |x - \xi| < y - \eta, \\ 0 & \text{otherwise.} \end{cases} \tag{15.4}$$

This is the characteristic function of the sector $S_{(\xi,\eta)}$ delimited by the two half-lines originating at (ξ, η),

$$\ell^+_{(\xi,\eta)} = \{x - y = \xi - \eta\} \bigcap \{x \geq \xi\},$$
$$\ell^-_{(\xi,\eta)} = \{x + y = \xi + \eta\} \bigcap \{x \leq \xi\}.$$

The exterior normal to such a sector is

$$\frac{(1, -1)}{\sqrt{2}} \quad \text{on } \ell^+_{(\xi,\eta)} \quad \text{and} \quad \frac{(-1, -1)}{\sqrt{2}} \quad \text{on } \ell^-_{(\xi,\eta)}.$$

Compute in $\mathcal{D}'(E)$

$$\left\{\frac{\partial^2 u}{\partial y^2} - \frac{\partial^2 u}{\partial x^2}\right\}[\varphi] = \iint_{\mathbb{R}^2} u\{\varphi_{yy} - \varphi_{xx}\}dxdy$$
$$= \frac{1}{2} \iint_{S_{(\xi,\eta)}} (\varphi_{yy} - \varphi_{xx})dxdy$$
$$= -\frac{1}{2\sqrt{2}} \int_{\ell^+_{(\xi,\eta)}} (\varphi_x + \varphi_y)ds$$
$$+ \frac{1}{2\sqrt{2}} \int_{\ell^-_{(\xi,\eta)}} (\varphi_x - \varphi_y)ds,$$

where s is the abscissa along $\ell^\pm_{(\xi,\eta)}$ and ds is the corresponding measure. Now on $\ell^\pm_{(\xi,\eta)}$, one computes

$$\varphi_x \pm \varphi_y = \sqrt{2}\varphi'(s).$$

Therefore,

$$\frac{\partial^2 u}{\partial y^2} - \frac{\partial^2 u}{\partial x^2} = \delta_{(\xi,\eta)} \quad \text{in } \mathcal{D}'(E).$$

15.2   The fundamental solution of the Laplace operator. For $x, y \in \mathbb{R}^N$ and $x \neq y$, set

$$F(x; y) = \begin{cases} \dfrac{1}{(N-2)\omega_N} \dfrac{1}{|x-y|^{N-2}} & \text{if } N \geq 3, \\[2mm] \dfrac{-1}{2\pi} \ln|x-y| & \text{if } N = 2, \end{cases} \tag{15.5}$$

where ω_N is the area of the unit sphere in $\mathbb{R}^N$. By direct calculation,

$$\nabla_y F(x; y) = \frac{1}{\omega_N} \frac{x-y}{|x-y|^N} \quad \text{for } x \neq y, \quad N \geq 2. \tag{15.6}$$

From this,

$$\Delta_y H(x; y) = \operatorname{div}_y \nabla_y F(x; y) = 0 \quad \text{for } x \neq y, \quad N \geq 2. \tag{15.7}$$

Proposition 15.1 (the Stokes formula[10]). *For all $\varphi \in C_o^\infty(\mathbb{R}^N)$ and all $x \in \mathbb{R}^N$,*

$$\varphi(x) = -\int_{\mathbb{R}^N} F(x; y)\Delta\varphi \, dy. \tag{15.8}$$

Proof. For a fixed $x \in \mathbb{R}^N$, the function $y \to F(x; y)$ is integrable about x. Therefore,

$$\int_E F(x; y)\Delta\varphi \, dy = \lim_{\varepsilon \to 0} \int_{E - B_\varepsilon(x)} F(x; y)\Delta\varphi \, dy, \tag{15.9}$$

where $B_\varepsilon(x)$ denotes the ball centered at x and of radius ε. The last integral is transformed by applying recursively the Gauss–Green theorem,

$$\begin{aligned}
\int_{E - B_\varepsilon(x)} F(x; y)\Delta\varphi \, dy &= \int_{|x-y|=\varepsilon} F(x; y)\nabla\varphi \cdot \frac{(x-y)}{\varepsilon} dy \\
&\quad - \int_{E - B_\varepsilon(x)} \nabla_y F(x; y) \cdot \nabla\varphi \, dy \\
&= \int_{|x-y|=\varepsilon} F(x; y)\nabla\varphi \cdot \frac{(x-y)}{\varepsilon} dy \\
&\quad - \int_{|x-y|=\varepsilon} \varphi\nabla_y F(x; y) \cdot \frac{(x-y)}{\varepsilon} dy \\
&\quad + \int_{E - B_\varepsilon(x)} \varphi\Delta_y F(x; y) \, dy \\
&= I_{1,\varepsilon} + I_{2,\varepsilon} + I_{3,\varepsilon}.
\end{aligned}$$

The last integral is zero for all $\varepsilon > 0$ in view of (15.7) since the domain of integration excludes the singular point $y = x$.

---

[10]This is a particular case of a more general Stokes representation formula when φ is not required to vanish on ∂E. See [11, Chapter II, pp. 55–56].

From (15.5) for $\{|x - y| = \varepsilon\}$,

$$
F(x; y) = \begin{cases} \dfrac{1}{\omega_N(N-2)} \dfrac{1}{\varepsilon^{N-2}} & \text{if } N \geq 3, \\[3mm] \dfrac{1}{2\pi} |\ln \varepsilon| & \text{if } N = 2. \end{cases}
$$

Therefore,

$$
\lim_{\varepsilon \to 0} I_{1,\varepsilon} = 0.
$$

The second integral is computed with the aid of (15.6) for $\{|x - y| = \varepsilon\}$ and gives

$$
\begin{aligned}
I_{2,\varepsilon} &= -\frac{1}{\omega_N \varepsilon^{N-1}} \int_{|x-y|=\varepsilon} \varphi \, dy \\
&= \frac{-1}{\omega_N \varepsilon^{N-1}} \int_{|x-y|=\varepsilon} (\varphi(y) - \varphi(x)) dy \\
&\quad - \frac{1}{\omega_N \varepsilon^{N-1}} \int_{|y-x|=\varepsilon} \varphi(x) dy.
\end{aligned}
$$

The last integral equals $\varphi(x)$ for all $\varepsilon > 0$, whereas the first integral tends to zero as $\varepsilon \to 0$ since

$$
\left| \frac{1}{\omega_N \varepsilon^{N-1}} \int_{|x-y|=\varepsilon} (\varphi(y) - \varphi(x)) dy \right| \leq \sup_{|x-y|=\varepsilon} |\varphi(y) - \varphi(x)|.
$$

Combining these calculations in (15.9) proves the Stokes formula (15.8). $\square$

The Stokes formula (15.8) can be rewritten in terms of distributions as

$$
-\Delta_y F(x; y) = \delta_x.
$$

Therefore, $F(\cdot; \cdot)$ given by (15.5) is the fundamental solution of the Laplace operator.

16   Weak derivatives and main properties

Let $u \in L^1_{\text{loc}}(E)$, and let α be a multiindex. If the distribution $D^\alpha u$ coincides a.e. with a function $w \in L^1_{\text{loc}}(E)$, we say that w is the *weak* D^α-*derivative* of u, and

$$
\int_E u D^\alpha \varphi \, dx = (-1)^{|\alpha|} \int_E w \varphi \, dx \quad \text{for all } \varphi \in \mathcal{D}(E).
$$

If $u \in C^{|\alpha|}_{\text{loc}}(E)$, then w coincides with the classical D^α derivative of u.

Let $1 \leq p \leq \infty$, and let m be a nonnegative integer. A function $u \in L^p(E)$ is said to be in $W^{m,p}(E)$ if all its weak derivatives $D^\alpha u$ for all $|\alpha| \leq m$ are in $L^p(E)$. Equivalently,[11]

$$W^{m,p}(E) = \left\{ \begin{array}{c} \text{the collection of all } u \in L^p(E) \text{ such that} \\ D^\alpha u \in L^p(E) \text{ for all } |\alpha| \leq m \end{array} \right\}. \tag{16.1}$$

A norm in $W^{m,p}(E)$ is

$$\|u\|_{m,p} = \sum_{|\alpha| \leq m} \|D^\alpha u\|_p. \tag{16.2}$$

If $m = 0$, then $W^{m,p}(E) = L^p(E)$ and $\|\cdot\|_{0,p} = \|\cdot\|_p$. Also, define

$$H^{m,p}(E) = \{\text{the closure of } C^\infty(E) \text{ with respect to } \|\cdot\|_{m,p}\}, \tag{16.3}$$
$$W_0^{m,p}(E) = \{\text{the closure of } C_0^\infty(E) \text{ with respect to } \|\cdot\|_{m,p}\}. \tag{16.4}$$

Proposition 16.1. $W^{m,p}(E)$ *is a Banach space.*

Proof. Let $\{u_n\}$ be a Cauchy sequence in $W^{m,p}(E)$. Then $\{u_n\}$ and $\{D^\alpha u_n\}$ are Cauchy sequences in $L^p(E)$ for all multiindices $0 < |\alpha| \leq m$. By the completeness of $L^p(E)$, there exist $u \in L^p(E)$ and functions $u_\alpha \in L^p(E)$ such that

$$u_n \to u \quad \text{and} \quad D^\alpha u_n \to u_\alpha \quad \text{in } L^p(E).$$

For all $\varphi \in \mathcal{D}(E)$,

$$\lim \int_E D^\alpha u_n \varphi dx = \lim (-1)^{|\alpha|} \int_E u_n D^\alpha \varphi dx$$
$$= (-1)^{|\alpha|} \int_E u D^\alpha \varphi dx.$$

Also,

$$\lim \int_E D^\alpha u_n \varphi dx = \int_E u_\alpha \varphi dx.$$

Therefore, u_α is the weak D^α-derivative of u.   □

Corollary 16.2. $H^{m,p}(E) \subset W^{m,p}(E)$.

Theorem 16.3 (Meyers–Serrin[12]). *Let* $1 \leq p < \infty$. *Then* $C^\infty(E)$ *is dense in* $W^{m,p}(E)$, *and as a consequence,* $H^{m,p}(E) = W^{m,p}(E)$.

---

[11] S. L. Sobolev, On a theorem of functional analysis, *Mat. Sb.*, **46** (1938), 471–496.
[12] N. Meyers and J. Serrin, $H = W$, *Proc. Nat. Acad. Sci.*, **51** (1964), 1055–1056.

Proof. Having chosen $u \in W^{m,p}(E)$ and $\varepsilon \in (0, 1)$, we exhibit a function $\varphi \in C^\infty(E)$ such that $\|u - \varphi\|_{m,p} < \varepsilon$. For $j = 1, 2, \ldots$, set

$$E_j = \left\{ x \in E \,|\, \operatorname{dist}\{x, \partial E\} > \frac{1}{j} \quad \text{and} \quad |x| < j \right\}.$$

Also set

$$E_o = E_{-1} = \emptyset \quad \text{and} \quad \mathcal{O}_j = E_{j+1} \bigcap \overline{E}^c_{j-1}.$$

The set $\mathcal{O}_j$ for $j \geq 2$ is the set of points of E such that

$$\frac{1}{j+1} < \operatorname{dist}\{x, \partial E\} < \frac{1}{j-1} \quad \text{and} \quad j - 1 < |x| < j + 1.$$

The sets $\mathcal{O}_j$ are open, and their collection $\mathcal{U}$ forms an open covering of E.

Let Φ be a partition of unity subordinate to $\mathcal{U}$, and let

$$\psi_j = \left\{ \begin{array}{c} \text{sum of the finitely many } \varphi \in \Phi \text{ whose} \\ \text{supports are contained in } \mathcal{O}_j \end{array} \right\}.$$

Then

$$\psi_j \in C_o^\infty(\mathcal{O}_j) \quad \text{and} \quad \sum \psi_j(x) = 1 \quad \text{for all } x \in E.$$

If ε_j are positive numbers satisfying

$$0 < \varepsilon_j < \frac{1}{(j+1)(j+2)},$$

the mollification $J_{\varepsilon_j} * (\psi_j u)$ has support in

$$E_{j+2} \bigcap \overline{E}^c_{j-2} \overset{\text{def}}{=} \mathcal{E}_j \subset E.$$

Since $\psi_j u \in W^{m,p}(E)$, one may choose ε_j so small that

$$\| J_{\varepsilon_j} * (\psi_j u) - \psi_j u \|_{m,p;\mathcal{E}_j} < \frac{1}{2^j} \varepsilon.$$

Set

$$\varphi = \sum J_{\varepsilon_j} * (\psi_j u),$$

and observe that within any compact subset of E, all of the terms in the sum vanish except at most finitely many. Therefore, $\varphi \in C^\infty(E)$. For $x \in E_j$,

$$\varphi(x) = \sum_{i=1}^{j+2} J_{\varepsilon_i} * (\psi_i u)(x) \quad \text{and} \quad u(x) = \sum_{i=1}^{j+2} \psi_i(x) u(x).$$

Therefore, for all $j = 1, 2, \ldots$,

$$\|u - \varphi\|_{m,p;E_j} \leq \sum_{i=1}^{j+2} \| J_{\varepsilon_i} * (\psi_i u) - \psi_i u \|_{m,p;E} \leq \varepsilon \sum \frac{1}{2^i}.$$

From this, by monotone convergence, $\|u - \varphi\|_{m,p} < \varepsilon$. $\square$

17   Domains and their boundaries

Let E denote an open set in $\mathbb{R}^N$, and let ∂E denote its boundary. We list here some assumptions that at times might be necessary to impose on the structure of ∂E.

A countable collection of open balls $\{B_\rho(x_j)\}$ centered at points $x_j \in \partial E$ and radius ρ is an open, *locally finite* covering of ∂E if $\partial E \subset \bigcup B_\rho(x_j)$ and there exists a positive integer k such that any $(k+1)$ distinct elements of $\{B_\rho(x_j)\}$ have empty intersection.

17.1 ∂E of class C^1. The boundary ∂E is said to be of class C^1 if there exist a positive ρ and an open, locally finite covering $\{B_\rho(x_j)\}$ of ∂E such that for all $x_j \in \partial E$, the portion of ∂E within the ball $B_\rho(x_j)$ can be represented in a local system of coordinates with the origin at x_j as the graph of function f_j of class class C^1 in a neighborhood of the origin of the new local coordinates and such that

$$f_j(0) = 0 \quad \text{and} \quad Df_j(0) = 0. \tag{17.1}$$

Denote by K_j the $(N-1)$-dimensional domain where f_j is defined, and set

$$\||\partial E\||_1 = \sup_j \max_{K_j} |Df_j|. \tag{17.2}$$

This quantity depends upon the choice of the covering $\{B_\rho(x_j)\}$. However, having fixed one such covering, if $\||\partial E\||_1$ is finite, it is invariant under homotetic transformations of the coordinates. In particular, it does not depend upon the size of E.

17.2   Positive geometric density. The boundary ∂E satisfies the property of *positive geometric density* with respect to the Lebesgue measure μ in $\mathbb{R}^N$ if there exists $\theta \in (0, 1)$ and $\rho_o > 0$ such that for all $x_o \in \partial E$ and every ball $B_\rho(x_o)$ centered at x_o and radius $\rho \leq \rho_o$,

$$\mu\left(E \bigcap B_\rho(x_o)\right) \leq (1 - \theta)\mu(B_\rho(x_o)). \tag{17.3}$$

17.3   The segment property. The boundary ∂E has the *segment property* if there exists a locally finite, open covering of ∂E with balls $\{B_t(x_j)\}$, a corresponding sequence of unit vectors $\mathbf{n}_j$, and a number $t^* \in (0, 1)$ such that

$$x \in \overline{E} \bigcap B_t(x_j) \Longrightarrow x + t\mathbf{n}_j \in E \quad \text{for all } t \in (0, t^*). \tag{17.4}$$

In some sense, such a requirement forces the domain E to lie locally on one side of its boundary.

The unit disc from which a radius is removed does not satisfy the segment property. For $x \in \mathbb{R}$, set

$$h(x) = \begin{cases} \sqrt{|x|}\sin^2\dfrac{1}{x} & \text{for } |x| > 0, \\ 0 & \text{for } x = 0. \end{cases} \tag{$17.5)_1$}$$

The bidimensional set

$$E = (-1, 1) \times \{y > h(x)\} \tag{$17.5)_1$}$$

satisfies the segment property.

17.4   The cone property. Let C_o denote a closed, circular, spherical cone of solid angle ω, height h, and vertex at the origin. Such a cone has volume[13]

$$\mu(C_o) = \frac{\omega}{N} h^N. \tag{17.6}$$

A domain E is said to have the *cone property* if there exist some C_o such that for all $x \in \overline{E}$, there exists a circular, spherical cone C_x with vertex at x and congruent to C_o, all contained in $\overline{E}$.

17.5   On the various properties of ∂E. The cone property does not imply the segment property. For example, the unit disc from which a radius is removed satisfies the cone property and does not satisfy the segment property. The segment property does not imply the cone property. For example, the set in $(17.5)_1$–$(17.5)_2$ does not satisfy the cone property.

The cone property does not imply the property of positive geometric density. For example, the unit disc from which a radius is removed satisfies the cone property and does not satisfy the property of positive geometric density.

The property of positive geometric density does not imply the cone property.

The segment property does not imply that ∂E is of class C^1. Conversely, ∂E of class C^1 does not imply the segment property. For example, let

$$E = \{x^2 + y^2 < 1\} - \left\{ x^2 + y^2 = \frac{1}{4} \right\}. \tag{17.7}$$

The boundary of E is regular, but ∂E does not satisfy the segment property.

18   More on smooth approximations

The approximations constructed in the proof of the Meyers–Serrin theorem might deteriorate near ∂E, and it is natural to ask whether a function in $W^{m,p}(E)$ can be approximated in the sense of $W^{m,p}(E)$ by functions that are smooth up to $\overline{E}$. This is, in general, not the case, as indicated by the following example. Let E as in (17.7), and set

$$u(x, y) = \begin{cases} 1 & \text{for } \dfrac{1}{4} < x^2 + y^2 < 1, \\ 0 & \text{for } x^2 + y^2 < \dfrac{1}{4}. \end{cases}$$

The function u is in $W^{m,p}(E)$, but there is no smooth function up to ∂E that approximates u in the norm of $W^{m,p}(E)$.

This example shows that such an approximation property is, in general, false for domains that do not satisfy the segment property.

---

[13]That is, the solid of intersection of a circular cone of solid angle ω with vertex at the origin with a sphere of radius h centered at the origin.

Proposition 18.1. *Let E be a bounded domain in $\mathbb{R}^N$ with boundary ∂E satisfying the segment property. Then $C_o^\infty(\mathbb{R}^N)$ is dense in $W^{m,p}(E)$ for $1 \le p < \infty$.*

Proof. Since ∂E is bounded, the open covering claimed by the segment property is finite, say, for example,

$$\{B_t(x_1), B_t(x_2), \dots, B_t(x_n)\} \quad \text{for some } n \in \mathbb{N} \tag{18.1}$$

for some $t > 0$. Denote by $\mathbf{n}_j$ the corresponding unit vectors pointing inside E and that realize the segment property. By reducing t if necessary, we may assume that

$$\text{for all } x \in \partial E \bigcap B_{8t}(x_j), \quad x + \tau \mathbf{n}_j \in E \quad \text{for all } \tau \in (0, 8t)$$

for all $j = 1, 2, \dots, n$. Consider the open set

$$B_o = E - \bigcup_{j=1}^n \overline{B}_t(x_j),$$

and construct an open covering for E by setting

$$\mathcal{U} = \{B_o, B_{2t}(x_1), B_{2t}(x_2), \dots, B_{2t}(x_n)\}. \tag{18.2}$$

Let Φ be a partition of unity subordinate to $\mathcal{U}$, and for $j = 1, 2, \dots, n$, let

$$\psi_j = \{\text{the sum of the finitely many } \varphi \in \Phi \text{ supported in } B_{2t}(x_j)\}.$$

For $j = 0$, define ψ_o analogously by replacing $B_{2t}(x_j)$ with B_o. Set

$$u_j(x) = \begin{cases} (u\psi_j)(x) & \text{for } x \in E, \\ 0 & \text{otherwise,} \end{cases} \quad j = 0, 1, 2, \dots, n.$$

Let Γ_j be the portion of ∂E within the ball $B_{4t}(x_j)$. By the definition of a weak derivative,

$$u_j \in W^{m,p}(\mathbb{R}^N - \Gamma_j).$$

To prove the proposition, having fixed $\varepsilon > 0$, it suffices to find functions $\varphi_j \in C_o^\infty(\mathbb{R}^N)$ such that

$$\|u_j - \varphi_j\|_{m,p} < \frac{\varepsilon}{n+1} \quad \text{for all } j = 0, 1, 2, \dots, n.$$

Indeed, putting $\varphi = \sum_{j=n}^n \varphi_j$ it would give

$$\|u - \varphi\|_{m,p} \le \sum_{j=0}^n \|u_j - \varphi_j\|_{m,p} < \varepsilon.$$

For $j = 0$, such a φ_o can be found by a mollification process since u_o is compactly supported in E.

Fix an index $j \in \{1, 2, \ldots, n\}$. To construct φ_j, move the portion of the boundary Γ_j towards the outside of E by setting

$$\Gamma_{j,\tau} = \Gamma_j - \tau \mathbf{n}(x_j), \quad \tau \in (0, 8t).$$

Then define

$$u_{j,\tau}(x) = u_j(x + \tau \mathbf{n}(x_j)) \quad \text{for all } x \in \mathbb{R}^N - \Gamma_{j,\tau}.$$

By the definition of a weak derivative, $u_{j,\tau} \in W^{m,p}(\mathbb{R}^N - \Gamma_{j,\tau})$ and

$$D^\alpha u_{j,\tau}(x) = D^\alpha u_j(x + \tau \mathbf{n}(x_j)) \quad \text{for all } x \in \mathbb{R}^N - \Gamma_{j,\tau}.$$

Since the translation operation is continuous in $L^p(E)$, there exists $\tau_\varepsilon \in (0, 4t)$ such that

$$\|u_{j,\tau} - u_j\|_{m,p;E} \leq \frac{\varepsilon}{2(n+1)}.$$

The function $\varphi_j \in C_o^\infty(\mathbb{R}^N)$ is constructed by the mollification

$$\varphi_j = J_\delta * u_{j,\tau},$$

where for a fixed $\tau \in (0, \tau_\varepsilon)$, the positive number δ is chosen so small that

$$\|J_\delta * u_{j,\tau} - u_{j,\tau}\|_{m,p;E} \leq \frac{\varepsilon}{2(n+1)}. \qquad \square$$

19 Extensions into $\mathbb{R}^N$

Proposition 19.1. *Let E be a bounded domain in $\mathbb{R}^N$ with boundary ∂E satisfying the segment property and of class C^1. Every function $u \in C^1(\overline{E})$ admits an extension $w \in C_o^1(\mathbb{R}^N)$ such that $w = u$ in $\overline{E}$. Moreover, there exists a constant C depending only upon N, p, and $\|\|\partial E\|\|_1$ and independent of u such that* [14]

$$\|w\|_{1,p;\mathbb{R}^N} \leq C\|u\|_{1,p;E}. \tag{19.1}$$

Proof. We first establish the proposition in a special case. For $R > 0$, let $\mathcal{B}_R$ denote the $(N-1)$-dimensional ball of radius R centered at the origin of $\mathbb{R}^{N-1}$; i.e.,

$$\mathcal{B}_R = \{|\overline{x}| < R\}, \quad \overline{x} = (x_1, x_2, \ldots, x_{N-1}).$$

Also set

$$Q_R^+ = \mathcal{B}_R \times [0, R), \qquad Q_R^- = \mathcal{B}_R \times (-R, 0], \qquad Q_R = Q_R^+ \bigcup Q_R^-.$$

---

[14]L. Lichenstein, *Eine elementare Bemerkung zur reellen Analyis*, Math. Z., **30** (1929), 794–795.

Assume that as a function of the first $(N-1)$ variables, $\overline{x} \to u(\overline{x}, x_N)$ is compactly supported in $\mathcal{B}_R$, and assume that $u(\overline{x}, R) = 0$. Thus u vanishes on the top and on the lateral boundary of Q_R^+, and it is of class C^1 up to $x_N = 0$.

The extension claimed by the proposition in such a particular case is

$$\widetilde{u}(\overline{x}, x_N) = -3u(\overline{x}, -x_N) + 4u\left(\overline{x}, -\frac{1}{2}x_N\right), \quad x_N \leq 0.$$

The function w defined as u within Q_R^+ and as $\widetilde{u}$ within Q_R^- satisfies the indicated requirements.

The general case is proved by a local flattening of ∂E so that the transformed function is recast in the special case indicated above.

Since ∂E is of class $C^{1,\lambda}$ and bounded, it admits a finite covering, which we take as as in (18.1), such that the portion of ∂E within each $B_t(x_j)$ can be represented as the graph of a smooth function.

Referring back to the proof of Proposition 18.1, consider the finite covering (18.2) of ∂E, and let Φ be the same partition of unity subordinate to $\mathcal{U}$. Then construct as before the functions ψ_j, renormalized so that their sum is 1. The functions ψ_j can be chosen so that

$$\sup_{B_{2t}(x_j)} |D\psi_j| \leq \frac{\gamma}{t},$$

for a positive constant γ. The portion $\partial E \cap B_t(x_j)$ can be represented in a local system of coordinates as the graph of

$$x_N = f_j(\overline{x}), \quad \overline{x} = (x_1, x_2, \ldots, x_{N-1}),$$

where f_j is of class C^1 within the $(N-1)$-dimensional ball $\mathcal{B}_t$. These coordinates are obtained by a rotation and translation of the original coordinates so that the point x_j will coincide with the origin of the new coordinates.

For each j fixed, we *flatten* $\partial E \cap B_t(x_j)$ by introducing the system of coordinates

$$(\xi_1, \xi_2, \ldots, \xi_{N-1}, \xi_N) = (\overline{x}, x_N - f_j(\overline{x})).$$

This maps $\partial E \cap B_t(x_j)$ into $\mathcal{B}_t$, and by taking t even smaller if necessary, $E \cap B_t(x_j)$ is mapped into the cylinders $Q_t^+ = \mathcal{B}_t \times [0, t)$.

The functions $\widetilde{u\psi_j}$ obtained from the $u\psi_j$ with these transformations of coordinates are of class C^1 in the closed cylinder $\overline{Q_t^+}$, and

$$\overline{x} \to \widetilde{u\psi_j}(\overline{x}, \xi_N) \text{ has compact support in } \mathcal{B}_t.$$

Next we extend $\widetilde{u\psi_j}$ with a function $\widetilde{w}_j$ of class C^1 in the whole cylinder $Q_t = \mathcal{B}_t \times (-t, t)$ by setting

$$\widetilde{w}_j(\overline{x}, \xi_N) = -3\widetilde{u\psi_j}(\overline{x}, -\xi_N) + 4\widetilde{u\psi_j}\left(\overline{x}, -\frac{1}{2}\xi_N\right) \quad \text{for } \xi_N \leq 0.$$

Let w_j denote the function obtained from $\tilde{w}_j$ by the change of variables that maps Q_t^+ back to $E \cap B_t(x_j)$. The extension claimed by the proposition is constructed by setting

$$w = u\psi_o + \sum_{j=1}^{n} w_j.$$

Indeed, for each $j = 1, 2, \ldots, n$,

$$\|\tilde{w}_j\|_{1,p;Q_t} \le \frac{2\gamma}{t}(1 + \|\partial E\|_1)\|u\|_{1,p;E\cap B_t(x_j)},$$

and each $B_t(x_j)$ overlaps at most finitely many balls $B_t(x_i)$.   □

Remark 19.1. The proof shows that the extension w satisfies

$$\|w\|_{p,\mathbb{R}^N} \le C\|u\|_{p,E}. \tag{19.2}$$

Remark 19.2. The boundedness of E is not essential. The only relevant requirement is that we may construct a *locally finite*, open covering of E. In such a case, however, we have to assume that $u \in W^{1,p}(E)$.

20   The chain rule

Proposition 20.1. *Let* $u \in W^{1,p}(E)$ *for some* $1 \le p < \infty$, *and let* $f \in C^1(\mathbb{R})$ *satisfy*

$$\sup |f'| \le M \quad \text{for some positive constant } M.$$

Then the composition $f(u)$ *belongs to* $W^{1,p}(E)$, *and*

$$Df(u) = f'(u)Du. \tag{20.1}$$

Proof. Let $\{u_n\}$ be a sequence of functions in $C^\infty(E)$ converging to u in $W^{1,p}(E)$. Then

$$\lim \|f(u_n) - f(u)\|_p \le M \lim \|u_n - u\|_p = 0.$$

Moreover,

$$\begin{aligned}
\lim \|Df(u_n) - f'(u)Du\|_p &= \lim \|f'(u_n)Du_n - f'(u)Du\|_p \\
&\le \lim M\|Du_n - Du\|_p \\
&\quad + \lim \||f'(u_n) - f'(u)||Du|\|_p.
\end{aligned}$$

The sequence

$$\{(f'(u_n) - f'(u))^p |Du|^p\}$$

tends to zero a.e. in E. Moreover, it is dominated by the integrable function $(2M)^p|Du|^p$. Therefore,

$$Df(u_n) \longrightarrow f'(u)Du \quad \text{in } L^p(E).$$

Also, for all $\varphi \in C_o^\infty(E)$,

$$\lim \int_E Df(u_n)\varphi dx = -\int_E f(u)D\varphi dx.$$

Thus $Df(u) = f'(u)Du$ in $\mathcal{D}'(E)$. □

Proposition 20.2. *Let $u \in W^{1,p}(E)$ for some $1 \le p < \infty$. Then u^+, u^-, and $|u|$ belong to $W^{1,p}(E)$, and*

$$Du^+ = \begin{cases} Du & \text{a.e. } [u > 0], \\ 0 & \text{a.e. } [u \le 0], \end{cases}$$

$$Du^- = \begin{cases} -Du & \text{a.e. } [u < 0], \\ 0 & \text{a.e. } [u \ge 0], \end{cases}$$

$$D|u| = \begin{cases} Du & \text{a.e. } [u > 0], \\ 0 & \text{a.e. } [u = 0], \\ -Du & \text{a.e. } [u < 0]. \end{cases}$$

Proof. For $\varepsilon > 0$, let

$$f_\varepsilon(u) = \begin{cases} \sqrt{u^2 + \varepsilon^2} - \varepsilon & \text{if } u > 0, \\ 0 & \text{if } u \le 0. \end{cases}$$

Then $f_\varepsilon \in C^1(\mathbb{R})$ and $|f_\varepsilon'| \le 1$. Therefore, for all $\varphi \in C_o^\infty(E)$,

$$\int_E f_\varepsilon(u)D\varphi dx = -\int_E Df_\varepsilon(u)\varphi dx$$

$$= -\int_{[u>0]} \frac{u\,Du}{\sqrt{u^2 + \varepsilon^2}}\varphi dx.$$

Letting $\varepsilon \to 0$,

$$\int_E u^+ D\varphi dx = -\int_{[u>0]} Du\varphi dx.$$

Thus the conclusion holds for u^+. The remaining statements follow from $u^- = (-u)^+$ and $|u| = (u^+ + u^-)$. □

Corollary 20.3. *Let $u \in W^{1,p}(E)$. Then $Du = 0$ a.e. on any level set of u.*

Corollary 20.4. *Let* $f, g \, W^{1,p}(E)$. *Then* $\max\{f; g\}$ *and* $\min\{f; g\}$ *are in* $W^{1,p}(E)$,
and

$$D \max\{f; g\} = \begin{cases} Df & \text{a.e. } [f > g], \\ Dg & \text{a.e. } [f < g], \\ 0 & \text{a.e. } [f = g]. \end{cases}$$

A similar formula holds for $\min\{f; g\}$.

Proof. The proof follows from Proposition 20.2 and the formulas

$$2 \max\{f; g\} = (f + g) + |f - g|,$$
$$2 \min\{f; g\} = (f + g) - |f - g|. \qquad \square$$

21   Steklov averagings

Given a function $u \in L^p(E)$, we regard it as defined in the whole $\mathbb{R}^N$ by setting
it to be zero outside E. For $h \neq 0$, set

$$u_{h,i}(x) = \frac{1}{h} \int_{x_i}^{x_i+h} u(x_1, \ldots, \xi_i, \ldots, x_N) d\xi_i.$$

These are the Steklov averages of u with respect to the variable x_i.

Proposition 21.1. *Let* $u \in L^p(E)$ *for some* $p \in [1, \infty)$. *Then* $u_{h,i}$ *tends to* u *in*
$L^p(E)$ *as* $h \to 0$.

Proof. For almost all $x \in E$,

$$u_{h,i}(x) - u(x) = \frac{1}{h} \int_{x_i}^{x_i+h} [u(x_1, \ldots, \xi_i, \ldots, x_N) - u(x)] d\xi_i$$

$$= \frac{1}{h} \int_0^h [u(x_1, \ldots, x_i + \sigma, \ldots, x_N) - u(x)] d\sigma$$

$$= \frac{1}{h} \int_0^h [T_{\sigma,i} u(x) - u(x)] d\sigma,$$

where $T_{\sigma,i}$ is the translation of a function in $L^p(E)$ with respect to the variable x_i
of an amount σ. From this, for almost all $x \in E$,

$$|u_{h,i}(x) - u(x)| \leq \frac{1}{h} \int_0^h |T_{\sigma,i} u(x) - u(x)| d\sigma$$

$$\leq \frac{1}{|h|^{1/p}} \left(\int_0^{|h|} |T_{\sigma,i} u(x) - u(x)|^p d\sigma \right)^{1/p}.$$

Taking the p-power and integrating in dx over E gives

$$\|u_{h,i} - u\|_p \leq \sup_{|\sigma| \leq |h|} \|T_{\sigma,i}u - u\|_p.$$

Thus $\|u_{h,i} - u\|_p \to 0$ since the translation operator $T_{\sigma,i}$ is continuous in $L^p(E)$. $\quad\square$

For $\delta > 0$, let E_δ be the subset of E defined by

$$E_\delta = \{x \in E \,|\, \text{dist}\{x; \partial E\} > \delta\}. \tag{21.1}$$

Assume that δ is so small that E_δ is not empty, and denote by $\mathbf{h}$ a vector in $\mathbb{R}^N$,

$$\mathbf{h} = (h_1, h_2, \dots, h_N) \quad \text{such that } |\mathbf{h}| < \delta.$$

For $x \in E_\delta$ and $0 < |h_i| < \delta$, set

$$
\begin{aligned}
w_{h_i,i}(x) &= \frac{u(x_1, \dots, x_i + h_i, \dots, x_N) - u(x_2, \dots, x_i, \dots, x_N)}{h_i} \\
&= \frac{\partial u_{h_i,i}}{\partial x_i} \quad \text{a.e. in } E_\delta.
\end{aligned}
\tag{21.2}
$$

If $h_i = 0$, we set $w_{0,i} = 0$ and denote by $\mathbf{w_h}$ the vector of components $w_{h_i,i}$, $i = 1, 2, \dots, N$.

Proposition 21.2. *Let $u \in L^p(E)$ for some $1 < p < \infty$ and assume that there exist positive constants C and δ_o such that*

$$\|\mathbf{w_h}\|_{p,E_\delta} \leq C \quad \text{for all } \delta \leq \delta_o \quad \text{and} \quad \text{for all } |\mathbf{h}| < \delta. \tag{21.3}$$

Then

$$u \in W^{1,p}(E) \quad \text{and} \quad \|Du\|_p \leq C.$$

If E is of finite measure, the conclusion continues to hold for $p = \infty$.

Proof. Fix $\delta \in (0, \delta_o)$ and $\varphi \in C_o^\infty(E_\delta)$. For any fixed $i \in \{1, 2, \dots, N\}$,

$$\lim_{h_i \to 0} \int_E w_{h_i,i}\varphi \, dx = -\lim_{h_i \to 0} \int_E u_{h_i,i} D_i\varphi \, dx = -\int_E u D_i\varphi \, dx.$$

The family $\{w_{h_i,i}\}$ for $0 < |h_i| < \delta$ is uniformly bounded in $L^p(E_\delta)$. Therefore, we may extract a subsequence, relabelled with h, converging weakly to some function $w_i \in L^p(E_\delta)$.[15] For such a subsequence,

$$\lim_{h_i \to 0} \int_E w_{h_i,i}\varphi \, dx = \int_E w_i\varphi \, dx \quad \text{for all } \varphi \in C_o^\infty(E_\delta).$$

---

[15] See Section 19 of Chapter V. This induces the restriction $p > 1$.

Thus $w_i = D_i u$ in E_δ. This identifies w_i as the distributional D_i-derivative of u in E_δ. Once the limit has been identified, the selection of subsequences is unnecessary, and the entire family $\{w_{h_i,i}\}$ converges to $D_i u$ weakly in $L^p(E_\delta)$.

Since the L^p-norm is weakly lower semicontinuous,

$$\|Du\|_p^p \leq \liminf_{\delta \to 0} \int_{E_\delta} |Du|^p dx$$

$$\leq \liminf_{\delta \to 0} \liminf_{h \to 0} \int_{E_\delta} |\mathbf{w_h}|^p \, dx \leq C^p.$$

If $p = \infty$ and E is of finite measure, the assumption (21.3) reads

$$\|\mathbf{w_h}\|_{\infty, E_\delta} \leq C_\infty \quad \text{for all } \delta \leq \delta_o \quad \text{and} \quad \text{for all } |\mathbf{h}| < \delta. \qquad (21.3)_\infty$$

Therefore, for all fixed $p \gg 1$,

$$\|\mathbf{w_h}\|_{p, E_\delta} \leq C_\infty \mu(E)^{1/p} \quad \text{for all } \delta \leq \delta_o \quad \text{and} \quad \text{for all } |\mathbf{h}| < \delta.$$

Now repeating the same arguments gives

$$\|Du\|_p \leq C_\infty \mu(E)^{1/p} \quad \text{for all } p \gg 1.$$

Letting $p \to \infty$ yields $u \in W^{1,\infty}(E)$ and $\|Du\|_\infty \leq C_\infty$. $\qquad \square$

22 Characterizing $W^{1,p}(E)$ for $1 < p < \infty$

Let E be a domain in $\mathbb{R}^N$, and let E_δ be defined as in (21.1), where δ is so small that E_δ is not empty. Continue to denote by $\mathbf{w_h}$ the vectors whose components are given by (21.2) for $0 < |h_i| < \delta$ and by $w_{0,i} = 0$ for $h_i = 0$.

Proposition 22.1. *A function u belongs to $W^{1,p}(E)$ for some $1 < p < \infty$ if and only if there exists a positive number δ_o and a vector-valued function $\mathbf{w} \in L^p(E)$ such that*[16]

$$\|u(\cdot + \mathbf{h}) - u - \mathbf{h} \cdot \mathbf{w}\|_{p, E_\delta} = o(|\mathbf{h}|) \quad \text{as } |\mathbf{h}| \to 0 \qquad (22.1)$$

for every positive number $\delta \leq \delta_o$ and every vector $|\mathbf{h}| < \delta$. In such a case, $\mathbf{w}$ is the distributional derivative Du.

Proof (sufficient condition). Let $u \in L^p(E)$ satisfy (22.1) and for $|\mathbf{h}| > 0$, let $\mathbf{w_h}$ denote its discrete gradient as in (21.2). Having fixed $\delta > 0$, choose $\mathbf{h}$ of the form $\mathbf{h} = (0, \ldots, h_i, \ldots, 0)$ with $0 < |h_i| < \delta$. For such a choice,

$$\|\mathbf{w_h}\|_{p, E_\delta} = \frac{1}{|\mathbf{h}|} \|u(\cdot + \mathbf{h}) - u\|_{p, E_\delta}$$

$$\leq \frac{1}{|\mathbf{h}|} \|u(\cdot + \mathbf{h}) - u - \mathbf{h} \cdot \mathbf{w}\|_{p, E_\delta} + \frac{1}{|\mathbf{h}|} \|\mathbf{h} \cdot \mathbf{w}\|_{p, E_\delta}$$

$$\leq 1 + \|\mathbf{w}\|_p.$$

---

[16] $\mathbf{w} \in L^p(E)$ means that each component of $\mathbf{w}$ is in $L^p(E)$.

Therefore, $u \in W^{1,p}(E)$ and $\mathbf{w} = Du$ by Proposition 21.2. $\qquad\square$

Proof (necessary condition). Let $u \in W^{1,p}(E)$ and fix some $\delta > 0$ so small that E_δ is not empty. For each $i \in \{1, 2, \ldots, N\}$ and $0 < |h_i| < \delta$,

$$u(x_1, \ldots, x_i + h_i, \ldots, x_N) - u(x_1, \ldots, x_i, \ldots, x_N)$$

$$= \int_0^{h_i} u_{x_i}(x_1, \ldots, x_i + \sigma, \ldots, x_N) d\sigma$$

$$= h_i (u_{x_i})_{h_i} \quad \text{a.e. in } E_\delta,$$

where $(u_{x_i})_{h_i}$ is the Steklov average of u_{x_i}. From this,

$$\|u(\cdot + \mathbf{h}) - u - \mathbf{h} \cdot Du\|_{p, E_\delta} \leq |\mathbf{h}| \|(Du)_\mathbf{h} - Du\|_{p, E_\delta} = o(\mathbf{h}). \qquad\square$$

22.1   Remarks on $W^{1,\infty}(E)$. Assume that (22.1) holds for $p = \infty$ for some $\mathbf{w} \in L^\infty(E)$. Then

$$\frac{\|u(\cdot + \mathbf{h}) - u\|_{\infty, E_\delta}}{|\mathbf{h}|} \leq (1 + \|\mathbf{w}\|_{\infty, E}) \tag{22.2}$$

for all $|\mathbf{h}| < \delta$. If E is of finite measure, this implies that $u \in W^{1,\infty}(E)$ and $Du = \mathbf{w}$.

If (22.1) holds for $p = \infty$, it follows from (22.2) that u is Lipschitz continuous in E_δ. Conversely, if u is Lipschitz continuous in some domain E, it also is in $W^{1,\infty}(E')$ for every subdomain $E' \subset E$ of finite measure. Indeed, the discrete gradient $\mathbf{w}_h$ of u is pointwise bounded above by the Lipschitz constant of u.

It remains to investigate whether a Taylor formula of the type (22.1) would hold for such functions. This is the content of the Rademacher theorem.

23   The Rademacher theorem

A continuous function $f : \mathbb{R}^N \to \mathbb{R}$ is differentiable at $x \in \mathbb{R}^N$ if there exists a vector $Df(x) \in \mathbb{R}^N$ such that

$$f(y) = f(x) + Df(x) \cdot (y - x) + o(|x - y|) \quad \text{as } y \to x.$$

If such a vector $Df(x)$ exists, then

$$Df(x) = (f_{x_1}(x), f_{x_2}(x), \ldots, f_{x_2}(x)) = \nabla f(x).$$

However, the existence of $\nabla f(x)$ does not imply that f is differentiable at x.

Let $\mathbf{u} = (u_1, u_2, \ldots, u_N)$ be a fixed unit vector, and for $x \in \mathbb{R}^N$, set

$$D'_\mathbf{u} f(x) = \liminf_{\tau \to 0} \frac{f(x + \tau \mathbf{u}) - f(x)}{\tau},$$

$$D''_\mathbf{u} f(x) = \limsup_{\tau \to 0} \frac{f(x + \tau \mathbf{u}) - f(x)}{\tau}.$$

Since f is continuous, the limit can be taken for τ ranging over the set $\{\tau_n\}$ of the rational numbers in $(-1, 1)$. Therefore, $D'_\mathbf{u} f$ and $D''_\mathbf{u} f$ are both measurable. Also set

$$D_\mathbf{u} f(x) = \lim_{\tau \to 0} \frac{f(x + \tau\mathbf{u}) - f(x)}{\tau},$$

provided that the limit exists.

A function $f : \mathbb{R}^N \to \mathbb{R}$ is *locally* Lipschitz continuous if for every compact subset $K \subset \mathbb{R}^N$, there exists a positive constant L_K such that

$$\left| \frac{f(x) - f(y)}{x - y} \right| \le L_K |x - y| \quad \text{for all } x, y \in K.$$

A Lipschitz-continuous function f in $\mathbb{R}^N$ is locally Lipschitz continuous. The converse is false.

Proposition 23.1. *Let* $f : \mathbb{R}^N \to \mathbb{R}$ *be locally Lipschitz continuous. Then* $D_\mathbf{u} f$ *exists a.e. in* $\mathbb{R}^N$. *In particular,* ∇f *exists a.e. in* $\mathbb{R}^N$, *and*

$$D_\mathbf{u} f = \mathbf{u} \cdot \nabla f \quad \text{a.e. in } x \in \mathbb{R}^N.$$

Proof. Let $E_\mathbf{u}$ denote the set where $D_\mathbf{u} f$ does not exist; i.e.,

$$E_\mathbf{u} = \{x \in \mathbb{R}^N | D'_\mathbf{u} f(x) < D''_\mathbf{u} f(x)\}. \tag{23.1}$$

Since $D'_\mathbf{u} f$ and $D''_\mathbf{u} f$ are measurable, $E_\mathbf{u}$ is measurable. For $x \in \mathbb{R}^N$ fixed, the function of one variable

$$\mathbb{R} \ni t \longrightarrow f(x + t\mathbf{u})$$

is absolutely continuous in every subinterval of $\mathbb{R}$, and hence it is a.e. differentiable in $\mathbb{R}$. Therefore, the intersection of $E_\mathbf{u}$ with any line parallel to $\mathbf{u}$ has one-dimensional Lebesgue measure zero. Thus by Fubini's theorem, $\mu(E_\mathbf{u}) = 0$.

Continue to denote by $\{\tau_n\}$ the sequence of the rationals in $(-1, 1)$, and let $\mathbf{e}_j$, $j = 1, 2, \ldots, N$, be the coordinate unit vectors of $\mathbb{R}^N$. Since f is Lipschitz continuous, for every $\zeta \in C_o^\infty(\mathbb{R}^N)$,

$$\left| \frac{f(x + \tau_n\mathbf{u}) - f(x)}{\tau_n} \zeta \right| \le L_\zeta |\zeta|,$$

where L_ζ is the Lipschitz constant of f over a sufficiently large ball containing the support of ζ. Therefore, by dominated convergence and an elementary change of variables,

$$
\begin{aligned}
\int_{\mathbb{R}^N} D_{\mathbf{u}} f \zeta \, dx &= \lim_{\tau_n \to 0} \int_{\mathbb{R}^N} \frac{f(x + \tau_n \mathbf{u}) - f(x)}{\tau_n} \zeta \, dx \\
&= -\lim_{\tau_n \to 0} \int_{\mathbb{R}^N} f \frac{\zeta(x + \tau_n \mathbf{u}) - \zeta(x)}{\tau_n} \, dx \\
&= -u_j \int_{\mathbb{R}^N} f \zeta_{x_j} \, dx \\
&= -u_j \lim_{\tau_n \to 0} \int_{\mathbb{R}^N} f \frac{\zeta(x + \tau_n \mathbf{e}_j) - \zeta(x)}{\tau_n} \, dx \\
&= u_j \lim_{\tau_n \to 0} \int_{\mathbb{R}^N} \frac{f(x + \tau_n \mathbf{e}_j) - f(x)}{\tau_n} \zeta \, dx \\
&= \int_{\mathbb{R}^N} \mathbf{u} \cdot \nabla f \zeta \, dx.
\end{aligned}
$$

This, in turn, implies that

$$
\int_{\mathbb{R}^N} (D_{\mathbf{u}} f - \mathbf{u} \cdot \nabla f) \zeta \, dx = 0
$$

for all $\zeta \in C_o^\infty(\mathbb{R}^N)$. $\qquad\qquad\square$

Theorem 23.2 (Rademacher[17]). *Let $f : \mathbb{R}^N \to \mathbb{R}$ be locally Lipschitz continuous. Then f is a.e. differentiable in $\mathbb{R}^N$.*

Proof. Let $\{\mathbf{u}_n\}$ be a countable dense subset of the unit sphere ∂B_1, and let $E_{\mathbf{u}_n}$ be defined as in (23.1). By Proposition 23.1, $\mu(\bigcup E_{\mathbf{u}_n}) = 0$ for all n. To prove the theorem, we will establish that f is differentiable in $\mathbb{R}^N - \bigcup E_{\mathbf{u}_n}$.

Having fixed $x \in \mathbb{R}^N - \bigcup E_{\mathbf{u}_n}$, select $y \in B_1(x)$ with $y \neq x$ and set

$$
\mathbf{u} = \frac{y - x}{|x - y|}, \quad t = |x - y|, \quad y = x + t\mathbf{u}.
$$

Also, let L_R be the Lipschitz constant of f in the ball B_R centered at the origin and radius $R = 2 \max\{|x|; 1\}$. Then for all $\mathbf{u}_j \in \{\mathbf{u}_n\}$,

$$
\begin{aligned}
|f(y) - f(x) - \nabla f(x) \cdot (y - x)| &\leq |f(x + t\mathbf{u}) - f(x) - t\mathbf{u} \cdot \nabla f(x)| \\
&= |f(x + t\mathbf{u}_j) - f(x) - t\mathbf{u}_j \cdot \nabla f(x)| \\
&\quad + |f(x + t\mathbf{u}) - f(x + t\mathbf{u}_j)| \\
&\quad + t|(\mathbf{u} - \mathbf{u}_j) \cdot \nabla f(x)| \\
&\leq t \left| \frac{f(x + t\mathbf{u}_j) - f(x)}{t} - \mathbf{u}_j \cdot \nabla f(x) \right| \\
&\quad + t 2 L_R |\mathbf{u} - \mathbf{u}_j|.
\end{aligned}
$$

[17]H. Rademacher, Über partielle und totale Differenzierbarkeit, *Math. Ann.*, **79** (1919), 340–359.

Having fixed $\varepsilon > 0$, fix $y \in B_\delta(x)$, where $\delta > 0$ is to be chosen. Then for $\mathbf{u}$ fixed, choose

$$\mathbf{u}_j \in \{\mathbf{u}_n\} \quad \text{such that } 2L_R|\mathbf{u} - \mathbf{u}_j| \le \frac{1}{2}\varepsilon.$$

Such a choice is independent of δ. For $\mathbf{u}_j$ fixed, there exists $\delta > 0$ such that

$$\left| \frac{f(x + t\mathbf{u}_j) - f(x)}{t} - \mathbf{u}_j \cdot \nabla f(x) \right| \le \frac{1}{2}\varepsilon \quad \text{for all } 0 < t < \delta.$$

Therefore, for all $\varepsilon > 0$, there exists $\delta > 0$ such that

$$|f(y) - f(x) - \nabla f(x) \cdot (y - x)| \le |x - y|\varepsilon$$

provided for all $y \in \mathbb{R}^N$ such that $|x - y| \le \delta$. $\square$

Remark 23.1. Rademacher's theorem continues to hold for a function f Lipschitz continuous on a subset $E \subset \mathbb{R}^N$ modulo a preliminary application of the Kirzbraun–Pucci theorem.

PROBLEMS AND COMPLEMENTS

6 CHARACTERIZING THE BOUNDED LINEAR FUNCTIONALS ON $C_o(\mathbb{R}^N)$

Let E be a bounded open set in $\mathbb{R}^N$. The next theorem can be proved by a minor variant of the arguments in Sections 2–6.

Theorem 6.1c. *Let $T \in C(\overline{E})^*$. There exist two finite Borel measures μ_1 and μ_2 in $\overline{E}$ such that*

$$T(f) = \int_E f d\mu_1 - \int_E f d\mu_2 \quad \text{for all } f \in C(\overline{E}). \tag{6.1c}$$

Moreover, $\|T\| = |\mu|(E)$, where $|\mu|$ is the total variation of the signed measure $\mu = \mu_1 - \mu_2$. Finally, the restriction of the measures μ_1 and μ_2 to the Borel subsets of $\overline{E}$ is unique.

13 CONTINUOUS MAPS AND FUNCTIONALS

13.1 POSITIVE DISTRIBUTIONS ARE MEASURES. Let E be a bounded open set in $\mathbb{R}^N$. A distribution $T \in \mathcal{D}'(E)$ is positive if

$$\langle T, f \rangle \ge 0 \quad \text{whenever } f \ge 0.$$

Theorem 13.1c. *Let $T \in \mathcal{D}'(E)$ be positive. There exists a finite Borel measure μ in E such that*

$$\langle T, f \rangle = \int_E f \, d\mu \quad \text{for all nonnegative } f \in \mathcal{D}'(E). \tag{13.1c}$$

The measure μ is uniquely determined on the Borel sets.

Proof. The proof is a minor variant of the arguments in Sections 2–6. □

15   FUNDAMENTAL SOLUTIONS

15.1   Denote by x, y the variables in $\mathbb{R}^N$ and by t, τ the variables in $\mathbb{R}$. Then the fundamental solution of the Laplace operator in $\mathbb{R}^{N+1} = \mathbb{R}^N \times \mathbb{R}$ takes the form

$$F((x, t); (y, \tau)) = \frac{1}{(N-1)\omega_{N+1}} \frac{1}{[|x-y|^2 + (t-\tau)^2]^{\frac{N-1}{2}}} \tag{15.1c}$$

for $(x, t) \neq (y, \tau)$. For $\tau = 0$, we set

$$F(x - y; t) = \frac{1}{(N-1)\omega_{N+1}} \frac{1}{[|x-y|^2 + t^2]^{\frac{N-1}{2}}}. \tag{15.1c}_o$$

This is referred to as the fundamental solution of the Laplace equation in $\mathbb{R}^{N+1}$ with pole at $(y, 0)$.

Let $H(x - y; t)$ be the mollifying kernel of Section 21.2 of the Problems and Complements of Chapter V. Verify that

$$H(x - y; t) = 2 \frac{\partial}{\partial t} F(x - y; t) \quad \text{for } t > 0. \tag{15.2c}$$

Therefore, since $(x, t) \to F(x - y; t)$ is harmonic in $\mathbb{R}^N \times \mathbb{R}^+$,

$$\frac{\partial}{\partial t} H = -2\Delta_x F. \tag{15.3c}$$

16   WEAK DERIVATIVES AND MAIN PROPERTIES

16.1   Let E be a bounded open set in $\mathbb{R}^N$ containing the origin. Compute the values of $m \in \mathbb{N} \cup \{0\}$ for which the function $f(x) = |x|$ belongs to $W^{m,p}(E)$.

16.2   Taking weak derivatives is a local operation in the following sense.

Proposition 16.1c. *Let E be a bounded open set in $\mathbb{R}^N$. Then $u \in W^{m,p}(E)$ if and only if for each $x \in \overline{E}$, there exists an open neighborhood U_x about x such that $u \in W^{m,p}(E \cap U_x)$.*

16.3   CHARACTERIZING $W^{1,p}(a, b)$. Let $AC(a, b)$ denote the space of absolutely continuous functions in some finite interval (a, b).

Proposition 16.2c. *A function u defined in (a, b) is in $W^{1,p}(a, b)$ if and only if*

$$u \in L^p(a, b) \bigcap AC(a, b) \quad and \quad u' \in L^p(a, b). \tag{16.1c}$$

Proof. The implication $(16.1c) \Longrightarrow u \in W^{1,p}(a, b)$ is trivial. The converse follows from

$$u(y) = u(y) + \int_y^x Du(t)dt \quad \text{for a.e. } x, y \in (a, b),$$

where Du is the distributional derivative of u.   □

19   EXTENSIONS INTO $\mathbb{R}^N$

19.1   Introduce the notion of ∂E of class C^m for some positive integer m. The extension Proposition 19.1 is a particular case of the following.

Proposition 19.1c. *Let E be a bounded domain in $\mathbb{R}^N$ with boundary ∂E of class C^m for some positive integer m. Every function $u \in C^m(\overline{E})$ admits an extension $w \in C_o^m(\mathbb{R}^N)$ such that $w = u$ in $\overline{E}$ and*

$$\|w\|_{m,p;\mathbb{R}^N} \le C\|u\|_{m,p;E}, \tag{19.1c}$$

where C is a constant depending only upon N, m, p, and $\||\partial E\||_m$.

Proof. Let E be the cylinder Q_R^+, and assume that $u \in C^m(\overline{Q}_R^+)$ and that $\overline{x} \to u(\overline{x}, x_N)$ has compact support in $\mathcal{B}_R$. Extend u in Q_R^- by setting

$$\tilde{u}(\overline{x}, x_N) = \sum_{j=1}^{m+1} c_j u\left(\overline{x}, -\frac{1}{j}x_N\right) \quad \text{for } x_N \le 0,$$

where the numbers c_j satisfy the linear system

$$\sum_{j=1}^{m+1} (-1)^h \frac{c_j}{j^h} = 1, \quad h = 0, 1, \dots, m. \quad □$$

19.2   Let E be the unit disc in $\mathbb{R}^2$ from which one of the coordinate diameters has been removed. There exists a function $u \in W^{1,2}(E)$ that cannot be approximated in the norm of $W^{1,2}(E)$ by a sequence of functions in $C^1(\overline{E})$.

VIII

Topics on Integrable Functions of Real Variables

1 Vitali-type coverings

Let μ be the Lebesgue measure in $\mathbb{R}^N$, and refer the notions of measurability and integrability to such a measure. Let $E \subset \mathbb{R}^N$ be measurable and of finite measure, and let $\mathcal{F}$ be a collection of cubes in $\mathbb{R}^N$ with faces parallel to the coordinate planes whose union covers E. Such a covering is a Vitali-type covering. The cubes making up $\mathcal{F}$ are not required to be open or closed.

The next theorem asserts that E can be covered in a measure-theoretical sense by a countable collection of pairwise-disjoint cubes in $\mathcal{F}$. The key feature of this theorem is that, unlike the Vitali or the Besicovitch measure-theoretical covering theorems, the covering $\mathcal{F}$ is not required to be fine.[1]

This limited information on the covering $\mathcal{F}$ results in a weaker covering; i.e., the measure-theoretical covering is realized through an *estimation* rather than the equality of the measure of the set E in terms of the measure of the selected cubes.

Theorem 1.1 (Wiener[2]). *Let $E \subset \mathbb{R}^N$ be of finite measure, and let $\mathcal{F}$ be a Vitali-type covering for E. There exists a countable collection $\{Q_n\}$ of pairwise-disjoint cubes in $\mathcal{F}$ such that*

$$\frac{1}{5^N}\mu(E) \leq \sum \mu(Q_n). \tag{1.1}$$

---

[1] See Sections 17 and 20 of Chapter II.

[2] N. Wiener, The ergodic theorem, *Duke Math. J.*, **5** (1939), 1–18. The theorem is more general in that the set E is not required to be measurable. Let μ_e be the outer measure associated with μ by the procedure of Section 16.1 of Chapter II. Then the conclusion continues to hold, provided that $\mu_e(E) < \infty$, by replacing the left-hand side of (1.1) with $5^{-N}\mu_e(E)$.

Remark 1.1. The set E is not required to be bounded. It is not claimed here that the union of the disjoint cubes Q_n, satisfying (1.1), covers E.

Proof of Theorem 1.1. Label $\mathcal{F}$ by $\mathcal{F}_1$ and set

$$2\rho_1 = \{\text{the supremum of the edges of the cubes in } \mathcal{F}_1\}.$$

If $\rho_1 = \infty$, we select a cube Q of edge so large that

$$\mu(Q) \geq \frac{1}{5^N}\mu(E).$$

If $\rho_1 < \infty$, select a cube $Q_1 \in \mathcal{F}_1$ of edge $\ell_1 > \rho_1$ and subdivide $\mathcal{F}_1$ into two subcollections of cubes $\mathcal{F}_2$ and $\mathcal{F}_2'$ by setting

$$\mathcal{F}_2 = \{\text{the collection of cubes } Q \in \mathcal{F}_1 \text{ that do not intersect } Q_1\},$$
$$\mathcal{F}_2' = \{\text{the collection of cubes } Q \in \mathcal{F}_1 \text{ that intersect } Q_1\}.$$

Denote by Q_1' the cube with the same center as Q_1 and edge $5\ell_1$. Then by construction,

$$\bigcup\{Q | Q \in \mathcal{F}_2'\} \subset Q_1'.$$

If $\mathcal{F}_2$ is empty, then

$$E \subset Q_1' \quad \text{and} \quad \frac{1}{5^N}\mu(E) \leq \mu(Q_1).$$

If $\mathcal{F}_2$ is not empty, set

$$2\rho_2 = \{\text{the supremum of the edges of the cubes in } \mathcal{F}_2\}$$

and select a cube $Q_2 \in \mathcal{F}_2$ of edge $\ell_2 > \rho_2$. Then subdivide $\mathcal{F}_2$ into the two subcollections

$$\mathcal{F}_3 = \{\text{the collection of cubes } Q \in \mathcal{F}_2 \text{ that do not intersect } Q_2\},$$
$$\mathcal{F}_3' = \{\text{the collection of cubes } Q \in \mathcal{F}_2 \text{ that intersect } Q_2\}.$$

Denote by Q_2' the cube with the same center as Q_2 and edge $5\ell_2$. Then by construction,

$$\bigcup\{Q | Q \in \mathcal{F}_3'\} \subset Q_2'.$$

If $\mathcal{F}_3$ is empty,

$$E \subset Q_1' \bigcup Q_2' \quad \text{and} \quad \frac{1}{5^N}\mu(E) \leq \mu(Q_1) + \mu(Q_2).$$

If $\mathcal{F}_3$ is not empty, we repeat the process to inductively define subfamilies of cubes $\{\mathcal{F}_n\}$, positive numbers $\{\rho_n\}$ and $\{\ell_n\}$, and cubes $\{Q_n\}$ and $\{Q'_n\}$ by the procedure

$\mathcal{F}_n = \{$collection of cubes in $\mathcal{F}_{n-1}$ that do not intersect $Q_{n-1}\}$,

$2\rho_n = \{$the supremum of the edges of the cubes in $\mathcal{F}_n\}$,

$Q_n = \{$a cube selected out of $\mathcal{F}_n$ of edge $\ell_n > \rho_n\}$,

$Q'_n = \{$a cube with the same center as Q_n and edge $5\ell_n\}$.

If $\mathcal{F}_{n+1}$ is empty for some $n \in \mathbb{N}$, then

$$E \subset Q'_1 \bigcup Q'_2 \bigcup \cdots \bigcup Q'_n \quad \text{and} \quad \frac{1}{5^N}\mu(E) \le \sum_{j=1}^{n} \mu(Q_j).$$

If $\mathcal{F}_n$ is not empty for all $n \in \mathbb{N}$, consider the series $\sum \mu(Q_n)$. If the series diverges, (1.1) is trivial. If the series converges, then $\rho_n \to 0$ as $n \to \infty$. In such a case, we claim that every cube $Q \in \mathcal{F}$ belongs to some Q'_n. Indeed, if not, Q must belong to all $\mathcal{F}_n$ and therefore its edge is zero. Thus

$$E \subset \bigcup Q'_n \quad \text{and} \quad \mu(E) \le \sum \mu(Q'_n) = 5^N \sum \mu(Q_n). \qquad \square$$

Corollary 1.2. *Let $E \subset \mathbb{R}^N$ be measurable and of finite measure, and let $\mathcal{F}$ be a collection of cubes in $\mathbb{R}^N$ with faces parallel to the coordinate planes and covering E. For every $\varepsilon > 0$, there exists a finite collection $\{Q_1, Q_2, \ldots, Q_m\}$ of disjoint cubes in $\mathcal{F}$ such that*

$$\frac{1}{5^N}\mu(E) - \varepsilon \le \sum_{j=1}^{m} \mu(Q_j). \tag{1.1}'$$

2 The maximal function

Let Q denote a cube centered at the origin and with faces parallel to the coordinate planes. For $x \in \mathbb{R}^N$, we let $Q(x)$ denote the cube centered at x and congruent to Q. The maximal function $M(f)$ of a function $f \in L^1_{\text{loc}}(\mathbb{R}^N)$ is defined by[3]

$$M(f)(x) = \sup_Q \frac{1}{\mu(Q)} \int_{Q(x)} |f(y)|dy. \tag{2.1}$$

From the definition, it follows that $M(f)$ is nonnegative and subadditive with respect to the argument f; i.e.,

$$M(f + g) \le M(f) + M(g).$$

Moreover, $M(\alpha f) = |\alpha| M(f)$ for all $\alpha \in \mathbb{R}$.

---

[3]For $N = 1$, the notion and the main properties were introduced and established in G. H. Hardy and J. E. Littlewood, A maximal theorem with function-theoretical applications, *Acta Math.*, **54** (1930), 81–116. The N-dimensional version is in N. Wiener, The ergodic theorem, *Duke Math. J.*, **5** (1939), 1–18.

Proposition 2.1. $M(f)$ *is measurable and lower semicontinuous. Moreover, if* f *is the characteristic function of a bounded measurable set* E, *there exists positive constants* C_o, C_1, *and* γ *depending only upon* E *such that*

$$\frac{C_o}{|x|^N} \le M(\chi_E)(x) \le \frac{C_1}{|x|^N} \quad \text{for all } |x| > \gamma. \tag{2.2}$$

Proof. If $|f| \equiv 0$, $M(f) \equiv 0$ also. Otherwise, $M(f)(x) > 0$ for all $x \in \mathbb{R}^N$. Let $c > 0$ and $x \in [M(f) > c]$. There exist $\varepsilon > 0$ and a cube Q such that

$$\frac{1}{\mu(Q)} \int_{Q(x)} |f| dx \ge M(f)(x) - \varepsilon > c.$$

By the absolute continuity of the integral, there exists $\delta > 0$ such that

$$M(f)(y) \ge \frac{1}{\mu(Q)} \int_{Q(y)} |f| dx > c$$

for all $|y - x| < \delta$. Thus $[M(f) > c]$ is open. This implies that $M(f)$ is lower semicontinuous and measurable. To prove (2.2), observe that

$$M(\chi_E)(x) = \sup_Q \frac{\mu\left(E \cap Q(x)\right)}{\mu(Q)}.$$

Since E is bounded, it is included in some cube Q_o.

Let d be the maximum distance of points in Q_o from the origin.

Fix $x \in \mathbb{R}^N$ such that $|x| \ge 2d/\sqrt{N}$, and let $Q(x)$ be the smallest cube centered at x and containing E. Then

$$M(\chi_E)(x) \ge \frac{\mu\left(E \cap Q(x)\right)}{\mu(Q)} = \frac{\mu(E)}{\mu(Q)} = \frac{C_o}{|x|^N}.$$

The bound above is estimated analogously. $\qquad\qquad\square$

Corollary 2.2. *Let* $f \in L^1(\mathbb{R}^N)$ *be of compact support in* $\mathbb{R}^N$ *and not identically zero. There exist constants* C_o, C_1, *and* γ *depending only upon* f *such that*

$$\frac{C_o}{|x|^N} \le M(f)(x) \le \frac{C_1}{|x|^N} \quad \text{for all } |x| > \gamma. \tag{2.2}'$$

It follows from (2.2)–(2.2)' that $M(f)$ is not in $L^1(\mathbb{R}^N)$ even if f is bounded and compactly supported unless $f \equiv 0$.

Proposition 2.3. *Let* $f \in L^1(\mathbb{R}^N)$. *Then for all* $t > 0$,

$$\mu([M(f) > t]) \le \frac{5^N}{t} \int_{\mathbb{R}^N} |f| dx. \tag{2.3}$$

Proof. Assume first that f is of compact support. Then $(2.2)'$ implies that the set $[M(f) > t]$ is of finite measure. For every $x \in [M(f) > t]$, there exists a cube $Q(x)$ centered at x and with faces parallel to the coordinate planes such that

$$\mu(Q) \leq \frac{1}{t} \int_{Q(x)} |f| dy. \tag{2.4}$$

The collection $\mathcal{F}$ of all such cubes is a covering of $[M(f) > t]$. Out of $\mathcal{F}$, we may extract a countable collection $\{Q_n\}$ of disjoint cubes such that

$$\frac{1}{5^N} \mu([M(f) > t]) \leq \sum \mu(Q_n).$$

From this and (2.4),

$$\mu([M(f) > t]) \leq 5^N \sum \mu(Q_n)$$
$$\leq \frac{5^N}{t} \sum \int_{Q_n} |f| dx \leq \frac{5^N}{t} \int_{\mathbb{R}^N} |f| dx.$$

If $f \in L^1(\mathbb{R}^N)$, we may assume that $f \geq 0$. Let $\{f_n\}$ be a nondecreasing sequence of compactly supported, nonnegative functions in $L^1(\mathbb{R}^N)$ converging to f a.e. in $\mathbb{R}^N$. Then $\{M(f_n)\}$ is a nondecreasing sequence of measurable functions converging to $M(f)$ a.e. in $\mathbb{R}^N$. Therefore, by monotone convergence,

$$\mu([M(f) > t]) = \lim \mu([M(f_n) > t])$$
$$\leq \frac{5^N}{t} \lim \int_{\mathbb{R}^N} f_n dx \leq \frac{5^N}{t} \int_{\mathbb{R}^N} f dx. \qquad \square$$

3 Strong L^p estimates for the maximal function

Proposition 3.1. *Let $f \in L^p(\mathbb{R}^N)$ for $p \in (1, \infty]$. Then $M(f) \in L^p(\mathbb{R}^N)$ and*

$$\|M(f)\|_p \leq \gamma_p \|f\|_p, \quad \text{where } \gamma_p^p = \frac{2^p p 5^N}{p-1}. \tag{3.1}$$

Proof. The estimate is obvious if $p = \infty$. Assuming that $p \in (1, \infty)$, fix $t > 0$ and set

$$g(x) = \begin{cases} f(x) & \text{if } |f(x)| \geq \frac{1}{2}t \\ 0 & \text{if } |f(x)| < \frac{1}{2}t. \end{cases}$$

Such a function g is in $L^1(\mathbb{R}^N)$. Indeed,

$$\int_{\mathbb{R}^N} |g| dx = \int_{[|f| \geq \frac{1}{2}t]} |f| dx \leq \left(\frac{2}{t}\right)^{p-1} \int_{\mathbb{R}^N} |f|^p dx.$$

Since $|f| \leq |g| + \frac{1}{2}t$,

$$M(f)(x) \leq \sup \frac{1}{\mu(Q)} \int_{Q(x)} |g| dy + \frac{1}{2}t = M(g)(x) + \frac{1}{2}t.$$

From this, one derives the inclusion

$$[M(f) > t] \subset \left[M(g) > \frac{1}{2}t \right].$$

Therefore, taking into account Proposition 2.3 applied to g and $M(g)$,

$$\mu([M(f) > t]) \leq \mu\left(\left[M(g) > \frac{1}{2}t \right] \right) \leq \frac{2 \cdot 5^N}{t} \int_{\mathbb{R}^N} |g| dx$$

$$= \frac{2 \cdot 5^N}{t} \int_{[|f| \geq \frac{1}{2}t]} |f| dx.$$

We now express the integral of $M(f)^p$ in terms of the distribution function of $M(f)$; i.e.,[4]

$$\int_{\mathbb{R}^N} M(f)^p dx = p \int_0^\infty t^{p-1} \mu([M(f) > t]) dt$$

$$\leq 2p5^N \int_0^\infty t^{p-2} \left(\int_{[|f| \geq \frac{1}{2}t]} |f| dx \right) dt.$$

Interchanging the order of integration by means of Fubini's theorem,

$$\int_{\mathbb{R}^N} M(f)^p dx \leq 2p5^N \int_{\mathbb{R}^N} |f| \left(\int_0^{2|f|} t^{p-2} dt \right) dx$$

$$= \frac{2^p p5^N}{p-1} \int_{\mathbb{R}^N} |f|^p dx. \qquad \square$$

3.1 Estimates of weak and strong type. Let $E \subset \mathbb{R}^N$ be measurable. A measurable function $g : E \to \mathbb{R}$ is in the space weak-$L^1(E)$, denoted by $L_w^1(E)$, if there exists a constant C depending only upon g such that

$$\mu([|g| > t]) \leq \frac{C}{t} \quad \text{for all } t > 0.$$

If $g \in L^1(E)$,

$$\mu([|g| > t]) \leq \frac{1}{t} \int_E |g| dx.$$

---

[4] See Section 15.1 of Chapter III and, in particular, $(15.4)_p$.

Therefore, $L^1(E)$ is included in weak-$L^1(E)$. The converse is false. For example, the function

$$g(x) = \begin{cases} \dfrac{1}{|x|^N} & \text{for } |x| > 0, \\ 0 & \text{for } |x| = 0 \end{cases}$$

belongs to weak-$L^1(\mathbb{R}^N)$ and not to $L^1(\mathbb{R}^N)$.

By Proposition 2.3, the maximal function $M(f)$ of a function $f \in L^1(\mathbb{R}^N)$ is in weak-$L^1(\mathbb{R}^N)$ and, in general, not in $L^1(\mathbb{R}^N)$.

Let T be a map acting on $L^1(E)$ and such that $T(f)$ is a real-valued, measurable function defined on E. The maximal function

$$T(f) = M(f) \tag{3.2}$$

is one such map. Another example can be constructed by setting

$$T(f) = (J_\varepsilon * f), \tag{3.3}$$

where J_ε are mollifying kernels.[5]

A map T is of weak type in $L^1(E)$ if $T(f)$ is in weak-$L^1(E)$ for every $f \in L^1(E)$. By Proposition 2.3, the map T in (3.2) is of weak type in $L^1(\mathbb{R}^N)$.

A map T is of strong type in $L^p(E)$ for some $1 \le p \le \infty$ if

$$f \in L^p(E) \implies T(f) \in L^p(E).$$

The convolution in (3.3) is of strong type in $L^p(\mathbb{R}^N)$ for all $p \in [1, \infty)$. By Proposition 3.1, the maximal function $M(f)$ is of strong type in $L^p(E)$ for $p \in (1, \infty)$.

4 The Calderón–Zygmund decomposition theorem

Theorem 4.1 (Calderón–Zygmund[6]). *Let f be a nonnegative function in $L^1(\mathbb{R}^N)$. Then for any fixed $\alpha > 0$, the whole $\mathbb{R}^N$ can be partitioned into two disjoint sets E and F such that the following hold:*

(i) *$f(x) \le \alpha$ almost everywhere in E.*

(ii) *F is the countable union of closed cubes Q_n with faces parallel to the coordinate planes and with pairwise disjoint interior.*

For each of these cubes,

$$\alpha < \frac{1}{\mu(Q_n)} \int_{Q_n} f(y)\,dy \le 2^N \alpha. \tag{4.1}$$

---

[5]See Section 21 and Section 21 of the Problems and Complements of Chapter V.

[6]A. P. Calderón and A. Zygmund, On the existence of certain singular integrals, *Acta. Math.*, **88** (1952), 85–139.

Proof. Let $\alpha > 0$ be fixed, and decompose $\mathbb{R}^N$ into closed cubes with pairwise-disjoint interior with faces parallel to the coordinate planes and of equal edge. Since $f \in L^1(\mathbb{R}^N)$, such a partition can be realized so that for every cube Q' of such a partition,

$$\frac{1}{\mu(Q')} \int_{Q'} f(y)dy \leq \alpha.$$

Having fixed one such cube Q', we partition it into 2^N equal cubes by bisecting Q' with hyperplanes parallel to the coordinate planes. Let Q'' be any one of these new cubes. Then either

$$\frac{1}{\mu(Q'')} \int_{Q''} f(y)dy \leq \alpha \tag{a}$$

or

$$\frac{1}{\mu(Q'')} \int_{Q''} f(y)dy > \alpha. \tag{b}$$

If (b) occurs, then Q'' is not subdivided further and is taken as one of the cubes of the collection $\{Q_n\}$ claimed by the theorem. Indeed,

$$\alpha < \frac{1}{\mu(Q'')} \int_{Q''} f(y)dy \leq \frac{2^N}{\mu(Q')} \int_{Q'} f(y)dy \leq 2^N \alpha.$$

If (a) occurs, we subdivide further Q'' into 2^N subcubes and on each of them repeat the same alternative.

For each of the cubes Q' of the initial partition of $\mathbb{R}^N$, we carry on this recursive partitioning process. The process terminates only if case (b) occurs. Otherwise, it is continued recursively.

Let $F = \bigcup Q_n$, where Q_n are cubes for which (b) occurs. By construction these are cubes with faces parallel to the coordinate planes and with pairwise-disjoint interior. Moreover, (4.1) holds for all of them.

Setting $E = \mathbb{R}^N - F$, it remains to prove (i). Let x be a Lebesgue point of f in F. There exists a sequence of cubes $\tilde{Q}_j$ with faces parallel to the coordinate planes and containing x resulting from the indicated recursive partition such that

$$\lim_{j \to \infty} \text{diam}\{\tilde{Q}_j\} = 0 \quad \text{and} \quad \frac{1}{\mu(\tilde{Q}_j)} \int_{\tilde{Q}_j} f(y)dy \leq \alpha.$$

The collection of cubes $\{\tilde{Q}_j\}$ forms a regular family $\mathcal{F}_x$ at x. Therefore,[7]

$$f(x) = \lim_{j \to \infty} \frac{1}{\mu(\tilde{Q}_j)} \int_{\tilde{Q}_j} f(y)dy \leq \alpha.$$

Since almost every $x \in F$ is a Lebesgue point, (i) follows. $\square$

---

[7]See Section 12 and, in particular, Proposition 12.1 of Chapter IV.

5   Functions of bounded mean oscillation

Let Q_o be a cube in $\mathbb{R}^N$ centered at the origin and with faces parallel to the coordinate planes. For a function $f \in L^1_{\text{loc}}(Q_o)$ and a cube $Q \subset Q_o$ with faces parallel to the coordinate planes, let f_Q denote the integral average of f in Q; i.e.,

$$f_Q = \frac{1}{\mu(Q)} \int_Q f(y) dy.$$

A function $f \in L^1(Q_o)$ is said to be of *bounded mean oscillation* if

$$|f|_o = \sup_{Q \subset Q_o} \frac{1}{\mu(Q)} \int_Q |f - f_Q| dy < \infty. \tag{5.1}$$

The collection of all functions $f \in L^1(Q_o)$ of bounded mean oscillation is denoted by $\text{BMO}(Q_o)$. One verifies that $\text{BMO}(Q_o)$ is a linear space and that

$$\|f\|_o = \|f\|_1 + |f|_o,$$

defines a norm on $\text{BMO}(Q_o)$. Moreover, from the definition and the completeness of $L^1(Q_o)$, it follows that $\text{BMO}(Q_o)$ is complete.

Theorem 5.1 (John–Nirenberg[8]). *There exist two positive constants C_1, C_2 depending only on N such that for every $f \in \text{BMO}(Q_o)$, for all cubes $Q \in Q_o$, and for all $t \geq 0$,*

$$\mu\{x \in Q \mid |f(x) - f_Q| > t\} \leq C_1 \exp\left\{-\frac{C_2 t}{|f|_o}\right\} \mu(Q). \tag{5.2}$$

Remark 5.1. Let $f \in L^1(Q_o)$ and assume that there exists positive constants γ_1 and γ_2 such that the inequality

$$\mu\{x \in Q \mid |f(x) - f_Q| > t\} \leq \gamma_1 \exp\{-\gamma_2 t\} \mu(Q) \tag{5.2}'$$

holds for all $t \geq 0$ and all subcubes $Q \subset Q_o$. Then

$$\int_Q |f - f_Q| dy = \int_0^\infty \mu\{x \in Q \mid |f(x) - f_Q| > t\} dt$$
$$\leq \gamma_1 \mu(Q) \int_0^\infty e^{-\gamma_2 t} dt \leq \frac{\gamma_1}{\gamma_2} \mu(Q) \tag{5.3}$$

for all subcubes $Q \subset Q_o$. Thus $f \in \text{BMO}(Q_o)$ and $|f|_o \leq \gamma_1/\gamma_2$.

This implies that (5.2)–(5.2)$'$ are necessary and sufficient for a function $f \in L^1(Q_o)$ to be of bounded mean oscillation in Q_o.

---

[8]F. John and L. Nirenberg, On functions of bounded mean oscillation, *Comm. Pure Appl. Math.*, **XIV** (1961), 415–426.

Remark 5.2. A function $f \in L^\infty(Q_o)$ is in $\mathrm{BMO}(Q_o)$. The converse is false as the function $x \to \ln|x|$ is in of bounded mean oscillation within the unit cube Q_o centered at the origin but is not bounded in Q_o.[9]

Remark 5.3. If $f \in \mathrm{BMO}(Q_o)$, then $(f - f_Q) \in L^p(Q)$ for all $1 \le p < \infty$ for every cube $Q \subset Q_O$. Indeed, from (5.2),

$$
\int_Q |f - f_Q|^p dy = p \int_0^\infty t^{p-1} \mu\{x \in Q | |f(x) - f_Q| > t\} dt
$$

$$
\le pC_1\mu(Q) \int_0^\infty t^{p-1} \exp\left\{-\frac{C_2 t}{|f|_o}\right\} dt
$$

$$
\le pC_1\mu(Q) \left(\frac{|f|_o}{C_2}\right)^p \int_0^\infty t^{p-1} e^{-t} dt
$$

$$
\le \gamma(N, p) |f|_o^p \mu(Q)
$$

for a constant $\gamma(N, p)$ depending upon only N and p. This in turn implies that $f \in L^p(Q)$ since

$$
\int_Q |f|^p dy \le \gamma(p) \left\{\int_Q |f - f_Q|^p dy + |f_Q|^p \mu(Q)\right\}
$$

$$
\le \gamma'(N, p)\mu(Q)\{|f|_o^p + \|f\|_{1,Q}^p\}
$$

$$
\le \gamma''(N, p)\mu(Q)\|f\|_o^p.
$$

6   Proof of Theorem 5.1

Having fixed some cube $Q \subset Q_o$, we may assume without loss of generality that $Q = Q_o$. Also, by possibly replacing f with $f/\|f\|_o$, we may assume that $\|f\|_o = 1$. Set

$$
f_{1,o}(x) = \begin{cases} |f(x) - f_{Q_o}| & \text{for } x \in Q_o, \\ 0 & \text{otherwise.} \end{cases}
$$

Since $f_{1,o} \in L^1(\mathbb{R}^N)$, having fixed some $\alpha > 1$, by the Calderón–Zygmund decomposition theorem, there exists a countable collection of closed cubes $\{Q_n^1\}$ with faces parallel to the coordinate planes and with pairwise-disjoint interior such that

$$
\alpha < \frac{1}{\mu(Q_n^1)} \int_{Q_n^1} |f - f_{Q_o}| dy \le 2^N \alpha \tag{6.1}
$$

and

$$
|f(x) - f_{Q_o}| \le \alpha \quad \text{for a.e. } x \in Q_o - \bigcup Q_n^1. \tag{6.2}_1
$$

---

[9]See Proposition 5.2c of the Problems and Complements.

It follows from $(6.1)_1$ and $\| f \|_o = 1$ that

$$\sum \mu(Q_n^1) \leq \frac{1}{\alpha} \int_{Q_o} |f - f_{Q_o}| dy \leq \frac{1}{\alpha} \mu(Q_o). \qquad (6.3)_1$$

Also, using again that $\| f \|_o = 1$,

$$|f_{Q_n^1} - f_{Q_o}| \leq \frac{1}{\mu(Q_n^1)} \int_{Q_n^1} |f - f_{Q_o}| dy \leq 2^N \alpha. \qquad (6.4)_1$$

Finally, since $\alpha > 1$ and $\| f \|_o = 1$,

$$\frac{1}{\mu(Q_n^1)} \int_{Q_n^1} |f(y) - f_{Q_n^1}| dy < \alpha \quad \text{for all cubes } Q_n^1. \qquad (6.5)$$

For $n \in \mathbb{N}$ fixed, set

$$f_{1,n}(x) = \begin{cases} |f(x) - f_{Q_n^1}| & \text{for } x \in Q_n^1, \\ 0 & \text{otherwise.} \end{cases}$$

We again apply the Calderón–Zygmund decomposition for the same $\alpha > 1$ to the function $f_{1,n}$ starting from the cube Q_n^1 and using (6.5). This generates a countable collection of cubes $\{Q_{n,m}^2\}$ with faces parallel to the coordinate planes and with pairwise-disjoint interior such that

$$\alpha < \frac{1}{\mu(Q_{n,m}^2)} \int_{Q_{n,m}^2} |f - f_{Q_n^1}| dy \leq 2^N \alpha \qquad (6.1)'$$

and

$$|f(x) - f_{Q_n^1}| \leq \alpha \quad \text{for a.e. } x \in Q_n^1 - \bigcup_m Q_{n,m}^2. \qquad (6.2)_1'$$

For such a collection, the analogue of $(6.3)_1$ is also satisfied; i.e.,

$$\sum_m \mu(Q_{n,m}^2) \leq \frac{1}{\alpha} \sum_m \int_{Q_{n,m}^2} |f - f_{Q_{n,m}^2}| dy$$

$$\leq \frac{1}{\alpha} \sum_m \mu(Q_{n,m}^2) \leq \frac{1}{\alpha} \mu(Q_n^1), \qquad (6.4)_1'$$

where we have used that $\| f \|_o = 1$. Next, we claim that

$$|f(x) - f_{Q_o}| \leq 2 \cdot 2^N \alpha \quad \text{for a.e. } x \in Q_o - \bigcup_{n,m} Q_{n,m}^2.$$

If $x \in Q_o - \bigcup Q_n^1$, this follows from $(6.2)_1$. If, on the other hand,

$$x \in \bigcup_n Q_n^1 - \bigcup_{n,m} Q_{n,m}^2,$$

then by $(6.4)_1$ and $(6.2)'_1$,

$$|f(x) - f_{Q_o}| \le |f(x) - f_{Q_n^1}| + |f_{Q_n^1} - f_{Q_o}| \le 2 \cdot 2^N \alpha.$$

Adding $(6.4)'_1$ with respect to n and taking into account $(6.3)_1$ gives

$$\sum_{n,m} \mu(Q_{n,m}^2) \le \frac{1}{\alpha^2} \mu(Q_o).$$

Also, using $(6.4)_1$ and $(6.1)'$,

$$|f_{Q_{n,m}^2} - f_{Q_o}| \le |f_{Q_{n,m}^2} - f_{Q_n^1}| + |f_{Q_n^1} - f_{Q_o}|$$

$$\le \frac{1}{\mu(Q_{n,m}^2)} \int_{Q_{n,m}^2} |f - f_{Q_n^1}| dy + |f_{Q_n^1} - f_{Q_o}|$$

$$\le 2 \cdot 2^N \alpha.$$

We now relabel $\{Q_{n,m}^2\}$ to obtain a countable collection $\{Q_n^2\}$ of closed cubes with faces parallel to the coordinate planes, with pairwise-disjoint interior, and such that

$$|f(x) - f_{Q_o}| \le 2 \cdot 2^N \alpha \quad \text{for a.e. } x \in Q_o - \bigcup Q_n^2, \tag{6.2}_2$$

$$\sum \mu(Q_n^2) \le \frac{1}{\alpha^2} \mu(Q_o). \tag{6.3}_2$$

$$|f_{Q_n^2} - f_{Q_o}| \le 2 \cdot 2^N \alpha. \tag{6.4}_2$$

Repeating the process k times generates a countable collection $\{Q_n^k\}$ of closed subcubes of Q_o with faces parallel to the coordinate planes, with pairwise-disjoint interior, and such that

$$|f(x) - f_{Q_o}| \le k \cdot 2^N \alpha \quad \text{for a.e. } x \in Q_o - \bigcup_n Q_n^k, \tag{6.2}_k$$

$$\sum_n \mu(Q_n^k) \le \frac{1}{\alpha^k} \mu(Q_o). \tag{6.3}_k$$

$$|f_{Q_n^k} - f_{Q_o}| \le k \cdot 2^N \alpha. \tag{6.4}_k$$

From this, for a fixed positive integer k,

$$\mu\{x \in Q_o | |f(x) - f_{Q_o}| > k2^N \alpha\} \le \sum_n \mu(Q_n^k) \le \frac{1}{\alpha^k} \mu(Q_o).$$

This inequality continues to hold for $k = 0$. Now fix $t > 0$ and let $k \ge 0$ be such that

$$k2^N \alpha < t \le (k+1)2^N \alpha.$$

Then

$$\mu\{x \in Q_o | |f(x) - f_{Q_o}| > t\} \leq \mu\{x \in Q_o | |f(x) - f_{Q_o}| > k2^N \alpha\}$$

$$\leq \frac{1}{\alpha^k} \mu(Q_o) \leq \alpha e^{-\gamma t} \mu(Q_o),$$

where $\gamma = (\ln \alpha)/2^N \alpha$.   □

7   The sharp maximal function

Continue to denote by Q_o and Q closed cubes in $\mathbb{R}^N$ with faces parallel to the coordinate planes. Given a function $f \in L^1(Q_o)$, we regard it as defined in the whole $\mathbb{R}^N$ by setting it to be zero outside Q_o.

For a cube Q whose center is in Q_o, let $f_{Q \cap Q_o}$ denote the integral average of f over $Q \cap Q_o$; i.e.,

$$f_{Q \cap Q_o} = \frac{1}{\mu(Q \cap Q_o)} \int_{Q \cap Q_o} f(y)dy. \tag{7.1}$$

Also set

$$|f|_{Q_o} = \frac{1}{\mu(Q_o)} \int_{Q_o} |f|dy. \tag{7.1}_o$$

The *sharp* maximal function $x \to f^\#(x)$ for $x \in Q_o$ is defined by

$$f^\#(x) = \sup_{Q \ni x} \frac{1}{\mu(Q \cap Q_o)} \int_{Q \cap Q_o} |f(y) - f_{Q \cap Q_o}|dy, \tag{7.2}$$

where the supremum is taken over all cubes Q containing x. This is also called the function of *maximal mean oscillation*.

It follows from (7.2) that $f \in \text{BMO}(Q_o)$ if and only if $f^\# \in L^\infty(Q_o)$. Also, from (7.2) and the definition of a maximal function $M(f)$,

$$f^\#(x) \leq 2^{N+1} M(f)(x) \quad \text{for a.e. } x \in Q_o.$$

By Proposition 2.3, this implies that if $f \in L^1(Q_o)$, then

$$\mu([f^\# > t]) \leq \frac{2 \cdot 10^N}{t} \int_{Q_o} |f|dy \quad \text{for all } t > 0. \tag{7.3}$$

Equivalently, $f^\# \in L_w^1(Q_o)$. If $f \in L^p(Q_o)$ for $p \in (1, \infty)$, by Proposition 3.1,

$$\|f^\#\|_p \leq 2^{N+1} \gamma_p \|f\|_p, \tag{7.4}$$

where the constant γ_p is the one appearing in (3.1).

The next theorem asserts the converse of (7.4); i.e., if $f^\# \in L^p(Q_o)$, then also $f \in L^p(Q_o)$.

Theorem 7.1 (Fefferman–Stein[10]). *Let $f \in L^1(Q_o)$ and assume that the corresponding sharp maximal function $f^\#$ is in $L^p(Q_o)$. Then $f \in L^p(Q_o)$ and there exists a positive constant $\gamma = \gamma(N, p)$ depending only upon N and p such that*

$$\|f\|_p \leq \gamma(N, p)\{\|f^\#\|_p + \mu(Q_o)|f|_{Q_o}\}. \tag{7.5}$$

8 Proof of the Fefferman–Stein theorem

Fix $t > |f|_{Q_o}$, and apply the Calderón–Zygmund decomposition to the function $|f|$ for $\alpha = t$. This generates a countable collection $\{Q_n^t\}$ of closed cubes with faces parallel to the coordinate planes, with pairwise-disjoint interior, and such that

$$t < \frac{1}{\mu(Q_n^t)} \int_{Q_n^t} |f| dy \leq 2^N t \tag{8.1}_t$$

and

$$|f| \leq t \quad \text{a.e. in } Q_o - \bigcup Q_n^t. \tag{8.2}_t$$

Without loss of generality, we may arrange that Q_o is part of the initial partition of $\mathbb{R}^N$ in the Calderón–Zygmund process. Therefore, the cubes Q_n^t result from repeated bisections starting from the parent cube Q_o.

Now let τ be a number satisfying

$$t > \tau > |f|_{Q_o}, \tag{8.3}$$

and let $\{Q_j^\tau\}$ be the corresponding decomposition for $\alpha = \tau$ satisfying the analogue of $(8.1)_t$ and $(8.2)_t$; i.e.,

$$\tau < \frac{1}{\mu(Q_j^\tau)} \int_{Q_j^\tau} |f| dy \leq 2^N \tau. \tag{8.1}_\tau$$

$$|f| \leq \tau \quad \text{a.e. in } Q_o - \bigcup Q_j^\tau. \tag{8.2}_\tau$$

By the Calderón–Zygmund recursive partition process and the previous remarks regarding Q_o, each of the cubes Q_n^t is a subcube of some Q_j^τ. Therefore,

$$M(t) = \sum \mu(Q_n^t) \leq \sum \mu(Q_j^\tau) = M(\tau).$$

Also, for any t satisfying (8.3),

$$\mu\{x \in Q_o | |f| > t\} \leq M(t). \tag{8.4}$$

$\square$

---

[10]C. Fefferman and E. M. Stein, H^p spaces of several variables, *Acta Math.*, **129** (1972), 137–193.

Lemma 8.1. *Let* $t > 2^{N+1}|f|_{Q_o}$. *Then*

$$M(t) \le \mu \left\{ x \in Q_0 | f^{\#}(x) > \frac{t}{C} \right\} + \frac{2}{C} M \left(\frac{t}{2^{N+1}} \right), \qquad (8.5)$$

where C *is an arbitrary positive constant.*

Proof. Set $\tau = t/2^{N+1}$, and determine the two countable families of cubes $\{Q_n^t\}$ and $\{Q_j^{\tau}\}$ satisfying (8.1)$_t$–(8.2)$_t$ and (8.1)$_{\tau}$–(8.2)$_{\tau}$, respectively. Fix one of the cubes Q_j^{τ}, and consider those cubes Q_n^t out of $\{Q_n^t\}$ that are contained in Q_j^{τ}. For any one of such cubes, either

$$Q_j^{\tau} \subset \left\{ x \in Q_0 | f^{\#}(x) > \frac{t}{C} \right\} \qquad (a)$$

or

$$Q_j^{\tau} \not\subset \left\{ x \in Q_0 | f^{\#}(x) > \frac{t}{C} \right\}. \qquad (b)$$

If (a) occurs,

$$\sum_{Q_n^t \subset Q_j^{\tau}} \mu(Q_n^t) \le \mu \left\{ x \in Q_j^{\tau} | f^{\#}(x) > \frac{t}{C} \right\}$$

$$= \mu \left(\left[f^{\#} > \frac{t}{C} \right] \cap Q_j^{\tau} \right). \qquad (8.6)$$

If (b) occurs, there exists some $x \in Q_j^{\tau}$ such that $f^{\#}(x) \le t/C$. From the definition of $f^{\#}$,

$$\frac{1}{\mu(Q_j^{\tau})} \int_{Q_j^{\tau}} |f - f_{Q_j^{\tau}}| dy \le \frac{t}{C}. \qquad (8.7)$$

By the lower bound in (8.1)$_t$ and the upper bound in (8.1)$_{\tau}$,

$$|f|_{Q_n^t} > t \quad \text{and} \quad |f|_{Q_j^{\tau}} \le 2^N \tau.$$

From this, for each of the cubes Q_n^t contained in the fixed cube Q_j^{τ},

$$\int_{Q_n^t} |f - f_{Q_j^{\tau}}| dy \ge (|f|_{Q_n^t} - |f|_{Q_j^{\tau}}) \mu(Q_n^t)$$

$$\ge (t - 2^N \tau) \mu(Q_n^t) = \frac{t}{2} \mu(Q_n^t).$$

Adding over all the cubes Q_n^t contained in Q_j^{τ} gives

$$\sum_{Q_n^t \subset Q_j^{\tau}} \mu(Q_n^t) \le \frac{2}{t} \sum_{Q_n^t \subset Q_j^{\tau}} \int_{Q_n^t} |f - f_{Q_j^{\tau}}| dy$$

$$\le \frac{2}{t} \int_{Q_j^{\tau}} |f - f_{Q_j^{\tau}}| dy \le \frac{2}{C} \mu(Q_j^{\tau}). \qquad (8.8)$$

where in the last step, we have used (8.7). Combining case (a), leading to (8.6), and case (b), leading to (8.8), gives

$$\sum_{Q_n^t \subset Q_j^\tau} \mu(Q_n^t) \le \mu\left(\left[f^\# > \frac{t}{C}\right] \cap Q_j^\tau\right) + \frac{2}{C}\mu(Q_j^\tau).$$

Adding over j proves the lemma.   □

Taking into account (8.4), the estimate (7.5) of the theorem will be derived from the limiting process

$$\int_{Q_o} |f|^p dy = \lim_{s \to \infty} p \int_0^s t^{p-1} \mu\{x \in Q_o | |f(x)| > t\} dt$$
$$\le \limsup_{s \to \infty} p \int_0^s t^{p-1} M(t) dt, \tag{8.9}$$

provided that the last limit is finite. To estimate such a limit, fix $s > 2^{N+1}|f|_{Q_o}$ and use (8.5) to compute

$$p \int_0^s t^{p-1} M(t) dt = p \int_0^{2^{N+1}|f|_{Q_o}} t^{p-1} M(t) dt$$
$$+ p \int_{2^{N+1}|f|_{Q_o}}^s t^{p-1} M(t) dt$$
$$\le (2^{N+1}|f|_{Q_o})^p \mu(Q_o)$$
$$+ p \int_0^\infty t^{p-1} \mu\left\{x \in Q_o | f^\#(x) > \frac{t}{C}\right\} dt$$
$$+ \frac{2p}{C} \int_0^s t^{p-1} M\left(\frac{t}{2^{N+1}}\right) dt$$
$$= (2^{N+1}|f|_{Q_o})^p \mu(Q_o) + C^p \|f^\#\|_p^p$$
$$+ \frac{2}{C} 2^{(N+1)p} p \int_0^s t^{p-1} M(t) dt.$$

Choosing $C = 4 \cdot 2^{(N+1)p}$ gives

$$p \int_0^s t^{p-1} M(t) dt \le 2(2^{N+1}|f|_{Q_o})^p \mu(Q_o) + 2C^p \|f^\#\|_p^p.$$

Putting this in (8.9) proves the theorem.   □

9   The Marcinkiewicz interpolation theorem

Let E be a measurable subset of $\mathbb{R}^N$, and let $1 \le p < \infty$. A measurable function $f : E \to \mathbb{R}$ is in the space weak-$L^p(E)$, denoted by $L_w^p(E)$, if there is a positive constant C such that

$$\mu([|f| > t]) \leq \frac{A^p}{t^p} \quad \text{for all } t > 0. \tag{9.1}$$

Set

$$\|f\|_{p,w} = \inf\{A| \text{ for which (9.1) holds}\}. \tag{9.2}$$

Let $f, g \in L_w^p(E)$, and let α, β be nonzero real numbers. Then for all $t > 0$,

$$[|\alpha f + \beta g| > t] \subset \left[|f| > \frac{t}{2|\alpha|}\right] \bigcup \left[|g| > \frac{t}{2|\beta|}\right]. \tag{9.3}$$

Thus $L_w^p(E)$ is a linear space. However, $\| \cdot \|_{p,w}$ is not a norm on $L_w^p(E)$.
   If $f \in L^p(E)$, then for all $t > 0$,

$$\mu([|f| > t]) \leq \frac{1}{t^p} \|f\|_p^p.$$

Therefore, $f \in L_w^p(E)$ and $\|f\|_p = \|f\|_{p,w}$. The converse is false as there exist
functions $f \in L_w^p(E)$ that are not in $L^p(E)$.[11]
   The space $L_w^\infty(E)$ is defined as the collection of measurable functions for
which (9.1) holds for some constant A, for all $p \geq 1$, and for all $t > 0$.
   If $t > A$, then $\mu([|f| > t]) = 0$. Thus if $f \in L_w^\infty(E)$, then $f \in L^\infty(E)$
and $\|f\|_\infty \leq \|f\|_{\infty,w}$. On the other hand, if $f \in L^\infty(E)$, then (9.1) holds for
$A = \|f\|_\infty$. Thus $L_w^\infty(E) = L^\infty(E)$ and $\| \cdot \|_{\infty,w} = \| \cdot \|_\infty$.

9.1   Quasi-linear maps and interpolation. A map $T : L^p(E) \to L^q(E)$ is
quasi-linear if there exists a positive constant C such that for all $f, g \in L^p(E)$,

$$|T(f + g)(x)| \leq C(|T(f)(x)| + |T(g)(x)|) \quad \text{for a.e. } x \in E. \tag{9.4}$$

If $T : L^p(E) \to L^q(E)$ is quasi-linear, then for all $t > 0$,

$$[|T(f + g)| > t] \subset \left[|T(f)| > \frac{t}{2C}\right] \bigcup \left[|T(g)| > \frac{t}{2C}\right]. \tag{9.5}$$

A quasi-linear map $T : L^p(E) \to L^q(E)$ is of *strong type* (p, q) if there exists a
positive constant $M_{p,q}$ such that

$$\|T(f)\|_q \leq M_{p,q}\|f\|_p \quad \text{for all } f \in L^p(E). \tag{9.6}$$

A quasi-linear map T defined in $L^p(E)$ and such that $T(f)$ is measurable for all
$f \in L^p(E)$ is of *weak type* (p, q) if there exists a positive constant $N_{p,q}$ such that

$$\|T(f)\|_{q,w} \leq N_{p,q}\|f\|_p \quad \text{for all } f \in L^p(E). \tag{9.7}$$

---

[11] Examples can be constructed for either bounded or unbounded domains starting from the function
$f(x) = |x|^{-N}$ defined for $|x| > 0$. As indicated in Section 3.1, such a function is in $L_w^1(\mathbb{R}^N)$ and not
in $L^1(\mathbb{R}^N)$.

When $p = q$, we set $M_{p,p} = M_p$ and $N_{p,p} = N_p$.[12]

Theorem 9.1 (Marcinkiewicz[13]). *Let T be a quasi-linear map defined both in $L^p(E)$ and $L^q(E)$ for some pair $1 \le p < q \le \infty$. Assume that T is is both of weak type (p, p) and of weak type (q, q); i.e., there exist positive constants N_p and N_q such that*

$$\begin{aligned}
\|T(f)\|_{p,w} &\le N_p \|f\|_p \quad \text{for all } f \in L^p(E), \\
\|T(f)\|_{q,w} &\le N_q \|f\|_p \quad \text{for all } f \in L^q(E).
\end{aligned} \tag{9.8}$$

Then T is of strong type (r, r) for every $p < r < q$, and

$$\|T(f)\|_r \le \gamma N_p^\delta N_q^{1-\delta} \|f\|_r \quad \text{for all } f \in L^r(E), \tag{9.9}$$

where

$$\delta = \begin{cases} \dfrac{p(q-r)}{r(q-p)} & \text{if } q < \infty, \\[2mm] \dfrac{p}{r} & \text{if } q = \infty \end{cases} \tag{9.9$'$}$$

and

$$\gamma = 2C \begin{cases} \left(\dfrac{r(q-p)}{(r-p)(q-r)} \right)^{1/r} & \text{if } q < \infty, \\[3mm] \left(\dfrac{r}{r-p} \right)^{1/r} & \text{if } q = \infty, \end{cases} \tag{9.9$''$}$$

where C is the constant appearing in (9.4).

10   Proof of the Marcinkiewicz theorem

Having fixed $r \in (p, q)$ and some $f \in L^r(E)$, decompose it as $f = f_1 + f_2$, where

$$\begin{aligned}
f_1(x) &= \begin{cases} f(x) & \text{for } |f(x)| > \lambda t, \\ 0 & \text{for } |f(x)| \le \lambda t, \end{cases} \\[2mm]
f_2(x) &= \begin{cases} 0 & \text{for } |f(x)| \ge \lambda t, \\ f(x) & \text{for } |f(x)| < \lambda t, \end{cases}
\end{aligned} \tag{10.1}$$

where $t > 0$ and λ is a positive constant to be chosen later.

---

[12] Examples of maps of strong and weak type are in Section 3.1. Further examples will arise from the Riesz potentials in Section 23.

[13] J. Marcinkiewicz, Sur l'interpolation d'opérations. *C. R. Acad. Sci. Paris*, **208** (1939), 1272–1273.

We claim that $f_1 \in L^p(E)$ and $f_2 \in L^q(E)$. Since $f \in L^r(E)$, the set $[|f| > \lambda t]$ has finite measure. Therefore, by Hölder's inequality,

$$\int_E |f_1|^p dy \leq \left(\int_E |f|^r dy\right)^{\frac{p}{r}} (\mu([|f| > \lambda t]))^{1-\frac{p}{r}}.$$

Thus $f_1 \in L^p(E)$. Moreover,

$$\int_E |f_2|^q dy \leq \int_E |f_2|^{q-r} |f|^r dy \leq (\lambda t)^{q-r} \int_E |f|^r dy.$$

This implies that $f_2 \in L^q(E)$. Assume first that $1 \leq p < q < \infty$. Then using the quasi-linear structure of T resulting in (9.5) and using the assumptions in (9.8),

$$
\begin{aligned}
\mu([|T(f)| > t]) &\leq \mu\left(\left[|T(f_1)| > \frac{t}{2C}\right]\right) \\
&\quad + \mu\left(\left[|T(f_2)| > \frac{t}{2C}\right]\right) \\
&\leq \frac{(2CN_p)^p}{t^p} \|f_1\|_p^p + \frac{(2CN_q)^q}{t^q} \|f_2\|_q^q \qquad (10.2) \\
&= \frac{(2CN_p)^p}{t^p} \int_{|f|>\lambda t} |f|^p dy \\
&\quad + \frac{(2CN_q)^q}{t^q} \int_{|f|\leq\lambda t} |f|^q dy.
\end{aligned}
$$

From this,

$$
\begin{aligned}
\int_E |T(f)|^r dy &= r \int_0^\infty t^{r-1} \mu([|T(f)| > t]) dt \\
&\leq r(2CN_p)^p \int_0^\infty t^{r-p-1} \int_{|f|>\lambda t} |f|^p dy \\
&\quad + r(2CN_q)^q \int_0^\infty t^{r-q-1} \int_{|f|\leq\lambda t} |f|^q dy.
\end{aligned}
$$

The integrals on the right-hand side are transformed by means of Fubini's theorem and give

$$
\begin{aligned}
\int_0^\infty t^{r-p-1} \int_{|f|>\lambda t} |f|^p dy &= \int_E |f|^p \left(\int_0^{|f|/\lambda} t^{r-p-1} dt\right) dy \\
&= \frac{1}{r-p} \frac{1}{\lambda^{r-p}} \int_E |f|^r dy, \\
\int_0^\infty t^{r-q-1} \int_{|f|\leq\lambda t} |f|^q dy &= \int_E |f|^q \left(\int_{|f|/\lambda}^\infty t^{r-q-1} dt\right) dy \\
&= \frac{1}{q-r} \lambda^{q-r} \int_E |f|^r dy.
\end{aligned}
$$

Combining these estimates,

$$\|T(f)\|_r^r \le r \left\{ \frac{(2C)^p}{r-p} \frac{N_p^p}{\lambda^{r-p}} + \frac{(2C)^q}{q-r} N_q^q \lambda^{q-r} \right\} \|f\|_r^r. \tag{10.3}$$

Minimizing the right-hand side with respect to λ proves (9.9) with the indicated value of δ.

Now turning to the case where $q = \infty$, we choose the parameter λ in (10.1) so large that

$$\mu \left(\left[|T(f_2)| > \frac{t}{2C} \right] \right) = 0. \tag{10.4}$$

This would occur if $\|T(f_2)\|_\infty > \frac{t}{2C}$. From this and the second part of (9.8) with $q = \infty$,

$$\lambda t \ge \|f_2\|_\infty \ge \frac{1}{N_\infty} \|T(f_2)\|_\infty \ge \frac{1}{N_\infty} \frac{t}{2C}.$$

Thus (10.4) is verified if $\lambda = (2C N_\infty)^{-1}$. We now proceed as before, starting from (10.2), where the terms involving f_2 are discarded. This gives an analogue of (10.3) without the terms involving q; i.e.,

$$\|T(f)\|_r^r \le r \frac{(2C)^p}{r-p} \frac{N_p^p}{\lambda^{r-p}} \|f\|_r^r, \quad \lambda = \frac{1}{2C N_\infty}. \qquad \square$$

11   Rearranging the values of a function

Let f be a nonnegative measurable function in $\mathbb{R}$. Its distribution function,

$$[0, \infty) \longrightarrow \lambda(t) = \mu([f > t]),$$

is nonincreasing and right continuous. If f is the characteristic function of a set E of infinite measure, then $\lambda(t) = \infty$ for all $t \in [0, 1)$. For $x \ge 0$, set

$$\Lambda(x) = \{t \ge 0 \mid \lambda(t) > x\}. \tag{11.1}$$

If f is identically zero, then $\Lambda(x) = \emptyset$ for all $x \ge 0$. If f is positive in a set of positive measure, then

$$\Lambda(0) = \begin{cases} [0, \text{ess sup } f) & \text{if } f \text{ is essentially bounded,} \\ [0, \infty) & \text{otherwise.} \end{cases}$$

If f is compactly supported in $\mathbb{R}$, then

$$\Lambda(x) = \emptyset \quad \text{for all } x \ge \mu(\text{supp}\{f\}).$$

In general, one has the inclusion

$$\Lambda(y) \subset \Lambda(x) \quad \text{for } x < y.$$

The decreasing rearrangement of f is defined by

$$[0, \infty) \longrightarrow f_*(x) = \begin{cases} \sup\{t \in \Lambda(x)\} & \text{if } \Lambda(x) \neq \emptyset, \\ 0 & \text{if } \Lambda(x) = \emptyset. \end{cases} \quad (11.2)$$

If f is the characteristic function of the interval $[\alpha, \beta]$, then f_* is the characteristic function of the interval $[0, \beta - \alpha)$. Thus rearranging such an f amounts to mapping the rectangular block $[\alpha, \beta] \times [0, 1]$ into the right-open block of unit height and same width, starting at the origin of the coordinate system. More generally, if f is the characteristic function of a bounded measurable set E, its rearrangement f_* is the characteristic function of the interval $[0, \mu(E))$. Now assume that f is of the form

$$f = f_1 \chi_{[\alpha_1, \beta_1]} + f_2 \chi_{[\alpha_2, \beta_2]}, \quad (11.3)$$

where $f_1 > f_2$ are two given positive numbers and the intervals $[\alpha_1, \beta_1]$ and $[\alpha_2, \beta_2]$ are disjoint. Then

$$f_* = f_1 \chi_{[0, a_1)} + f_2 \chi_{[a_1, a_2)}, \quad (11.3)'$$

where

$$a_1 = (\beta_1 - \alpha_1) \quad \text{and} \quad (a_2 - a_1) = (\beta_2 - \alpha_2).$$

Thus the two rectangular blocks

$$[\alpha_1, \beta_1] \times [0, f_1] \quad \text{and} \quad [\alpha_2, \beta_2] \times [0, f_2]$$

are rearranged into right-open blocks of decreasing height and same width, one next to the other, starting from the origin. More generally, suppose that f is a simple function with canonical representation

$$f = \sum_{i=1}^{n} f_i \chi_{E_i}. \quad (11.4)$$

Then the rearrangement of f takes each of the blocks $E_i \times [0, f_i]$ and transforms them into rectangular, right-open blocks of base $\mu(E_i)$ and height f_i rearranged in decreasing height, one next to the other, starting at the origin; i.e.,

$$f_* = \sum_{j=1}^{n} f_j \chi_{[a_j, b_j)} \quad \text{with } f_j > f_{j+1} \quad (11.4)'$$

and

$$a_1 = 0, \qquad b_j = a_{j+1}, \qquad (b_j - a_j) = \mu(E_j).$$

A nonnegative measurable function f is the pointwise limit of an increasing sequence of simple functions. Therefore, one might interpret f_*, roughly speaking, as the operation of rearranging the values of f and the sets where such values are taken into rectangular blocks ordered in decreasing height starting at the origin.

12 Basic properties of rearrangements

Proposition 12.1. *Let f be nonnegative and measurable in $\mathbb{R}$. Then we have the following:*

If f vanishes for $x < 0$ and is continuous and decreasing for $x \geq 0$, then $f_* = f$.

The decreasing rearrangement $x \to f_*$ is lower semicontinuous.

The function f and its rearrangement f_* are equimeasurable; i.e.,

$$\mu([f > t]) = \mu([f_* > t]) \quad \text{for all } t \geq 0. \tag{12.1}$$

Moreover, for all $0 < t_1 \leq t_2$,

$$\mu([t_1 < f \leq t_2]) = \mu([t_1 < f_* \leq t_2]). \tag{12.2}$$

As a consequence,

$$\int_{\mathbb{R}} f dx = \int_{\mathbb{R}} f_* dx. \tag{12.3}$$

Decreasing rearrangements preserve the ordering; i.e.,

$$f \leq g \quad \text{implies that} \quad f_* \leq g_*. \tag{12.4}$$

Let f be a nonnegative function in $L^1(\mathbb{R})$, and let $\{f_n\}$ be a countable collection of simple functions such that $f_n \leq f_{n+1}$ and converging to f pointwise in $\mathbb{R}$. Then $\{f_{*n}\}$ converges to f_* monotonically in $\mathbb{R}$.

For any measurable subset $E \subset \mathbb{R}$,

$$\int_E f dx \leq \int_0^{\mu(E)} f_* dx. \tag{12.5}$$

In particular,

$$\left| \int_0^x f dt \right| \leq \int_0^{|x|} f_* dt. \tag{12.6}$$

Proof. All of the assertions up to (12.5) are a direct consequence of the definition by working, for example, with simple functions and by a limiting process. To prove (12.5), set $g = f \chi_E$ and observe that g_* vanishes outside the interval $[0, \mu(E))$. Then by (12.3) and (12.4),

$$\int_E f dx = \int_{\mathbb{R}} g dx = \int_{\mathbb{R}} g_* dx$$

$$= \int_0^{\mu(E)} g_* dx \leq \int_0^{\mu(E)} f_* dx. \qquad \square$$

Proposition 12.2. *A nonnegative simple function f in* $\mathbb{R}$ *can be represented in the form*

$$f = \sum_{j=1}^{n} \varphi_j \chi_{F_j}, \tag{12.7}$$

where φ_i *are positive numbers and* F_j *are measurable sets satisfying*

$$F_j \supset F_{j+1}, \quad j = 1, 2, \ldots, n. \tag{12.7}'$$

Moreover,

$$f_* = \sum_{j=1}^{n} \varphi_j (\chi_{F_j})_*. \tag{12.8}$$

Proof. A simple function can be written in the form (11.4), where the sets E_i are mutually disjoint and the numbers f_i are mutually distinct. After a possible reordering, we may assume that

$$0 < f_1 < f_2 < \cdots < f_n.$$

Then set

$$\varphi_1 = f_1, \qquad \varphi_2 = (f_2 - f_1), \qquad \ldots, \qquad \varphi_n = (f_n - f_{n-1})$$

and

$$F_j = [f \geq f_j], \quad j = 1, 2, \ldots, n.$$

This proves (12.7)–(12.7)'. To establish (12.8), first assume that

$$f = f_1 \chi_{E_1} + f_2 \chi_{E_2}, \quad f_1 < f_2.$$

Then according to (11.4)',

$$\begin{aligned} f_* &= f_2 \chi_{[0,a_2)} + f_1 \chi_{[a_2,a_2+a_1)} \\ &= f_1 \chi_{[0,a_1+a_2)} + (f_2 - f_1) \chi_{[0,a_2)}, \quad a_i = \text{meas } E_i. \end{aligned}$$

On the other hand, by (12.7)–(12.7)',

$$f = f_1 \chi_{E_1 \cup E_2} + (f_2 - f_1) \chi_{E_2}.$$

Moreover,

$$(\chi_{E_1 \cup E_2})_* = \chi_{[0,a_1+a_2)} \quad \text{and} \quad (\chi_{E_1})_* = \chi_{[0,a_2)}.$$

The general case follows from this by induction.   □

Proposition 12.3. *Let f and g be any two nonnegative measurable functions defined in $\mathbb{R}^+$. Then for all $t > 0$,*

$$\int_0^t fg\,dx \le \int_0^t f_*g_*\,dx. \tag{12.9}$$

Proof. Assume first that f and g are simple and take only the values 0 and 1. Then set

$$E = [f = 1], \qquad E_* = [f_* = 1],$$
$$G = [g = 1], \qquad G_* = [g_* = 1].$$

From the definition of rearrangement,

$$\mu(E) = \mu(E_*) \quad \text{and} \quad E_* = [0, \mu(E)),$$
$$\mu(G) = \mu(G_*) \quad \text{and} \quad G_* = [0, \mu(G)).$$

From this, for every $t > 0$,

$$\mu\left(E \bigcap (0, t)\right) \le \mu\left(E_* \bigcap (0, t)\right),$$
$$\mu\left(G \bigcap (0, t)\right) \le \mu\left(G_* \bigcap (0, t)\right).$$

Now compute

$$\int_0^t fg\,dx = \mu\left(\left[E \bigcap G\right] \bigcap (0, t)\right)$$
$$= \mu\left(\left[E \bigcap (0, t)\right] \bigcap \left[G \bigcap (0, t)\right]\right)$$
$$\le \min\left\{\mu\left(\left[E \bigcap (0, t)\right]\right); \mu\left(\left[G \bigcap (0, t)\right]\right)\right\}$$
$$\le \min\left\{\mu\left(\left[E_* \bigcap (0, t)\right]\right); \mu\left(\left[G_* \bigcap (0, t)\right]\right)\right\}$$
$$= \mu\left(\left[E_* \bigcap G_*\right] \bigcap (0, t)\right)$$
$$= \int_0^t f_*g_*\,dx.$$

By making use of Proposition 12.2 and linear combinations and iterations, the inequality holds for simple functions. By approximation, it continues to hold for nonnegative integrable functions in $\mathbb{R}$. □

13 Symmetric rearrangements

Let f be a nonnegative measurable function in $\mathbb{R}$, and let f_* be its decreasing rearrangement. The *symmetric* rearrangement of f is defined by

$$f^*(x) = \begin{cases} f_*(2x) & \text{if } x \ge 0, \\ f_*(-2x) & \text{if } x < 0. \end{cases} \tag{13.1}$$

If f is the characteristic function of the interval $[\alpha, \beta]$, then f^* is the characteristic function of the interval

$$\left(-\frac{1}{2}(\beta - \alpha), \frac{1}{2}(\beta - \alpha)\right).$$

Thus f^* distributes a rectangular block of unit height symmetrically on the left and on the right of the origin by keeping its total width unchanged.

If f is the characteristic function of a bounded measurable set E, then f^* rearranges the block $E \times [0, 1]$ into a rectangular block of base equal to $\mu(E)$, of unit height, and symmetric about the origin. Thus

$$\chi_E^* = \chi_{E^*}, \quad \text{where } E^* = \left(-\frac{1}{2}\mu(E), \frac{1}{2}\mu(E)\right).$$

Now suppose that f is made out of two blocks of the form (11.3). Then f^* first distributes the block of largest height $[\alpha_1, \beta_1] \times [0, f_1]$ into a block of the same height f_1, symmetric about the origin, and of total width $(\beta_1 - \alpha_1)$; i.e.,

$$f_1\chi_{(-a_1,a_1)}, \quad \text{where } a_1 = \frac{1}{2}(\beta_1 - \alpha_1). \tag{13.2}$$

Then f^* distributes the block $[\alpha_2, \beta_2] \times [0, f_2]$ into two blocks, symmetric about the origin, each of height f_2 and width $\frac{1}{2}(\beta_2 - \alpha_2)$, adjacent, one on the left and one on the right of the first symmetric block in (13.2); i.e.,

$$f_2\chi_{(-a_2,-a_1]} + f_2\chi_{[a_1,a_2)}, \quad \text{where } a_2 = a_1 + \frac{1}{2}(\beta_2 - \alpha_2).$$

More generally, if f is a simple function of the form (11.4), then

$$f^* = \sum_{j=1}^{n} f_j\chi_{(-b_j,-a_j]} + \sum_{j=1}^{n} f_j\chi_{[a_j,b_j)},$$

where the values f_j are arranged in decreasing order $f_j > f_{j+1}$ and the intervals $[a_j, b_j]$ are given by $a_1 = 0$ and

$$b_j = a_{j+1}, \quad (b_j - a_j) = \frac{1}{2}\mu(E_j), \quad j = 1, 2, \ldots, n.$$

Since a nonnegative measurable function f is the pointwise limit of a nondecreasing sequence of simple functions $\{f_n\}$, one might interpret f^*, roughly speaking, to be the operation of rearranging the values of f and the sets where such values are taken into rectangular blocks ordered in a decreasing and symmetric way starting at the origin.

It follows from the definition that f^* is symmetric and equimeasurable with f and f_*. As consequence,

$$\int_{\mathbb{R}} f\, dx = \int_{\mathbb{R}} f_*\, dx = \int_{\mathbb{R}} f^*\, dx.$$

Proposition 13.1. *Let f be a nonnegative simple function in* $\mathbb{R}$. *Then f can be represented as in* (12.7)–(12.7)' *and, moreover,*

$$f^* = \sum_{i=1}^{n} \varphi_j \chi_{F_j}^*.$$

Proof. If $n = 2$, this is established by direct verification along the lines of Proposition 12.1. For general $n \in \mathbb{N}$, it follows by induction.   □

Let f be a nonnegative function in $L^1(\mathbb{R})$, and let $\{f_n\}$ be a countable collection of simple functions such that $f_n \leq f_{n+1}$ and converging to f pointwise in $\mathbb{R}$. Then $\{f_n^*\}$ converges to f^* monotonically in $\mathbb{R}$.

14   A convolution inequality for rearrangements

Theorem 14.1 (F. Riesz[14]). *Let f, g, and h be nonnegative measurable functions in* $\mathbb{R}$, *and let* f^*, g^*, *and* h^* *denote their symmetric decreasing rearrangements. Then*

$$\mathcal{I} = \int_{\mathbb{R}} \int_{\mathbb{R}} f(x)g(y)h(x-y)dxdy \tag{14.1}$$
$$\leq \int_{\mathbb{R}} \int_{\mathbb{R}} f^*(x)g^*(y)h^*(x-y)dxdy = \mathcal{I}^*.$$

14.1   Approximations by simple functions. It suffices to prove the theorem for characteristic functions of measurable sets. Indeed, having fixed nonnegative, measurable functions f, g, and h in $L^1(\mathbb{R})$, these are the pointwise limit of nondecreasing sequences of simple functions $\{f_n\}$, $\{g_n\}$, and $\{h_n\}$, each having the representation

$$f_n = \sum_{j=1}^{n} \varphi_j \chi_{F_j}, \quad F_j \supset F_{j+1}, \quad j = 1, 2, \dots, n,$$

$$g_m = \sum_{s=1}^{m} \gamma_s \chi_{G_s}, \quad G_s \supset G_{s+1}, \quad s = 1, 2, \dots, m,$$

$$h_k = \sum_{\ell=1}^{k} \theta_\ell \chi_{H_\ell}, \quad H_\ell \supset H_{\ell+1}, \quad \ell = 1, 2, \dots, k.$$

By Proposition 13.1, their symmetric rearrangements are

$$f_n^* = \sum_{j=1}^{n} \varphi_j \chi_{F_j}^*, \qquad g_m^* = \sum_{s=1}^{m} \gamma_s \chi_{G_s}^*, \qquad h_k^* = \sum_{\ell=1}^{k} \theta_\ell \chi_{H_\ell}^*.$$

---

[14]F. Riesz, Sur une inégalité intégrale, *J. London Math. Soc.*, 5 (1930), 162–168; also in A. Zygmund, On an integral inequality, *J. London Math. Soc.*, 8 (1933), 175–178.

Now assume that (14.1) holds for characteristic functions of measurable sets. Then

$$
\begin{aligned}
\mathcal{I} &= \int_{\mathbb{R}} \int_{\mathbb{R}} f(x)g(y)h(x - y)dxdy \\
&= \lim_{n,m,k\to\infty} \int_{\mathbb{R}} \int_{\mathbb{R}} f_n(x)g_m(y)h_k(x - y)dxdy \\
&= \lim_{n,m,k\to\infty} \sum_{j,s,\ell} \varphi_j \gamma_s \theta_\ell \int_{\mathbb{R}} \int_{\mathbb{R}} \chi_{F_j}(x)\chi_{G_s}(y)\chi_{H_\ell}(x - y)dxdy \\
&\leq \lim_{n,m,k\to\infty} \sum_{j,s,\ell} \varphi_j \gamma_s \theta_\ell \int_{\mathbb{R}} \int_{\mathbb{R}} \chi_{F_j}^*(x)\chi_{G_s}^*(y)\chi_{H_\ell}^*(x - y)dxdy \\
&\leq \lim_{n,m,k\to\infty} \int_{\mathbb{R}} \int_{\mathbb{R}} f_n^*(x)g_m^*(y)h_k^*(x - y)dxdy \\
&= \int_{\mathbb{R}} \int_{\mathbb{R}} f^*(x)g^*(y)h^*(x - y)dxdy.
\end{aligned}
$$

Thus in what follows, we may assume that

$$
f = \chi_F, \qquad g = \chi_G, \qquad h = \chi_H, \tag{14.2}
$$

where F, G, and H are measurable subsets of $\mathbb{R}$. In addition, we may assume that each of these sets has finite measure; i.e.,

$$
\mu(F), \mu(G), \mu(H) < \infty. \tag{14.2$'$}
$$

If any two of them, say, for example, F and G, have infinite measure, then f^* and g^* would both be identically equal to 1. This would imply that $\mathcal{I}^* = \infty$ and (14.1) is trivial.

If only one of these sets, say, for example, F, has infinite measure, then $f^* \equiv 1$ in $\mathbb{R}$. In such a case, introduce the truncations

$$
f_n = f\chi_{[-n,n]}.
$$

Then assuming that (14.1) holds for functions of the form (14.2)–(14.2$'$),

$$
\begin{aligned}
\mathcal{I} &= \int_{\mathbb{R}} \int_{\mathbb{R}} f(x)g(y)h(x - y)dxdy \\
&= \lim_{n\to\infty} \int_{\mathbb{R}} \int_{\mathbb{R}} f_n(x)g(y)h(x - y)dxdy \\
&\leq \lim_{n\to\infty} \int_{\mathbb{R}} \int_{\mathbb{R}} f_n^*(x)g^*(y)h^*(x - y)dxdy \\
&\leq \int_{\mathbb{R}} \int_{\mathbb{R}} f^*(x)g^*(y)h^*(x - y)dxdy.
\end{aligned}
$$

15   Reduction to a finite union of intervals

Since F is measurable and of finite measure, for every $\varepsilon > 0$, there exists an open set $F_{o,\varepsilon}$ containing F and such that[15]

$$\mu(F_{o,\varepsilon} - F) < \frac{1}{2}\varepsilon.$$

Such an open set $F_{o,\varepsilon}$ is the countable union of mutually disjoint open intervals $\{I_n\}$. Since $F_{o,\varepsilon}$ is of finite measure, there exists a positive integer n_ε such that

$$\sum_{j>n_\varepsilon} \mu(I_j) < \frac{1}{2}\varepsilon.$$

Setting

$$F_\varepsilon = \bigcup_{j=1}^{n_\varepsilon} I_j, \qquad F_{1,\varepsilon} = F \cap \bigcup_{j>n_\varepsilon} I_j, \qquad F_{2,\varepsilon} = (F_\varepsilon - F),$$

the set F can be represented as

$$F = F_\varepsilon \bigcup F_{1,\varepsilon} - F_{2,\varepsilon} \quad \text{with } \mu(F_{1,\varepsilon}) + \mu(F_{2,\varepsilon}) < \varepsilon,$$

where F_ε is the finite union of open disjoint intervals. Moreover, since $F_{2,\varepsilon} \subset F_\varepsilon$ and F_ε and $F_{1,\varepsilon}$ are disjoint,

$$\chi_F = \chi_{F_\varepsilon} + \chi_{F_{1,\varepsilon}} - \chi_{F_{2,\varepsilon}}.$$

Similar decompositions hold for G and H. It is also clear that sets of arbitrarily small measure give arbitrarily small contributions in the integrals $\mathcal{I}$ and $\mathcal{I}^*$. Therefore, in proving (14.1), one may assume that F, G, and H are finite unions of disjoint, open intervals. Moreover, by changing ε if necessary, we may assume that the endpoints of the intervals making up F and, respectively, G and H are rational.

Thus the proof of (14.1) reduces to the case when f, g, and h are of the form (14.2)–(14.2)' and, in addition, F, G, and H are unions of nonoverlapping intervals whose endpoints are rationals. In such a case, in the integral $\mathcal{I}$, we may introduce a change of variables by rescaling x and y of a multiple equal to the minimum, common denominator of the endpoints of the intervals making up F, G, and H. This reduces (14.1) to the case when each of the sets F, G, and H is the finite union of intervals of the type $(i, i + 1)$ for integral i.

Finally, we may assume that the number of intervals making up each of the sets F, G, and H is even. This can be realized by bisecting each of these intervals and by effecting a further change of variables.

---

[15]See Proposition 12.3 of Chapter II.

Thus in proving Theorem 14.1, we may assume that f, g, and h are of the form (14.2)–(14.2)' and, in addition,

$$F = \bigcup_{i=1}^{2R}(m_i, m_i + 1) \qquad \text{for positive integers } m_i \text{ and some positive integer } R,$$

$$G = \bigcup_{j=1}^{2S}(n_j, n_j + 1) \qquad \text{for positive integers } n_j \text{ and some positive integer } S, \qquad (15.1)$$

$$H = \bigcup_{\ell=1}^{2T}(k_\ell, k_\ell + 1) \qquad \text{for positive integers } k_\ell \text{ and some positive integer } T.$$

From this,

$$\mu(F) = 2R, \qquad \mu(G) = 2S, \qquad \mu(H) = 2T.$$

Moreover,

$$f^* = \chi_{F^*}, \qquad g^* = \chi_{G^*}, \qquad h^* = \chi_{H^*}, \qquad (15.2)$$

where

$$F^* = (-R, R), \qquad G^* = (-S, S), \qquad H^* = (-T, T). \qquad (15.3)$$

From the definition of symmetric, decreasing rearrangement, it follows that for all $x \in \mathbb{R}$,

$$h^*(\cdot - x) = \chi_{H_x^*}, \qquad \text{where } H_x^* = (x - T, x + T). \qquad (15.4)$$

With this notation, we rewrite $\mathcal{I}$ as

$$\mathcal{I} = \int_{\mathbb{R}} f(x)\Gamma(x)dx, \qquad (15.5)$$

where

$$\Gamma(x) = \int_{\mathbb{R}} g(y)h(y - x)dy. \qquad (15.6)$$

We also rewrite $\mathcal{I}^*$ as

$$\mathcal{I}^* = \int_{\mathbb{R}} f^*(x)\Gamma^*(x)dx, \qquad (15.5^*)$$

where

$$\Gamma^*(x) = \int_{\mathbb{R}} g^*(y)h^*(y - x)dy. \qquad (15.6^*)$$

Moreover,

$$\mathcal{I}^* = \int_{-R}^{R} \Gamma^*(x)dx. \qquad (15.7)$$

16   Proof of Theorem 14.1: The case where $T + S \leq R$

Without loss of generality, we may assume that

$$\mu(H) \leq \mu(G); \quad \text{i.e.,} \quad T \leq S.$$

Indeed, we may always reduce to such a case by interchanging the role of g and h and effecting a suitable change of variables in the integral $\mathcal{I}$.

Starting from (15.5), estimate and compute

$$\mathcal{I} \leq \int_{\mathbb{R}} \int_{\mathbb{R}} h(y - x) dy dx$$

$$= \int_{\mathbb{R}} g(y) dy \int_{\mathbb{R}} h(\eta) d\eta$$

$$= \mu(G) \mu(H) = 4ST.$$

Next, we show that $\mathcal{I}^* = 4ST$. From (15.4) and the definition of Γ^*,

$$\Gamma^*(x) = \mu \left((-S, S) \bigcap (x - T, x + T) \right)$$

$$= \begin{cases} 0 & \text{for } x \leq -(S + T), \\ x + (S + T) & \text{for } -(S + T) \leq x \leq -(S - T), \\ 2T & \text{for } -(S - T) \leq x \leq (S - T), \\ (S + T) - x & \text{for } (S - T) \leq x \leq (S + T), \\ 0 & \text{for } (S + T) \leq x. \end{cases} \quad (16.1)$$

Now assuming that $(S + T) \leq R$, we compute

$$\mathcal{I}^* = \int_{-R}^{R} \Gamma^*(x) dx = \int_{-(S+T)}^{(S+T)} \Gamma^*(x) dx = 4ST.$$

Thus far, no use has been made of the structure (15.1) of the sets F, G, and H. Such a structure will be employed in examining the case $S + T > R$.

17   Proof of Theorem 14.1: The case where $S + T > R$

Since S, T, and R are positive integers, the difference $(S + T) - R$ is a positive integer; i.e.,

$$\frac{1}{2}\mu(G) + \frac{1}{2}\mu(H) - \frac{1}{2}\mu(F) = S + T - R = n \quad (17.1)$$

for some integer n. The arguments of the previous section show that the theorem holds for $n \leq 0$. We show by induction that if it holds for some integer $(n - 1) \geq 0$, then it continues to hold for n. Set

$$G_1 = \bigcup_{j=1}^{2S-1} (n_j, n_j + 1), \qquad \begin{array}{l} \text{i.e., the set } G \text{ from which the last} \\ \text{interval on the right has been removed,} \end{array}$$

$$H_1 = \bigcup_{\ell=1}^{2T-1} (k_\ell, k_\ell + 1), \qquad \begin{array}{l} \text{i.e., the set } H \text{ from which the last} \\ \text{interval on the right has been removed.} \end{array}$$

By construction,

$$\frac{1}{2}\mu(G_1) + \frac{1}{2}\mu(H_1) - \frac{1}{2}\mu(F)$$
$$= \left(S - \frac{1}{2}\right) + \left(T - \frac{1}{2}\right) - R = n - 1 \geq 0. \tag{17.2}$$

Also set

$$g_1 = \chi_{G_1} \qquad h_1 = \chi_{H_1}.$$

From the definitions, it follows that

$$g_1^* = \chi_{G_1^*}, \quad \text{where } G_1^* = \left(-S + \frac{1}{2}, S - \frac{1}{2}\right),$$

$$h_1^* = \chi_{H_1^*}, \quad \text{where } H_1^* = \left(-T + \frac{1}{2}, T - \frac{1}{2}\right).$$

Moreover,

$$h_1^*(\cdot - x) = \chi_{H_{1,x}^*},$$

where

$$H_{1,x}^* = \left(x - T + \frac{1}{2}, x + T - \frac{1}{2}\right).$$

Taking into account (17.2), the induction hypothesis is that

$$\mathcal{I}_1 = \int_{\mathbb{R}} f(x) \int_{\mathbb{R}} g_1(y) h_1(y - x) dy dx$$
$$= \int_{\mathbb{R}} f(x) \Gamma_1(x) dx$$
$$\leq \int_{\mathbb{R}} f^*(x) \int_{\mathbb{R}} g_1^*(y) h_1^*(y - x) dy dx$$
$$= \int_{\mathbb{R}} f^*(x) \Gamma_1^*(x) dx = \mathcal{I}_1^*.$$

Next, we observe that Γ_1^* is defined by formula (16.1) with S and T replaced by, respectively, $S - \frac{1}{2}$ and $T - \frac{1}{2}$; i.e.,

$$\Gamma_1^*(x) = \begin{cases} 0 & \text{for } x \le -(S+T-1), \\ x+(S+T-1) & \text{for } -(S+T-1) \le x \le -(S-T), \\ 2T-1 & \text{for } -(S-T) \le x \le (S-T), \\ (S+T-1)-x & \text{for } (S-T) \le x \le (S+T-1), \\ 0 & \text{for } (S+T-1) \le x. \end{cases} \qquad (17.3)$$

From this and (16.1), one verifies that

$$\Gamma^*(x) - \Gamma_1^*(x) = 1 \quad \text{for all } |x| \le (S+T-1).$$

In particular, this holds true for all $|x| \le R$ since by (17.2), $R \le (S+T-1)$. Using these remarks, we compute

$$\begin{aligned} \mathcal{I}^* - \mathcal{I}_1^* &= \int_{\mathbb{R}} f^*(x)\Gamma^*(x)dx - \int_{\mathbb{R}} f^*(x)\Gamma_1^*(x)dx \\ &= \int_{-R}^{R} \{\Gamma^*(x) - \Gamma_1^*(x)\}dx = 2R. \end{aligned} \qquad (17.4)$$

Next, we examine the structure of the function

$$x \longrightarrow \Gamma(x) - \Gamma_1(x) = \mu(\{G \cap H_x\}) - \mu(G_1 \cap H_{1,x}).$$

Here by H_x and $H_{1,x}$, we have denoted the sets H and H_1 shifted by x.

Lemma 17.1. $0 \le \Gamma(x) - \Gamma_1(x) \le 1$ for all $x \in \mathbb{R}$.

Assuming the lemma for the moment, we compute

$$\begin{aligned} \mathcal{I} - \mathcal{I}_1 &= \int_{\mathbb{R}} f(x)\Gamma(x)dx - \int_{\mathbb{R}} f(x)\Gamma_1(x)dx \\ &= \int_{\mathbb{R}} f(x)\{\Gamma(x) - \Gamma_1(x)\}dx \\ &\le \int_{\mathbb{R}} \chi_F dx = 2R. \end{aligned}$$

This and (17.4) now give

$$\mathcal{I} - \mathcal{I}_1 \le \mathcal{I}^* - \mathcal{I}_1^*.$$

This, in turn, implies the theorem since by the induction hypothesis, $\mathcal{I}_1 \le \mathcal{I}_1^*$. $\qquad \square$

17.1   Proof of Lemma 17.1. It is apparent that such a function is affine within any interval of the form $(n, n+1)$ for integral n. Therefore, it must take its extrema for some integral value of x. If x is an integer, the set H_x is the finite union of unit intervals whose endpoints are integers. Now, still for integral x, the set $H_{1,x}$ is precisely H_x from which the last interval on the right has been removed. Set

$$I_G = \{\text{the rightmost interval of } G\},$$
$$I_{H_x} = \{\text{the rightmost interval of } H_x\}.$$

If I_G coincides with I_{H_x}, then removing them both amounts to removing a single interval of unit length out of $G \cap H_x$. Therefore,

$$\mu\left(G \bigcap H_x\right) - \mu\left(G_1 \bigcap H_{1,x}\right) = 1.$$

If I_{H_x} is on the right with respect to I_G, then removing it has no effect on the intersection $G \cap H_x$; i.e.,

$$G \bigcap H_x = G \bigcap H_{1,x}.$$

Now, by removing I_G out of G, the two sets $G \cap H_x$ and $G_1 \cap H_{1,x}$ differ by at most one interval of unit length. Thus

$$\mu\left(G \bigcap H_x\right) - \mu\left(G_1 \bigcap H_{1,x}\right) \le 1.$$

Finally, if I_{H_x} is on the left with respect to I_G, we arrive at the same conclusion by interchanging the role of G and H_x.   □

18   Hardy's inequality

Proposition 18.1. *Let* $f \in L^p(\mathbb{R}^+)$, $p > 1$, *be nonnegative. Then*[16]

$$\int_0^\infty \frac{1}{x^p} \left(\int_0^x f(t)dt\right)^p dx \le \left(\frac{p}{p-1}\right)^p \int_0^\infty f^p dx. \qquad (18.1)$$

Proof. Fix $0 < \xi < \eta < \infty$. Then by integration by parts,

$$\int_\xi^\eta \frac{1}{x^p} \left(\int_0^x f(t)dt\right)^p dx$$

$$= \frac{-1}{p-1} \int_\xi^\eta \left(\int_0^x f(t)dt\right)^p \frac{d}{dx} x^{1-p} dx$$

$$= \frac{\xi^{1-p}}{p-1} \left(\int_0^\xi f(t)dt\right)^p - \frac{\eta^{1-p}}{p-1} \left(\int_0^\eta f(t)dt\right)^p$$

$$+ \frac{p}{p-1} \int_\xi^\eta x^{1-p} f(x) \left(\int_0^x f(t)dt\right)^{p-1} dx.$$

---

[16]H. H. Hardy, Note on a theorem of Hilbert, *Math. Z.*, **6** (1920), 314–317.

The second term on the right-hand side is nonpositive, and it is discarded. The first term tends to zero as $\xi \to 0$. Indeed, by Hölder's inequality,

$$\xi^{1-p} \left(\int_0^\xi f(t)dt \right)^p \le \int_0^\xi f^p(t)dt.$$

Therefore, letting $\xi \to 0$ and applying Hölder's inequality in the resulting inequality gives

$$\int_0^\eta \frac{1}{x^p} \left(\int_0^x f(t)dt \right)^p dx$$

$$\le \frac{p}{p-1} \int_0^\eta x^{1-p} f(x) \left(\int_0^x f(t)dt \right)^{p-1} dx$$

$$\le \frac{p}{p-1} \left\{ \int_0^\eta \frac{1}{x^p} \left(\int_0^\eta f(t)dt \right)^p dx \right\}^{\frac{p-1}{p}} \left(\int_0^\eta f^p dx \right)^{\frac{1}{p}}. \qquad \square$$

The constant on the right-hand side of (18.1) is the best possible as it can be tested for the family of functions

$$f_\varepsilon(x) = \begin{cases} x^{-\frac{1}{p}-\varepsilon} & \text{for } x \ge 1, \\ 0 & \text{for } 0 \le x < 1 \end{cases}$$

for $\varepsilon > 0$. Assume that (18.1) were to hold for a smaller constant, say, for example,

$$\left(\frac{p}{p-1} \right)^p (1-\delta)^p \quad \text{for some } \delta \in (0,1).$$

If (18.1) were applied to f_ε, it would give

$$\int_1^\infty \frac{1}{x^p} \left(\int_1^x t^{-\frac{1}{p}-\varepsilon}dt \right)^p dx \le \left(\frac{p}{p-1} \right)^p (1-\delta)^p \frac{1}{p\varepsilon}. \tag{18.2}$$

To estimate the the left-hand side below, set

$$A_\varepsilon = \left(\frac{p}{p-1-\varepsilon p} \right)^p, \qquad B_{\varepsilon,\rho} = \left(1 - \frac{1}{(1+\rho)^{1-\frac{1}{p}-\varepsilon}} \right)^p,$$

where $\varepsilon > 0$ is so small that $(p-1-\varepsilon p) > 0$ and ρ is an arbitrary positive number. Then

$$\int_1^\infty \frac{1}{x^p} \left(\int_1^x t^{-\frac{1}{p}-\varepsilon}dt \right)^p dx = A_\varepsilon \int_1^\infty \frac{1}{x^p} \left(x^{1-\frac{1}{p}-\varepsilon} - 1 \right)^p dx$$

$$\ge A_\varepsilon B_\varepsilon \int_{1+\rho}^\infty x^{-1-p\varepsilon} dx$$

$$= A_\varepsilon B_\varepsilon \frac{1}{p\varepsilon} \frac{1}{(1+\rho)^{\varepsilon p}}.$$

Putting this in (18.2), multiplying by $p\varepsilon$, and letting $\varepsilon \to 0$ gives

$$1 - \frac{1}{(1+\rho)^{\frac{p-1}{p}}} \le (1 - \delta).$$

Since $\rho > 0$ is arbitrary, this is a contradiction.

19 A convolution-type inequality

Theorem 19.1. *Let f and g be nonnegative measurable functions in $\mathbb{R}$, and let $p, q > 1$ and $\sigma \in (0, 1)$ be linked by*[17]

$$\frac{1}{p} + \frac{1}{q} > 1 \quad \text{and} \quad 2 - \frac{1}{p} - \frac{1}{q} = \sigma. \tag{19.1}$$

There exists a constant C depending only upon p, q, and σ such that

$$\int_{\mathbb{R}} \int_{\mathbb{R}} \frac{f(x)g(y)}{|x - y|^\sigma} dx dy \le C \|f\|_p \|g\|_q. \tag{19.2}$$

Remark 19.1. The constant $C(p, q, \sigma)$ can be computed explicitly as

$$C(p, q, \sigma)) = \frac{4}{1 - \sigma} \left\{ \left(\frac{p}{p-1} \right)^{p\frac{q-1}{q}} + \left(\frac{q}{q-1} \right)^{q\frac{p-1}{p}} \right\}. \tag{19.3}$$

Thus $C(p, q, \sigma)$ tends to infinity as either $\sigma \to 1$ or $p \to 1$. Also, $q = 1$ is not permitted in (19.2).

19.1 Some reductions. Assume that (19.2) holds for nonnegative and symmetrically decreasing functions. Then for general nonnegative functions f and g, by Theorem 14.1 applied to the triple f, g, and $h(x) = |x|^{-\sigma}$,

$$\int_{\mathbb{R}} \int_{\mathbb{R}} \frac{f(x)g(y)}{|x - y|^\sigma} dx dy \le \int_{\mathbb{R}} \int_{\mathbb{R}} \frac{f^*(x)g^*(y)}{|x - y|^\sigma} dx dy$$
$$\le C \|f^*\|_p \|g^*\|_q.$$

Now $(f^*)^p$ and $(g^*)^p$ are the decreasing, symmetric rearrangements of f^p and g^p, respectively. Therefore, $\|f^*\|_p = \|f\|_p$ and $\|g^*\|_q = \|g\|_q$. Thus it suffices to prove (19.2) for nonnegative and symmetrically decreasing functions f and g.

Next, it suffices to prove (19.2) in the seemingly weaker form

$$\int_0^\infty \int_0^\infty \frac{f(x)g(y)}{|x - y|^\sigma} dx dy \le C_0 \|f\|_p \|g\|_q \tag{19.2'}$$

---

[17]G. H. Hardy, J. E. Littlewood, and G. Pólya, The maximum of a certain bilinear form, *Proc. London Math. Soc.*, 2-25 (1926), 265–282; G. H. Hardy and J. E. Littlewood, Some properties of fractional integrals I, *Math. Z.*, **27** (1928), 565–606.

for a constant C_o depending only upon p and q. For this, divide the domain of integration in (19.2) into the four coordinate quadrants. By changing the sign of both variables, one verifies that the contributions of the first and third quadrants to the integral in (18.1) are equal. The contribution of the second quadrant is majorized by the contribution of the first quadrant. Indeed, by changing x into $-x$,

$$\int_0^\infty \int_{-\infty}^0 \frac{f(x)g(y)}{|x-y|^\sigma} dx dy = \int_0^\infty \int_0^\infty \frac{f(x)g(y)}{|x+y|^\sigma} dx dy$$
$$\leq \int_0^\infty \int_0^\infty \frac{f(x)g(y)}{|x-y|^\sigma} dx dy.$$

Similarly, the contribution of the fourth quadrant is majorized by the contribution of the first quadrant. We conclude that

$$\int_{\mathbb{R}} \int_{\mathbb{R}} \frac{f(x)g(y)}{|x-y|^\sigma} dx dy \leq 4 \int_0^\infty \int_0^\infty \frac{f(x)g(y)}{|x-y|^\sigma} dx dy. \qquad \square$$

20 Proof of Theorem 19.1

Divide the first quadrant further into the two octants $\{x \geq y\}$ and $\{y > x\}$ and write

$$\int_0^\infty \int_0^\infty \frac{f(x)g(y)}{|x-y|^\sigma} dx dy = \int_0^\infty f(x) \left(\int_0^x \frac{g(y)}{(x-y)^\sigma} dy \right) dx$$
$$+ \int_0^\infty g(y) \left(\int_0^y \frac{f(x)}{(y-x)^\sigma} dx \right) dy$$
$$= J_1 + J_2.$$

We estimate the first of these integrals in terms of the right-hand side of (19.2), the estimation of the second being similar.

Lemma 20.1. *Let $t \to u(t)$, $v(t)$ be nonnegative and measurable on a measurable subset $E \subset \mathbb{R}$ of finite measure. Assume, in addition, that u is nondecreasing and v is nonincreasing. Then*

$$\int_E uv dt \leq \frac{1}{\mu(E)} \left(\int_E u dt \right) \left(\int_E v dt \right).$$

Proof. By the stated monotonicity of u and v,

$$\int_E \int_E (u(x) - u(y))(v(y) - v(x)) dx dy \geq 0.$$

From this,

$$\int_E \int_E u(x)v(y) dx dy + \int_E \int_E u(y)v(x) dx dy$$
$$\geq \int_E \int_E u(x)v(x) dx dy + \int_E \int_E u(y)v(y) dx dy. \qquad \square$$

Applying the lemma with

$$E = (0, x), \quad u(t) = (x - t)^{-\sigma}, \quad v(t) = g(t)$$

gives

$$\int_0^x \frac{g(y)}{(x - y)^\sigma} dy \le \frac{1}{(1 - \sigma)x^\sigma} \int_0^x g(y)dy. \tag{20.1}$$

Using Hölder's inequality, estimate

$$G(x) = \int_0^x g(y)dy \le \left(\int_0^x g^q(y)dy \right)^{\frac{1}{q}} x^{\frac{q-1}{q}}.$$

Now return to J_1. Using (20.1), the expression of $G(x)$, and Hölder's inequality,

$$J_1 \le \frac{1}{1 - \sigma} \int_0^\infty f(x)x^{-\sigma} G(x)dx$$

$$\le \frac{1}{1 - \sigma} \|f\|_p \left(\int_0^\infty x^{-\sigma \frac{p}{p-1}} G(x)^{\frac{p}{p-1}} dx \right)^{\frac{p-1}{p}}$$

$$= \frac{1}{1 - \sigma} \|f\|_p \left(\int_0^\infty x^{-\sigma \frac{p}{p-1}} G(x)^q G(x)^{\frac{p}{p-1}-q} dx \right)^{\frac{p-1}{p}}$$

$$\le \frac{1}{1 - \sigma} \|f\|_p \|g\|_q^{1-q\frac{p-1}{p}} \left(\int_0^\infty x^{-\sigma \frac{p}{p-1}} x^{\frac{q-1}{q}(\frac{p}{p-1}-q)} G(x)^q dx \right)^{\frac{p-1}{p}}$$

$$\le \frac{1}{1 - \sigma} \|f\|_p \|g\|_q^{1-q\frac{p}{p-1}} \left(\int_0^\infty \frac{1}{x^q} \left(\int_0^x g(y)dy \right)^q dx \right)^{\frac{p-1}{p}}.$$

The last integral is estimated by means of Hardy's inequality and gives

$$\left(\int_0^\infty \frac{1}{x^q} \left(\int_0^x g(y)dy \right)^q dx \right)^{\frac{p-1}{p}} \le \left(\frac{q}{q-1} \right)^{q\frac{p-1}{p}} \|g\|_q^{q\frac{p-1}{p}}.$$

Putting this in the previous inequality proves the theorem.   □

21   An equivalent form of Theorem 19.1

Theorem 21.1. *Let f be a nonnegative function in $L^p(\mathbb{R})$ for some $p \in (1, \infty)$, and set*

$$h(y) = \int_\mathbb{R} \frac{f(x)}{|x - y|^\sigma} dx \quad \text{for some } \sigma \in \left(\frac{p-1}{p}, 1 \right). \tag{21.1}$$

There exists a constant C depending only upon σ and p such that

$$\|h\|_{p^*} \le C\|f\|_p, \quad \text{where } \frac{1}{p^*} = \frac{1}{p} + \sigma - 1. \tag{21.2}$$

Proof. From the range of σ and the second part of (21.2), it follows that $p^* > 1$. If $q > 1$ denotes the Hölder conjugate of p^*,

$$\frac{1}{q} = 1 - \frac{1}{p^*}, \qquad \sigma = 2 - \frac{1}{p} - \frac{1}{q}.$$

Thus $p. q > 1$ and $\sigma \in (0, 1)$ satisfy (19.1). Since q and p^* are conjugate,[18]

$$\|h\|_{p^*} = \sup_{\substack{g \in L^q(\mathbb{R}) \\ \|g\|_q = 1}} \int_{\mathbb{R}} hg \, dy. \tag{21.3}$$

Majorizing the right-hand side by Theorem 19.1 implies (21.2).   □

Remark 21.1. Theorems 19.1 and 21.1 are equivalent. The previous argument shows that Theorem 19.1 implies Theorem 21.1. On the other hand, assuming that (21.2) holds, (21.3) implies Theorem 19.1.

Remark 21.2. The constant C in (21.2) is the same as the constant $C(p, q, \sigma)$ in (19.2) with $p^* = \frac{q}{q-1}$. From the explicit form (19.3), it follows that the values $p = 1$ and $p^* = \infty$ are not permitted in (21.2).

22   An N-dimensional version of Theorem 21.1

For a function $f \in L^p(\mathbb{R}^N)$, $p > 1$, define its potential of order δ as

$$\mathcal{U}(f)(x) = \int_{\mathbb{R}^N} \frac{f(y)}{|x - y|^\delta} dy, \quad \delta \in (0, N). \tag{22.1}$$

Proposition 22.1. *There exists a constant C depending only upon N, p, and δ such that*

$$\|\mathcal{U}(f)\|_{p^*} \le C\|f\|_p, \quad \text{where } \frac{1}{p^*} = \frac{1}{p} + \frac{\delta}{N} - 1. \tag{22.2}$$

Proof. The arithmetic mean of N positive numbers is more than the corresponding geometric mean.[19] Therefore,

$$|x - y| = \left(\sum_{i=1}^N (x_i - y_i)^2 \right)^{\frac{1}{2}} \ge \sqrt{N} \prod_{i=1}^N |x_i - y_i|^{\frac{1}{N}}.$$

From this and Fubini's theorem,

$$|\mathcal{U}(f)(x)| \le \frac{1}{N^{\delta/2}} \int_{\mathbb{R}} \frac{dy_1}{|x_1 - y_1|^{\delta/N}} \int_{\mathbb{R}} \frac{dy_2}{|x_2 - y_2|^{\delta/N}}$$
$$\cdots \int_{\mathbb{R}} \frac{|f(y)|}{|x_N - y_N|^{\delta/N}} dy_N.$$

---

[18] See Proposition 4.1 of Chapter V.
[19] See Section 14.1 of the Problems and Complements of Chapter IV.

By Theorem 21.1 and the continuous version of Minkowski's inequality,[20]

$$\left(\int_{\mathbb{R}} |\mathcal{U}(f)(x_1, \ldots, x_{N-1}, x_N)|^{p^*} dx_N \right)^{1/p^*}$$

$$\leq \frac{1}{N^{\delta/2}} \int_{\mathbb{R}} \frac{dy_1}{|x_1 - y_1|^{\delta/N}} \int_{\mathbb{R}} \frac{dy_2}{|x_2 - y_2|^{\delta/N}}$$

$$\cdots \left(\int_{\mathbb{R}} |f(y_1, \ldots, y_{N-1}; y_N)|^p dy_N \right)^{1/p}.$$

The proposition is proved by repeated application of (22.2).   □

Remark 22.1. The constant C is given by (19.3) with $p^* = \frac{q}{q-1}$ and $\sigma = \frac{\delta}{N}$. Thus C tends to infinity as $\delta \to N$ or $p \to 1$ or $q \to 1$. In particular, $p^* = \infty$ is not permitted in (22.2).

23 L^p estimates of Riesz potentials

Let E be a Lebesgue-measurable subset of $\mathbb{R}^N$, and let $f \in L^p(E)$ for some $1 \leq p \leq \infty$. The Riesz potential generated by f in $\mathbb{R}^N$ is defined by[21]

$$\mathcal{V}_f(x) = \int_E \frac{f(y)}{|x - y|^{N-1}}. \tag{23.1}$$

The definition is formal, and it is natural to ask whether such a potential is well defined as an integrable function.

One may regard f as a function in $L^p(\mathbb{R}^N)$ by extending it to be zero outside E. By such an extension, one might regard the domain of integration in (23.1) as the whole $\mathbb{R}^N$. By Proposition 22.1 with $\delta = (N - 1)$, we have the following.

Theorem 23.1. *Let* $f \in L^p(E)$ *for* $1 < p < N$. *There exists a constant* C *depending only upon* N *and* p *such that*

$$\|\mathcal{V}_f\|_{p^*} \leq C\|f\|_p, \quad \text{where } p^* = \frac{Np}{N - p}. \tag{23.2}$$

Remark 23.1. The constant C in (23.2) tends to infinity if either $p \to 1$ or $p \to N$. Following Remark 22.1, this can be traced back to the form (19.3) of the constant C with $\sigma = \frac{N-1}{N}$ and $p^* = \frac{q}{q-1}$.

Similar estimates for the limiting cases $p = 1$ and $p \geq N$ require a preliminary estimation of the potential generated by a function f, constant on a set E of finite measure.

---

[20]See Section 4.3 of Chapter V.

[21]F. Riesz, *Sur les fonctions subharmoniques et leur rapport à la théorie du potentiel*, *Acta Math.*, **48** (1926), 329–343 and **54** (1930), 162–168. A motivation for the notion of Riesz potentials and Theorem 23.1 below is provided in Section 23.4 of the Problems and Complements.

Proposition 23.2. *Let E be of finite measure. For every $r \in [1, \frac{N}{N-1})$,*

$$\sup_{x \in E} \int_E \frac{dy}{|x - y|^{(N-1)r}} \le \frac{\kappa_N^{\frac{N-1}{N}r}}{1 - \frac{N-1}{N}r} \mu(E)^{1 - \frac{N-1}{N}r}, \tag{23.3}$$

where κ_N is the volume of the unit ball in $\mathbb{R}^N$.

Proof. Fix $x \in E$, and choose $\rho > 0$ such that $\mu(E) = \mu(B_\rho(x))$. Then

$$\int_E \frac{dy}{|x - y|^{(N-1)r}} = \int_{E \cap B_\rho(x)} \frac{dy}{|x - y|^{(N-1)r}} + \int_{E - B_\rho(x)} \frac{dy}{|x - y|^{(N-1)r}}$$

$$\le \int_{B_\rho(x)} \frac{dy}{|x - y|^{(N-1)r}}$$

$$= \frac{N\kappa_N}{N - (N-1)r} \rho^{N - (N-1)r}. \qquad \square$$

Proposition 23.3. *Let E be of finite measure, and let $f \in L^1(E)$. Then $\mathcal{V}_f \in L^q(E)$ for all $q \in [1, \frac{N}{N-1})$, and*

$$\|\mathcal{V}_f\|_q \le \frac{\kappa_N^{\frac{N-1}{N}}}{\left(1 - \frac{N-1}{N}q\right)^{\frac{1}{q}}} \mu(E)^{\frac{1}{q} - \frac{N-1}{N}} \|f\|_1. \tag{23.4}$$

Proof. Fix $q \in (1, \frac{N}{N-1})$, and write

$$\frac{|f(y)|}{|x - y|^{N-1}} = |f(y)|^{1 - \frac{1}{q}} \frac{|f(y)|^{\frac{1}{q}}}{|x - y|^{N-1}}.$$

Then by Hölder's inequality,

$$\mathcal{V}_f(x) \le \|f\|_1^{1 - \frac{1}{q}} \left(\int_E \frac{|f(y)|}{|x - y|^{(N-1)q}} dy \right)^{\frac{1}{q}}.$$

Take the q power and integrate in dx over E to obtain

$$\|\mathcal{V}_f\|_q^q \le \|f\|_1^{q-1} \int_E |f(y)| \left(\int_E \frac{dx}{|x - y|^{(N-1)q}} \right) dy$$

$$\le \|f\|_1^q \sup_{y \in E} \int_E \frac{dx}{|x - y|^{(N-1)q}}$$

$$\le \frac{\kappa_N^{\frac{N-1}{N}q}}{1 - \frac{N-1}{N}q} \mu(E)^{1 - \frac{N-1}{N}q} \|f\|_1^q. \qquad \square$$

Remark 23.2. The limiting integrability $q = \frac{N}{N-1}$ is not permitted in (23.4).

Proposition 23.4. *Let E be of finite measure, and let $f \in L^p(E)$ for some $p > N$. Then $\mathcal{V}_f \in L^\infty(E)$ and*

$$\|\mathcal{V}_f\|_\infty \le C(N, p)\mu(E)^{\frac{p-N}{Np}}\|f\|_p, \tag{23.5}$$

where

$$C(N, p) = \kappa_N^{\frac{N-1}{N}}\left(\frac{N(p-1)}{p-N}\right)^{\frac{p-1}{p}}. \tag{23.5}'$$

Proof. By Hölder's inequality and the definition of $\mathcal{V}_f$,

$$\|\mathcal{V}_f\|_\infty \le \|f\|_p\left(\sup_{x \in E}\int_E \frac{dy}{|x - y|^{(N-1)\frac{p}{p-1}}}\right)^{\frac{p-1}{p}}.$$

The estimate (23.5) and the form $(23.5)'$ of the constant $C(N, p)$ now follow from Proposition 23.2. $\qquad\square$

24   The limiting case $p = N$

The value $p = N$ is not permitted in either Theorem 23.1 or Proposition 23.4. The next theorem indicates that the potential $\mathcal{V}_f$ of a function $f \in L^N(E)$ belongs to some intermediate space between some $L^q(E)$ for large q and $L^\infty(E)$.

Theorem 24.1. *Let E be of finite measure, and let $f \in L^N(E)$. There exist constants C_1 and C_2 depending only upon N such that*[22]

$$\int_E \exp\left(\frac{|\mathcal{V}_f|}{C_1\|f\|_N}\right)^{\frac{N}{N-1}} dx \le C_2\mu(E). \tag{24.1}$$

Proof. For any $q > N$ and $1 < r < \frac{N}{N-1}$ satisfying

$$\frac{1}{r} = 1 + \frac{1}{q} - \frac{1}{N}, \tag{24.2}$$

formally write

$$\frac{|f(y)|}{|x - y|^{N-1}} = \frac{|f(y)|^{\frac{N}{q}}}{|x - y|^{(N-1)\frac{r}{q}}}\frac{|f(y)|^{1-\frac{N}{q}}}{|x - y|^{(N-1)\frac{r(N-1)}{N}}}.$$

Then by Hölder's inequality applied with the conjugate exponents

$$\frac{q - N}{Nq} + \frac{1}{q} + \frac{N - 1}{N} = 1,$$

---

[22]N. S. Trudinger, On embeddings into Orlicz spaces and some applications, *J. Math. Mech.*, **17** (1967), 473–483.

we obtain, at least formally,[23]

$$|V_f(x)| \leq \|f\|_N^{1-\frac{N}{q}} \left(\int_E \frac{|f(y)|^N}{|x-y|^{(N-1)r}} dy \right)^{\frac{1}{q}} \left(\int_E \frac{dy}{|x-y|^{(N-1)r}} \right)^{\frac{N-1}{N}}.$$

Take the q power of both sides and integrate in dx over E to obtain

$$\|V_f\|_q \leq \|f\|_N^{1-\frac{N}{q}} \left\{ \int_E |f(y)|^N \left(\int_E \frac{dx}{|x-y|^{(N-1)r}} \right) dy \right\}^{\frac{1}{q}}$$

$$\times \sup_{x \in E} \left(\int_E \frac{dy}{|x-y|^{(N-1)r}} \right)^{\frac{N-1}{N}}$$

$$\leq \|f\|_N \sup_{x \in E} \left(\int_E \frac{dy}{|x-y|^{(N-1)r}} \right)^{\frac{1}{r}}.$$

These formal calculations become rigorous provided that the last term is finite. By Proposition 23.2, this occurs if $r < \frac{N}{N-1}$, and we estimate

$$\sup_{x \in E} \left(\int_E \frac{dy}{|x-y|^{(N-1)r}} \right)^{\frac{1}{r}} \leq \frac{\kappa_N^{\frac{N-1}{N}}}{\left(1 - \frac{N-1}{N}r\right)^{\frac{1}{r}}} \mu(E)^{\frac{1}{q}}$$

$$= \kappa_N^{\frac{N-1}{N}} \left(\frac{q}{r}\right)^{1+\frac{1}{q}-\frac{1}{N}} \mu(E)^{\frac{1}{q}}$$

$$\leq \kappa_N^{\frac{N-1}{N}} q^{\frac{N-1}{N}+\frac{1}{q}} \mu(E)^{\frac{1}{q}}$$

for all $q > N$. Therefore, for all such q,

$$\|V_f\|_q \leq \kappa_N^{\frac{N-1}{N}} q^{\frac{N-1}{N}+\frac{1}{q}} \mu(E)^{\frac{1}{q}} \|f\|_N. \tag{24.3}$$

Set $q = \frac{N}{N-1}s$ and let s range over the positive integers larger than $N-1$. Then from (24.3), after we take the q power, we derive

$$\int_E \left\{ \left(\frac{|V_f|}{\|f\|_N}\right)^{\frac{N}{N-1}} \right\}^s dx \leq \frac{N}{N-1}\mu(E) \left(\frac{N\kappa_N}{N-1} \right)^s s^{s+1}.$$

Let C be a constant to be chosen. Divide both sides by $C^{\frac{Ns}{N-1}}s!$ and add for all integers $s = N, N+1, \dots, k$. This gives

$$\int_E \sum_{s=N}^{k} \frac{1}{s!} \left\{ \left(\frac{|V_f|}{C\|f\|_N}\right)^{\frac{N}{N-1}} \right\}^s dx$$

$$\leq \frac{N}{N-1}\mu(E) \sum_{s=0}^{\infty} \left(\frac{N\kappa_N}{(N-1)C^{\frac{N}{N-1}}} \right)^s \frac{s^s}{(s-1)!}$$

---

[23]This justifies the limitation $q > N$.

for all $k \geq N$. The right-hand side is convergent provided that we choose C so large that

$$\frac{N\kappa_N}{(N-1)C^{\frac{N}{N-1}}} < \frac{1}{e}.$$

Making use of (24.3), it is readily seen that the sum on the left-hand side can be taken for $s = 0, 1, 2, \ldots$ by possibly modifying the various constants on the right-hand side. Letting $k \to \infty$ and using the monotone convergence theorem proves (24.1). $\qquad \square$

PROBLEMS AND COMPLEMENTS

5 FUNCTIONS OF BOUNDED MEAN OSCILLATION

We continue to denote by Q_o a cube centered at the origin and with faces parallel to the coordinate planes. We denote by Q a subcube of Q_o with faces parallel to the coordinate planes.

Proposition 5.1c. *Let* $f \in L^1(Q_o)$, *and assume that for every subcube* $Q \subset Q_o$, *there is a constant* γ_Q *such that*

$$\mu\{x \in Q || f(x) - \gamma_Q| > t\} \leq \overline{\gamma}_1 \exp\{-\overline{\gamma}_2 t\}\mu(Q) \qquad (5.1c)$$

for all $t > 0$, *for two given constants* $\overline{\gamma}_1$ *and* $\overline{\gamma}_2$ *independent of* Q *and* t. *Then* f *is of bounded mean oscillation in* Q_o.

Proof. By an argument similar to the one leading to (5.3),

$$\sup_{Q \in Q_o} \frac{1}{\mu(Q)} \int_Q |f - \gamma_Q| dy \leq \frac{\overline{\gamma}_1}{\overline{\gamma}_2}.$$

From this, for every $Q \subset Q_o$,

$$\frac{1}{\mu(Q)} \int_Q |f - f_Q| dy \leq \frac{2}{\mu(Q)} \int_Q |f - \gamma_Q| dy$$

$$\leq 2\frac{\overline{\gamma}_1}{\overline{\gamma}_2}. \qquad \square$$

Proposition 5.2c. *The function* $x \to \ln |x|$ *is of bounded mean oscillation in the unit cube* Q_o *centered at the origin of* $\mathbb{R}^N$.

Proof. Having fixed $Q \in Q_o$, let ξ be the element of largest Euclidean length in Q and set $\gamma_Q = \ln |\xi|$. Then

$$\Sigma_t = \{x \in Q || \ln |x| - \gamma_Q| > t\}$$

$$= \left\{x \in Q | \ln \frac{|\xi|}{|x|} > t\right\}$$

$$= \{x \in Q || x| < |\xi| e^{-t}\}.$$

Let h be the edge of Q, and denote by η an element of least Euclidean length in Q. If Σ_t is not empty, it must contain η. Therefore,

$$|\xi| e^{-t} \geq |\eta| \geq |\xi| - |\xi - \eta| \geq |\xi| - \sqrt{N} h.$$

Therefore,

$$|\xi| \leq \frac{\sqrt{N} h}{1 - e^{-t}},$$

and Σ_t is contained in the ball

$$|x| < \frac{\sqrt{N} h}{1 - e^{-t}} e^{-t}.$$

This, in turn, implies that

$$\mu(\Sigma_t) \leq \left(\frac{\omega_N \sqrt{N}}{N(e^t - 1)} \right)^N \mu(Q).$$

If $t \geq 1$, this gives (5.1c) for $\overline{\gamma}_2 = N$ and a suitable constant $\overline{\gamma}_1$. On the other hand, if $0 < t < 1$, inequality (5.1c) is still satisfied by possibly modifying $\overline{\gamma}_1$.   □

11   REARRANGING THE VALUES OF A FUNCTION

Here we provide an alternative, equivalent definition of the rearrangement f_* of a nonnegative, measurable function $f : \mathbb{R} \to \mathbb{R}$. First, for a measurable set $E \subset \mathbb{R}$, define the rearrangement E_* of E as the interval

$$E_* = \begin{cases} \emptyset & \text{if } \mu(E) = 0, \\ [0, \mu(E)) & \text{if } 0 < \mu(E) < \infty, \\ [0, \infty) & \text{if } \mu(E) = \infty. \end{cases}$$

The rearrangement of the characteristic function of E is defined as

$$(\chi_E)_* = \chi_{[0, \mu(E))} = \chi_{E_*}.$$

In particular, if $E = [f > t]$ for some $t \geq 0$,

$$(\chi_{[f>t]})_* = \begin{cases} 0 & \text{if } \lambda(t) = \mu([f > t]) = 0, \\ \chi_{[0, \lambda(t))} & \text{if } 0 < \lambda(t) < \infty, \\ \chi_{[0, \infty)} & \text{if } \lambda(t) = \infty. \end{cases}$$

Then for all $x \geq 0$,

$$f_*(x) \overset{\text{def}}{=} \int_0^\infty \chi_{[f>t]_*}(x) dt. \tag{11.1c}$$

This definition is equivalent to (11.2), and the basic properties of rearrangements can be established directly from (11.1c).

11.1   The decreasing rearrangement of f could be also defined from

$$[0, \infty) \longrightarrow \bar{\lambda}(t) = \mu([f \geq t])$$

by setting

$$[0, \infty) \longrightarrow \bar{f}(x) = \sup\{t \geq 0 | \bar{\lambda}(t) \geq x\}. \qquad (11.2c)$$

Introduce the necessary changes to generate from this a well-defined function $\bar{f}$, and prove that $\bar{f} = f_*$ a.e. in $\mathbb{R}$. Moreover, $\bar{f}$ is upper semicontinuous.

13   SYMMETRIC REARRANGEMENTS

The form (11.1c) of f_* permits one to give a direct definition of symmetric rearrangement f^* of a nonnegative and measurable function $f : \mathbb{R} \to \mathbb{R}$. First for a measurable set $E \subset \mathbb{R}$, define E^* as

$$E^* = \begin{cases} \emptyset & \text{if } \mu(E) = 0, \\ \left(-\dfrac{1}{2}\mu(E), \dfrac{1}{2}\mu(E)\right) & \text{if } 0 < \mu(E) < \infty, \\ \mathbb{R} & \text{if } \mu(E) = \infty. \end{cases}$$

The symmetric rearrangement of the characteristic function of a measurable set $E \subset \mathbb{R}$ is

$$\chi_E^* = \chi_{E^*}.$$

In particular, if $E = [f > t]$ for some $t \geq 0$,

$$\chi_{[f>t]}^* = \begin{cases} 0 & \text{if } \lambda(t) = \mu[f > t]) = 0, \\ \chi_{\left(-\frac{1}{2}\lambda(t), \frac{1}{2}\lambda(t)\right)} & \text{if } 0 < \lambda(t) < \infty, \\ 1 & \text{if } \lambda(t) = \infty. \end{cases}$$

Then for all $x \in \mathbb{R}$,

$$f^*(x) = \int_0^\infty \chi_{[f>t]}^*(x)dt = \begin{cases} f_*(2x) & \text{if } x \geq 0, \\ f_*(-2x) & \text{if } x < 0. \end{cases} \qquad (13.1c)$$

13.1   **REARRANGING THE VALUES OF A FUNCTION OF N VARIABLES.** The definition (13.1c) of f^* in terms of an integral can be used to introduce a notion of rearrangement of a nonnegative, measurable function $f : \mathbb{R}^N \to \mathbb{R}$. Let E be a Lebesgue-measurable subset of $\mathbb{R}^N$. If $\mu(E) = 0$, we set $E^* = \emptyset$. If $0 < \mu(E) < \infty$, let E^* be the open ball B_R centered at the origin and of volume equal to $\mu(E)$; i.e.,

$$\mu(E^*) = \kappa_N R^N = \mu(E).$$

If $\mu(E) = \infty$, we set $E^* = \mathbb{R}^N$. The N-dimensional, symmetric rearrangement of the characteristic function of E is defined as

$$\chi_E^* = \chi_{E^*} = \begin{cases} 0 & \text{if } \mu(E) = 0, \\ \chi_{B_R} & \text{if } 0 < \mu(E) < \infty, \\ 1 & \text{if } \mu(E) = \infty. \end{cases}$$

In particular, if $E = [f > t]$ for some $t \geq 0$, then

$$\chi_{[f>t]}^* = \begin{cases} 0 & \text{if } \lambda(t) = 0, \\ \chi_{B_R}\, R = \left(\dfrac{\lambda(t)}{\kappa_N}\right)^{\frac{1}{N}} & \text{if } 0 < \lambda(t) < \infty, \\ 1 & \text{if } \lambda(t) = \infty. \end{cases}$$

The symmetric rearrangement of f is then defined as

$$\mathbb{R} \ni x \longrightarrow f^*(x) \stackrel{\text{def}}{=} \int_0^\infty \chi_{[f>t]}^*(x)dt.$$

Proposition 13.1c. *Let f^* be the symmetric rearrangement of a nonnegative, measurable function $f : \mathbb{R}^N \to \mathbb{R}$. Then*

(i) *f^* is a decreasing function of $|x|$;*

(ii) *f^* is lower semicontinuous;*

(iii) *for all $1 \leq p \leq \infty$, $\|f\|_p = \|f^*\|_p$.*

23 L^p ESTIMATES OF RIESZ POTENTIALS

23.1   The value $p = 1$ is not permitted in Theorem 23.1. To construct a counterexample, let $J_{1/n}$ be the Friedrichs mollifying kernels.[24] If (23.2) were to hold for $p = 1$,

$$\int_{\mathbb{R}^N} \left(\int_{\mathbb{R}^N} \frac{J_{1/n}(y)}{|x-y|^{N-1}} dy \right)^{\frac{N}{N-1}} dx \leq C^{\frac{N}{N-1}}$$

for a constant C independent of n. Letting $n \to \infty$,

$$\int_{\mathbb{R}^N} \frac{1}{|x|^N} dx \leq C.$$

---

[24] See Section 21 of Chapter V.

23.2 The values $p = N$ and $p^* = \infty$ are not permitted in Theorem 23.1, as indicated by the following counterexample. For $0 < \varepsilon \ll 1$, set

$$f(x) = \begin{cases} \dfrac{1}{|x|} \dfrac{1}{|\ln|x||^{\frac{1+\varepsilon}{N}}} & \text{for } |x| \le \dfrac{1}{2}, \\ 0 & \text{for } |x| > \dfrac{1}{2}. \end{cases}$$

One verifies that $f \in L^N(\mathbb{R}^N)$ and that the corresponding potential $\mathcal{V}_f(x)$ is unbounded for x near the origin since

$$\mathcal{V}_f(0) = \int_{|x|<\frac{1}{2}} \frac{1}{|x|^N} \frac{1}{|\ln|x||^{\frac{1+\varepsilon}{N}}} dx = \infty$$

for ε sufficiently small.

23.3 GENERALIZED RIESZ POTENTIALS. Let E be a domain in $\mathbb{R}^N$, and for $\alpha \in (0, N)$ and $f \in L^p(E)$, consider the potentials

$$\mathcal{V}_{\alpha, f}(x) = \int_E \frac{f(y)}{|x - y|^{N-\alpha}} dy. \tag{23.1c}$$

If $\alpha = 1$, these coincide with the Riesz potentials.

Theorem 23.1c. *Let* $f \in L^p(E)$ *for* $1 < p < \frac{N}{\alpha}$. *There exists a constant* C *depending upon only* N, p, *and* α *such that*

$$\|\mathcal{V}_{\alpha, f}\|_q \le C \|f\|_p, \quad \text{where } q = \frac{Np}{N - \alpha p}. \tag{23.2c}$$

Remark 23.1c. The constant C in (23.1c) tends to infinity as either $p \to 1$ or $p \to \frac{N}{\alpha}$.

Proposition 23.2c. *For every* $\alpha \in (0, N)$ *and every* $1 \le r < \frac{N}{N-\alpha}$,

$$\sup_{x \in E} \int_E \frac{dy}{|x - y|^{(N-\alpha)r}} \le \frac{\kappa_N^{\frac{N-\alpha}{N}r}}{1 - \frac{N-\alpha}{N}r} \mu(E)^{1 - \frac{N-\alpha}{N}r}. \tag{23.3c}$$

Proposition 23.3c. *Let* E *be of finite measure, and let* $f \in L^1(E)$. *Then* $\mathcal{V}_{\alpha, f} \in L^q(E)$ *for all* $q \in [1, \frac{N}{N-\alpha})$ *and*

$$\|\mathcal{V}_{\alpha, f}\|_q \le \frac{\kappa_N^{\frac{N-\alpha}{N}}}{\left(1 - \frac{N-\alpha}{N}q\right)^{\frac{1}{q}}} \mu(E)^{\frac{1}{q} - \frac{N-\alpha}{N}} \|f\|_1. \tag{23.4c}$$

Remark 23.2c. The limiting integrability $q = \frac{N}{N-\alpha}$ is not permitted in (23.4c).

Proposition 23.4c. *Let E be of finite measure, and let $f \in L^p(E)$ for some $p > \frac{N}{\alpha}$. Then $\mathcal{V}_{\alpha,f} \in L^\infty(E)$ and*

$$\|\mathcal{V}_{\alpha,f}\|_\infty \leq C(N, p, \alpha)\mu(E)^{\frac{\alpha p - N}{Np}}\|f\|_p, \tag{23.5c}$$

where

$$C(N, p, \alpha) = \kappa_N^{\frac{N-\alpha}{N}}\left(\frac{N(p-1)}{\alpha p - N}\right)^{\frac{p-1}{p}}. \tag{23.5c}'$$

23.4 MOTIVATING RIESZ POTENTIALS AS EMBEDDINGS. Consider the Stokes formula (15.8) of Chapter VII written for a function $\varphi \in C_o^\infty(E)$. After an integration by parts,

$$\varphi(x) = \int_E \nabla_y F(x; y) \cdot \nabla\varphi \, dy.$$

Using (15.6) of Chapter VII,

$$|\varphi(x)| \leq \frac{1}{\omega_N} \int_E \frac{|\nabla\varphi|}{|x - y|^{N-1}} dy. \tag{23.6c}$$

Now given a function $u \in W_o^{1,p}(E)$, there is a sequence $\{\varphi_n\}$ of functions in $C_o^\infty(E)$ such that $\varphi_n \to u$ in the norm of $W_o^{1,p}(E)$.[25] Thus up to a limiting process, (23.6c) continues to hold for functions $\varphi \in W_o^{1,p}(E)$.

Given that $|\nabla\varphi| \in L^p(E)$, it is natural to ask what the order of integrability of φ is. This motivates Theorem 23.1 and Proposition 23.3. Theorems of this kind are called *embedding theorems* and are systematically treated in Chapter IX.

24 THE LIMITING CASE $p = \frac{N}{\alpha}$

The value $\alpha p = N$ is not permitted in either Theorem 23.1c or in Proposition 23.4c.

Theorem 24.1c. *Let E be of finite measure, and let $f \in L^p(E)$ for $p = \frac{N}{\alpha}$. There exist constants C_1 and C_2 depending only upon N such that*

$$\int_E \exp\left(\frac{|\mathcal{V}_{\alpha,f}|}{C_1\|f\|_p}\right)^{\frac{N}{N-\alpha}} dx \leq C_2\mu(E). \tag{24.1c}$$

---

[25]This follows from the definition of $W_o^{1,p}(E)$ in Section 16 of Chapter VII.

IX

Embeddings of $W^{1,p}(E)$ into $L^q(E)$

1 Multiplicative embeddings of $W_o^{1,p}(E)$

Let E be a bounded, open set in $\mathbb{R}^N$. An embedding from $W^{1,p}(E)$ into $L^q(E)$ is an estimate of the $L^q(E)$-norm of a function $u \in W^{1,p}(E)$ in terms of its $W^{1,p}(E)$-norm. The structure of such an estimate and the various constants involved should not depend on the particular function $u \in W^{1,p}(E)$ nor on the size of E, although they might depend on the structure of ∂E.

Since typically $q > p$, an embedding estimate amounts to, roughly speaking, an improvement on the order of integrability of u. Also, if p is sufficiently large, one might expect a function $u \in W^{1,p}(E)$ to possess some local regularity beyond a higher degree of integrability.

The backbone of such embeddings is that of $W_o^{1,p}(E)$ into $L^q(E)$ in view of its relative simplicity. Functions in $W_o^{1,p}(E)$ are limits of functions in $C_o^\infty(E)$ in the norm of $W_o^{1,p}(E)$, and in this sense, they vanish on ∂E. This permits embedding inequalities in a multiplicative form, such as (1.1) below.[1] The proof is based only on calculus ideas and applications of Hölder's inequality.

Theorem 1.1 (Gagliardo–Nirenberg[2]). *Let* $u \in W_o^{1,p}(E) \cap L^r(E)$ *for some* $r \geq 1$. *There exists a constant* C *depending upon* N, p, r *such that*

$$\|u\|_q \leq C \|Du\|_p^\theta \|u\|_r^{1-\theta}, \tag{1.1}$$

---

[1] Such an inequality would be false for functions not vanishing, in some sense, on a subset of $\overline{E}$. For example, a constant, nonzero function would not satisfy (1.1).

[2] E. Gagliardo, Proprietà di alcune funzioni in n variabili, *Ricerche Mat.*, **7** (1958), 102–137; L. Nirenberg, On elliptic partial differential equations, *Ann. Scuola Norm. Sup. Pisa*, **3**-13 (1959), 115–162.

where $\theta \in [0, 1]$ and $p, q \geq 1$ are linked by

$$\theta = \left(\frac{1}{r} - \frac{1}{q}\right)\left(\frac{1}{N} - \frac{1}{p} + \frac{1}{r}\right)^{-1} \tag{1.2}$$

and their admissible range is

$$\begin{cases} \text{if } N = 1, \quad \text{then } q \in [r, \infty] \quad \text{and} \\ \theta \in \left[0, \dfrac{p}{p + r(p - 1)}\right], \qquad C = \left(1 + \dfrac{p - 1}{rp}\right)^{\theta}; \end{cases} \tag{1.3}$$

$$\begin{cases} \text{if } N > p \geq 1, \quad \text{then} \\ q \in \left[r, \dfrac{Np}{N - p}\right] \quad \text{if } r \leq \dfrac{Np}{N - p} \quad \text{and} \\ q \in \left[\dfrac{Np}{N - p}, r\right] \quad \text{if } r \geq \dfrac{Np}{N - p}; \quad \text{moreover,} \\ \theta \in [0, 1] \quad \text{and} \quad C = \left[\dfrac{p(N - 1)}{N - p}\right]^{\theta}; \end{cases} \tag{1.4}$$

$$\begin{cases} \text{if } p > N > 1, \quad \text{then } q \in [r, \infty] \quad \text{and} \\ \theta \in \left[0, \dfrac{Np}{Np + r(p - N)}\right]; \\ \text{if } p = N, \quad \text{then } q \in [r, \infty) \quad \text{and} \quad \theta \in [0, 1). \end{cases} \tag{1.5}$$

Moreover, the constant C is given explicitly in (6.1) below.

By taking $\theta = r = 1$ in (1.4) yields the following embedding.

Corollary 1.2. *Let $u \in W_o^{1,p}(E)$, and assume that $1 \leq p < N$. Then*[3]

$$\|u\|_{p^*} \leq \frac{p(N - 1)}{N - p}\|Du\|_p, \quad \text{where } p^* = \frac{Np}{N - p}. \tag{1.1}'$$

Since $C_o^\infty(E)$ is dense in $W_o^{1,p}(E)$, in proving Theorem 1.1, we may assume that $u \in C_o^\infty(E)$. Also, since E is bounded, we may assume, possibly after a translation, that it is contained in a cube centered at the origin and faces parallel to the coordinate planes, say, for example,

$$Q = \left\{x \in \mathbb{R}^N \mid \max_{1 \leq i \leq N} |x_i| < M\right\} \quad \text{for some } M > 0. \tag{1.6}$$

We will assume that E coincides with such a cube and that $u \in C_o^\infty(Q)$. The embedding constant C in (1.1) is independent of M.

---

[3]The constant $\gamma = \frac{p(N-1)}{N-p}$ in (1.1)$'$ is not optimal. The best constant $\gamma = \gamma(N, p)$ is computed in G. Talenti, Best constants in Sobolev inequalities, *Ann. Mat. Pura Appl.*, **110** (1976), 353–372. When $p = 1$, one has $\gamma(N, 1) = \frac{1}{N}(\frac{N}{\omega_N})^{1/N}$, where ω_N is the area of the unit sphere in $\mathbb{R}^N$.

2 Proof of Theorem 1.1 for $N = 1$

Assume first $p > 1$ and $q < \infty$. For $r \geq 1, s > 1, q \geq r$, and all $x \in Q$,

$$
\begin{aligned}
|u(x)|^q &= |u(x)|^r (|u(x)|^s)^{\frac{q-r}{s}} \\
&= |u(x)|^r \left(\int_{-\infty}^{x} D|u(\xi)|^s d\xi \right)^{\frac{q-r}{s}} \\
&\leq |u(x)|^r \left(\int_E s|u(\xi)|^{s-1} |Du(\xi)| d\xi \right)^{\frac{q-r}{s}}.
\end{aligned}
$$

Integrate in dx over E and apply Hölder's inequality to the last integral to obtain

$$
\|u\|_q^q \leq \|u\|_r^r \left(s\|Du\|_p \|u\|_{\frac{p(s-1)}{p-1}}^{s-1} \right)^{\frac{q-r}{s}}.
$$

Choose s from

$$
\frac{p(s-1)}{p-1} = r, \quad \text{i.e.,} \quad s = 1 + \frac{r(p-1)}{p},
$$

and set

$$
\theta = \frac{q-r}{qs} = \left(\frac{1}{r} - \frac{1}{q} \right) \left(\frac{1}{N} - \frac{1}{p} + \frac{1}{r} \right)^{-1}.
$$

Then

$$
\|u\|_q \leq C\|Du\|_p^\theta \|u\|_r^{1-\theta}, \quad \text{where } C = \left(1 + \frac{r(p-1)}{p} \right)^\theta.
$$

The limiting cases $p = 1$ and $q = \infty$ are established similarly. $\square$

3 Proof of Theorem 1.1 for $1 \leq p < N$

The proof for the case (1.4) uses two auxiliary lemmas.

Lemma 3.1. $\|u\|_{\frac{N}{N-1}} \leq \|Du\|_1$.

Proof. It suffices to establish the stronger inequality

$$
\|u\|_{\frac{N}{N-1}} \leq \prod_{i=1}^{N} \|u_{x_i}\|_1^{1/N}. \tag{3.1}
$$

Such inequality holds for $N = 2$. Indeed,

$$\iint_E u^2(x_1, x_2)dx_1dx_2 = \iint_E u(x_1, x_2)u(x_1, x_2)dx_1dx_2$$

$$\leq \iint_E \max_{x_2} u(x_1, x_2) \max_{x_1} u(x_1, x_2)dx_1dx_2$$

$$= \int_{\mathbb{R}} \max_{x_2} u(x_1, x_2)dx_1 \int_{\mathbb{R}} \max_{x_1} u(x_1, x_2)dx_2$$

$$\leq \left(\iint_E |u_{x_1}|dx \right) \left(\iint_E |u_{x_2}|dx \right).$$

The lemma is now proved by induction; i.e., if (3.1) holds for $N \geq 2$, it continues to hold for $N + 1$. Set

$$x = (\overline{x}, x_{N+1}) \quad \text{and} \quad \overline{x} = (x_1, x_2, \ldots, x_N).$$

Then by Hölder's inequality,

$$\|u\|_{\frac{N+1}{N}}^{\frac{N+1}{N}} = \int_{\mathbb{R}} \int_{\mathbb{R}^N} |u(\overline{x}, x_{N+1})|^{\frac{N+1}{N}} d\overline{x}dx_{N+1}$$

$$= \int_{\mathbb{R}} dx_{N+1} \int_{\mathbb{R}^N} |u(\overline{x}, x_{N+1})||u(\overline{x}, x_{N+1})|^{\frac{1}{N}} d\overline{x}$$

$$\leq \int_{\mathbb{R}} dx_{N+1} \left(\int_{\mathbb{R}^N} |u(\overline{x}, x_{N+1})|d\overline{x} \right)^{\frac{1}{N}}$$

$$\times \left(\int_{\mathbb{R}^N} |u(\overline{x}, x_{N+1})|^{\frac{N}{N-1}} d\overline{x} \right)^{\frac{N-1}{N}}$$

Observe that

$$\int_{\mathbb{R}^N} |u(\overline{x}, x_{N+1})|d\overline{x} \leq \int_{\mathbb{R}} \int_{\mathbb{R}^N} |u_{x_{N+1}}(\overline{x}, x_{N+1})|dx_{N+1}d\overline{x}$$

$$= \int_E |u_{x_{N+1}}|dx.$$

Moreover, by the induction hypothesis,

$$\left(\int_{\mathbb{R}^N} |u(\overline{x}, x_{N+1})|^{\frac{N}{N-1}} d\overline{x} \right)^{\frac{N-1}{N}} \leq \prod_{i=1}^{N} \left(\int_{\mathbb{R}^N} |u_{x_i}(\overline{x}, x_{N+1})|d\overline{x} \right)^{\frac{1}{N}}.$$

Therefore,

$$\int_E |u|^{\frac{N+1}{N}} dx \leq \int_{\mathbb{R}} \prod_{i=1}^{N} \left(\int_{\mathbb{R}^N} |u_{x_i}(\overline{x}, x_{N+1})|d\overline{x} \right)^{\frac{1}{N}} dx_{N+1}$$

$$\times \left(\int_E |u_{x_{N+1}}|dx \right)^{\frac{1}{N}}.$$

By the generalized Hölder inequality,

$$\int_{\mathbb{R}} \prod_{i=1}^{N} \left(\int_{\mathbb{R}^N} |u_{x_i}(\bar{x}, x_{N+1})| d\bar{x} \right)^{\frac{1}{N}} dx_{N+1}$$

$$\leq \prod_{i=1}^{N} \left(\int_{\mathbb{R}} \int_{\mathbb{R}^N} |u_{x_i}(\bar{x}, x_{N+1})| d\bar{x} dx_{N+1} \right)^{\frac{1}{N}}$$

$$= \left(\prod_{i=1}^{N} \int_{E} |u_{x_i}| dx \right)^{\frac{1}{N}}.$$

Combining these last two inequalities proves the lemma.   □

Lemma 3.2. $\|u\|_{\frac{Np}{N-p}} \leq \frac{p(N-1)}{N-p} \|Du\|_p.$

Proof. Write

$$\|u\|_{\frac{Np}{N-p}} = \left(\int_{E} \left(|u|^{\frac{p(N-1)}{N-p}} \right)^{\frac{N}{N-1}} dx \right)^{\frac{N-1}{N} \frac{N-p}{p(N-1)}}$$

and apply (3.1) to the function $w = |u|^{p(N-1)/(N-p)}$. This gives

$$\|u\|_{\frac{Np}{N-p}} \leq \left[\prod_{i=1}^{N} \left(\int_{E} \left| \left(|u|^{\frac{p(N-1)}{N-p}} \right)_{x_i} \right| dx \right)^{\frac{1}{N}} \right]^{\frac{N-p}{p(N-1)}}$$

$$= \gamma \prod_{i=1}^{N} \left(\int_{E} |u|^{\frac{p(N-1)}{N-p} - 1} |u_{x_i}| dx \right)^{\frac{N-p}{Np(N-1)}},$$

where

$$\gamma = \left(\frac{p(N-1)}{N-p} \right)^{\frac{N-p}{p(N-1)}}.$$

Now for all $i = 1, 2, \ldots, N$, by Hölder's inequality,

$$\int_{E} |u|^{\frac{p(N-1)}{N-p} - 1} |u_{x_i}| dx \leq \left(\int_{E} |u_{x_i}|^p dx \right)^{\frac{1}{p}} \left(\int_{E} |u|^{\frac{Np}{N-p}} dx \right)^{\frac{p-1}{p}}$$

and

$$\prod_{i=1}^{N} \left(\int_{E} |u|^{\frac{p(N-1)}{N-p} - 1} |u_{x_i}| dx \right)^{\frac{N-p}{Np(N-1)}}$$

$$= \prod_{i=1}^{N} (\|u_{x_i}\|_p)^{\frac{N-p}{Np(N-1)}} \|u\|_{\frac{Np}{N-p}}^{\frac{p-1}{p(N-1)}}$$

$$\leq \|Du\|_p^{\frac{N-p}{p(N-1)}} \|u\|_{\frac{Np}{N-p}}^{\frac{p-1}{p} \frac{N}{N-1}}.$$

Combining these inequalities proves the lemma.   □

4   Proof of Theorem 1.1 for $1 \leq p < N$, concluded

According to (1.2), the case $\theta = 0$ corresponds to $q = r$, and therefore (1.1) is trivial. The case where $\theta = 1$ corresponds to $q = \frac{Np}{N-p}$ and is contained in Lemma 3.2.

Let $r \in [1, \infty)$ and $p < N$ be fixed, and choose $\theta \in (0, 1)$ and q such that

$$\min \left\{ r; \frac{Np}{N-p} \right\} < q < \max \left\{ r; \frac{Np}{N-p} \right\} \quad \text{and} \quad \theta q < \frac{Np}{N-p}. \tag{4.1}$$

Then by Hölder's inequality,

$$\int_E |u|^q dx = \int_E |u|^{\theta q} |u|^{(1-\theta)q} dx$$

$$\leq \left(\int_E |u|^{\frac{Np}{N-p}} dx \right)^{\frac{N-p}{Np} \theta q} \left(\int_E |u|^r dx \right)^{\frac{(1-\theta)q}{r}},$$

where we have set

$$(1-\theta)q \frac{Np}{Np - (N-p)\theta q} = r. \tag{4.2}$$

From this, by Lemma 3.2,

$$\|u\|_q \leq C \|Du\|_p^{\theta} \|u\|_r^{1-\theta}, \quad \text{where } C = \left(\frac{p(N-1)}{N-p} \right)^{\theta}.$$

By direct calculation, one verifies that (4.2) is exactly (1.2). Moreover, the ranges indicated in (1.4) correspond to the compatibility of (4.1) and (4.2).

5   Proof of Theorem 1.1 for $p \geq N > 1$

Let $F(x; y)$ be the fundamental solution of the Laplace operator.[4] Then for $u \in C_o^\infty(E)$,

$$u(x) = -\int_{\mathbb{R}^N} F(x; y) \Delta u(y) dy$$

$$= \frac{1}{\omega_N} \int_{\mathbb{R}^N} Du(y) \cdot \frac{(x-y)}{|x-y|^N} dy. \tag{5.1}$$

---

[4] See Section 15.2 and, in particular, the Stokes formula (15.8) of Chapter VII.

Such an identity holds for all $x \in E$. Fix $\rho > 0$ and rewrite it as

$$\omega_N u(x) = \int_{|x-y|<\rho} Du(y) \cdot \frac{(x-y)}{|x-y|^N} dy$$
$$+ \int_{|x-y|>\rho} Du(y) \cdot \frac{x-y}{|x-y|^N} dy. \tag{5.2}$$

The second integral can be computed by an integration by parts as

$$\int_{|x-y|>\rho} Du(y) \cdot \frac{x-y}{|x-y|^N} dy$$
$$= \frac{1}{N-2} \int_{|x-y|>\rho} Du(y) \cdot D \frac{1}{|x-y|^{N-2}} dy$$
$$= \frac{1}{\rho^{N-1}} \int_{|x-y|=\rho} u(y) d\sigma$$

since $F(x; y)$ is harmonic in $\mathbb{R}^N - \{x\}$. Here $d\sigma$ denotes the surface measure on the sphere $\{|x - y| = \rho\}$. Put this in (5.2), multiply by $N\rho^{N-1}$, and integrate in $d\rho$ over $(0, R)$, where R is a positive number to be chosen later. This gives

$$\omega_N R^N |u(x)| \leq N \int_0^R \left(\int_{|x-y|<\rho} \frac{|Du(y)|}{|x-y|^{N-1}} dy \right) \rho^{N-1} d\rho$$
$$+ N \int_0^R \left(\int_{|x-y|=\rho} |u(y)| d\sigma \right) d\rho.$$

From this, for all $x \in E$,

$$\omega_N |u(x)| \leq \int_{B_R(x)} \frac{|Du(y)|}{|x-y|^{N-1}} dy + \frac{N}{R^N} \int_{B_R(x)} |u(y)| dy \tag{5.3}$$
$$= I_1(x, R) + N I_2(x, R),$$

where $B_R(x)$ is the ball of radius R centered at x.

5.1 Estimate of $I_1(x, R)$. Choose two positive numbers $a, b < N$ such that

$$\frac{a}{q} + b \left(1 - \frac{1}{p} \right) = N - 1. \tag{5.4}$$

Since

$$\frac{N}{q} + N \left(1 - \frac{1}{p} \right) > N \left(1 - \frac{1}{N} \right) = N - 1, \tag{5.5}$$

such a choice can be made. Now write

$$\frac{|Du(y)|}{|x-y|^{N-1}} = |Du|^{p(\frac{1}{p}-\frac{1}{q})} \frac{|Du|^{\frac{p}{q}}}{|x-y|^{\frac{a}{q}}} \frac{1}{|x-y|^{b(1-\frac{1}{p})}},$$

and apply the generalized Hölder inequality with the conjugate exponents

$$\left(\frac{1}{p} - \frac{1}{q}\right) + \frac{1}{q} + \left(1 - \frac{1}{p}\right) = 1.$$

This gives

$$I_1(x, R) \leq \|Du\|_p^{1-\frac{p}{q}} \left(\int_{B_R(x)} \frac{|Du(y)|^p}{|x - y|^a} dy\right)^{\frac{1}{q}} \left(\int_{B_R(x)} \frac{1}{|x - y|^b} dy\right)^{1-\frac{1}{p}}.$$

Taking the q power and integrating over E gives

$$\|I_1(R)\|_q \leq \frac{\omega_N^{1-\frac{1}{p}+\frac{1}{q}} R^{N\left(\frac{1}{N} - \frac{1}{p} + \frac{1}{q}\right)}}{(N - a)^{\frac{1}{q}} (N - b)^{1-\frac{1}{p}}} \|Du\|_p.$$

5.2   Estimate of $I_2(x, R)$. For all $x \in \mathbb{R}^N$,

$$I_2(x, R) \leq R^{-N} \left(\int_{|x-y|<R} |u(y)|^r dy\right)^{\frac{1}{r}} \left(\int_{|x-y|<R} 1 dy\right)^{1-\frac{1}{r}}$$

$$\leq \left(\frac{\omega_N}{N}\right)^{1-\frac{1}{r}} R^{-\frac{N}{r}} \|u\|_r^{1-\frac{r}{q}} \left(\int_{|\xi|<R} |u(x + \xi)|^r d\xi\right)^{\frac{1}{q}}.$$

Take the q power and integrate in dx over $\mathbb{R}^N$ to obtain

$$\|I_2\|_q \leq \left(\frac{\omega_N}{N}\right)^{1+\frac{1}{q}-\frac{1}{r}} R^{-N\left(\frac{1}{r}-\frac{1}{q}\right)} \|u\|_r.$$

6   Proof of Theorem 1.1 for $p \geq N > 1$, concluded

Combining these estimates into (5.3) gives

$$\|u\|_q \leq \frac{\omega_N^{\frac{1}{q}-\frac{1}{p}}}{(N - a)^{\frac{1}{q}} (N - b)^{1-\frac{1}{p}}} \|Du\|_p R^{N\left(\frac{1}{N} - \frac{1}{p} + \frac{1}{q}\right)}$$

$$+ \left(\frac{\omega_N}{N}\right)^{\frac{1}{q}-\frac{1}{r}} \|u\|_r R^{-N\left(\frac{1}{r}-\frac{1}{q}\right)}.$$

Setting

$$A = \frac{\omega_N^{\frac{1}{q}-\frac{1}{p}} \|Du\|_p}{(N - a)^{\frac{1}{q}} (N - b)^{1-\frac{1}{p}}}, \qquad B = \left(\frac{\omega_N}{N}\right)^{\frac{1}{q}-\frac{1}{r}} \|u\|_r,$$

the previous inequality takes the form

$$\|u\|_q \leq AR^\beta + BR^{-\bar{\beta}}.$$

where

$$\beta = N\left(\frac{1}{N} - \frac{1}{p} + \frac{1}{q}\right), \qquad \overline{\beta} = N\left(\frac{1}{r} - \frac{1}{q}\right).$$

Minimizing the right-hand side with respect to the parameter $R \in (0, \infty)$, we find

$$\|u\|_q \leq \left\{\left(\frac{\overline{\beta}}{\beta}\right)^{\frac{\beta}{\beta+\overline{\beta}}} + \left(\frac{\beta}{\overline{\beta}}\right)^{\frac{\overline{\beta}}{\beta+\overline{\beta}}}\right\} A^{\frac{\overline{\beta}}{\beta+\overline{\beta}}} B^{\frac{\beta}{\beta+\overline{\beta}}}.$$

Setting

$$\frac{\overline{\beta}}{\beta + \overline{\beta}} = \theta, \qquad \frac{\beta}{\beta + \overline{\beta}} = 1 - \theta \quad \text{so that} \quad \frac{\beta}{\overline{\beta}} = \frac{1 - \theta}{\theta}$$

proves (1.1)–(1.2), with the constant C given explicitly by

$$C = \left\{\left(\frac{\theta}{1-\theta}\right)^{1-\theta} + \left(\frac{1-\theta}{\theta}\right)^{\theta}\right\} \left(\frac{\omega_N}{N}\right)^{(\frac{1}{q}-\frac{1}{r})(1-\theta)}$$

$$\times \left(\frac{\omega_N^{\frac{1}{q}-\frac{1}{p}}}{(N-a)^{\frac{1}{q}}(N-b)^{1-\frac{1}{p}}}\right)^{\theta}. \tag{6.1}$$

The ranges indicated in (1.5) depend on the possibility of choosing numbers a and b as in (5.4).   □

7   On the limiting case $p = N$

A function $u \in W_o^{1,p}(E)$ for $p > N$ belongs to $L^\infty(E)$ by the embedding (1.1), (1.5). However, the constant $C = C(N, p)$ in (6.1) deteriorates as $p \to N$. Indeed, as $p \to N$, the number b in (5.4) tends to N, and consequently $C(N, p) \to \infty$. On the other hand, if $p = N$, the same embedding implies that u is $L^q(E)$ for all $1 \leq q < \infty$.

The embedding is rather precise, as there exist functions in $W_o^{1,N}(E)$ for $N > 1$ that are not essentially bounded.[5] It does not, however, provide the precise embedding space for $W_o^{1,N}(E)$.

A sharp embedding for $p = N$ can be derived from the limiting estimates of the Riesz potentials.[6]

Theorem 7.1. *Let $u \in W_o^{1,N}(E)$. There exist constants C_1 and C_2 depending only upon N such that*

$$\int_E \exp\left\{\frac{|u|}{C_1\|Du\|_N}\right\}^{\frac{N}{N-1}} dx \leq C_1\mu(E).$$

---

[5]The function $u(x) = \ln|\ln|x||$ is not bounded near the origin and belongs to $W^{1,N}(B_{1/e})$.
[6]See Section 24 of Chapter VIII.

Proof. We may assume that $u \in C_o^\infty(E)$. From the representation formula (5.1),

$$|u(x)| \leq \frac{1}{\omega_N} \int_E \frac{|Du|}{|x-y|^{N-1}} dy. \tag{7.1}$$

The embedding now follows from the limiting potential estimates.   □

Remark 7.1. The same potential estimates starting from the Riesz potentials in (7.1) contain the results of Corollary 1.2. This is, however, weaker than the multiplicative embedding of Theorem 1.1. On the other hand, if $p = N$, the limiting embedding of Theorem 7.1 is not covered by Theorem 1.1.

8 Embeddings of $W^{1,p}(E)$

Embedding inequalities for functions in $W^{1,p}(E)$ depend, in general, on the structure of ∂E. A minimal requirement is that E satisfy the cone property.[7] The embedding constants are independent of E and its size and depend on the structure of ∂E only through the cone property.

Let κ_N denote the volume of the unit ball in $\mathbb{R}^N$, and denote by $C(N, p)$ a positive constant depending only on N and p and independent of E.

Theorem 8.1. *Assume that E satisfies the cone condition for a fixed circular cone C_o of solid angle ω, height h, and vertex at the origin. Let $u \in W^{1,p}(E)$. If $1 < p < N$, then $u \in L^{p^*}(E)$, where $p^* = \frac{Np}{N-p}$ and*

$$\|u\|_{p^*} \leq \frac{C(N, p)}{\omega} \left\{ \frac{1}{h} \|u\|_p + \|Du\|_p \right\}, \quad p^* = \frac{Np}{N-p}. \tag{8.1}$$

The constant $C(N, p)$ in (8.1) tends to infinity as either $p \to 1$ or $p \to N$.
If $p = 1$ and $\mu(E) < \infty$, then $u \in L^q(E)$ for all $q \in [1, \frac{N}{N-1})$ and

$$\|u\|_q \leq \frac{C(N, q)}{\omega} \mu(E)^{\frac{1}{q}-\frac{N-1}{N}} \left\{ \frac{1}{h} \|u\|_1 + \|Du\|_1 \right\}, \tag{8.2}$$

where

$$C(N, q) = \frac{\kappa_N^{\frac{N-1}{N}}}{\left(1 - \frac{N-1}{N}q\right)^{1/q}}. \tag{8.2$'$}$$

If $p > N$, then $u \in L^\infty(E)$ and

$$\|u\|_\infty \leq \frac{C(N, p, \omega)}{\mu(C_o)^{1/p}} \{ \|u\|_p + h\|Du\|_p \}. \tag{8.3}$$

---

[7] See Section 17.4 of Chapter VII.

where[8]

$$C(N, p\omega) = \frac{1}{N} \left(\frac{\omega_N}{\omega}\right)^{\frac{N-1}{N}} \left(\frac{N(p-1)}{p-N}\right)^{\frac{p-1}{p}}. \tag{8.3}'$$

If $p > N$ and, in addition, E is convex, the function u is Hölder continuous in $\overline{E}$, and for every $x, y \in \overline{E}$ such that $|x - y| < h$,[9]

$$|u(x) - u(y)| \leq \frac{C(N, p)}{\omega} |x - y|^{1 - \frac{N}{p}} \|Du\|_p, \tag{8.4}$$

where

$$C(N, p) = 2^{N+1} \kappa_N^{\frac{p-1}{p}} \frac{Np}{p-N}. \tag{8.4}'$$

Remark 8.1. The estimates exhibit an explicit dependence on the height h and the solid angle ω of the cone C_o. They deteriorate as either h or ω tend to zero.

Remark 8.2. Estimate (8.4) depends on the height h of the cone C_o through the requirement that $|x - y| < h$.

Remark 8.3. The value $q = 1^* = \frac{N}{N-1}$ is not permitted in (8.2). Such a limiting value is permitted for embeddings of $W_o^{1,p}(E)$, as indicated in Corollary 1.2.

Remark 8.4. The limiting case $p = N$, not permitted in either (8.3) or (8.4), is given a sharp form in Section 13.

Remark 8.5. If $p > N$ and E is not convex, the estimate (8.4) can be applied locally. Thus if $p > N$, a function $u \in W^{1,p}(E)$ is locally Hölder continuous in E.

9   Proof of Theorem 8.1

Proof of (8.1) *and* (8.2)[10]. It suffices to prove the various assertions for $u \in C^\infty(E)$.[11] Fix $x \in E$, and let $C_x \subset \overline{E}$ be a cone congruent to C_o and claimed by the cone property. Then

$$|u(x)| = \left|\int_0^h \frac{\partial}{\partial\rho} \left(1 - \frac{\rho}{h}\right) u(\rho\mathbf{n}) d\rho\right|$$

$$\leq \int_0^h |Du(\rho\mathbf{n})| d\rho + \frac{1}{h} \int_0^h |u(\rho\mathbf{n})| d\rho,$$

---

[8] ω_N is the area of the unit sphere in $\mathbb{R}^N$. The volume of the unit ball in $\mathbb{R}^N$ is $\kappa_N = N^{-1}\omega_N$.

[9] That is, the equivalence class u has a representative that is Hölder continuous in $\overline{E}$.

[10] S. L. Sobolev and S. M. Nikol'skii, Embedding theorems, *Izdat. Akad. Nauk SSSR Leningrad*, 1963, 227–242.

[11] By the density Theorem 16.3 of Chapter VII.

where **n** denotes an arbitrary unit vector ranging on the same solid angle as C_x. Integrating over such a solid angle,

$$\omega|u(x)| \leq \int_{C_x} \frac{|Du(y)|}{|x-y|^{N-1}}dy + \frac{1}{h}\int_{C_x}\frac{|u(y)|}{|x-y|^{N-1}}dx$$
$$\leq \int_E \frac{|Du(y)|}{|x-y|^{N-1}}dy + \frac{1}{h}\int_E \frac{|u(y)|}{|x-y|^{N-1}}dx. \tag{9.1}$$

The right-hand side of (9.1) is the sum of two Riesz potentials. Therefore, inequality (8.1) follows from the second part of (9.1) and the $L^{p^*}(E)$-estimates of the Riesz potentials.[12] Analogously, (8.2) and the form of the constant $C(N,p)$ follows from the $L^q(E)$-estimates of the Riesz potentials.

To establish (8.3), start from the first part of (9.1) and estimate

$$\omega|u(x)| \leq \sup_{z\in C_x}\int_{C_x}\frac{|Du(y)|}{|z-y|^{N-1}}dy + \frac{1}{h}\sup_{z\in C_x}\int_{C_x}\frac{|u(y)|}{|z-y|^{N-1}}dx.$$

Therefore, by the $L^\infty(E)$-estimates of the Riesz potentials,[13]

$$\omega|u(x)| \leq C(N,p)\mu(C_o)^{\frac{p-N}{Np}}\left\{\frac{1}{h}\|u\|_{p,C_x} + \|Du\|_{p,C_x}\right\}$$
$$\leq \frac{C(N,p)}{h}\mu(C_o)^{\frac{p-N}{Np}}\{\|u\|_p + h\|Du\|_p\}. \qquad \square$$

Proof of (8.4)[14]. For $x \in \overline{E}$, let C_x be a circular, spherical cone congruent to C_o and all contained $\overline{E}$. Then for all $0 < \rho \leq h$, the circular, spherical cone $C_{x,\rho}$ of vertex at x and radius ρ, coaxial with C_x, and with the same solid angle ω is contained in $\overline{E}$. Its volume is

$$\mu\left(C_{x,\rho}\right) = \frac{\omega}{N}\rho^N.$$

Denote by $(u)_{x,\rho}$ the average of u over $C_{x,\rho}$; i.e.,

$$(u)_{x,\rho} = \frac{1}{\mu(C_{x,\rho})}\int_{C_{x,\rho}}u(\xi)d\xi. \qquad \square$$

Lemma 9.1. *For every pair* $x, y \in E$ *such that* $|x - y| = \rho \leq h$,

$$|u(y) - (u)_{x,\rho}| \leq 2^{N+1-\frac{N}{p}}\frac{\kappa_N^{\frac{p-1}{p}}}{\omega}\frac{Np}{p-N}\rho^{1-\frac{N}{p}}\|Du\|_p.$$

---

[12] See Section 23 of Chapter VIII. The behavior of the constant $C(N,p)$ follows from Remark 23.1 there.

[13] See Proposition 23.4 of Chapter VIII with E replaced by C_x. The form of the constant $C(N,p)$ is given by (23.5)' there. The form (8.3)' of the constant $C(N,p,\omega)$ is computed from this and the expression of the volume of C_o given in (17.6) of Chapter VII.

[14] C. B. Morrey, On the solutions of quasilinear elliptic partial differential equations, *Trans. Amer. Math. Soc.*, **43** (1938), 126–166.

Proof. Fix $x, y \in E$ such that $|x - y| = \rho \leq h$. Since E is convex, for all $\xi \in C_{x,\rho}$,

$$|u(y) - u(\xi)| = \left| \int_0^1 \frac{\partial}{\partial t} u(y + t(\xi - y)) dt \right|.$$

First, integrate in $d\xi$ over $C_{x,\rho}$, and then in the resulting integral, perform the change of variables $y + t(\xi - y) = \eta$. The Jacobian is t^{-N}, and the new domain of integration is transformed in those η given by

$$|y - \eta| = t|\xi - y| \quad \text{as } \xi \text{ ranges over } C_{x,\rho}.$$

Therefore, such a transformed domain is contained in the ball $B_{2\rho t}(y)$ of center y and radius $2\rho t$. These operations give

$$\frac{\omega}{N} \rho^N |u(y) - (u)_{x,\rho}| \leq \int_0^1 \left(\int_{C_{x,\rho}} |\xi - y| |Du(y + t(\xi - y))| d\xi \right) dt$$

$$\leq \int_0^1 t^{-(N+1)} \int_{E \cap B_{2\rho t}(y)} |\eta - y| |Du(\eta)| d\eta dt$$

$$\leq \kappa_N^{\frac{p-1}{p}} \int_0^1 t^{-(N+1)} (2\rho t)^{N(1-\frac{1}{p})+1} \|Du\|_p dt. \qquad \square$$

To conclude the proof of (8.4), fix $x, y \in E$ and let

$$z = \frac{1}{2}(x + y), \qquad \rho = |x - z| = |y - z| = \frac{1}{2}|x - y|.$$

Then

$$|u(x) - u(y)| \leq |u(x) - (u)_{z,\rho}| + |u(y) - (u)_{z,\rho}|$$

$$\leq \frac{2^{N+1}}{\omega} \kappa_N^{\frac{p-1}{p}} \frac{Np}{p - N} |x - y|^{1-\frac{N}{p}} \|Du\|_p. \qquad \square$$

10   Poincaré inequalities

The multiplicative inequalities of Theorem 1.1 cannot hold for functions in $W^{1,p}(E)$, as can be verified by taking $u = \text{const.}$ In general, an integral norm of u cannot be controlled in terms of some integral norm of its gradient unless one has some information on the values of the function in some subset of $\overline{E}$.

10.1   The Poincaré inequality. Let E be a bounded domain in $\mathbb{R}^N$, and for $u \in L^1(E)$, let

$$u_E = \fint_E u dx = \frac{1}{\mu(E)} \int_E u dx$$

denote the integral average of u over E.

Theorem 10.1. *Assume that E is bounded and convex, and let $u \in W^{1,p}(E)$ for some $1 < p < N$. There exists a constant C depending only upon N and p such that*

$$\|u - u_E\|_{p^*} \le C \frac{(\operatorname{diam} E)^N}{\mu(E)} \|Du\|_p, \quad \text{where } p^* = \frac{Np}{N - p}, \tag{10.1}$$

$$\|u - u_E\|_1 \le C(\operatorname{diam} E)^N \|Du\|_N. \tag{10.2}$$

Proof of (10.1). Having fixed $x, y \in E$, denote by $R(x, y)$ the distance from x to ∂E along the direction of $(y - x)$, and write

$$|u(x) - u(y)| \le \int_0^{R(x,y)} \left| \frac{\partial}{\partial \rho} u(x + \rho \mathbf{n}) \right| d\rho, \quad \mathbf{n} = \frac{(y - x)}{|y - x|}.$$

Integrate in dy over E to obtain

$$\mu(E)|u(x) - u_E| \le \int_E \left(\int_0^{R(x,y)} |Du(x + \rho \mathbf{n})| d\rho \right) dy.$$

The integral in dy is calculated by introducing polar coordinates with pole at x. Therefore, if $\mathbf{n}$ is the angular variable spanning the sphere $|\mathbf{n}| = 1$, the right-hand side is majorized by

$$(\operatorname{diam} E)^{N-1} \int_0^{\operatorname{diam} E} \int_{|\mathbf{n}|=1} \int_0^{R(x,y)} \rho^{N-1} \frac{|Du(x + \rho \mathbf{n})|}{|x - y|^{N-1}} d\rho \, d\mathbf{n} \, dr$$

$$\le (\operatorname{diam} E)^N \int_E \frac{|Du|}{|x - y|^{N-1}} dy.$$

Therefore,

$$|u(x) - u_E| \le \frac{(\operatorname{diam} E)^N}{\mu(E)} \int_E \frac{|Du(y)|}{|x - y|^{N-1}} dy. \tag{10.3}$$

The proof of (10.1) now follows from this and the $L^{p^*}(E)$-estimates of the Riesz potentials.[15]

Inequality (10.2) follows from (10.1) and Hölder's inequality. Indeed, having fixed $1 < p < N$,

$$\|u - u_E\|_1 \le \|u - u_E\|_{p^*} \mu(E)^{1 - \frac{N-p}{Np}}$$

$$\le C\mu(E)^{1 - \frac{N-p}{Np}} \frac{(\operatorname{diam} E)^N}{\mu(E)} \|Du\|_p$$

$$\le C(\operatorname{diam} E)^N \|Du\|_N. \qquad \square$$

---

[15] See Theorem 23.1 of Chapter VIII. Following Remark 23.1 there, the constant C in (10.1) tends to infinity as either $p \to 1$ or $p \to N$.

The estimate depends upon the structure of the convex set E through the ratio $(\operatorname{diam} E)^N/\mu(E)$. If E is a ball, then such a ratio is $2^N/\kappa_N$, where κ_N is the volume of the unit ball in $\mathbb{R}^N$. In general, if R is the radius of the smallest ball containing E and ρ is the radius of the largest ball contained in E,

$$\frac{(\operatorname{diam} E)^N}{\mu(E)} \le \frac{2^N}{\kappa_N} \left(\frac{R}{\rho}\right)^N.$$

10.2 Multiplicative Poincaré inequalities.

Proposition 10.2. *Let E be a bounded and convex subset of $\mathbb{R}^N$, and let $u \in W^{1,p}(E)$ for some $1 < p < N$. There exists a constant C depending upon only N and p such that*

$$\|u - u_E\|_q \le C^\theta \left(\frac{(\operatorname{diam} E)^N}{\mu(E)}\right)^\theta \|Du\|_p^\theta \|u - u_E\|_r^{1-\theta}, \qquad (10.4)$$

where the numbers $r > 1$, $1 < p < N$, and $\theta \in [0, 1]$ are linked by

$$\theta = \left(\frac{1}{r} - \frac{1}{q}\right)\left(\frac{1}{N} - \frac{1}{p} + \frac{1}{r}\right)^{-1}. \qquad (10.4)'$$

Proof. The case $\theta = 0$ corresponds to $q = r$, and (10.4) is trivial. The case where $\theta = 1$ corresponds to $q = \frac{Np}{N-p}$ and coincides with (10.1). Let $r \in (1, \infty)$ and $p \in (1, N)$ be fixed, and choose $\theta \in (0, 1)$ and q such that

$$\min\left\{r; \frac{Np}{N-p}\right\} < q < \max\left\{r; \frac{Np}{N-p}\right\} \quad \text{and} \quad \theta q < \frac{Np}{N-p}.$$

Then by Hölder's inequality,

$$\int_E |u - u_E|^q\, dx = \int_E |u - u_E|^{\theta q} |u - u_E|^{(1-\theta)q}\, dx$$

$$\le \left(\int_E |u - u_E|^{\frac{Np}{N-p}}\, dx\right)^{\frac{N-p}{Np}\theta q} \left(\int_E |u - u_E|^r\, dx\right)^{\frac{(1-\theta)q}{r}},$$

where we have set

$$(1-\theta)q \frac{Np}{Np - (N-p)\theta q} = r.$$

The inequality follows from this and (10.1). $\square$

Remark 10.1. It follows from (10.4) that the multiplicative embedding inequality (1.1) continues to hold for functions $u \in W^{1,p}(E)$ of zero average over a convex domain E.

11   The discrete isoperimetric inequality

Proposition 11.1. *Let E be a bounded convex open set in $\mathbb{R}^N$, and let $u \in W^{1,1}(E)$. Assume that the set where u vanishes has positive measure. Then*

$$\|u\|_1 \le \kappa_N^{\frac{N-1}{N}} \frac{(\operatorname{diam} E)^N \mu(E)^{\frac{1}{N}}}{\mu([u=0])} \|Du\|_1. \tag{11.1}$$

where κ_N is the volume of the unit ball in $\mathbb{R}^N$.

Proof. For $x, y \in E$, let

$$\mathbf{n} = \frac{(y-x)}{|x-y|}$$

denote the unit vector ranging over the unit ball centered at x. Then for almost all $x \in E$ and almost all $y \in [u=0]$,

$$|u(x)| = \left| \int_0^{|y-x|} \frac{\partial}{\partial \rho} u(x+\mathbf{n}\rho) d\rho \right| \le \int_0^{|y-x|} |Du(x+\mathbf{n}\rho)| d\rho.$$

Integrating in dx over E and in dy over $[u=0]$ gives

$$\mu([u=0]) \|u\|_1 \le \int_E \left\{ \int_{[u=0]} \int_0^{|y-x|} |Du(x+\mathbf{n}\rho)| d\rho dy \right\} dx.$$

The integral over $[u=0]$ is computed by introducing polar coordinates with center at x.

Denoting by $R(x, y)$ the distance from x to ∂E in the direction of $\mathbf{n}$,

$$\int_{[u=0]} \int_0^{|y-x|} |Du(x+\mathbf{n}\rho)| d\rho dy$$

$$\le \int_0^{R(x,y)} s^{N-1} ds \int_{|\mathbf{n}|=1} \int_0^{R(x,y)} |Du(x+\mathbf{n}\rho)| d\rho d\mathbf{n}.$$

Combining these remarks, we arrive at

$$\mu([u=0]) \|u\|_1 \le \frac{1}{N} (\operatorname{diam} E)^N \int_E \int_E \frac{|Du(y)|}{|x-y|^{N-1}} dy dx.$$

Inequality (11.1) follows from this since[16]

$$\sup_{y \in E} \int_E \frac{dx}{|x-y|^{N-1}} \le N \kappa_N^{\frac{N-1}{N}} \mu(E)^{\frac{1}{N}}. \qquad \square$$

---

[16]See Proposition 23.2 of Chapter VIII with $r=1$.

If E is the ball B_R of radius R centered at the origin,

$$\|u\|_1 \le 2^N \kappa_N \frac{R^{N+1}}{\mu([|u| = 0])} \|Du\|_1. \tag{11.2}$$

For a real number ℓ and $u \in W^{1,1}(E)$, set

$$u_\ell = \begin{cases} \ell & \text{if } u > \ell, \\ u & \text{if } u \le \ell. \end{cases}$$

For $k < \ell$, the function $(u_\ell - k)_+$ belongs to $W^{1,1}(E)$.[17] Putting such a function in (11.1) gives[18]

$$(\ell - k)\mu([u > \ell]) \le \kappa_N^{\frac{N-1}{N}} \frac{(\text{diam } E)^N \mu(E)^{\frac{1}{N}}}{\mu([u < k])} \int_{[k < u < \ell]} |Du|\,dx. \tag{11.3}$$

This is referred to as a discrete version of the isoperimetric inequality.[19]

12   Morrey spaces

A function $f \in L^1(E)$ belongs to the Morrey space $M^p(E)$ for some $p \ge 1$ if there exists a constant C_f depending upon f such that[20]

$$\sup_{x \in E} \int_{B_\rho(x) \cap E} |f(y)|\,dy \le C_f \rho^{N\left(1 - \frac{1}{p}\right)} \quad \text{for all } \rho > 0. \tag{12.1}$$

If $f \in M^p(E)$, we set

$$\|f\|_{M^p} = \inf\left\{C_f \text{ for which (12.1) holds}\right\}. \tag{12.2}$$

It follows from the definition that $L^p(E) \subset M^p(E)$ and

$$\|f\|_p \ge \kappa_N^{\frac{1}{p} - 1} \|f\|_{M^p}. \tag{12.3}$$

Moreover,

$$L^1(E) = M^1(E) \quad \text{and} \quad \|f\|_1 = \|f\|_{M^1},$$

$$L^\infty(E) = M^\infty \quad \text{and} \quad \|f\|_\infty = \frac{1}{\kappa_N}\|f\|_{M^\infty}.$$

---

[17]See Proposition 20.2 of Chapter VII.

[18]E. De Giorgi, Sulla differenziabilità e l'analiticità delle estremali degli integrali multipli regolari, *Mem. Accad. Sci. Torino Cl. Sci. Mat. Fis. Nat.* (3), **3** (1957), 25–43.

[19]A continuous version is in W. H. Fleming and R. Rischel, An integral formula for total gradient variation, *Arch. Math.*, **XI** (1960), 218–222.

[20]See [41]; also in S. Campanato, Proprietà di inclusione per spazi di Morrey, *Ricerche Mat.*, **12** (1963), 67–86 and Proprietà di Hölderianità di alcune classi di funzioni, *Ann. Scuola Norm. Sup. Pisa* (3), **17** (1963), 175–188.

We regard f as being defined in the whole $\mathbb{R}^N$ by setting it to be zero outside E. Thus if $f \in M^p(E)$,

$$\|f\|_1 \le \|f\|_{M^p}(\operatorname{diam} E)^{N\left(1-\frac{1}{p}\right)}. \tag{12.4}$$

12.1 Embeddings for functions in the Morrey spaces. Let $\alpha \in (0, N)$, and consider the potential $V_{\alpha, f}$ generated by some $f \in M^p(E)$. Precisely,

$$V_{\alpha, f}(x) = \int_E \frac{f(y)}{|x - y|^{N-\alpha}} dy. \tag{12.5}$$

If $\alpha = 1$, this is the Riesz potential generated by f. If $p > 1$, such a potential is well defined as a function in $L^q(E)$ for $q = \frac{N}{N-\alpha p}$.[21]

The next proposition gives an estimate of $\|V_{\alpha, f}\|_\infty$ in terms of $\|f\|_{M^p}$, provided that $p > \frac{N}{\alpha}$ and E is bounded.

Proposition 12.1. *Let* $f \in M^p(E)$ *for some* $p > \frac{N}{\alpha}$. *Then*[22]

$$\|V_{\alpha, f}\|_\infty \le \frac{N(p-1)}{\alpha p - N}(\operatorname{diam} E)^{\frac{\alpha p - N}{p}}\|f\|_{M^p}. \tag{12.6}$$

Proof. For x and y in $\mathbb{R}^N$, let $\mathbf{n} = \frac{y-x}{|x-y|}$. Then using of polar coordinates, compute

$$\frac{d}{d\rho}\int_{B_\rho(x)}|f(y)|dy = \frac{d}{d\rho}\int_{|\mathbf{n}|=1}\int_0^\rho r^{N-1}|f(x + \mathbf{n}r)|dr d\mathbf{n}$$

$$= \int_{|\mathbf{n}|=1}\rho^{N-1}|f(x + \mathbf{n}\rho)|d\mathbf{n}.$$

Using this calculation in the definition of $V_{\alpha, f}$,

$$|V_{\alpha, f}(x)| \le \int_E \frac{|f(y)|}{|x - y|^{N-\alpha}}dy$$

$$\le \int_{|\mathbf{n}|=1}\int_0^{\operatorname{diam} E}\frac{1}{\rho^{N-\alpha}}\rho^{N-1}|f(x + \mathbf{n}\rho)|d\mathbf{n}d\rho$$

$$= \int_0^{\operatorname{diam} E}\frac{1}{\rho^{N-\alpha}}\left(\frac{d}{d\rho}\int_{B_\rho(x)}|f(y)|dy\right)d\rho$$

$$= \frac{\|f\|_1}{(\operatorname{diam} E)^{N-\alpha}} - \lim_{\varepsilon \to 0}\frac{1}{\varepsilon^{N-\alpha}}\int_{B_\varepsilon(x)}|f(y)|dy$$

$$+ (N - \alpha)\int_0^{\operatorname{diam} E}\frac{1}{\rho^{N+1-\alpha}}\int_{B_\rho(x)}|f(y)|dyd\rho.$$

To prove (12.6), estimate each of these terms by making use of the definition (12.1)–(12.2) of $\|f\|_{M^p}$. $\qquad\square$

---

[21] See Theorem 23.1c of the Problems and Complements of Chapter VIII.

[22] Compare with Proposition 23.4c of the Problems and Complements of Chapter VIII.

The main interest of inequality (12.6) is for $\alpha = 1$ when applied to the Riesz potential of the type of the right-hand side of (10.3). One of the assertions of the embedding Theorem 8.1—namely, inequality (8.4)—is that if $p > N$ and if E is convex, then a function in $W^{1,p}(E)$ is Hölder continuous with Hölder exponent $\eta = \frac{p-N}{p}$. The next theorem asserts that the embedding in (8.4) holds for functions whose weak gradient is in $M^p(E)$ for $p > N$.

Theorem 12.2. *Let $u \in W^{1,1}(E)$, and assume that $|Du| \in M^p(E)$ for some $p > N$. Then $u \in C^\eta(E)$ with $\eta = \frac{p-N}{p}$. Moreover, for every ball $B_\rho(x) \subset E$,*

$$\operatorname*{ess\,osc}_{B_\rho(x)} u \leq \frac{2^{N+1}}{\kappa_N} \frac{N(p-1)}{p-N} \|Du\|_{M^p} \rho^\eta. \tag{12.7}$$

Proof. Consider the potential inequality (10.3) written for the ball $B_\rho(x)$ replacing E. It gives[23]

$$|u(x) - u_{x,\rho}| \leq \frac{2^N}{\kappa_N} \int_{B_\rho(x)} \frac{|Du(y)|}{|x-y|^{N-1}} dy.$$

This implies (12.7) in view of Proposition 12.1 applied with $\alpha = 1$.   □

Since $L^p(E) \subset M^p(E)$, the embedding (12.7) generalizes the embedding (8.4).

13   Limiting embedding of $W^{1,N}(E)$

Let f be in the Morrey space $M^N(E)$, and consider its Riesz potential,

$$\mathcal{V}_f(x) = \int_E \frac{f(y)}{|x-y|^{N-1}} dy.$$

Proposition 13.1. *Let $f \in M^N(E)$. There exist constants C_1 and C_2 depending upon only N such that*

$$\int_E \exp\left\{\frac{|\mathcal{V}_f|}{C_1 \|f\|_{M^N}}\right\} dy \leq C_2(\operatorname{diam} E)^N. \tag{13.1}$$

Proof. Let $q > 1$, and rewrite the integrand in the definition of the potential $\mathcal{V}_f$ as

$$\frac{|f(y)|^{\frac{1}{q}}}{|x-y|^{\left(N-\frac{1}{q}\right)\frac{1}{q}}} \frac{|f(y)|^{1-\frac{1}{q}}}{|x-y|^{\left(N-\frac{q+1}{q}\right)\left(1-\frac{1}{q}\right)}}.$$

Then by Hölder's inequality,

$$|\mathcal{V}_f(x)| \leq \left(\int_E \frac{|f(y)|}{|x-y|^{N-\frac{1}{q}}} dy\right)^{\frac{1}{q}} \left(\int_E \frac{|f(y)|}{|x-y|^{N-\frac{q+1}{q}}} dy\right)^{1-\frac{1}{q}}. \tag{13.2}$$

---

[23] $u_{x,\rho}$ denotes the integral average of u over the ball centered at x and of radius ρ.

The second integral is estimated by Proposition 12.1 with $\alpha = \frac{q+1}{q}$ and $p = N$. This gives

$$\int_E \frac{|f(y)|}{|x-y|^{N-\frac{q+1}{q}}}\,dy \le (N-1)q(\text{diam } E)^{\frac{1}{q}}\|f\|_{M^N}.$$

We put this in (13.2), take the q power of both sides, and integrate over E in dx to obtain

$$\int_E |\nabla_f|^q\,dx \le \left\{(N-1)q(\text{diam } E)^{\frac{1}{q}}\|f\|_{M^N}\right\}^{q-1}$$

$$\times \int_E |f(y)| \left(\int_E \frac{dx}{|x-y|^{N-\frac{1}{q}}}\right)dy.$$

The last double integral is estimated by using (12.4) with $p = N$ as[24]

$$\int_E |f(y)| \left(\int_E \frac{dx}{|x-y|^{N-\frac{1}{q}}}\,dx\right)dy \le \|f\|_1 \sup_{y\in E} \int_E \frac{dx}{|x-y|^{N-\frac{1}{q}}}$$

$$\le \|f\|_1 \omega_N q(\text{diam } E)^{\frac{1}{q}}$$

$$\le \omega_N q(\text{diam } E)^{N-1+\frac{1}{q}}\|f\|_{M^N}.$$

Combining these estimates yields

$$\int_E |\nabla_f|^q\,dx \le \omega_N(N-1)^{q-1}q^q(\text{diam } E)^N\|f\|_{M^N}^q.$$

Since $q > 1$ is arbitrary, we write this inequality for $q = 2, 3, \ldots$. Then[25]

$$\int_E \sum_{q=0}^k \frac{1}{q!}\left\{\frac{|\nabla_f|}{C_1\|f\|_N}\right\}^q\,dx \le \frac{\omega_N(\text{diam } E)^N}{(N-1)} \sum_{q=0}^\infty \left(\frac{N-1}{C_1}\right)^q \frac{q^q}{q!}$$

for all $k \in \mathbb{N}$ and all $C_1 > 0$.   □

Let $u \in W^{1,1}(E)$ be such that $|Du| \in M^N(E)$; i.e.,

$$\int_{B_\rho(x)} |Du|\,dy \le \|Du\|_{M^N}\rho^{N-1} \quad \text{for all } B_\rho(x) \subset E. \tag{13.3}$$

---

[24]Also use Proposition 23.2c of the Problems and Complements of Chapter VIII with $\alpha = \frac{1}{q}$ and $r = 1$. Note that $N\kappa_N = \omega_N$.

[25]In a manner similar to the proof of Theorem 24.1 of Chapter VIII.

Theorem 13.2. *Let $u \in W^{1,1}(E)$, and let* (13.3) *hold. There exist constants C_1 and C_2 depending only upon N such that*[26]

$$\int_{B_\rho(x)} \exp\left\{\frac{|u - (u)_{x,\rho}|}{C_1 \|Du\|_{M^N}}\right\} dy \leq C_2 \mu(B_\rho). \tag{13.4}$$

Proof. The proof follows from Proposition 13.1 starting from the potential inequality (10.3). □

This estimate can be regarded as a limiting case of the Hölder estimates of Theorem 12.2 when $p \to N$.

14 Compact embeddings

Let E be a bounded domain in $\mathbb{R}^N$ with the cone property, and let K be a bounded set in $W^{1,p}(E)$; i.e.,

$$K = \{u \in W^{1,p}(E) | \|u\|_{1,p} \leq C \text{ for some positive constant } C\}.$$

If $1 \leq p < N$, by the embedding Theorem 8.1, K is a bounded subset of $L^{p^*}(E)$, where $p^* = \frac{Np}{N-p}$. Since E is of finite measure, K is also a bounded set of $L^q(E)$ for all $1 \leq q < p^*$. The next theorem asserts that K is a compact subset of $L^q(E)$ for all $1 \leq q < p^*$.

Theorem 14.1 (Reillich[27]). *Let E be a bounded domain in $\mathbb{R}^N$ with the cone property, and let $1 \leq p < N$. Then the embedding of $W^{1,p}(E)$ into $L^q(E)$ is compact for all $1 \leq q < p^*$.*

Proof. The proof consists of verifying that a bounded subset of $W^{1,p}(E)$ satisfies the conditions for a subset of $L^q(E)$ to be compact.[28] For $\delta > 0$, let E_δ be defined as

$$E_\delta = \{x \in E | \operatorname{dist}\{x; \partial E\} > \delta\},$$

and let δ be so small that E_δ is not empty. For $q \in [1, p^*)$ and $u \in W^{1,p}(E)$,

$$\|u\|_{q, E-E_\delta} \leq \|u\|_p^* \mu(E - E_\delta)^{\frac{1}{q} - \frac{1}{p^*}}.$$

---

[26]F. John and L. Nirenberg, On functions of bounded mean oscillation, *Comm. Pure Appl. Math.*, **14** (1961), 415–426.

[27]R. Reillich, Ein Satz über mittlere Konvergenz, *Nachr. Akad. Wiss. Göttingen Math.-Phys. Kl.*, 1930, 30–35; V. I. Kondrachov, Sur certaines propriétés des fonctions dans l'espace L^p, *C. R. (Dokl.) Acad. Sci. USSR (N.S.)*, **48** (1945), 535–539.

[28]See Section 22 of Chapter V.

Next, for a vector $h \in \mathbb{R}^N$ of length $|h| < \delta$, compute

$$\int_{E_\delta} |u(x+h) - u(x)| dx \le \int_{E_\delta} \left(\int_0^1 \left| \frac{d}{dt} u(x+th) \right| dt \right) dx$$

$$\le |h| \int_0^1 \left(\int_{E_\delta} |Du(x+th)| dx \right) dt$$

$$\le |h| \mu(E)^{\frac{p-1}{p}} \|Du\|_p.$$

Therefore, for all $\sigma \in (0, \frac{1}{q})$,

$$\int_{E_\delta} |T_h u - u|^q dx = \int_{E_\delta} |T_h u - u|^{q\sigma + q(1-\sigma)} dx$$

$$\le \left(\int_{E_\delta} |T_h u - u| dx \right)^{q\sigma} \left(\int_{E_\delta} |T_h u - u|^{\frac{q(1-\sigma)}{1-q\sigma}} dx \right)^{1-q\sigma}.$$

Choose σ so that

$$\frac{(1-\sigma)q}{1-q\sigma} = p^*; \quad \text{i.e.,} \quad \sigma q = \frac{p^* - q}{p^* - 1}.$$

Such a choice is possible if $1 < q < p^*$. Applying the embedding Theorem 8.1 gives

$$\int_{E_\delta} |T_h u - u|^q dx \le \gamma^{(1-\sigma)q} \left(\int_{E_\delta} |T_h u - u| dx \right)^{\frac{p^*-q}{p^*-1}} \|u\|_{1,p}^{(1-\sigma)q}$$

for a constant γ depending only upon N, p, and the geometry of the cone property of E.[29] Combining these estimates, we arrive at

$$\|T_h u - u\|_{q,E_\delta} \le \gamma_1 |h|^\sigma \|u\|_{1,p},$$

where the constant γ_1 is given by

$$\gamma_1 = \gamma^{\frac{1}{q}-\sigma} \mu(E)^{\frac{p-1}{p}\sigma}.$$

The assertion of the theorem is now a consequence of the characterization of the precompact subsets in $L^q(E)$. $\qquad\square$

Corollary 14.2. *Let E be a bounded domain in $\mathbb{R}^N$ with the cone property, and let $1 \le p < N$. Then every sequence $\{f_n\}$ of functions equibounded in $W^{1,p}(E)$ has a subsequence $\{f_{n'}\}$ strongly convergent in $L^q(E)$ for all $1 \le q < p^*$.*

---

[29] This constant is the one occurring in (8.1).

15   Fractional Sobolev spaces in $\mathbb{R}^N$

Let $s \in (0, 1)$, $p \geq 1$, and $N \geq 1$, and consider the set of all functions $u \in L^p(\mathbb{R}^N)$ with finite seminorm

$$\|u\|_{s,p} = \left(\int_{\mathbb{R}^N} \int_{\mathbb{R}^N} \frac{|u(x) - u(y)|^p}{|x - y|^{N+sp}} dxdy \right)^{\frac{1}{p}}. \tag{15.1}$$

They form a linear subspace of $L^p(\mathbb{R}^N)$, which is a Banach space when equipped with the norm

$$\|u\|_{s,p} = \|u\|_p + \|u\|_{s,p}.$$

Such a Banach space is denoted with $W^{s,p}(\mathbb{R}^N)$, and it is called the fractional Sobolev space of order s.

Proposition 15.1. $W^{1,p}(\mathbb{R}^N)$ *is continuously embedded into* $W^{s,p}(\mathbb{R}^N)$ *for all* $s \in (0, 1)$. *Moreover,*

$$\|u\|_{s,p} \leq \left(\frac{2\omega_N}{ps(1-s)} \right)^{\frac{1}{p}} \|Du\|_p^s \|u\|_p^{1-s}. \tag{15.2}$$

Proof. A change of variable in the definition of the seminorm in (15.1) gives

$$\|u\|_{s,p}^p = \int_{\mathbb{R}^N} |\xi|^{-(N+sp)} d\xi \int_{\mathbb{R}^N} |u(x + \xi) - u(x)|^p dx$$

$$= \int_{|\xi|<\lambda} |\xi|^{-(N+sp)} d\xi \int_{\mathbb{R}^N} |u(x + \xi) - u(x)|^p dx$$

$$+ \int_{|\xi|>\lambda} |\xi|^{-(N+sp)} d\xi \int_{\mathbb{R}^N} |u(x + \xi) - u(x)|^p dx,$$

where λ is a positive parameter to be chosen later. The second integral is majorized by

$$2^p \frac{\omega_N}{sp} \frac{1}{\lambda^{sp}} \|u\|_p^p.$$

The first integral is estimated by

$$\int_{|\xi|<\lambda} |\xi|^{-(N+sp)} d\xi \int_{\mathbb{R}^N} \left| \int_0^1 \frac{\partial}{\partial t} u(x + t\xi) dt \right|^p dx$$

$$\leq \int_{|\xi|<\lambda} |\xi|^{p-(N+sp)} d\xi \int_{\mathbb{R}^N} |Du|^p dx$$

$$\leq \frac{\omega_N}{p(1-s)} \lambda^{p(1-s)} \|Du\|_p^p.$$

Combining these estimates, we arrive at

$$\|u\|_{s,p}^p \leq \frac{2^p \omega_N}{sp} \frac{1}{\lambda^{sp}} \|u\|_p^p + \frac{\omega_N}{p(1-s)} \lambda^{p(1-s)} \|Du\|_p^p,$$

valid for every $\lambda > 0$. To prove the proposition, minimize the right-hand side with respect to λ.   □

Proposition 15.2. $C_o^\infty(\mathbb{R}^N)$ *is dense in* $W^{s,p}(\mathbb{R}^N)$.

Proof. For $\varepsilon > 0$, let ζ_ε denote a nonnegative piecewise smooth cutoff function in $\mathbb{R}^N$ satisfying

$$
\begin{cases}
\zeta_\varepsilon = 1 & \text{in the ball } \left\{ |x| < \dfrac{1}{\varepsilon} \right\}, \\[2mm]
\zeta_\varepsilon = 0 & \text{outside the ball } \left\{ |x| < \dfrac{2}{\varepsilon} \right\}, \\[2mm]
|D\zeta_\varepsilon| \le \varepsilon.
\end{cases}
$$

One verifies that for every $u \in W^{s,p}(\mathbb{R}^N)$,

$$
\|u\zeta_\varepsilon\|_{s,p} \le \|u\|_{s,p} + \varepsilon\gamma\|u\|_{s,p},
$$

where γ depends only upon N and p, and that

$$
\|u - u\zeta_\varepsilon\|_{s,p} \longrightarrow 0 \quad \text{as } \varepsilon \to 0.
$$

The functions $J_\varepsilon * (u\zeta_\varepsilon)$ are in $C_o^\infty(\mathbb{R}^N)$, and as u ranges over $W^{s,p}(\mathbb{R}^N)$ and ε ranges over $(0, 1)$, they span a dense subset of $W^{s,p}(\mathbb{R}^N)$. For this, we have only to verify that

$$
\|\|J_\varepsilon * u - u\|\|_{s,p} \longrightarrow 0 \quad \text{as } \varepsilon \to 0.
$$

For a.e. pair $x, y \in \mathbb{R}^N$, write

$$
\frac{|[(J_\varepsilon * u)(x) - u(x)] - [(J_\varepsilon * u)(y) - u(y)]|^p}{|x - y|^{N+sp}}
$$

$$
\le \int_{|\xi|<\varepsilon} J_\varepsilon(\xi) \left| \frac{u(x+\xi) - u(y+\xi)}{|(x+\xi) - (y+\xi)|^{\frac{N}{p}+s}} - \frac{u(x) - u(y)}{|x - y|^{\frac{N}{p}+s}} \right|^p d\xi.
$$

Integrating in $dxdy$ over $\mathbb{R}^N \times \mathbb{R}^N$ gives

$$
\|\|J_\varepsilon * u - u\|\|_{s,p}^p
$$

$$
\le \sup_{|\xi| \le \varepsilon} \iint_{\mathbb{R}^N \times \mathbb{R}^N} \left| \frac{u(x+\xi) - u(y+\xi)}{|(x+\xi) - (y+\xi)|^{\frac{N}{p}+s}} - \frac{u(x) - u(y)}{|x - y|^{\frac{N}{p}+s}} \right|^p dxdy.
$$

If $u \in W^{s,p}(\mathbb{R}^N)$, by the definition (15.1),

$$
w(x, y) = \frac{u(x) - u(y)}{|x - y|^{\frac{N}{p}+s}} \in L^p(\mathbb{R}^N \times \mathbb{R}^N).
$$

Therefore, if T_ξ is the translation operator in $L^p(\mathbb{R}^N \times \mathbb{R}^N)$,

$$\||J_\varepsilon * u - u\||_{s,p}^p \leq \sup_{|\xi| < \varepsilon} \|T_\xi w - w\|_{p,\mathbb{R}^N \times \mathbb{R}^N}.$$

By the continuity of the translation in $L^p(\mathbb{R}^N \times \mathbb{R}^N)$, the right-hand side tends to zero as $\varepsilon \to 0$.[30]   □

16 Traces

Let $\mathbb{R}_+^N$ denote the upper half-space $\{x_N > 0\}$ whose coordinates we denote by $(\bar{x}, x_N)$; i.e.,

$$\mathbb{R}_+^N = \mathbb{R}^{N-1} \times \mathbb{R}^+, \qquad x = (\bar{x}, x_N), \qquad \bar{x} = (x_1, \ldots, x_{N-1}).$$

If u is a function in $W^{1,p}(\mathbb{R}_+^N)$ that is continuous in $\mathbb{R}_+^N$ up to $\{x_N = 0\}$, the trace of u on the hyperplane $\{x_N = 0\}$ is defined by

$$tr(u) = u(\bar{x}, 0), \quad u \in W^{1,p}(\mathbb{R}_+^N) \cap C(\{x_N \geq 0\}).$$

Proposition 16.1. *Let $u \in W^{1,p}(\mathbb{R}_+^N)$ be continuous in $\overline{\mathbb{R}}_+^N$. Then*

$$\|u(\cdot, 0)\|_{r,\mathbb{R}^{N-1}}^r \leq r \|Du\|_{p,\mathbb{R}_+^N} \|u\|_{q,\mathbb{R}_+^N}^{r-1} \tag{16.1}$$

for all $q, r \geq 1$ such that $q(p-1) = p(r-1)$, provided that $u \in L^q(\mathbb{R}_+^N)$.

Proof. We may assume that $u \in C_o^\infty(\mathbb{R}^N)$. Then for all $\bar{x} \in \mathbb{R}^{N-1}$ and all $r \geq 1$,

$$|u(\bar{x}, 0)|^r \leq r \int_0^\infty |u(\bar{x}, x_N)|^{r-1} \left| \frac{\partial}{\partial x_N} u(\bar{x}, x_N) \right| dx_N.$$

To conclude the proof, integrate both sides in $d\bar{x}$ over $\mathbb{R}^{N-1}$ and apply Hölder's inequality to the resulting integral on the right-hand side.   □

If $u \in W^{1,p}(\mathbb{R}_+^N)$, there exist a sequence of functions $\{u_n\}$ in $C_o^\infty(\mathbb{R}^N)$ converging to u in $W^{1,p}(\mathbb{R}_+^N)$.[31] By (16.1) with $r = p$,

$$\|u_n(\cdot, 0) - u_m(\cdot, 0)\|_{p,\mathbb{R}^{N-1}} \leq \|u_n - u_m\|_{1,p,\mathbb{R}_+^N}.$$

Therefore, $\{u_n\}$ is a Cauchy sequence in $L^p(\mathbb{R}^{N-1})$ converging to some function $tr(u) \in L^p(\mathbb{R}^{N-1})$. We define such a function as the trace of $u \in W^{1,p}(\mathbb{R}_+^N)$ on the hyperplane $\{x_N = 0\}$. We will use the perhaps improper but suggestive symbolism $tr(u) = u(\cdot, 0)$.

---

[30] See Section 20 of Chapter V.
[31] See Sections 18 and 19 of Chapter VII.

Proposition 16.2. *Let* $u \in W^{1,p}(\mathbb{R}^N_+)$ *for some* $p \geq 1$. *Then*

$$\|u(\cdot,0)\|_{p,\mathbb{R}^{N-1}} \leq p^{\frac{1}{p}} \|u\|_{p,\mathbb{R}^N_+}^{1-\frac{1}{p}} \|Du\|_{p,\mathbb{R}^N_+}^{\frac{1}{p}}. \tag{16.2}$$

If $1 \leq p < N$, *then* $u(\cdot,0) \in L^{p\frac{N-1}{N-p}}(\mathbb{R}^{N-1})$ *and*

$$\|u(\cdot,0)\|_{p\frac{N-1}{N-p},\mathbb{R}^{N-1}} \leq \frac{p(N-1)}{N-p} \|Du\|_{p,\mathbb{R}^N_+}. \tag{16.3}$$

If $p > N$, *the equivalence class* $u(\cdot,0)$ *has a Hölder-continuous representative, which we continue to denote by* $u(\cdot,0)$, *and there exists a constant* $\gamma(N,p)$ *depending upon only* N *and* p *such that*

$$\|u(\cdot,0)\|_{\infty,\mathbb{R}^{N-1}} \leq \gamma \|u\|_{p,\mathbb{R}^N_+}^{1-\frac{N}{p}} \|Du\|_{p,\mathbb{R}^N_+}^{\frac{N}{p}}, \tag{16.4}$$

$$|u(\bar{x},0) - u(\bar{y},0)| \leq \gamma |\bar{x} - \bar{y}|^{1-\frac{N}{p}} \|Du\|_{p,\mathbb{R}^N_+} \tag{16.5}$$

for all $\bar{x}, \bar{y} \in \mathbb{R}^{N-1}$.

Proof. Inequality (16.2) follows from (16.1) with $r = p$. The domain $\mathbb{R}^N_+$ satisfies the cone condition with cone C_o of solid angle $\frac{1}{2}\omega_N$ and height $h \in (0,\infty)$. Then (16.5) follows from (8.4) of Theorem 8.1, whereas (16.4) follows from (8.3) by minimizing over $h \in (0,\infty)$.[32]

To prove (16.3), let $\{u_n\}$ be a sequence of functions in $C_o^\infty(\mathbb{R}^N)$ converging to u in $W^{1,p}(\mathbb{R}^N_+)$. For these, by $(1.1)'$ of Corollary 1.2,

$$\|u_n\|_{\frac{Np}{N-p},\mathbb{R}^N} \leq \frac{p(N-1)}{N-p} \|Du_n\|_{p,\mathbb{R}^N}.$$

Then from (16.1) with $r = p\frac{N-1}{N-p}$,

$$\|u_n(\cdot,0)\|_{r,\mathbb{R}^{N-1}} \leq r^{\frac{1}{r}} \|u_n\|_{\frac{Np}{N-p},\mathbb{R}^N_+}^{1-\frac{1}{r}} \|Du_n\|_{p,\mathbb{R}^N_+}^{\frac{1}{r}}$$

$$\leq r \|Du_n\|_{p,\mathbb{R}^N_+}. \qquad \square$$

17 Traces and fractional Sobolev spaces

Set $\mathbb{R}^{N+1}_+ = \mathbb{R}^N \times \mathbb{R}_+$ and denote the coordinates in $\mathbb{R}^{N+1}_+$ by (x,t), where $x \in \mathbb{R}^N$ and $t \geq 0$. If $u \in W^{1,p}(\mathbb{R}^{N+1}_+)$ for some $p > 1$, we will describe the regularity of its trace on the hyperplane $\{t = 0\}$ in terms of the fractional Sobolev spaces $W^{s,p}(\mathbb{R}^N)$, where $s = 1 - 1/p$. We will adopt the symbolism

$$D_N = \left(\frac{\partial}{\partial x_1}, \frac{\partial}{\partial x_2}, \dots, \frac{\partial}{\partial x_N}\right), \quad D = \left(D_N, \frac{\partial}{\partial t}\right).$$

---

[32] The constant $\gamma(N,p)$ can be computed explicitly from (8.3)–(8.4) of Theorem 8.1. This shows that $\gamma(N,p) \to \infty$ as $p \to N$.

Proposition 17.1. *Let* $u \in W^{1,p}(\mathbb{R}_+^{N+1})$ *for some* $p > 1$. *Then the trace of* u *on the hyperplane* $\{t = 0\}$ *belongs to the fractional Sobolev space* $W^{1-\frac{1}{p},p}(\mathbb{R}^N)$. *Moreover,*

$$\|u(\cdot, 0)\|_{1-\frac{1}{p}, p; \mathbb{R}^N} \leq \{2(p-1)\}^{1/p} \left(\frac{p}{p-1}\right)^2 \|u_t\|_{p, \mathbb{R}_+^{N+1}}^{\frac{1}{p}} \|D_N u\|_{p, \mathbb{R}_+^{N+1}}^{1-\frac{1}{p}}.$$
$$(17.1)$$

Proof. For every pair $x, y \in \mathbb{R}^N$, set

$$2\xi = x - y,$$

and consider the point $z \in \mathbb{R}_+^{N+1}$ of coordinates

$$z = \left(\frac{1}{2}(x + y), \lambda|\xi|\right),$$

where λ is a positive parameter to be chosen later. Then

$$|u(x, 0) - u(y, 0)| \leq |u(z) - u(x, 0)| + |u(z) - u(y, 0)|$$

$$\leq |\xi| \int_0^1 |D_N u(x - \rho\xi, \lambda\rho|\xi|)|d\rho$$

$$+ |\xi| \int_0^1 |D_N u(y + \rho\xi, \lambda\rho|\xi|)|d\rho$$

$$+ \lambda|\xi| \int_0^1 |u_t(x - \rho\xi, \lambda\rho|\xi|)|d\rho$$

$$+ \lambda|\xi| \int_0^1 |u_t(y + \rho\xi, \lambda\rho|\xi|)|d\rho.$$

From this,

$$\frac{|u(x, 0) - u(y, 0)|^p}{|x - y|^{N+(p-1)}} \leq \frac{1}{2^p} \left(\int_0^1 \frac{|D_N u(x - \rho\xi, \lambda\rho|\xi|)|}{|x - y|^{\frac{N-1}{p}}} d\rho\right)^p$$

$$+ \frac{1}{2^p} \left(\int_0^1 \frac{|D_N u(y + \rho\xi, \lambda\rho|\xi|)|}{|x - y|^{\frac{N-1}{p}}} d\rho\right)^p$$

$$+ \frac{1}{2^p}\lambda^p \left(\int_0^1 \frac{|u_t(x - \rho\xi, \lambda\rho|\xi|)|}{|x - y|^{\frac{N-1}{p}}} d\rho\right)^p$$

$$+ \frac{1}{2^p}\lambda^p \left(\int_0^1 \frac{|u_t(y + \rho\xi, \lambda\rho|\xi|)|}{|x - y|^{\frac{N-1}{p}}} d\rho\right)^p.$$

Next, integrate both sides over $\mathbb{R}^N \times \mathbb{R}^N$. In the resulting inequality, take the $\frac{1}{p}$ power and estimate the various integrals on the right-hand side by the continuous version of Minkowski's inequality.[33] This gives

$$\||u(\cdot,0)\||_{1-\frac{1}{p},\mathbb{R}^N} \le \int_0^1 \left(\int_{\mathbb{R}^N} \int_{\mathbb{R}^N} \frac{|D_N u(x - \rho\xi, \lambda\rho|\xi|)|^p}{|x-y|^{N-1}} dxdy \right)^{\frac{1}{p}} d\rho$$
$$+ \lambda \int_0^1 \left(\int_{\mathbb{R}^N} \int_{\mathbb{R}^N} \frac{|u_t(x - \rho\xi, \lambda\rho|\xi|)|^p}{|x-y|^{N-1}} dxdy \right)^{\frac{1}{p}} d\rho.$$

Compute the first integral by first integrating in dy, and perform such an integration in polar coordinates with pole at x. Denoting with $\mathbf{n}$ the unit vector spanning the unit sphere in $\mathbb{R}^N$ and recalling that $2|\xi| = |x - y|$, we obtain

$$\int_{\mathbb{R}^N} \int_{\mathbb{R}^N} \frac{|D_N u(x - \rho\xi, \lambda\rho|\xi|)|^p}{|x-y|^{N-1}} dxdy$$
$$= 2 \int_{|\mathbf{n}|=1} d\mathbf{n} \int_0^\infty d|\xi| \int_{\mathbb{R}^N} |D_N u(x + \rho\mathbf{n}|\xi|, \lambda\rho|\xi|)|^p dx$$
$$= 2 \frac{\omega_N}{\lambda\rho} \int_{\mathbb{R}^{N+1}_+} |D_N u|^p dx.$$

Compute the second integral in a similar fashion, and combine them into

$$\||u(\cdot,0)\||_{1-\frac{1}{p},p;\mathbb{R}^N} \le 2^{1/p} \lambda^{-\frac{1}{p}} \|D_N u\|_{p,\mathbb{R}^{N+1}_+} \int_0^1 \rho^{-\frac{1}{p}} d\rho$$
$$+ 2^{1/p} \lambda^{1-\frac{1}{p}} \|u_t\|_{p,\mathbb{R}^{N+1}_+} \int_0^1 \rho^{-\frac{1}{p}} d\rho$$
$$= 2^{1/p} \frac{p}{p-1} \left(\lambda^{-\frac{1}{p}} \|D_N u\|_{p,\mathbb{R}^{N+1}_+} + \lambda^{1-\frac{1}{p}} \|u_t\|_{p,\mathbb{R}^{N+1}_+} \right).$$

The proof is completed by minimizing with respect to λ. $\square$

Remark 17.1. Proposition 17.1 admits a converse; i.e., a function

$$u \in W^{1-\frac{1}{p},p}(\mathbb{R}^N)$$

is the trace on the hyperplane $\{t = 0\}$ of a function in $W^{1,p}(\mathbb{R}^{N+1}_+)$.[34]

Thus a measurable function u defined in $\mathbb{R}^N$ is in $W^{1-\frac{1}{p},p}(\mathbb{R}^N)$ if and only if it is the trace on $\{t = 0\}$ of a function in $W^{1,p}(\mathbb{R}^{N+1}_+)$.

18   Traces on ∂E of functions in $W^{1,p}(E)$

Let E be a bounded domain in $\mathbb{R}^N$ with boundary ∂E of class C^1 and with the segment property. There exists a finite covering of ∂E with open balls $B_t(x_j)$ of

---

[33] See Section 4.3 of Chapter V.
[34] See Section 17 of the Problems and Complements.

radius $t > 0$ and center $x_j \in \partial E$ such that the portion of ∂E within $B_t(x_j)$ can be represented in a local system of coordinates as the graph of a function f_j of class C^1 in a neighborhood of the origin of the local coordinate system. Now consider the covering of E given by

$$\mathcal{U} = \{B_o, B_t(x_1), \ldots, B_t(x_n)\}, \quad \text{where } B_o = E - \bigcup_{j=1}^{n} \overline{B}_{\frac{1}{2}t}(x_j).$$

Let Φ be a partition of unity subordinate to $\mathcal{U}$, and construct functions $\psi_j \in C_o^\infty(B_t(x_j))$ satisfying

$$\sum_{j=1}^{n} \psi_j(x) = 1 \quad \text{for all } x \in \overline{E} \quad \text{and} \quad |D\psi_j| \leq \frac{2}{t}.$$

For each x_j fixed, introduce a local system of coordinates

$$\xi = (\bar{\xi}, \xi_N), \quad \text{where } \bar{\xi} = (\xi_1, \xi_2, \ldots, \xi_{N-1})$$

such that $E \cap B_t(x_j)$ is mapped into the cylinder

$$Q_t^+ = \widetilde{B}_t \times [0, t], \quad \text{where } \widetilde{B}_t = \{|\bar{\xi}| < t\}$$

and $\partial E \cap B_t(x_j)$ is mapped into the portion of the hyperplane

$$\{\xi_N = 0\} \bigcap \{|\bar{\xi}| < t\}.$$

If f is a measurable function defined in E, denote by $\widetilde{f}$ the transformed of f by the new coordinate system. In these new coordinates,

$$\widetilde{u\psi_j} \in W^{1,p}\left(\mathbb{R}^{N-1} \times \mathbb{R}^+\right),$$

and its trace on $\{\xi_N = 0\}$ can be defined as in Section 16. In particular, (16.1) and Proposition 16.1 hold for it.

If $\widetilde{u\psi_j}(\cdot, 0)$ is such a trace, we define the trace of $u\psi_j$ on $\partial E \cap B_t(x_j)$ as the function obtained from $\widetilde{u\psi_j}(\cdot, 0)$ upon returning to the original coordinates. With perhaps improper but suggestive notation, we denote it by $u\psi_j|_{\partial E}$. We then define the trace of u on ∂E as

$$tr(u) = \sum_{j=1}^{n} u\psi|_{\partial E}, \quad \text{and denote it by } u|_{\partial E}.$$

Applying (16.1) to each $\widetilde{u\psi_j}$, we obtain

$$\left(\int_{\{|\bar\xi|<t\}} |\widetilde{u\psi_j}|^r d\bar\xi \right)^{\frac{1}{r}} \le r^{\frac{1}{r}} \left(\iint_{Q_t^+} |D\widetilde{u\psi_j}|^p d\xi \right)^{\frac{1}{rp}}$$

$$\times \left(\iint_{Q_t^+} |\widetilde{u\psi_j}|^q d\xi \right)^{\frac{1}{q}\left(1-\frac{1}{r}\right)}$$

$$\le \gamma \|D\widetilde{u}\|_{p,Q_t^+}^{\frac{1}{r}} \|\widetilde{u}\|_{q,Q_t^+}^{1-\frac{1}{r}} + \gamma t^{-\frac{1}{r}} \|\widetilde{u}\|_{p,Q_t^+}^{\frac{1}{r}} \|\widetilde{u}\|_{q,Q_t^+}^{1-\frac{1}{r}}$$

for all $q, r \ge 1$ such that $q(r-1) = p(r-1)$, provided that $u \in L^q(E)$.

Next, we return to the original coordinates and add up the resulting inequalities for $j = 1, 2, \ldots, n$. Recalling that only finitely many $B_t(x_j)$ have nonempty mutual intersection, we deduce that

$$\|u\|_{r,\partial E} \le \gamma (\|Du\|_p + \|u\|_p)^{\frac{1}{r}} \|u\|_q^{1-\frac{1}{r}}. \tag{18.1}$$

Remark 18.1. In deriving (18.1), the requirement that E be bounded can be eliminated. It is necessary only that the open covering of ∂E be locally finite, whence we observe that the notion of trace of u on ∂E is of a local nature. We conclude that the trace on ∂E of a function $u \in W^{1,p}(E)$ is well defined for any domain $E \subset \mathbb{R}^N$ with boundary of class C^1 and with the segment property. Moreover, such a trace satisfies (18.1).

Proceeding as before, we may derive a counterpart of the embedding Proposition 16.2, namely, the following.

Proposition 18.1. *Let $u \in W^{1,p}(E)$, and assume that ∂E is of class C^1 and with the segment property. There exists a constant γ that can be determined a priori only in terms of N, p and the structure of ∂E such that for all $p \ge 1$ and for all $\varepsilon > 0$,*

$$\|u\|_{p,\partial E} \le \varepsilon^{p-1} \|Du\|_p + \gamma \left(1 + \frac{1}{\varepsilon}\right) \|u\|_p. \tag{18.2}$$

If $1 \le p < N$, the trace $u|_{\partial E}$ belongs to $L^{p\frac{N-1}{N-p}}(\partial E)$, and

$$\|u\|_{p\frac{N-1}{N-p},\partial E} \le \gamma \|u\|_{1,p}. \tag{18.3}$$

If $p > N$, the equivalence class $u \in W^{1,p}(E)$ has a representative that is Hölder continuous in $\overline{E}$, and

$$\|u\|_{\infty,\partial E} \le \gamma\varepsilon\|Du\| + \gamma \left(1 + \frac{1}{\varepsilon}\right) \|u\|_p, \tag{18.4}$$

$$|u(x) - u(y)| \le \gamma|x-y|^{1-\frac{N}{p}} \|u\|_{1,p} \quad \text{for all } x, y \in \overline{E}. \tag{18.5}$$

Proof. Inequality (18.2) follows from (18.1) with $q = r = p$ and an application of Young's inequality. To prove the remaining inequalities, we apply Proposition 16.2 to the functions $u\widetilde{\psi}_j$ introduced earlier. Then we return to the original coordinates and add over $j = 1, 2, \ldots, n$. $\square$

18.1 Traces and fractional Sobolev spaces. Let E be a bounded domain in $\mathbb{R}^N$ with boundary ∂E of class C^1 and with the segment property. Motivated by Proposition 17.1, we may introduce the notion of a fractional Sobolev space $W^{s,p}(\partial E)$ for $s \in (0, 1)$ on the $(N - 1)$-dimensional domain ∂E. A function $u \in L^p(\partial E)$ belongs to $W^{s,p}(\partial E)$ if the seminorm

$$\|u\|_{s,p;\partial E} = \left(\int_{\partial E} \int_{\partial E} \frac{|u(x) - u(y)|^p}{|x - y|^{(N-1)+sp}} d\sigma(x)d\sigma(y) \right)^{\frac{1}{p}} \tag{18.6}$$

is finite. Here $d\sigma(\cdot)$ is the surface measure on ∂E. A norm in $W^{s,p}(\partial E)$ is given by

$$\|u\|_{s,p;\partial E} = \|u\|_{p,\partial E} + \|u\|_{s,p;\partial E}.$$

Statements concerning $W^{s,p}(\partial E)$ can be derived from Section 17 by working with the functions $u\widetilde{\psi}_j$, returning to the original coordinates, and adding over $j = 1, 2, \ldots, n$.

Theorem 18.2. *Let $u \in W^{1,p}(E)$ for $p > 1$, and let ∂E be of class C^1 and with the segment property. Then the trace of u on ∂E belongs to $W^{s,p}(\partial E)$, where $s = 1 - \frac{1}{p}$, and*

$$\|u\|_{1-\frac{1}{p},p;\partial E} \leq \gamma \|u\|_{1,p} \tag{18.7}$$

for a constant γ depending only upon N, p and the structure of ∂E.

Remark 18.2. Theorem 18.2 admits a converse; i.e., a function

$$u \in W^{1-\frac{1}{p},p}(\partial E)$$

is the trace on ∂E of a function in $W^{1,p}(E)$.[35] Thus a measurable function u defined in ∂E is in $W^{1-\frac{1}{p},p}(\partial E)$ if and only if it is the trace on ∂E of a function in $W^{1,p}(E)$.

19 Multiplicative embeddings of $W^{1,p}(E)$

Multiplicative embeddings in the form (1.1) hold for functions in $W_o^{1,p}(E)$ and are, in general, false for functions in $W^{1,p}(E)$.

The Poincaré inequalities of Section 10 recover a multiplicative form of the embedding of $W^{1,p}(E)$ for functions of zero integral average on E.

---

[35]See Theorems 18.1c and 18.2c of the Problems and Complements.

The discrete form of the isoperimetric inequality (11.1) would be vacuous if u were a nonzero constant. It is meaningful only if the measure of the set $[u = 0]$ is positive.

These remarks imply that a multiplicative embedding of $W^{1,p}(E)$ into $L^q(E)$ is possible only if some information is available on the values of u on some subset of $\overline{E}$.

The next theorem provides a multiplicative embedding in terms of the trace of u on some subset Γ of ∂E, provided that E is convex.

Theorem 19.1. *Let E be a bounded, open, convex subset of $\mathbb{R}^N$, $N \geq 2$, and let $\Gamma \subset \partial E$ be open in the relative topology of ∂E. There exist constants γ and C_Γ such that for every $u \in W^{1,p}(E)$,*[36]

$$\|u\|_{q,E} \leq \gamma C_\Gamma^{\frac{N}{q}} (\|u\|_{m,E}^{1-\alpha} \|u\|_{s,\Gamma}^{\alpha} + C_\Gamma^{2\theta} \|u\|_{r,E}^{1-\theta} \|\nabla u\|_{p,E}^{\theta}), \tag{19.1}$$

where the parameters $\{\alpha, \theta, m, s, r, p, q\}$ satisfy

$$m, r \geq 1, \qquad sp > 1, \qquad q \geq \max\{m; r\}, \qquad \alpha, \theta \in [0, 1], \tag{19.2}_1$$

and, in addition, the two sets of parameters $\{\alpha, m, s, q, N\}$ and $\{\theta, r, p, q, N\}$ are linked by

$$\theta = \left(\frac{1}{r} - \frac{1}{q}\right)\left(\frac{1}{N} - \frac{1}{p} + \frac{1}{r}\right)^{-1},$$

$$\alpha = \left(\frac{1}{m} - \frac{1}{q}\right)\left(\frac{1}{Ns} - \frac{1}{s} + \frac{1}{m}\right)^{-1} \tag{19.2}_2$$

and their admissible range is restricted by

$$s \geq \max\left\{1; m\frac{N-1}{N}\right\}, \tag{19.2}_3$$

$$r \leq \frac{(s-1)p}{p-1} \leq q,$$
$$q < \infty \qquad \text{if } p \geq N,$$
$$q \leq \frac{Np}{N-p} \qquad \text{if } p < N. \tag{19.2}_4$$

The constant γ depends upon only the parameters $\{\alpha, \theta, m, s, r, p, q\}$, and it is independent of u. The constant C_Γ depends upon only the geometry of E and Γ.

---

[36]E. DiBenedetto and D. J. Diller, On the rate of drying in a photographic film, *Adv. Differential Equuations*, 1-6 (1996), 989–1003; E. DiBenedetto and D. J. Diller, A new form of the Sobolev multiplicative inequality and applications to the asymptotic decay of solutions to the Neumann problem for quasilinear parabolic equations with measurable coefficients, *Proc. Nat. Acad. Sci. Ukraine*, **8** (1997), 81–88.

Remark 19.1. If u vanishes on Γ in the sense of the traces, then (19.1) coincides with the multiplicative embedding of Theorem 1.1; i.e.,

$$\|u\|_q \le \gamma C_\Gamma^{\frac{N}{q}+2\theta}\|u\|_r^{1-\theta}\|Du\|_p^\theta. \tag{19.1}'$$

Taking $\alpha = 0$ in the second part of (19.2)$_2$ and using the arbitrariness of m and s yields the same range as in (1.3)–(1.5) for the parameters $\{\theta, r, q, p, N\}$. Thus the multiplicative embedding (1.1) continues to hold for functions in $W^{1,p}(E)$ vanishing only on a nontrivial portion of ∂E.

Remark 19.2. The difference between (19.1)$'$ and the embedding (1.1) is the presence of the constant C_Γ in the right-hand side of (19.1)$'$. This is because u is known to vanish only on $\Gamma \subset \partial E$, as opposed to (1.1), where u vanishes on the whole ∂E in the sense of $W_o^{1,p}(E)$.

Remark 19.3. Since $q \ge m$, it follows from (19.2)$_2$ that $\alpha, \theta \in [0, 1]$. The links (19.2)$_2$ arise naturally from the proof. Through a rescaling argument, one can see that these are the only possible links between the two sets of parameters

$$\{\theta, r, p, q, N\} \quad \text{and} \quad \{\alpha, m, s, r, N\}.$$

For example, let E be the ball B_R of radius R centered at the origin of $\mathbb{R}^N$. By rescaling R, inequality (19.1) is independent of the radius of E only if (19.2)$_2$ holds.

Remark 19.4. The constant γ in (19.1) depends upon only the indicated parameters. In particular, it is independent of u and possible rotations, translations, and dilations of E and Γ.

Remark 19.5. The constant C_Γ depends on the geometry of E and Γ in the following manner. Let $\bar{x} \in \Gamma$, and for $\varepsilon > 0$, let $B_\varepsilon(\bar{x})$ be the N-dimensional ball centered at $\bar{x}$ and radius ε. The number ε is chosen as the largest radius for which

$$B_\varepsilon(\bar{x}) \bigcap \partial E \subset \Gamma. \tag{19.3}$$

Then choose $x_o \in B_\varepsilon(\bar{x}) \cap E$, and let $\rho > 0$ be the largest radius for which $B_\rho(x_o) \subset E$. Finally, let $R > 0$ be the smallest radius for which $E \subset B_R(x_o)$. Thus

$$B_\rho(x_o) \subset B_\varepsilon(\bar{x}) \quad \text{and} \quad E \subset B_R(x_o). \tag{19.4}$$

Then the constant C_Γ is the smallest value of the ratio (R/ρ) for all the possible choices of $\bar{x} \in \Gamma$ and $x_o \in B_\varepsilon(\bar{x})$.

Remark 19.6. Choosing $\theta = \alpha = m = r = 1$ in (19.2)$_2$ gives

$$\|u\|_{p^*, E} \le \gamma C_\Gamma^{N/p^*}(\|u\|_{s^*,\Gamma}^\alpha + C_\Gamma^2\|\nabla u\|_{p,E}), \tag{19.1}''$$

where

$$p^* = \frac{Np}{N-p} \quad \text{and} \quad s^* = \frac{(N-1)p}{N-p}.$$

Remark 19.7. It is an open question to establish the embedding (19.1) for nonconvex domains.

20 Proof of Theorem 19.1: A special case

First, assume that E is the unit cube in $\mathbb{R}^N$, $N \geq 2$, with edges on the positive coordinate semiaxes; i.e.,

$$Q = \left\{ x \in \mathbb{R}^N \mid \max_{1 \leq i \leq N} x_i < 1 \right\}, \qquad N \geq 2.$$

As the portion Γ_o of ∂Q, we take the union of the faces of the cube lying on the coordinate planes; i.e.,

$$\Gamma_o = \partial Q \cap \bigcup_{i=1}^{N} \{x_i = 0\}.$$

To establish (19.1) for such a domain and such a Γ_o, we may assume that u is nonnegative and in $C^1(\overline{Q})$. Set

$$\overline{x}_i = (x_1, \ldots, x_{i-1}, 0, x_{i+1}, \ldots, x_N),$$
$$(\overline{x}_i, t) = (x_1, \ldots, x_{i-1}, t, x_{i+1}, \ldots, x_N), \qquad i = 1, \ldots, N-1.$$

Then for all $s \geq 1$ and all $x \in Q$,

$$
\begin{aligned}
u^s(x) &= u^s(\overline{x}_i) + s \int_0^{x_i} u^{s-1}(\overline{x}_i, t) \frac{\partial u(\overline{x}_i, t)}{\partial x_i} dt \\
&\leq u^s(\overline{x}_i) + s \int_0^1 u^{s-1}(\overline{x}_i, t) \left| \frac{\partial u(\overline{x}_i, t)}{\partial x_i} \right| dt \stackrel{\text{def}}{=} w_i(\overline{x}_i).
\end{aligned}
\tag{20.1}
$$

Choose numbers $q \geq \ell \geq 0$ and all $s \geq 1$ satisfying

$$(N-1)(q-\ell) < Ns, \tag{20.2}$$

and set

$$k = \frac{Ns\ell}{Ns - (N-1)(q-\ell)}, \tag{20.3$_1$}$$

or, equivalently,

$$\ell = \frac{(Ns - (N-1)q)k}{Ns - (N-1)k}. \tag{20.3$_2$}$$

One verifies that the requirement $0 \leq \ell \leq q$ is equivalent to $0 \leq k \leq q$.
    Having fixed $m, r \geq 1$, we choose k to satisfy

$$\max\{r; m\} \leq k \leq q. \tag{20.4}$$

Then from (20.1), for all $x \in Q$,

$$u^q(x) = u^\ell(x)\{u^{Ns}(x)\}^{\frac{q-\ell}{Ns}} = u^\ell(x) \prod_{i=1}^{N} w_i^{\frac{q-\ell}{Ns}}(\overline{x}_i).$$

Integrating this in dx_1 over $(0, 1)$ and applying Hölder's inequality,

$$\int_0^1 u^q(x)dx_1 \leq w_1(\overline{x}_1) \int_0^1 u^\ell(x) \prod_{i=2}^N w_i^{\frac{q-\ell}{Ns}}(\overline{x}_i)dx_1$$

$$\leq w_1(\overline{x}_1) \left(\int_0^1 u^k(x)dx_1\right)^{\frac{\ell}{k}} \prod_{i=2}^N \left(\int_0^1 w_i(\overline{x}_i)dx_1\right)^{\frac{q-\ell}{Ns}}.$$

In view of $(20.3)_1$, Hölder's inequality can be applied.

Next, integrate this in dx_2 over $(0, 1)$ and apply Hölder's inequality with the same conjugate exponents. Proceeding by induction to exhaust all the variables x_j yields

$$\int_Q u^q(x)dx \leq \left(\int_Q u^k(x)dx\right)^{\frac{\ell}{k}} \prod_{i=1}^N \left(\int_{Q_{N-1}^i} w_i(\overline{x}_i)d\overline{x}_i\right)^{\frac{q-\ell}{Ns}},$$

where $Q_{N-1}^i = Q \cap \{x_i = 0\}$ is the $(N-1)$-dimensional cube excluding the ith coordinate. By Hölder's inequality and the definition of $w_i(\overline{x}_i)$,

$$\int_{Q_{N-1}^i} w_i(\overline{x}_i)d\overline{x}_i \leq \int_{\Gamma_o} u^s(x)d\sigma$$

$$+ s\left(\int_Q |Du|^p dx\right)^{\frac{1}{p}} \left(\int_Q u^{\frac{(s-1)p}{p-1}} dx\right)^{\frac{p-1}{p}},$$

where $d\sigma$ is the surface measure on Γ_o. Combining these estimates, we arrive at

$$\|u\|_{q,Q} \leq \gamma \|u\|_{k,Q}^{\frac{\ell}{q}} \|u\|_{s,\Gamma_o}^{\frac{q-\ell}{q}}$$

$$+ \gamma \|u\|_{k,Q}^{\frac{\ell}{q}} \|u\|_{\frac{(s-1)p}{p-1},Q}^{\frac{s-1}{s}\frac{q-\ell}{q}} \|Du\|_{p,Q}^{\frac{q-\ell}{qs}}. \tag{20.5}$$

If either $m = q$ or $r = q$, it follows from (20.4) that $k = q$, and then $(20.3)_1$–$(20.3)_2$ imply that $k = \ell = q$. In such a case, the multiplicative inequality (19.1) follows from (20.5), and it is vacuous.[37]

Assume that $\max\{m; r\} < q$ and choose k satisfying

$$\max\{m; r\} < k < q.$$

By Hölder's inequality,

$$\|u\|_{k,Q} \leq \|u\|_{m,Q}^{\frac{m(q-k)}{k(q-m)}} \|u\|_{q,Q}^{\frac{q(k-m)}{k(q-m)}},$$

$$\|u\|_{k,Q} \leq \|u\|_{r,Q}^{\frac{r(q-k)}{k(q-r)}} \|u\|_{q,Q}^{\frac{q(k-r)}{k(q-r)}}. \tag{20.6}$$

---

[37]Thus in the definition $(19.2)_2$ of θ, it is stipulated that $\theta = 0$ if $r = q$. A similar stipulation holds for α.

We may also assume that the second part of $(19.2)_3$ holds with strict inequalities. Then by Hölder's inequality,[38]

$$\|u\|_{\frac{(s-1)p}{(p-1)},Q} \le \|u\|_{r,Q}^{\frac{r[q(p-1)-p(s-1)]}{p(s-1)(q-r)}} \|u\|_{q,Q}^{\frac{q[p(s-1)-r(p-1)]}{p(s-1)(q-r)}}. \tag{20.7}$$

Assume that of the two terms on the right-hand side of (20.5), the first majorizes the second. Then using the first part of (20.6),

$$\|u\|_{q,Q} \le 2\gamma \|u\|_{q,Q}^{\frac{t(k-m)}{k(q-m)}} \|u\|_{m,Q}^{\frac{m(q-k)}{k(q-m)}} \|u\|_{s,\Gamma_o}^{\frac{q-t}{q}}.$$

By reducing the powers of $\|u\|_{q,Q}$, this gives

$$\|u\|_{q,Q} \le \gamma' \|u\|_{m,Q}^{1-\alpha} \|u\|_{s,\Gamma_o}^{\alpha},$$

where α is given by the second part of $(19.2)_2$ and γ' is a constant depending upon only the set of parameters $\{N, p, q, s, m, r\}$.

If of the two terms on the right-hand side of (20.5), the second majorizes the first, using the second term of (20.6) and (20.7) gives

$$\|u\|_{q,Q} \le 2\gamma \|u\|_{q,Q}^{\frac{t(k-r)}{k(q-r)} + \frac{(q-t)[(s-1)p-r(p-1)]}{s(q-r)}}$$
$$\times \|u\|_{r,Q}^{\frac{tr(q-k)}{qk(q-r)} + \frac{r(q-t)[q(p-1)-p(s-1)]}{qsp(q-r)}} \|Du\|_{p,Q}^{\frac{q-t}{qs}}.$$

By reducing the powers of $\|u\|_{q,Q}$, this gives

$$\|u\|_{q,Q} \le \gamma'' \|u\|_{r,Q}^{1-\theta} \|Du\|_{p,Q}^{1-\theta},$$

where θ is given by the first term of $(19.2)_2$ and γ'' is a constant depending upon only the set of parameters $\{N, p, q, s, m, r\}$.

21   Constructing a map between E and Q: Part 1

Let E be a bounded, convex subset of $\mathbb{R}^N$, and let $\Gamma \subset \partial E$ be open in the relative topology of ∂E. Having fixed $\bar{x} \in \Gamma$, construct the ball $B_\varepsilon(\bar{x})$ as in (19.3). Then pick $x_o \in B_\varepsilon(\bar{x}) \cap E$ and construct the balls $B_\rho(x_o)$ and $B_R(x_o)$ as in (19.4).

Proposition 21.1. *There exists a map* $\mathcal{F} : \overline{E} \to \overline{Q}$ *and positive, absolute constants* $C > c > 0$ *independent of the geometry of* ∂E *and* Γ *such that*[39]

$$Q = \mathcal{F}^{-1}(E), \qquad \Gamma_o \subset \mathcal{F}^{-1}(\Gamma)$$

---

[38] If the second part of $(19.2)_3$ holds with some equality, the arguments are similar and indeed simpler.

[39] Q and $\Gamma_o \subset \partial Q$ are as in Section 20. The map $\mathcal{F}$ is not claimed to be unique.

and

$$c\rho|x - y| \leq |\mathcal{F}(x) - \mathcal{F}(y)| \leq CR\frac{R}{\rho}|x - y|. \tag{21.1}$$

Moreover, the Jacobian $J_{\mathcal{F}}$ satisfies

$$c\rho^N \leq J_{\mathcal{F}}(x) \leq CR^N \quad \text{for all } x \in E. \tag{21.2}$$

Proof. Let $\mathbf{n}$ be the unit vector in $\mathbb{R}^N$ ranging over the unit sphere S_1, and consider the map $\phi_{x_o,E} : S_1 \to \partial E$ defined by

$$\phi_{x_o,E}(\mathbf{n}) = x_o + t\mathbf{n}, \tag{21.3}$$

where t is the unique positive number such that $x_o + t\mathbf{n} \in \partial E$. Such a map is well defined since E is bounded and convex.

Lemma 21.2. *There exists a constant C_o depending on x_o and E such that*

$$\rho|\mathbf{n}_1 - \mathbf{n}_2| \leq |\phi_{x_o,E}(\mathbf{n}_1) - \phi_{x_o,E}(\mathbf{n}_2)| \leq 4R\frac{R}{\rho}|\mathbf{n}_1 - \mathbf{n}_2|. \tag{21.4}$$

Proof. Up to a translation, we may assume that $x_o = 0$ and set $\phi_{x_o,E} = \phi$. Having fixed $\mathbf{n}_1$ and $\mathbf{n}_2$ on S_1, the bound below in (21.4) follows from the definition (21.3) since the sphere S_ρ centered at the origin is contained in the interior of E. For the bound above, by intersecting E with the hyperplane through x_o and containing $\mathbf{n}_1$ and $\mathbf{n}_2$, it suffices to assume $N = 2$. Denoting by $\mathbf{i}$ and $\mathbf{j}$ the coordinate unit vectors in $\mathbb{R}^2$, we may assume up to a possible rotation that $\mathbf{n}_1 = \mathbf{i}$. Then

$$\mathbf{n}_2 = \mathbf{n} = (\cos\lambda, \sin\lambda) \quad \text{for some } \lambda \in (-\pi, \pi).$$

Therefore, the bound above in (21.4) reduces to

$$|\phi(\mathbf{n}) - \phi(\mathbf{i})| \leq 4R\frac{R}{\rho}\sqrt{1 - \cos\lambda}. \tag{21.4'}$$

Let $\Sigma_\mathbf{n}$ be the line segment joining $\phi(\mathbf{i})$ and $\phi(\mathbf{n})$. Also let $L_\mathbf{n}$ be the line through $\phi(\mathbf{i})$.

Case 1: $\Sigma_\mathbf{n}$ intersects B_ρ. Since $L_\mathbf{n}$ intersects B_ρ, the point x_* on the line $L_\mathbf{n}$ that minimizes the distance from $L_\mathbf{n}$ to the origin 0 is in B_ρ and thus on the line segment $\Sigma_\mathbf{n}$. Let λ_1 and λ_2 be the angles

$$\lambda_1 = \widehat{\phi(\mathbf{i})0x_*}, \qquad \lambda_2 = \widehat{x_*0\phi(\mathbf{n})}.$$

Then $\lambda_1 + \lambda_2 = \lambda$, and by elementary trigonometry,

$$|\phi(\mathbf{n}) - \phi(\mathbf{i})| = |\phi(\mathbf{i})|\sin\lambda_1 + |\phi(\mathbf{n})|\sin\lambda_2$$
$$\leq R(\sin\lambda_1 + \sin\lambda_2).$$

At least one of the λ_i is less than $\frac{1}{2}\lambda$—say, for example, λ_1. Then since $\frac{1}{2}\lambda \in (0, \frac{\pi}{2})$,

$$\sin \lambda_1 + \sin \lambda_2 \leq \sin \frac{1}{2}\lambda + \sin(\lambda - \lambda_1)$$

$$\leq \sin \frac{1}{2}\lambda + \sin \lambda \cos \lambda_1 - \cos \lambda \sin \lambda_1$$

$$\leq 2 \sin \frac{1}{2}\lambda + \sin \lambda$$

$$\leq \frac{4}{\sqrt{2}}\sqrt{1 - \cos \lambda}.$$

Case 2: Σ_n does not intersect B_ρ. Without loss of generality, by possibly interchanging the roles of $\mathbf{i}$ and $\mathbf{n}$, we may assume that $|\phi(\mathbf{i})| > |\phi(\mathbf{n})|$. Let $\lambda_o = \widehat{\phi(\mathbf{n})\phi(\mathbf{i})0}$. Then by the law of sines,

$$\frac{\sin \lambda}{|\phi(\mathbf{n}) - \phi(\mathbf{i})|} = \frac{\sin \lambda_o}{|\phi(\mathbf{n})|}.$$

From this,

$$|\phi(\mathbf{n}) - \phi(\mathbf{i})| = |\phi(\mathbf{n})|\frac{\sin \lambda}{\sin \lambda_o} \leq R\frac{\sin \lambda}{\sin \lambda_o}.$$

Since the line segment Σ_n does not intersect B_ρ, the smallest λ_o could possibly be is if the line L_n is tangent to B_ρ. In such a case,

$$\sin \lambda_o \geq \frac{\rho}{|\phi(\mathbf{i})|} \geq \frac{\rho}{R}.$$

Combining these estimates proves the lemma. $\square$

22   Constructing a map between E and Q: Part 2

Next, extend the map $\phi_{x_o,E}$ to a map $\varphi_{x_o,E}$ from the whole unit ball B_1 onto E by

$$\varphi_{x_o,E}(x) = x_o + |x|t(\mathbf{n}_x)\mathbf{n}_x, \tag{22.1}$$

where $|x| \leq 1$,

$$\mathbf{n}_x = \begin{cases} \dfrac{x}{|x|} & \text{if } x \neq 0, \\ 0 & \text{if } x = 0, \end{cases}$$

and $t(\mathbf{n}_x)$ is defined as in (21.3) in correspondence of the unit vector $\mathbf{n}_x$.

Lemma 22.1. *For all $x, y \in B_1$,*

$$\frac{1}{4}\rho\frac{\rho}{R}|x - y| \leq |\varphi_{x_o,E}(x) - \varphi_{x_o,E}(y)| \leq 5R\frac{R}{\rho}|x - y|. \tag{22.2}$$

Moreover, denoting by J_φ the Jacobian of $\varphi_{x_o, E}$,

$$J_\varphi(x) = |t(\mathbf{n}_x)|^N \quad \text{for all } x \in B_1.\tag{22.3}$$

Finally, from the definition of $t(\mathbf{n}_x)$, it follows that

$$\rho^N \leq J_\varphi(x) \leq R^N \quad \text{for all } x \in B_1.\tag{22.4}$$

Proof. Assume that $x_o = 0$, and set $\varphi_{x_o, E} = \varphi$. Fix any two nonzero vectors $x, y \in B_1$. By intersecting E with the hyperplane through the origin and containing x and y, it suffices to consider the case where $N = 2$. Assume, for example, that $|y| \leq |x|$. Then by elementary plane geometry,

$$\left| \frac{x}{|x|} - \frac{y}{|y|} \right| \leq \frac{1}{|y|}|x - y|.$$

From this and using the upper estimate in Lemma 21.2,

$$|\varphi(x) - \varphi(y)| \leq R|x - y| + 4R\frac{R}{\rho}|y|\left| \frac{x}{|x|} - \frac{y}{|y|} \right|$$

$$\leq 5R\frac{R}{\rho}|x - y|.$$

For the lower estimate in (22.2), assume, for example, that $|x| \geq |y|$. First, suppose that

$$|x| - |y| \leq \frac{1}{4}\frac{\rho}{R}|x - y|.$$

Then by the lower estimate of Lemma 21.2,

$$|\varphi(x) - \varphi(y)| = ||x|t(\mathbf{n}_x)\mathbf{n}_x - |y|t(\mathbf{n}_y)\mathbf{n}_y|$$

$$\geq |x||t(\mathbf{n}_x)\mathbf{n}_x - t(\mathbf{n}_y)\mathbf{n}_y| - R(|x| - |y|)$$

$$\geq \rho|x|\left| \frac{x}{|x|} - \frac{y}{|y|} \right| - \frac{1}{4}\rho|x - y|$$

$$\geq \rho|x|\left| \frac{x - y}{|x|} + \frac{y}{|x|} - \frac{y}{|y|} \right| - \frac{1}{4}\rho|x - y|$$

$$\geq \rho|x - y| - \rho(|x| - |y|) - \frac{1}{4}\rho|x - y|$$

$$\geq \frac{1}{2}\rho|x - y|.$$

If, on the other hand,

$$|x| - |y| > \frac{1}{4}\frac{\rho}{R}|x - y|,$$

then by elementary plane geometry,

$$|\varphi(x) - \varphi(y)| = ||x|t(\mathbf{n}_x)\mathbf{n}_x - |y|t(\mathbf{n}_y)\mathbf{n}_y|$$
$$\geq \rho(|x| - |y|) \geq \frac{1}{4}\rho\frac{\rho}{R}|x - y|.$$

To establish (22.3), we express the Lebesgue measure dv of B_1 in polar coordinates as

$$dv = r^{N-1}drd\mathbf{n} \quad \text{for } r \in (0, 1)$$

and $\mathbf{n}$ ranging over the unit sphere of $\mathbb{R}^N$. Likewise, the Lebesgue measure of E in polar coordinates is

$$d\mu = \tau^{N-1}d\tau d\mathbf{n} \quad \text{for } \tau \in (0, t(\mathbf{n})),$$

where $t(\mathbf{n})$ is the polar representation of ∂E with a pole at $x_o = 0$.

From the definition (22.1), it follows that $\tau = rt(\mathbf{n})$. Therefore,

$$d\mu = \tau^{N-1}d\tau d\mathbf{n} = |t(\mathbf{n})|^N r^{N-1}drd\mathbf{n} = |t(\mathbf{n})|^N dv.$$

This proves (22.3)–(22.4).   □

The transformation $\varphi_{x_o,E}^{-1}$ is a one-to-one Lipschitz map between E and B_1 with Lipschitz-continuous inverse. The boundary of E is mapped into ∂B_1, and the portion $\Gamma \subset \partial E$ is mapped into a subset $\Gamma_1 \subset \partial B_1$ open in the relative topology of the unit sphere S_1. The Lipschitz constants and the Jacobian of such a transformation are controlled by (22.2)–(22.4). Therefore, the proof of Proposition 21.1 reduces to the case where

$$E = B_1 \quad \text{and} \quad \Gamma_1 \subset \partial B_1.$$

Up to a rotation, we may assume that $(0, \ldots, 0, 1) \in \Gamma_1$. Since Γ_1 is open, there is $\varepsilon > 0$ such that

$$\{x \in S_1 | x_N > 1 - \varepsilon\} \subset \Gamma_1.$$

Set $x_\varepsilon = (0, \ldots, 0, 1 - \varepsilon)$ and construct the map

$$\varphi_{x_\varepsilon, B_1} : B_1 \longrightarrow B_1.$$

Such a map satisfies estimates analogous to (22.2)–(22.4) with ρ replaced by ε and $R = 2$. Moreover, Γ_1 is mapped into an open portion $\Gamma_2 \in \partial B_1$ such that

$$\varphi_{x_\varepsilon, B_1}(\{x \in S_1 | x_N > 0\}) \subset \Gamma_2.$$

Thus we have reduced to the case when

$$E = B_1 \quad \text{and} \quad \Gamma_2 = S_1 \bigcap \{x_N > 0\}.$$

Then after an appropriate rotation, the map

$$\varphi_{x_{\frac{1}{2}}, Q} : B_1 \longrightarrow B_1, \quad \text{where } x_{\frac{1}{2}} = \left(\frac{1}{2}, \frac{1}{2}, \dots, \frac{1}{2} \right)$$

maps the cube Q onto B_1 and Γ_o onto Γ_2.

The map $\mathcal{F}$ claimed by Proposition 21.1 is obtained by composing the maps occurring in the various steps of the construction.   □

23 Proof of Theorem 19.1, concluded

Let E be a bounded, convex subset of $\mathbb{R}^N$, and let $\Gamma \subset \partial E$ be open in the relative topology of ∂E. Having fixed $\overline{x} \in \Gamma$, construct the ball $B_\varepsilon(\overline{x})$ as in (19.3). Then pick $x_o \in B_\varepsilon(\overline{x}) \cap E$ and construct the balls $B_\rho(x_o)$ and $B_R(x_o)$ as in (19.4). Also let $\mathcal{F}$ be the map claimed by Proposition 21.1 for these choices of $\overline{x}$, ρ, and R. The Jacobian is estimated by (21.2), whereas by (21.1),

$$|D\mathcal{F}| \le CR\frac{R}{\rho}.$$

Now let $u \in W^{1,p}(E)$. By Theorem 19.1 applied for the unit cube Q,

$$\|u\|_{q,E} = \left(\int_Q |u(\mathcal{F})|^q J_{\mathcal{F}} dy \right)^{1/q} \le \gamma R^{N/q} \|u(\mathcal{F})\|_{q,Q}$$

$$\le \gamma R^{N/q} (\|u(\mathcal{F})\|_{m,Q}^{1-\alpha} \|u(\mathcal{F})\|_{s,\Gamma_o}^{\alpha} + \|u(\mathcal{F})\|_{r,Q}^{1-\theta} \|D_y u(\mathcal{F})\|_{p,Q}^{\theta})$$

$$\le \gamma R^{N/q} \frac{1}{\rho^{\frac{N(1-\alpha)}{m}}} \|u\|_{m,E}^{1-\alpha} \frac{1}{\rho^{\frac{(N-1)\alpha}{s}}} \|u\|_{s,\Gamma}^{\alpha}$$

$$+ \gamma R^{N/q} \left(\frac{R^2}{\rho} \right)^\theta \frac{1}{\rho^{\frac{N(1-\theta)}{r}}} \|u\|_{r,E}^{1-\theta} \frac{1}{\rho^{\frac{N\theta}{p}}} \|D_x u\|_{p,E}^{\theta}$$

$$\le \gamma \left(\frac{R}{\rho} \right)^{N/q} \left\{ \|u\|_{m,E}^{1-\alpha} \|u\|_{s,\Gamma}^{\alpha} + \left(\frac{R}{\rho} \right)^{2\theta} \|u\|_{r,E}^{1-\theta} \|D_x u\|_{p,E}^{\theta} \right\}.$$

Here γ denotes a positive constant depending upon only the set of parameters $\{N, p, q, m, s, r\}$ and independent of the geometry of ∂E and Γ. The dependence on such a geometry is traced by R and ρ.

PROBLEMS AND COMPLEMENTS

1 MULTIPLICATIVE EMBEDDINGS OF $W_o^{1,p}(E)$

1.1 The following two propositions are established by a minor variant of the arguments in Sections 2–4. Their significance is that u is not required to vanish, in some sense, on ∂E. Let Q be a cube in $\mathbb{R}^N$; i.e.,

$$Q = \prod_{j=1}^{N}(a_j, b_j), \quad -\infty < a_j < b_j < \infty, \quad j = 1, 2, \ldots, N.$$

Proposition 1.1c. *Let $u \in W^{1,p}(Q)$ for some $p \in [1, N)$. Then $u \in L^{p^*}(Q)$, and*

$$\|u\|_q \leq \sum_{j=1}^{N} \left\{ \frac{1}{N(b_j - a_j)} \|u\|_p + \frac{p(N-1)}{N(N-p)} \|u_{x_j}\|_p \right\}. \tag{1.1c}$$

Inequality (1.1c) continues to hold if $(b_j - a_j) = \infty$ for some j, provided that we set $(b_j - a_j)^{-1} = 0$.

Proposition 1.2c. *Let $u \in W^{1,p}(\mathbb{R}^N)$ for some $p \in [1, N)$. Then $u \in L^{p^*}(\mathbb{R}^N)$, and*

$$\|u\|_{p^*} \leq \frac{p(N-1)}{N(N-p)} \sum_{j=1}^{N} \|u_{x_j}\|_p. \tag{1.2c}$$

1.2 There exists function $u \in W_o^{1,N}(E)$ that is not essentially bounded.

1.3 The functional $u \to \|Du\|_p$ is a seminorm in $W^{1,p}(E)$ and a norm in $W_o^{1,p}(E)$. Such a norm is equivalent to $\|u\|_{1,p}$; i.e. there exists a constant γ depending only upon N and p such that

$$\gamma^{-1}\|Du\|_p \leq \|u\|_{1,p} \leq \gamma\|Du\|_p \quad \text{for all } u \in W_o^{1,p}(E).$$

14 COMPACT EMBEDDINGS

14.1 Theorem 14.1 is false if E is unbounded. To construct a counterexample, consider a sequence of balls $\{B_{\rho_j}(x_j)\}$ all contained in E such that $|x_j| \to \infty$ as $j \to \infty$. Then construct a function $\varphi \in W_o^{1,p}(B_{\rho_o}(x_o))$ such that its translated and rescaled copies φ_j all satisfy

$$\|\varphi_j\|_{1,p;B_{\rho_j}(x_j)} = 1 \quad \text{for all } j = 1, 2, \ldots.$$

The sequence $\{\varphi_j\}$ does not have a subsequence strongly convergent in any $L^q(E)$ for all $q \geq 1$.

14.2 Theorem 14.1 is false and $q = p^*$ for all $p \in [1, N)$. Construct an example.

17   TRACES AND FRACTIONAL SOBOLEV SPACES

17.1   CHARACTERIZING FUNCTIONS IN $W^{1-\frac{1}{p},p}(\mathbb{R}^N)$ AS TRACES.

Proposition 17.1c. *Every function* $\varphi \in W^{1-\frac{1}{p},p}(\mathbb{R}^N)$ *has an extension* $u \in W^{1,p}(\mathbb{R}^{N+1}_+)$ *such that the trace of* u *on the hyperplane* $\{t = 0\}$ *coincides with* φ *and*

$$\|u(\cdot, t)\|_{p,\mathbb{R}^N} \le \|\varphi\|_{p,\mathbb{R}^N} \quad \text{for all } t > 0, \tag{17.1c}$$

$$\|Du\|_{p,\mathbb{R}^{N+1}_+} \le \gamma \|\!|\varphi|\!\|_{1-\frac{1}{p},p;\mathbb{R}^N}, \tag{17.2c}$$

where γ *depends upon only* N *and* p.

Proof. Assume that $N \ge 2$, and let

$$F(x - y; t) = \frac{1}{(N-1)\omega_{N+1}} \frac{1}{\left(|x - y|^2 + t^2\right)^{\frac{N-1}{2}}}$$

be the fundamental solution of the Laplace equation in $\mathbb{R}^{N+1}$ with a pole at $(x, 0)$. Also, let[40]

$$\begin{aligned}
\Phi(x, t) &= \frac{2}{\omega_{N+1}} \int_{\mathbb{R}^N} \frac{t}{\left(|x - y|^2 + t^2\right)^{\frac{N+1}{2}}} \varphi(y)dy \\
&= 2 \int_{\mathbb{R}^N} \frac{\partial}{\partial t} F(x - y; t)\varphi(y)dy
\end{aligned}$$

be the Poisson integral of φ in $\mathbb{R}^{N+1}_+$. Since[41]

$$2 \int_{\mathbb{R}^N} \frac{\partial}{\partial t} F(x - y; t)dy = 1 \quad \text{for all } t > 0 \tag{17.3c}$$

and $(x, t) \to F(x, t)$ is harmonic in $\mathbb{R}^N \times \mathbb{R}^+$, we may regard $2F_t$ as a mollifying kernel following the parameter t. Therefore, (17.1c) follows from the properties of the mollifiers.[42] By (17.3c),

$$\frac{\partial}{\partial t} \int_{\mathbb{R}^N} \frac{\partial}{\partial t} F(x - y; t)dy = 0,$$

and

$$\frac{\partial}{\partial x_i} \int_{\mathbb{R}^N} \frac{\partial}{\partial t} F(x - y; t)dy = 0, \quad i = 1, 2, \ldots, N.$$

---

[40]See Section 15.1 of the Problems and Complements of Chapter VII. The Poisson integral of φ is introduced in Section 21.2 of the Problems and Complements of Chapter V.

[41]See (21.12c) of the Problems and Complements of Chapter V and (15.2c) of the Problems and Complements of Chapter VII.

[42]See Proposition 21.2c of the Problems and Complements of Chapter V.

Therefore, denoting with η any one of the components of (x, t),

$$\frac{\partial}{\partial \eta} \Phi(x, t) = \int_{\mathbb{R}^N} \frac{\partial^2}{\partial \eta \partial t} F(x - y; t) [\varphi(y) - \varphi(x)] dy,$$

and by direct calculation,

$$|D\Phi(x, t)| \leq \gamma \int_{\mathbb{R}^N} \frac{1}{(|x - y| + t)^{N+1}} |\varphi(x) - \varphi(y)| dy$$

for a constant γ depending only upon N. Integrate in dy by introducing polar coordinates with pole at x and radial variable ρt. If $\mathbf{n}$ denotes the unit vector spanning the unit sphere in $\mathbb{R}^N$,

$$|D\Phi(x, t)| \leq \gamma \int_{|\mathbf{n}|=1} d\mathbf{n} \int_0^\infty \frac{\rho^{N-1}}{(1 + \rho)^{N+1}} \frac{|\varphi(x + \rho t \mathbf{n}) - \varphi(x)|}{t} d\rho.$$

By the continuous version of Minkowski's inequality

$$\|D\Phi\|_{p, \mathbb{R}_+^{N+1}} \leq \gamma \int_{|\mathbf{n}|=1} d\mathbf{n} \int_0^\infty \frac{\rho^{N-1}}{(1 + \rho)^{N+1}} d\rho$$
$$\times \left(\int_0^\infty \frac{\|\varphi(\cdot + \rho t \mathbf{n}) - \varphi(\cdot)\|_{p, \mathbb{R}^N}^p}{t^p} dt \right)^{\frac{1}{p}}.$$

The last integral is computed by the change of variable $r = \rho t$. This gives

$$\|D\Phi\|_{p, \mathbb{R}_+^{N+1}} \leq \gamma \left(\int_0^\infty \frac{\rho^{N - \frac{1}{p}}}{(1 + \rho)^{N+1}} d\rho \right)$$
$$\times \int_{|\mathbf{n}|=1} \left(\int_0^\infty r^{N-1} \frac{\|\varphi(\cdot + r\mathbf{n}) - \varphi(\cdot)\|_{p, \mathbb{R}^N}^p}{r^{N+p-1}} \right)^{\frac{1}{p}} d\mathbf{n}$$
$$\leq \gamma(N, p) \left(\int_{\mathbb{R}^N} \int_{\mathbb{R}^N} \frac{|\varphi(x) - \varphi(y)|^p}{|x - y|^{N+p-1}} dx dy \right)^{\frac{1}{p}}$$
$$= \gamma(N, p) \|\varphi\|_{1 - \frac{1}{p}, p; \mathbb{R}^N}.$$

The extension claimed by the proposition can be taken to be

$$u(x, t) = e^{-t/p} \Phi(x, t).$$

To prove that φ is the trace of u, consider a sequence of functions $\{\varphi_n\}$ in $C_o^\infty(\mathbb{R}^N)$ that approximate φ in the norm of $W^{s,p}(\mathbb{R}^N)$ for $s = 1 - 1/p$. Such a sequence exists by virtue of Proposition 15.2. Then construct the corresponding Poisson integrals Φ_n, and let

$$u_n = e^{-t/p} \Phi_n.$$

Writing

$$u_n - u = 2e^{-t/p} \int_{\mathbb{R}^N} \frac{\partial}{\partial t} F(x - y; t)[\varphi_n(y) - \varphi(y)]dy$$

and applying (17.1c)–(17.2c) proves that $u_n \to u$ in $W^{1,p}(\mathbb{R}_+^{N+1})$. By the definition of trace, this implies that $u(\cdot, 0) = \varphi$. $\qquad \square$

Combining Proposition 16.1 and this extension procedure, we derive the following characterization of the traces.

Theorem 17.2c. *A function φ defined and measurable in $\mathbb{R}^N$ belongs to $W^{s,p}(\mathbb{R}^N)$ for some $s \in (0, 1)$ if and only if it is the trace on the hyperplane $\{t = 0\}$ of a function $u \in W^{1,p}(\mathbb{R}_+^{N+1})$, where $p = 1/(1 - s)$.*

18   TRACES ON ∂E OF FUNCTIONS IN $W^{1,p}(E)$

Theorem 18.1c. *Let E be a bounded domain in $\mathbb{R}^N$ with boundary ∂E of class C^1 and with the segment property. A given function $\varphi \in W^{1-\frac{1}{p},p}(\partial E)$ for some $p > 1$ admits an extension $u \in W^{1,p}(E)$ such that the trace of u on ∂E is φ.*

Theorem 18.2c. *Let E be a bounded domain in $\mathbb{R}^N$ with boundary ∂E of class C^1 and with the segment property. A function φ defined and measurable on ∂E belongs to $W^{s,p}(\partial E)$ for some $s \in (0, 1)$ if and only if it is the trace on ∂E of a function $u \in W^{1,p}(E)$, where $p = 1/(1 - s)$.*

18.1   TRACES ON A SPHERE. The embedding inequalities of Proposition 18.1 might be simplified if E is of relatively simple geometry, such as a ball or a cube in $\mathbb{R}^N$. Let B_R the ball of radius R about the origin of $\mathbb{R}^N$, and let $S_R = \partial B_R$.

Proposition 18.3c. *Let $u \in W^{1,p}(B_R)$, $p \in [1, \infty)$. Then*

$$\|u\|_{p,S_R}^p \le \frac{N}{R} \|u\|_p^p + p\|u\|_p^{p-1} \left\| \frac{\partial u}{\partial |x|} \right\|_p. \tag{18.1c}$$

If $p \in [1, N)$, then

$$\|u\|_{m,S_R}^m \le N\kappa_N^{\frac{1}{N}} \|u\|_{p*}^m + m\|u\|_{p*}^{m-1} \left\| \frac{\partial u}{\partial |x|} \right\|_p, \tag{18.2c}$$

where

$$p^* = \frac{Np}{N - p} \quad \text{and} \quad m = \frac{N - 1}{N} p^*.$$

Proof. We may assume that $u \in C^1(\overline{B}_R)$. Having fixed $q \ge p \ge 1$, set

$$\theta = \max\left\{ p; 1 + q\left(1 - \frac{1}{p}\right) \right\} \quad \text{so that } p \le \theta \le q.$$

Then for any unit vector $\mathbf{n}$,

$$R^N|u(R\mathbf{n})|^\theta = \int_0^R \frac{\partial}{\partial\rho}\left(\rho^N|u(\rho\mathbf{n})|^\theta\right)d\rho$$

$$= N\int_0^R \rho^{N-1}|u(\rho\mathbf{n})|^\theta d\rho$$

$$+ \theta\int_0^R \rho^N|u(\rho\mathbf{n})|^{\theta-1}\frac{\partial u(\rho\mathbf{n})}{\partial\rho}\,\text{sign}\,u(\rho\mathbf{n})d\rho.$$

Integrating in $d\mathbf{n}$ over the unit sphere $|\mathbf{n}| = 1$ gives

$$R\int_{\mathbf{n}=1} |u(R\mathbf{n})|^\theta R^{N-1}d\mathbf{n} = N\int_{B_R} |u|^\theta dx$$

$$+ \theta\int_{B_R} |x||u|^{\theta-1}\frac{\partial u}{\partial|x|}\,\text{sign}\,u\,dx$$

$$\leq N\mu(B_R)^{1-\frac{\theta}{q}}\left(\int_{B_R} |u|^q dx\right)^{\frac{\theta}{q}}$$

$$+ \theta R\left(\int_{B_R} |u|^{(\theta-1)\frac{p}{p-1}}dx\right)^{\frac{p-1}{p}}\left\|\frac{\partial u}{\partial|x|}\right\|_p.$$

Inequalities (18.1c) and (18.2c) follow from this for $q = p$ and $q = p^*$.   □

References

[1] R. A. Adams, *Sobolev Spaces*, Academic Press, New York, 1975.

[2] S. Banach, *Thèorie des Opérations Linéaires*, 2nd ed., Monografie Matematyczne, Warsaw, 1932; reprinted by Chelsea, New York, 1963.

[3] E. Borel, *Leçons sur la Théorie des Fonctions*, Gauthier-Villars, Paris, 1898.

[4] H. Brézis, *Analyse Fonctionnelle, Théorie et Applications*, Masson, Paris, 1983.

[5] H. Cartan, *Fonctions Analytiques d'une Variable Complexe*, Dunod, Paris, 1961.

[6] C. Carathéodory, *Vorlesungen über reelle Funktionen*, Chelsea, New York, 1946 (reprint).

[7] C. Carathéodory, *Algebraic Theory of Measure and Integration*, Chelsea, New York, 1963 (reprint).

[8] S. Campanato, *Sistemi ellittici in Forma di Divergenza*, Quaderni, Scuola Normale Superiore, Pisa, 1980.

[9] M. M. Day, *Normed Linear Spaces*, 3rd ed., Springer-Verlag, New York, 1973.

[10] B. Dacorogna and P. Marcellini, *Implicit Partial Differential Equations*, Birkäuser, Boston, 1999.

[11] E. DiBenedetto, *Partial Differential Equations*, Birkhäuser, Boston, 1995.

[12] J. Dieudonné, *Eléments d'Analyse*, Vol. I, Gauthier-Villars, Paris, 1969.

[13] U. Dini, *Fondamenti per la Teorica delle Funzioni di una Variabile Reale*, Pisa, 1878; reprint of Unione Matematica Italiana, Bologna.

[14] N. Dunford and J. T. Schwartz, *Linear Operators*, Part I, Wiley–Interscience, New York, 1958.

[15] I. Ekeland and R. Temam, *Analyse Convexe et Problèmes Variationnels*, Dunod, Paris, 1974.

[16] L. C. Evans and R. F. Gariepy, *Measure Theory and Fine Properties of Functions*, CRC Press, Boca Raton, 1992.

[17] H. Federer, *Geometric Measure Theory*, Springer-Verlag, New York, 1969.

[18] G. B. Folland, *Real Analysis*, Wiley–Interscience, New York, 1984.

[19] A. Friedman, *Foundations of Modern Analysis*, Dover, New York, 1982.

[20] E. Giusti, *Functions of Bounded Variation*, Birkhäuser, Basel, 1983.

[21] H. Hahn, *Theorie der reellen Funktionen*, Berlin, 1921.

[22] P. R. Halmos, *Measure Theory*, Springer-Verlag, New York, 1974.

[23] P. R. Halmos, *Introduction to Hilbert Spaces*, Chelsea, New York, 1957.

[24] G. H. Hardy, J. E. Littlewood, and G. Pólya, *Inequalities*, Cambridge University Press, Cambridge, U.K., 1963.

[25] F. Hausdorff, *Grundzüge der Mengenlehre*, Leipzig, 1914; reprinted by Chelsea, New York, 1955.

[26] C. Jordan, *Course d'Analyse*, Gauthier-Villars, Paris, 1893.

[27] J. L. Kelley, *General Topology*, Van Nostrand, New York, 1961.

[28] J. L. Kelley and I. Namioka, *Linear Topological Spaces*, Van Nostrand, New York, 1963.

[29] O. D. Kellogg, *Foundations of Potential Theory*, Dover, New York, 1953.

[30] K. Kuratowski, *Topology*, Vol. 1, Academic Press, New York, 1966.

[31] O. A. Ladyzhenskaya and N. N. Ural'tzeva, *Linear and Quasilinear Elliptic Equations*, Academic Press, New York, 1968.

[32] O. A. Ladyzhenskaya, V. A. Solonnikov, and N. N. Ural'tzeva, *Linear and Quasilinear of Parabolic Type*, Translations of Mathematical Monographs 23, American Mathematical Society, Providence, RI, 1968.

[33] H. Lebesgue, *Leçons sur l'Integration et la Recherche des Fonctions Primitives*, 2nd ed., Gauthier-Villars, Paris, 1928.

[34] E. H. Lieb and M. Loss, *Analysis*, Graduate Studies in Mathematics 14, American Mathematical Society, Providence, RI, 1996.

[35] J. L. Lions and E. Magenes, *Non-Homogeneous Boundary Value Problems and Applications*, Vols. 1–3, Springer-Verlag, Berlin, 1972.

[36] J. E. Littlewood, *Lectures on the Theory of Functions*, Oxford University Press, London, 1944.

[37] L. A. Liusternik and V. J. Sobolev, *Elements of Functional Analysis*, Fredrick Ungar, New York, 1967.

[38] G. Lorentz, *Approximation of Functions*, Chelsea, New York, 1986.

[39] V. G. Mazja, *Sobolev Spaces*, Springer-Verlag, New York, 1985.

[40] S. G. Mikhlin, *Integral Equations*, Vol. 4, Pergamon Press, New York, 1957.

[41] C. B. Morrey, *Multiple Integrals in the Calculus of Variations*, Springer-Verlag, New York, 1966.

[42] M. E. Munroe, *Measure and Integration*, Addison–Wesley, Reading, MA, 1971.

[43] I. P. Natanson, *Theory of Functions of a Real Variable*, Vols. 1 and 2, Fredrick Ungar, New York, 1955, 1960.

[44] F. Riesz and B. Nagy, *Leçons d'Analyse Fontionnelle*, 6th ed., Akadémiai Kiadó, Budapest, 1972.

[45] C. A. Rogers, *Hausdorff Measures*, Cambridge University Press, Cambridge, U.K., 1970.

[46] H. L. Royden, *Real Analysis*, 3rd ed., Macmillan, New York, 1988.

[47] W. Rudin, *Real and Complex Analysis*, 3rd Ed., McGraw–Hill, New York, 1986.

[48] S. Saks, *Theory of the Integral*, Monografje Matematyczne 7, 2nd revised ed., Hafner, New York, 1937 (English translation by L. C. Young, with two additional notes by S. Banach).

[49] L. Schwartz, *Théorie des Distributions*, Hermann and Cie, Paris, 1966.

[50] E. Stein, *Singular Integrals and Differentiability Properties of Functions*, Princeton University Press, Princeton, NJ, 1970.

[51] G. Talenti, *Calcolo delle Variazioni*, Quaderni, Unione Matematica Italiana, Roma, 1977.

[52] E. C. Tischmarsh, *The Theory of Functions*, 2nd ed., Oxford University Press, London, 1939.

[53] F. Tricomi, *Integral Equations*, Dover, New York, 1957.

[54] R. L. Wheeden and A. Zygmund, *Measure and Integral*, Marcel Dekker, New York, 1977.

[55] K. Yoshida, *Functional Analysis*, Springer-Verlag, New York, 1974.

Index

Birkhäuser Advanced Texts
Basler Lehrbücher

Series Editors:
H. Amann, University of Zürich
S. Kumar, University of North Carolina at Chapel Hill

This series presents, at an advanced level, introductions to some of the fields of current interest in mathematics. Starting with basic concepts, fundamental results and techniques are covered, and important applications and new developments discussed. The textbooks are suitable as an introduction for students and non-specialists, and they can also be used as background material for advanced courses and seminars.

We encourage preparation of manuscripts in TeX for delivery in camera-ready copy which leads to rapid publication, or in electronic form for interfacing with laser printers or typesetters. Proposals should be sent directly to the editors or to:
Birkhäuser Boston, 675 Massachusetts Ave., Cambridge, MA 02139
or
Birkhäuser Publishers, 40-44 Viadukstrasse, Ch-4010 Basel, Switzerland

M. Brodmann, **Algebraische Geometrie**
1989. Hardcover. ISBN 3-7643-1779-5

E.B. Vinberg, **Linear Representations of Groups**
1989. Harcover. ISBN 3-7643-2288-8

K. Jacobs, **Discrete Stochastics**
1991. Harcover. ISBN 3-7643-2591-7

S.G. Krantz, H.R. Parks, **A Primer of Real Analytic Functions**
1992. Hardcover. ISBN 3-7643-2768-5

L. Conlon, **Differentiable Manifolds: A First Course**
1992. Harcover. *First edition, 2nd revised printing.*
ISBN 0-8176-3626-9

M. Artin, **Algebra**
1993. Hardcover. ISBN 3-7643-2927-0

H. Hofer, E. Zehnder, **Symplectic Invariants and Hamiltonian Dynamics**
1994. Hardcover. ISBN 3-7643-5066-0

M. Rosenblum, J. Rovnyak, **Topics in Hardy Classes and Univalent Functions**
1994. Hardcover. ISBN 0-8176-5111-X

P. Gabriel, **Matrizen, Geometrie, Lineare Algebra**
1996. Hardcover. ISBN 3-7643-5376-7

M. Artin, **Algebra**
1998. Softcover. ISBN 3-7643-5938-2

K. Bichteler, **Integration—A Functional Approach**
1998. Hardcover. ISBN 3-7643-5936-6

S.G. Krantz, H.R. Parks, **The Geometry of Domains in Space**
1998. Harcover. ISBN 0-8176-4097-5

M.G. Nadkarni, **Spectral Theory of Dynamical Systems**
1998. Hardcover. ISBN 3-7643-5817-3

M.G. Nadkarni, **Basic Ergodic Theory**
1998. Hardcover. ISBN 3-7643-5816-5

V.S. Sunder, **Functional Analysis**
1998. Hardcover. ISBN 3-7643-5892-0

M. Holz, K. Steffens, E. Weitz, **Introduction to Cardinal Arithmetic**
1999. Hardcover. ISBN 3-7643-6124-7

M. Chipot, **Elements of Nonlinear Analysis**
2000. Harcover. ISBN 3-7643-6406-8

J.M. Gracia-Bondía, J.C. Várilly, H. Figueroa, **Elements of Noncommutative Geometry**
2000. Hardcover. ISBN 0-8176-4124-6

L. Conlon, **Differentiable Manifolds**
2001. Harcover. *Second Edition*
ISBN 0-8176-4134-3

H. Sohr, **The Navier–Stokes Equations: An Elementary Functional Analytic Approach**
2001. Hardcover. ISBN 3-7643-6545-5

M. Chipot, **goes to plus infinity**
2001. Hardcover. ISBN 3-7643-6646-X

E. DiBenedetto, **Real Analysis**
2002. Hardcover. ISBN 0-8176-4231-5

R. Estrada, R.P. Kanwal, **A Distributional Approach to Asymptotics: Theory and Applications**
2002. Hardcover. *Second Edition*
ISBN 0-8176-4142-4

CPSIA information can be obtained at www.ICGtesting.com
Printed in the USA

243707LV00007B/1/A

9 780817 642310